Biodiesel Production

# Biodiesel Production: Feedstocks, Catalysts, and Technologies

Edited by

*Dr. Samuel Lalthazuala Rokhum*
*National Institute of Technology Silchar*
*Department of Chemistry*
*NIT Road, Fakiratilla*
*Silchar, Assam*
*India*

*Prof. Gopinath Halder*
*National Institute of Technology Durgapur*
*Department of Chemical Engineering*
*National Institute Technology*
*Durgapur*
*India*

*Prof. Suttichai Assabumrungrat*
*Chulalongkorn University*
*Center of Excellence in Catalysis and Catalytic Reaction Engineering*
*and Bio-Circular-Green-economy Technology & Engineering Center (BCGeTEC)*
*Department of Chemical Engineering*
*Bangkok, Thailand*

*Assoc. Prof. Kanokwan Ngaosuwan*
*Rajamangala University of Technology Krungthep*
*Chemical Engineering Devision*
*Bangkok*
*Thailand*

WILEY

This edition first published 2022


*Registered Offices*
John Wiley & Sons, Inc., 111 River Street, Hoboken, NJ 07030, USA
John Wiley & Sons Ltd, The Atrium, Southern Gate, Chichester, West Sussex, PO19 8SQ, UK

*Editorial Office*
The Atrium, Southern Gate, Chichester, West Sussex, PO19 8SQ, UK

For details of our global editorial offices, customer services, and more information about Wiley products visit us at www.wiley.com.

Wiley also publishes its books in a variety of electronic formats and by print-on-demand. Some content that appears in standard print versions of this book may not be available in other formats.



For general information on our other products and services or for technical support, please contact our Customer Care Department within the United States at (800) 762-2974, outside the United States at (317) 572-3993 or fax (317) 572-4002. Wiley also publishes its books in a variety of electronic formats. Some content that appears in print may not be available in electronic formats. For more information about Wiley products, visit our web site at ww.wiley.com.

*Library of Congress Cataloging-in-Publication Data*

Names: Rokhum, Samuel Lalthazuala, editor. | Halder, Gopinath, editor. | Assabumrungrat, Suttichai, editor. | Ngaosuwan, Kanokwan, editor.
Title: Biodiesel production : feedstocks, catalysts, and technologies / [edited by] Prof. Samuel Lalthazuala Rokhum, National Institute of Technology Silchar, Department of Chemistry, NIT Road, Fakiratilla, Sichar, Assam, India, Prof. Gopinath Halder, Department of Chemical Engineering, National Institute Technology, Durgapur, India, Prof. Suttichai Assabumrungrat, Chulalongkorn University, Department of Chemical Engineering, Bangkok, Thailand, Dr. Kanokwan Ngaosuwan, Rajamangala University Technology Krungthep, Chemical Engineering Division, Krungthep, Bangkok, Thailand.
Other titles: Biodiesel production (John Wiley & Sons)
Description: Hoboken, NJ : Wiley, 2022. | Includes bibliographical references and index.
Identifiers: LCCN 2021056412 (print) | LCCN 2021056413 (ebook) | ISBN 9781119771333 (cloth) | ISBN 9781119771340 (adobe pdf) | ISBN 9781119771357 (epub)
Subjects: LCSH: Biodiesel fuels.
Classification: LCC TP359.B46 B5634 2022 (print) | LCC TP359.B46 (ebook) | DDC 665/.37–dc23/eng/20220228
LC record available at https://lccn.loc.gov/2021056412
LC ebook record available at https://lccn.loc.gov/2021056413

Cover Design: Wiley
Cover Image: © Julija Vidjajeva/Shutterstock

Set in 9.5/12.5pt STIXTwoText by Straive, Pondicherry, India

10 9 8 7 6 5 4 3 2 1

# Contents

# Preface

This book *Biodiesel Production: Feedstocks, Catalysts, and Technologies* includes the contribution of leading researchers in the fields of biodiesel, which will serve as a valuable source of information for scientists, researchers, graduate students, and professionals alike. It focusses on several aspects of biodiesel productions, technologies employed, and sustainability. It consists of 20 chapters, grouped together in three parts, in different technological aspects as follows.

The utilization of conventional and novel feedstocks for biodiesel production will be presented in Chapters 1–4.

Chapter 1 emphasizes on the conversion of several edible vegetable oils and animal fats to biodiesel. Different catalysts used and several factors that affect the overall biodiesel production are comprehensively discussed.

Chapter 2 provides the perspective of the biodiesel production from waste cooking oil via the conventional and modern technologies to bolster competitiveness of biodiesel with petrodiesel.

Chapter 3 addresses the state of the art and future perspectives of nonedible oils for biodiesel production. It provides several important aspects such as cultivation information, fatty acid composition, extraction, and conversion method for biodiesel production.

Chapter 4 proposes the important strategy of microalgae cultivation for the large-scale optimization of lipid accumulation as a potential sustainable approach for biodiesel production.

In the next part, Chapters 5–14, the various types of homogeneous and heterogeneous catalysts for biodiesel production will be discussed.

Chapter 5 reviews the utilization of homogeneous catalysts for various feedstocks under optimum conditions to serve the growing demand of biodiesel as a cost-effective production process.

Chapter 6 summarizes the development of metal oxide catalysts from various sources for biodiesel production. The reaction mechanistic pathways and causes of catalyst deactivation are discussed.

Chapter 7 focuses on the catalytic activity enhancement of metal oxides with particular focus on the role of supporters, their synthesis methods, and physicochemical properties to achieve eco-friendly and economically viable processes of biodiesel production.

Chapter 8 highlights the development of new mixed metal oxides with a variety of novel acidic, basic, and bifunctional catalysts from various feedstocks for enhancing their catalytic performance. Their stability, catalyst regeneration techniques, and recommendation for full scale biodiesel production are also addressed.

Chapter 9 presents the outlook of using nanotechnology-based catalysts for the development of more efficient, economically viable, durable, and stable nanocatalysts, targeting at achieving higher biodiesel quality and yields.

Chapter 10 reveals the advantages and issues of using ion-exchange resins catalysts for both cation and anion exchange resins especially in continuous biodiesel production.

Chapter 11 discusses the solid bifunctional catalysts with acid–base and Lewis–Brønsted functionalities. The preparation methods, their characterization results, and the optimum condition for biodiesel production were addressed.

Chapter 12 proposes the green concept for biodiesel production using catalysts derived from renewable resources. Essential information on their preparation methods, physicochemical properties, and catalytic activities as well as the challenges are discussed.

Chapter 13 exploits the usage of the promising ionic liquid catalyst to replace homogeneously catalyzed biodiesel production concurrently with techno-economic analysis, life cycle assessment, environmental impact assessment, and scale-up technologies.

Chapter 14 demonstrates the effective acid/base metal–organic frameworks (MOFs) catalysts for both transesterification and esterification reactions to intensify biodiesel production based on their catalytic synergy.

The strategies in terms of upstream, mainstream, and downstream process to fulfill the economical and sustainable for biodiesel production will be addressed in Chapters 15–21.

Chapter 15 scrutinizes the strategies for upstream biodiesel production dealing with the advanced feedstocks like waste cooking oil, waste animal fats, nonedible oils, or genetically engineered oils based on the appropriate catalyst, reaction conditions, and the following downstream processes.

Chapter 16 approaches the operating key parameters of mainstream strategies in terms of the novel reactor for biodiesel production based on the scientific and practical viewpoints to achieve efficiency and sustainable concept.

Chapter 17 discloses the downstream strategies to accomplish biodiesel standard as well as operating cost reduction using methanol recovery and glycerol by-product refining. The integration of bioenergy systems to produce antioxidant additives for improving biodiesel quality is also encouraged.

Chapter 18 addresses the conversion of bio-glycerol to value-added chemicals especially acrylic acid to boost alternative sustainable routes for biodiesel production.

Chapter 19 introduces the sustainability in the production and use of biodiesel, which is mainly dependent on the types of feedstocks and government policy as the incentives of using biodiesel.

Chapter 20 discusses the advanced, sustainable technology with respect to the diversified feedstock and design of the novel efficient catalytic system for production of biodiesel and its commercialization.

# List of Contributors

***Pratibha Agrawal***
Department of Applied Chemistry
Laxminarayan Institute of Technology
RTM Nagpur University, Nagpur
Maharashtra, India

***Weerinda Appamana***
Department of Chemical and Materials Engineering, Faculty of Engineering
Rajamangala University of Technology
Thanyaburi, Pathum Thani, Thailand

***Arumugam Arumugam***
Department of Chemical Engineering
School of Chemical and Biotechnology
Center for Bioenergy, SASTRA Deemed to Be University, Thanjavur, India

***Suttichai Assabumrungrat***
Center of Excellence on Catalysis and Catalytic Reaction Engineering
Department of Chemical Engineering
Faculty of Engineering, Chulalongkorn University, Bangkok, Thailand
Bio-Circular-Green-economy Technology & Engineering Center, BCGeTEC
Department of Chemical Engineering
Faculty of Engineering, Chulalongkorn University, Bangkok, Thailand

***Vasudeva Rao Bakuru***
Materials science and catalysis division
Poornaprajna Institute of Scientific Research, Bangalore Rural, India

***G. Baskar***
Department of Biotechnology, St. Joseph's College of Engineering, Chennai, India

***Bidangshri Basumatary***
Department of Chemistry, Bodoland University, Kokrajhar, Assam, India

***Sanjay Basumatary***
Department of Chemistry, Bodoland University, Kokrajhar, Assam, India

***Rahul Bhagat***
Department of Biotechnology, Government Institute of Science, Aurangabad
Maharashtra, India

***Trinath Biswal***
Department of Chemistry, Veer Surendra Sai University of Technology, Burla.
Sambalpur, Odisha. India

***Bishwajit Changmai***
Department of Chemistry, National Institute of Technology Silchar
Silchar, India

***Narita Chanthon***
Center of Excellence on Catalysis and Catalytic Reaction Engineering
Department of Engineering, Faculty of Engineering, Chulalongkorn University
Bangkok, Thailand

***Michael Van Lal Chhandama***
Department of Biotechnology, School of Sciences (Block-I), JAIN (Deemed-to-be University), Bengaluru, Karnataka, India

***Valeria D'Ambrosio***
Istituto di Ricerca Sulle Acque, Consiglio Nazionale delle Ricerche (IRSA-CNR) Bari, Italy

***Yosvany Díaz-Domínguez***
Faculty of Chemical Engineering Universidad Tecnológica de la Habana Havana, Cuba

***Marilyn Esclance DMello***
Materials science and catalysis division Poornaprajna Institute of Scientific Research, Bangalore Rural, India

***Apiluck Eiad-ua***
College of Nanotechnology, King Mongkut's Institute of Technology Ladkrabang, Bangkok, Thailand

***Jabbar Gardy***
School of Chemical and Process Engineering, University of Leeds Leeds, UK

***Gopinath Halder***
Department of Chemical Engineering National Institute of Technology Durgapur, India

***Ali Hassanpour***
School of Chemical and Process Engineering, University of Leeds, Leeds, UK

***Balkis Hazmi***
Institute of Nanoscience and Nanotechnology (ION2) Universiti Putra Malaysia, Serdang Selangor, Malaysia

***Kousuke Hiromori***
Department of Chemical Engineering Tohoku University, Sendai, Japan

***Avinash P. Ingle***
Biotechnology Centre, Department of Agricultural Botany, Dr. Panjabrao Deshmukh Agricultural University, Akola Maharashtra, India

***Pavan Narayan Kalbande***
Catalysis and Inorganic Chemistry Division, CSIR-National Chemical Laboratory, Pune, India
Academy of Scientific and Innovative Research (AcSIR), CSIR-National Chemical Laboratory, Pune, India

***Suresh Babu Kalidindi***
Department of Inorganic and Analytical Chemistry, School of Chemistry, Andhra University Visakhapatnam, India

***Aleksandra Sander***
Department of Mechanical and Thermal Process Engineering, University of Zagreb Faculty of Chemical Engineering and Technology, Zagreb, Croatia

***Bisheswar Karmakar***
Department of Chemical Engineering National Institute of Technology Durgapur, India

***Rupam Kataki***
Department of Energy, Tezpur University Tezpur, Assam. India

***Worapon Kiatkittipong***
Department of Chemical Engineering Faculty of Engineering and Industrial Technology, Silpakorn University, Nakhon Pathom, Thailand

***Hui Li***
School of Thermal Engineering,
Shandong Jianzhu University, Jinan,
PR China

***Brandon Lowe***
School of Chemical and Process
Engineering, University of Leeds
Leeds, UK

***Mangesh P. Moharil***
Biotechnology Centre, Department
of Agricultural Botany, Dr. Panjabrao
Deshmukh Agricultural University, Akola
Maharashtra, India

***Biswajit Nath***
Department of Chemistry,
Bodoland University, Kokrajhar,
Assam, India
Department of Chemistry, Science
College Kokrajhar, Assam, India

***Kanokwan Ngaosuwan***
Division of Chemical Engineering, Faculty
of Engineering, Rajamangala University
of Technology Krungthep, Bangkok,
Thailand

***Carlo Pastore***
Istituto di Ricerca Sulle Acque, Consiglio
Nazionale delle Ricerche (IRSA-CNR)
Bari, Italy

***Nattawat Petchsoongsakul***
Center of Excellence on Catalysis
and Catalytic Reaction Engineering
Department of Engineering, Faculty of
Engineering, Chulalongkorn University
Bangkok, Thailand

***Ana Petračić***
Department of Mechanical and Thermal
Process Engineering, University of Zagreb
Faculty of Chemical Engineering and
Technology, Zagreb, Croatia

***Ramón Piloto-Rodríguez***
Faculty of Chemical Engineering
Universidad Tecnológica de la Habana
Havana, Cuba

***Armando T. Quitain***
Faculty of Advanced Science and
Technology, Kumamoto University
Kumamoto, Japan
Center for International Education
Kumamoto University, Kumamoto, Japan

***Umer Rashid***
Institute of Nanoscience and
Nanotechnology (ION2), Universiti Putra
Malaysia, Serdang Selangor, Malaysia

***Samuel Lalthazuala Rokhum***
Hamid Yusuf Department of Chemistry
University of Cambridge, Cambridge, UK
Department of Chemistry, National Institute
of Technology, Silchar, Assam, India

***B. Sangeetha***
Department of Biotechnology, St. Joseph's
College of Engineering, Chennai, India

***Kumudini Belur Satyan***
Department of Biotechnology, School of
Sciences (Block-I), JAIN (Deemed-to-be
University), Bengaluru, Karnataka, India

***Shreshtha Saxena***
Biotechnology Centre, Department
of Agricultural Botany, Dr. Panjabrao
Deshmukh Agricultural University, Akola
Maharashtra, India

***Enrico Scelsi***
Istituto di Ricerca Sulle Acque, Consiglio
Nazionale delle Ricerche (IRSA-CNR),
Bari, Italy

***Krushna Prasad Shadangi***
Department of Chemical Engineering,
Veer Surendra Sai University of Technology,
Burla. Sambalpur, Odisha. India

***Naomi Shibasaki-Kitakawa***
Department of Chemical Engineering, Tohoku University, Sendai, Japan

***Nittan Singh***
Catalysis and Inorganic Chemistry Division, CSIR-National Chemical Laboratory, Pune, India
Academy of Scientific and Innovative Research (AcSIR), CSIR-National Chemical Laboratory, Pune, India

***Atthapon Srifa***
Department of Chemical Engineering Faculty of Engineering, Mahidol University, Nakhon Pathom, Thailand

***Putla Sudarsanam***
Catalysis and Inorganic Chemistry Division, CSIR-National Chemical Laboratory, Pune, India
Academy of Scientific and Innovative Research (AcSIR), CSIR-National Chemical Laboratory, Pune, India

***Pothiappan Vairaprakash***
Department of Chemistry, School of Chemical and Biotechnology, Center for Bioenergy, SASTRA Deemed to Be University, Thanjavur, India

***Chhangte Vanlalveni***
Department of Botany, Mizoram University, Aizawl, Mizoram, India

***Andrew E.H. Wheatley***
Hamid Yusuf Department of Chemistry University of Cambridge, Cambridge, UK

***Doonyapong Wongsawaeng***
Department of Nuclear Engineering Faculty of Engineering, Chulalongkorn University, Bangkok, Thailand

***Kejun Wu***
School of Chemical and Process Engineering, University of Leeds Leeds, UK
School of Chemical and Biological Engineering, Zhejiang University Hangzhou, P.R. China

## An Overview of Biodiesel Production

The advent of the industrial revolution had many benefits such as increases in wealth of the average masses, upgrade in living standards, and vast improvements in production of goods (both in quality and in quantity), which reduced prices drastically. Technological advancements also occurred in the transport sector, which enabled ease in travel, while the use of coal and petroleum skyrocketed: an example of this would be the 20-fold increase in coal imports between 1550 and 1700 in Newcastle, England. Consequently, a proportional increase in mining of these fossilized reserves had to be done as far as from the early nineteenth century. Since then, the energy demand per capita has increased manifold to the point where current consumption trends cannot be supported without exhausting the remaining global reserves – alternative energy sources must be sought. Additionally, large areas of forest land had been cleared for fuelwood, which served as the primary energy source for cooking and heating in rural households. Widespread deforestation led to a rapid rise in global temperature since less trees are available for climate modulation. Also, upon using wood and other fossilized sources as fuel, huge amounts of particulate matter, smoke, and other noxious gases ($NO_X$s, $SO_X$s, CO, and $CO_2$) are emitted, and thus their continued emission for the last few centuries has led to global warming, harmful impacts on terrestrial and aquatic life (through acid rain, aquatic pollution resulting in eutrophication), and changes in weather patterns, which has even impacted the overall health and life expectancy of humans (lung diseases caused by air pollution, water pollution leading to chronic diseases, etc.).

In order to combat or gradually reverse the effects of such a global situation where arable land and potable water are scarce, alternative energy sources that have no or negligible environmental impacts must be sought. Thus, renewable energy research over the last few decades has been steadily increasing and is now capable of changing an entire country's energy consumption trend. A good example is Brazil, which runs entirely on "sustainable" fuels, having produced 26.1% (a staggering 26.72 billion liters) of the global ethanol being used as fuel in 2017, and many countries have tried to replicate the so-called "Brazilian ethanol model." Among the wide variety of renewable energy sources available, feedstock for biofuels such as biodiesel and bioethanol are limited to a few varieties. Vegetable oils (edible or nonedible) cannot be directly used in engines due to their incompatible physicochemical properties. This had been tested by Dr. Rudolph Diesel who used peanut oil for his internal combustion (IC) engine and reported many problems in required performance when run for extended durations. Thus, such oils are converted into esters that are the main component of biodiesel, a fuel suitable for use in diesel engines with minor modifications. To convert vegetable oils as well as other potential feedstock such as microalgal lipids, animal fats and greases, waste oils, and other miscellaneous sources, various approaches may be used with different conversion efficiencies. The most efficient conversion process, however, is transesterification, which may or may not be coupled with an esterification pretreatment stage depending on the free fatty acid content of the oil.

For both esterification and transesterification, the reactants are the feedstock and an alcohol, which in the presence of a catalyst are converted into their esters, producing either water or glycerol as by-products. Depending on the reaction conditions (based on the approach used), catalysts may not be required, although a multitude of catalysts have been developed and tested with varying degrees of efficiency. Such catalysts range from the simplest mineral acids, enzymes, or bases, which are added for achieving a homogeneous system and discarded with every use to simple heterogeneous catalysts that rely on solid metal oxides or the use of inert carbonaceous or siliceous biomass doped with the required catalytic groups (including transition metals) or enzymes, as well as nanocatalysts that have increased efficiency (when compared with inert microporous support-based catalysts), while specially designed catalysts based on resin supports or metal organic frameworks have also been developed and can be very efficient but may be difficult to commercialize due to high development costs and unavoidable losses in each cycle of use. Strangely, processes such as supercritical fluid technology or superheated vapor technology can function reliably even without the use of catalysts, although the use of catalysts can augment the process, which may require a cost-to-benefit analysis before commercialization.

The process of biodiesel commercialization does not simply end at its production, since there are many stages that need to be considered for downstream processing as well as the consideration for treatment of hazardous materials generated (such as biodiesel wastewater that contains spent catalyst or leached ions) and the recovery of spent alcohol and the valorization of generated glycerol. Additionally, the produced fuel must have an acceptably long shelf life, and since biodiesel is prone to auto-oxidation (it contains high oxygen content that helps in reducing pollution due to complete fuel combustion), such additives are essential for storage. Such processes generally increase the cost of available fuel, which has made it necessary to consider these hurdles that are yet to be overcome before the complete utilization of biodiesel is feasible as an environment-friendly and affordable alternative to petrodiesel.

Editors: Samuel Lalthazuala Rokhum, Gopinath Halder, Kanokwan Ngaosuwan, Suttichai Assabumrungrat

# Part 1

# Biodiesel Feedstocks

# 1

# Advances in Production of Biodiesel from Vegetable Oils and Animal Fats

*Umer Rashid and Balkis Hazmi*

*Institute of Nanoscience and Nanotechnology (ION2), Universiti Putra Malaysia, Serdang, Selangor, Malaysia*

## 1.1 Introduction

Currently, the energy requirements of the world are mainly met through fossil fuel resources, such as gasoline, petroleum-based diesel, and natural gas. Such fossil-derived resources are too limited to fulfill the future energy demands and meet the challenges of rapid human population growth coupled with technological developments [1]. Presently, research is progressively more directed toward exploration of alternative renewable fuels. Several types of biofuels, such as vegetable oil/animal fat (raw, processed, or used), methyl esters from vegetable oil/animal fat, and ethanol or liquid fuels from biomass (bioethanol and biomethanol), have been investigated as a replacement for gasoline and petrodiesel [2].

At present over 197.97 million metric tons of 10 major vegetable oils are produced worldwide [3]. Vegetable oils are commonly derived from various oilseed crops. In a vegetable oil, almost 90–95% is glycerides, which are basically esters of glycerol and fatty acids (FAs) [4]. The vegetable oils can be considered as a feasible alternative for diesel fuel as the heating value of vegetable oils is comparable to that of diesel fuel [5, 6]. However, the uses of vegetable oils in direct injection diesel engines are restricted due to some unfavorable physical properties, particularly the viscosity. The viscosity of vegetable oil is roughly 10 times higher than the diesel fuel. Therefore, the use of vegetable oil in direct injection diesel engines creates poor fuel atomization, incomplete combustion, and carbon deposition on the injector [7, 8].

Several techniques are employed to bring down the physical and thermal properties of vegetable oils close to mineral diesel, by which these oils and fats can be used in internal combustion engines as fuel. This mainly requires improvement in viscosity of the vegetable oil. The possible treatments employed to improve the oil viscosity includes dilution with a suitable solvent, microemulsification, pyrolysis, and transesterification [9, 10].

The uses of biodiesel (BD) as a renewable, biodegradable, nontoxic, and eco-friendly neat diesel fuel or in blends with petroleum-based fuels are fascinating [11, 12]. "Biodiesel," termed as the monoalkyl esters of long-chain FAs, is derived from vegetable oils or animal fats.

*Biodiesel Production: Feedstocks, Catalysts, and Technologies*, First Edition. Edited by Samuel Lalthazuala Rokhum, Gopinath Halder, Suttichai Assabumrungrat, and Kanokwan Ngaosuwan.

Numerous types of conventional and nonconventional vegetable oils and animal fats including those of used oils from the frying industry, soybean oil, rapeseed oil, tallow, rubber seed oil, and palm oil have been investigated to produce BD [13–15]. The production of BD involves the conversion of vegetable oils/animal fats using methanol or ethanol and a catalyst to produce fatty acid methyl esters (FAMEs) and crude glycerin as by-product through a process termed as "transesterification" [16].

The transesterification process is accomplished by reacting vegetable oil with alcohol in the presence of alkaline or acidic catalyst. A catalyst is typically used to accelerate the reaction rate and yield. The stoichiometric equation requires 1 mol of triglyceride and 3 mol of alcohol to form 3 mol of methyl ester and 1 mol of glycerol [17]. Since the reaction is reversible, excess alcohol is used to shift the reaction equilibrium to the product's side. The most preferred catalysts are sulfuric, sulfonic, and hydrochloric acids as acidic catalysts and sodium hydroxide, sodium methoxide, and potassium hydroxide as alkaline catalysts [18]. The product, fatty esters, have improved viscosity and volatility relative to the triglycerides. A dense, liquid phase rich in glycerol is the coproduct of this process. The separated fatty esters have cetane number and heating value close to that of the conventional diesel. The transesterification process for converting vegetable oils to BD is shown in Figure 1.1.

The "R" groups are the FAs, which are usually 12–22 carbons in length. The large vegetable oil molecule is reduced to about one third of its original size, lowering the viscosity and making it like diesel fuel. The resulting fuel can work like diesel fuel in an engine. The by-product "glycerin" produced in this process is valuable due to its diverse industrial applications [19].

Technically, BD is a fuel comprising of monoalkyl esters of long-chain FAs derived from vegetable oils or animal fat, which meets current EN 14214 and ASTM D 6751 BD standards of Europe and the United States, respectively. These standards are frequently employed as references to evaluate and compare the properties of other fuels.

Presently, the BD is commonly produced using a base-catalyzed transesterification reaction because it involves low temperature and pressure processing, high conversions, no intermediate steps, and lower costs of processing materials [20]. Alkoxides and hydroxides of potassium and sodium are often used as catalysts in the transesterification of refined oils and low FA greases and fats. However, acid esterification followed by transesterification of high free fatty acid (FFA) fats and oils is also applicable. The base catalysts have better efficiency than the acid catalysts [21]. The base-catalyzed transesterification reaction can be carried out at lower temperature, yet at room temperature, with the base catalysts, whereas acid catalysis required higher temperature (100 °C) and longer reaction time. During the process, basic catalyst breaks the FAs from the glycerin one by one. When a methanol molecule contacts an FA molecule, it will bond and form BD molecule. The hydroxyl group

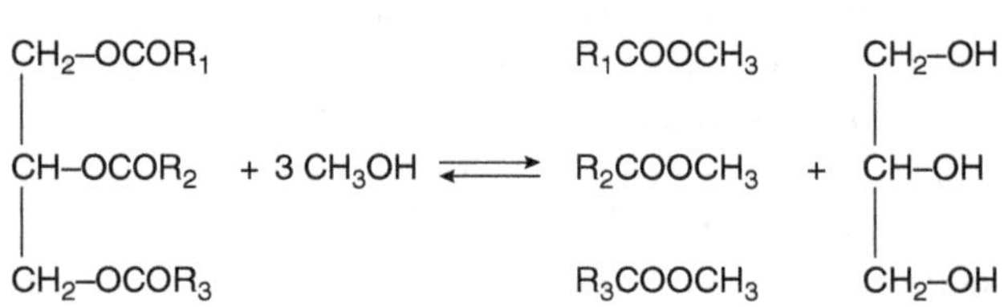


**Figure 1.1** General reaction for transesterification of vegetable oil.

from the catalyst alleviates the glycerol formation. The resulting product named as methyl esters (BD) has appreciably lower viscosity and increased volatility relative to the triglycerides present in vegetable oils [22–24].

The second usual method of producing BD involves the use of an acid as a substitute of a base catalyst. Any mineral acid can be employed to catalyze the process; the most used acids are sulfuric acid and sulfonic acid. Although yield is high, the acids, being corrosive, may cause damage to the equipment, and the reaction rate is also observed to be relatively low [9, 21]. Oil feedstocks containing more than 4% FFAs must pass through an acid esterification process to increase the BD yield [25]. Such feedstocks are filtered and preprocessed to remove water and contaminants and then fed to the acid esterification process. The catalyst (sulfuric acid) is dissolved in methanol and then mixed with the pretreated oil [26].

The alcohols employed in the transesterification are generally short-chain alcohols such as methanol, ethanol, propanol, and butanol producing esters named as methyl-, ethyl-, propyl-, and butyl-esters, respectively [9, 10]. It is reported that when transesterification of soybean oil using methanol, ethanol, and butanol was performed, 96–98% of ester's yield could be obtained after an hour of reaction [27]. Though utilizing different alcohols presents little differences with regard to the kinetic of reaction, the final yield of esters remains unchanged. Thus, assortment of the alcohol is based on cost and performance consideration. Generally, reaction temperature is set at near the boiling point of the alcohol used [28].

Due to the reality that many vegetable oils, including soybean, canola (rapeseed) oil, and rice bran oil, have a major number of FAs with double bonds, oxidative stability is a problem, particularly when storing BD for longer period of time [29, 30]. This problem becomes severe due to improper storage conditions, which may include exposure to air and/or light, temperatures above ambient, and presence of extraneous materials (contaminants) with catalytic effect on oxidation. Some additives such as antioxidants might control the oxidation.

Characterization of BD fuel properties and evaluation of its quality are the matters of great concern for the successful commercialization of this fuel. A high fuel value with no operational problems is a condition for market acceptance of BD. Accordingly, the analysis of BD and the monitoring of the transesterification reaction have been the subject of numerous publications [31, 32]. The constraints, which are used to define the quality of BD, can be divided in two groups [33]. One of them is also used for mineral diesel, and the second illustrates the composition and purity of fatty esters. The former includes, for example, density, viscosity, flash point, sulfur percentage, carbon residue, sulfated ash percentage, cetane number, and acid number. The latter comprises, for example, methanol, free glycerol, total glycerol, phosphorus contents, water, and esters content. Chromatography and spectroscopy are the mainly used analytical methods for BD analyses, but procedures based on physical properties are also available [34]. Furthermore, it is important to mention that in most chromatographic analyses, mainly gas chromatography (GC) has been applied to methyl and not to ethyl esters [29].

As the demand for vegetable oils for food has increased tremendously in recent years, hence, the contribution of nonedible oils such as jatropha, *Moringa oleifera*, rice bran oils, etc. can play an important role for BD production. In view of the limited petro-oil resources and rapidly growing energy demands of the world, there is an extensive need to take immediate initiatives for exploring alternative energy sources to meet the domestic needs and reduce the dependence on imported fossil fuels. In view of the future perspectives of biofuels, the present book chapter was designed with the main purposes to assess the feasibility of BD production from multi-feedstock vegetable oil sources.

## 1.2 History of the Use of Vegetable Oil in Biodiesel

The idea to use vegetable oils as fuels for diesel engines dates back to more than one hundred years. Historically, Rudolf Diesel, the inventor of diesel engine, at the Paris Exhibition in 1900, conducted engine tests, for the first time, on peanut oil [22, 35]. At that moment Diesel said, "The use of vegetable oils for engine fuels may seem insignificant today. However, such oils may in course of time be as important as petroleum and the coal tar products of the present time." Today, over a century later, the scientific community is working to fulfill his dream by considering potential benefits of BD as an alternative fuel to petrodiesel for future uses.

## 1.3 Feedstocks for Biodiesel Production

All over the world, the usual lipid feedstocks for BD production are refined vegetable oils. In this group, the oil of choice varies with location according to availability; the most abundant lipid is generally the most common feedstock. The bases for this are not only the desire to have an ample supply of product fuel but also because of the inverse relation between supply and cost. Refined oils can be comparatively costly under the best of conditions, compared with petroleum products, and the choice of oil for BD production depends on local availability and corresponding affordability. The four oil crops clearly dominate the feedstock sources used for worldwide BD production. With a share of nearly 85%, rapeseed oil is by far leading the field, followed by sunflower seed oil, soybean oil, and palm oil [36]. Apart from the "great four" – rapeseed oil, sunflower seed oil, soybean oil, and palm oil in BD production – other edible plant oils have also successfully been transesterified to produce BD.

The choice of raw material used for BD production in a specific region mainly depends on the respective climatic conditions. Thus, rapeseed and sunflower oils are mainly used in the European Union [37], palm oil predominates in BD production in tropical countries [38, 39], and soybean oil [40] and animal fats are the major feedstocks in the United States. FA ester production has also been demonstrated from a variety of other feedstocks, including the oils of coconut [41], rice bran [42], *Thespesia populnea* [43], safflower [44], palm kernel [45], *M. oleifera* [46], *Citrus reticulata* (mandarin orange) [47], *Jatropha curcas* [48], Ethiopian mustard [13], *Cynara cardunculus* [49], *Hibiscus esculentus* [50], maize [51], *Cyperus esculentus* (Barminas et al. [52]), *Prunus mahaleb* [53], kapok [54], tobacco [55], milkweed [7], *Yucca aloifolia* [56], *Oleum papaveris seminis* [57], *Pongamia* [58], *Brassica napus* [59], *Citrullus colocynthis* [53], rubber seed oils [60], palm FA distillate [61], the animal fats, tallow [7, 62], lard [63], and waste oils [64, 65]. As such, any animal or plant lipid should be a ready substrate for the production of BD. Such features as supply, cost, storage properties, and engine performance will determine whether a particular potential feedstock is actually acceptable for commercial fuel production.

One way of reducing the production costs for BD fuels is the use of nonedible oils, which tend to be considerably cheaper than edible vegetable oils [66]. A number of plant oils contain substances that make them unsuitable for human consumption. In some cases, these substances can be removed by refining. For example, gossypol contained in cottonseeds can effectively be eliminated from the oil and the press cake to allow utilization as a cooking oil and animal feed, respectively [55]. Sometimes harmful ingredients can also be eliminated by breeding, as was the case with glucosinolates and erucic acid in rapeseed. In many cases, however, the removal of toxic components from the fatty material has not been accomplished or even attempted yet.

### 1.3.1 Generations of Biodiesel

BD production has proven to be sustainable because of the wide coverage of raw material availability, estimated at more than 350 types of oilseed crops worldwide. The feedstocks are generally easily accessible but vary depending on geographic location, weather conditions, land type, and agricultural practices in any country. In addition, the BD feedstocks portray 75% of total manufacturing cost; thus it is essential to choose appropriate feedstock to ensure the BD production feasibility. Typically, BD raw materials can be categorized as first-generation, second-generation, and third-generation BD, accordingly to the material used for the BD synthesizing as shown in Table 1.1 [15].

### 1.3.2 First-Generation Biodiesel

First-generation BD may be defined as edible oils from agricultural products such as palm oil, olive, sunflower, coconut, canola, rapeseed soybeans, etc. [71, 72]. Today, derived BD from edible oils has reached approximately more than 95% and has raised many issues,

**Table 1.1** Main feedstocks of biodiesel.

| First-generation oil | Second-generation oil | Third-generation oil |
|---|---|---|
| Soybean | Rubber seed | *Nannochloropsis oculata* |
| Canola | Cotton seed | *Chlamydomonas pitschmannii* |
| Palm | Tobacco seed | *Isochrysis* sp. |
| Rapeseed | Karanja | *Chlorella vulgaris* |
| Coconut | Jojoba oil | *Monoraphidium* sp. |
| Olive | Neem | |
| Sunflower | Moringa | |
| Peanut | Jatropha | |
| Sesame | Coffee ground | |
| Mahua | Used cooking oil | |
| Barley | Tallow | |
| Wheat | Fish oil | |
| | Chicken fat | |
| | Bitter almond oil | |

*Source:* Adapted from [67–70].

especially the competition between food supply and oil demand crisis, deforestation, and soil destruction for feedstock plantation purposes [73]. In the past decade, the price of edible oils has escalated, while the production demand for BD conversion is continuously increasing, resulting in edible-based BD being less economically feasible [74]. Given these circumstances, the exploitation of first-generation BD as a replacement for diesel fuel has put the world's stock of edibles in jeopardy.

### 1.3.3 Second-Generation Biodiesel

Currently, BD derived from inedible oil has sparked scientists' interest in replacing reliance on edible oil-based diesel. The inedible oil crops can be planted in wasteland or fallow areas without intensive agriculture, which can produce high oil yields [39]. Besides, waste oils and animal fats can be categorized as second-generation BD. The use of waste as a BD feedstock may reduce the problem of waste disposal and the cost of BD production [75]. Both waste oils and animal fats are most likely to contain water and a slightly higher FFA value compared with virgin oils, which result in lower oil quality. Different types of raw materials will produce different quantities of yield and characteristics of the oil, so the selection of raw materials is very important because the cost of producing BD is very expensive.

### 1.3.4 Third-Generation Biodiesel

Recently, the use of microalgae-based BD has gained immense awareness and prospects for meeting the growing supply of BD feedstocks. The microalgae-based BD has the advantage of growing at a faster rate under photoautotrophic condition and is able to produce high yield of oil than edible and nonedible crop oil [76]. Also in the future, microalgae may make a significant contribution to addressing the issue of food production versus BD production and reducing competition for farmland [77]. Moreover, a study discovered that the algae-based BD has lower carbon footprint, which is beneficial to the ecosystem [67]. Nevertheless, it is important to study the production cost and energy output of algae-based BD so that it is much more feasible and cost-effective for mass production as an alternative to fossil fuel sources.

## 1.4 Basics of the Transesterification Reaction

The direct use of a vegetable oil in diesel engines is problematic because of its high viscosity (about 11–17 times higher than petrodiesel fuel), which reduces the fuel atomization leading to high engine deposits, thickening of lubricating oil, and lower volatilities that cause the formation of deposits in engines due to incomplete combustion [78, 79]. The extremely high flash points of vegetable oils and their tendency for thermal or oxidative polymerization aggravate the situation, leading to the formation of deposits on the injector nozzles, a gradual dilution and degradation of the lubricating oil, and the sticking of piston rings. As a consequence, long-term operation on neat plant oils or on mixtures of plant oils and fossil diesel fuel inevitably results in engine breakdown [80].

Chemically, the vegetable oils/animal fats consist of triglyceride molecules of three long-chain FAs that are ester bonded to a single glycerol molecule. These FAs differ by the length of carbon chains and by the number, orientation, and position of double bonds in these chains. Thus, BD refers to alkyl esters of long-chain FAs, which are synthesized either by transesterification with alcohols or by esterification of FAs. The latter strategy aimed at modifying plant oils by various technologies to produce fuels that approximate the properties and performance of fossil diesel. Four methods to decrease the high viscosity of vegetable oils to enable their use in common diesel engines without operational problems such as engine deposits have been investigated: blending with petrodiesel, pyrolysis, microemulsification, and transesterification [81]. Transesterification is by far the most common method that leads to the products commonly known as BD, which are alkyl esters of vegetable oils or animal fats.

Pyrolysis denotes thermal decomposition reactions, usually brought about in the absence of oxygen. Pyrolysis of vegetable and fish oils, optionally in the presence of metallic salts as catalysts, was conducted as a means of producing emergency fuels during the Second World War, as various Chinese, Japanese, or Brazilian publications show [13]. In addition, the technology has occasionally found entry into the more recent literature as well [82, 83]. This treatment results in a mixture of alkanes, alkenes, alkadienes, aromatics, and carboxylic acids, which are similar to hydrocarbon-based diesel fuels in many respects. The cetane number of plant oils is increased by pyrolysis, and the concentrations of sulfur, water, and sediment for the resulting products are acceptable. However, according to modern standards, the viscosity of the fuels is considered as too high, ash and carbon residue far exceed the values for fossil diesel, and the cold flow properties of pyrolyzed vegetable oils are poor [81]. Moreover, it is argued that the removal of oxygen during thermal decomposition eliminates one of the main ecological benefits of oxygenated fuels, namely, more complete combustion due to higher oxygen availability in the combustion chamber [22].

Microemulsification is the formation of thermodynamically stable dispersions of two usually not miscible liquids, brought about by one or more surfactants. Drop diameters in microemulsions typically range from 100 to 1000 Å [84]. Various investigators have studied the microemulsification of vegetable oils with methanol, ethanol, or 1-butanol [85, 86]. They arrived at the conclusion that microemulsions of vegetable oils and alcohols cannot be recommended for long-term use in diesel engines for similar reasons as applicable to neat vegetable oils. Moreover, microemulsions display considerably lower volumetric heating values as compared to hydrocarbon-based diesel fuel due to their high alcohol contents [84], and these have also been assessed insufficient in terms of cetane number and cold temperature behavior [87].

Transesterification with lower alcohols, however, has emerged to be an ideal modification, so that the term "biodiesel" is now only used to denote products obtained by this process. The reaction between triglycerides and lower alcohols, yielding free glycerol and the FA esters of the respective alcohol, was first described in 1852 [88]. In the 1930s and 1940s, this reaction was frequently applied in the fat and soap industry. The Belgian patent on the production of palm oil ethyl esters by acid-catalyzed transesterification describes the first use of a fuel, which would now be referred to as "biodiesel" [7].

Transesterification is one of the reversible reactions and proceeds essentially by mixing the animal fat/vegetable oil with an alcohol (usually methanol). However, the presence of a catalyst (usually a base) accelerates the conversion to produce the corresponding alkyl esters (or for methanol, the methyl esters) of the FA mixture that is found in the parent vegetable oil or animal fat [79]. The general scheme of the transesterification reaction is given in Figure 1.1.

The production of BD by transesterification has been the focus of many research studies. Several reviews on the production of BD by transesterification have been published [71, 89–92]. Usually, transesterification can proceed by base or acid catalysis. However, in homogeneous catalysis, alkali catalysis (sodium or potassium hydroxide; or the corresponding alkoxides) is a much more rapid process than acid catalysis [92, 93].

## 1.5 Variables Affecting Transesterification Reaction

The process of transesterification is affected by various variables depending upon the reaction conditions employed. The most pertinent variables for this kind of reaction are described as follows:

- Catalyst type and concentration
- Molar ratio of alcohol to vegetable oil
- Reaction temperature
- Agitation intensity
- Reaction time
- Water and FFA contents

## 1.6 Alkaline-Catalyzed Transesterification

Conventionally, transesterification reactions are alkali catalyzed. Alkaline catalysts, such as sodium hydroxide, sodium methoxide, potassium hydroxide, and potassium methoxide, are more effective and most commonly used for BD production [43, 94]. When compared with acid or other type of catalysts, basic ones show a high conversion under mild temperature conditions and in short reaction times [95]. For transesterification giving maximum yield, the alcohol should be free of moisture, and the FFA content of the oil should be <0.5% [96]. The absence of moisture in the transesterification reaction is important because according to the equation (as shown for methyl esters next), the hydrolysis of the formed alkyl esters to FFAs can occur.

$$\text{R-COOCH}_3 + \text{H}_2\text{O} \rightarrow \text{R-COOH} + \text{CH}_3\text{OH} \quad (\text{R} = \text{alkyl})$$

Similarly, because triacylglycerols are also esters, the reaction of the triacylglycerols with water can form FFA.

The ester yields are lower with crude oil in the presence of gums and extraneous material. The parameters (60 °C reaction temperature and 6 : 1 methanol:oil molar ratio) have almost become a standard for methanol-based transesterification. Other alcohols (ethanol and butanol) require higher temperatures (75 and 114 °C, respectively) for optimum conversion [21]. Alkoxides in solution with the corresponding alcohol have the advantage over hydroxides that the water forming reaction according to the equation

$$R'OH + XOH \rightarrow R'OX + H_2O \quad (R' = \text{alkyl; } X = \text{Na or K})$$

cannot occur in the reaction system, thus ensuring that the transesterification reaction system remains as water-free as possible. This reaction, though, is the one forming the transesterification causing alkoxide when using NaOH or KOH as catalysts.

Effects, similar to those discussed earlier, were observed in studies on transesterification of beef tallow [15]. FFA and especially water should be kept as low as possible [97]. NaOH reportedly was more effective than the alkoxide [98]; however, this may have been a result of the reaction conditions. Mixing was significant due to the immiscibility of NaOH/MeOH with beef tallow, with smaller NaOH/MeOH droplets resulting in faster transesterification [15]. Ethanol is more soluble in beef tallow, which increased yield [99], an observation that should hold for other feedstocks as well.

Several studies reveal the use of both NaOH and KOH in the transesterification of rapeseed oil [100]. At 32 °C, transesterification was 99% completed in 4 h when using an alkaline catalyst (NaOH or NaOMe). At >60 °C, using an alcohol:oil molar ratio of at least 6 : 1 and fully refined oils, the reaction was completed in 1 h yielding methyl, ethyl, or butyl esters [100]. Current work on producing BD from waste frying oils employed KOH. With the reaction conducted at ambient pressure and temperature, conversion rates of 80–90% were achieved within 5 min, even when stoichiometric amounts of methanol were employed [101]. Within two transesterifications (with more MeOH/KOH steps added to the methyl esters after the first step), the ester yields were 99%. It was concluded that FFA content up to 3% in the feedstock did not affect the process negatively, and phosphatides up to 300 ppm phosphorus contents were acceptable. The resultant methyl ester met the quality requirements for Austrian and European BD without further treatment.

In another study, similar to previous work on the transesterification of soybean oil, it was concluded that KOH is more effective than NaOH in the transesterification of safflower oil of Turkish origin [102]. In this experiment, the optimal conditions offering 97.7% methyl ester yields were as follows: 1.0% KOH catalyst (by weight), 69 ± 1 °C reaction temperature, 7 : 1 alcohol:vegetable oil molar ratio, and 18 min reaction time. Depending upon the vegetable oil and its constituent FAs influencing FFA content, adjustments to the alcohol:oil molar ratio and the amount of catalyst may be required as was reported for the alkaline transesterification of *Brassica carinata* oil [103]. However, advantages of NaOH over KOH as a catalyst are that sodium hydroxide-catalyzed transesterifications tend to be completed faster [104], and sodium hydroxide is cheaper. It was [34] reported that the esters yield is affected by methanol/oil molar ratio, catalyst type, and its concentration and reaction temperature. They observed that BD with the best properties was obtained using an optimum methanol/oil molar ratio (6 : 1), potassium hydroxide as catalyst (1%), and 65 °C reaction temperature.

Although the alkali hydroxides are the catalysts of choice for methanolysis, these cannot be applied in transesterifications with higher and secondary alcohols, as the reactivity between KOH or NaOH and the alcohol to form the respective alkoxide anion dramatically decreases with increasing chain length [43]. So, if higher or branched-chain esters are to be produced by alkaline catalysis, only the use of pure sodium or potassium is feasible, although under much higher reaction temperatures than those sufficient for methanolysis. Table 1.2 depicts the homogeneous alkaline catalysts at different reaction conditions for transesterification process.

The reaction mechanism of alkali-catalyzed transesterification has long been known. The actual catalytic species is the respective alcoholate anion (i.e. methoxide for methanolysis). Transesterification is started by a nucleophilic attack of the alkoxide ion on the carbonyl carbon atom of the triglyceride molecule, resulting in a tetrahedral intermediate. In a second step, this intermediate splits into the desired methyl ester and the anion of the diglyceride. The latter reacts with methanol to from a diglyceride molecule, which will analogously be converted into monoglyceride and glycerol, and a methoxide ion, which can start another catalytic cycle.

The optimum concentration of homogeneous alkaline catalysts ranges from 0.5 to 1.0% by weight of the oil [111]. A high amount of FFAs in the reaction mixture can partly be compensated by the addition of more catalyst [22]. However, it was reported that higher catalyst concentrations increase the solubility of the methyl esters in the glycerol phase, so that a significant amount of esters remains in the lower phase even after separation [112]. Therefore, several investigators suggested for calculating the optimum amount of KOH or NaOH necessary to facilitate transesterification and at the same time neutralize the acidity of the oils (Table 1.2).

In principle, transesterification is a reversible reaction, although in the production of vegetable oil alkyl esters, i.e. BD, the back reaction does not occur or is negligible largely because the glycerol formed is not miscible with the product, leading to a two-phase system. The transesterification of soybean oil with methanol or 1-butanol was reported to proceed [35] with pseudo-first-order or second-order kinetics, depending on the molar ratio of alcohol to soybean oil (30 : 1 pseudo-first order, 6 : 1 second order; NaOBu catalyst), whereas the reverse reaction was second order [65].

The methanolysis of sunflower oil at a molar ratio of methanol:sunflower oil of 3 : 1 was reported to begin with second-order kinetics, but then the rate decreased due to the formation of glycerol [108]. A force reaction (a reaction in which all three positions of the triacylglycerol react virtually simultaneously to give three alkyl ester molecules and glycerol), originally proposed as part of the forward reaction, has shown that second-order kinetics are not followed and miscibility phenomena [113] can play a significant role. The cause is that the vegetable oil starting material and methanol are not well miscible. The development of glycerol from triacylglycerols proceeds stepwise via the di- and monoacylglycerols, with an FA alkyl ester molecule being formed in each step. From the fact that diacylglycerols reach their maximum concentration before the monoacylglycerols, it was concluded that the last step, formation of glycerol from monoacylglycerols, proceeds more rapidly than the formation of monoacylglycerols from diacylglycerols [114].

The count of cosolvents such as tetrahydrofuran (THF) or methyl *tert*-butyl ether (MTBE) for methanolysis reaction was reported to notably accelerate the methanolysis of vegetable oils as a result of solubilizing methanol in the oil to a rate comparable to that of the faster butanolysis [115, 116]. This is to prevail over the limited miscibility of alcohol and oil at the early reaction stage, creating a single phase. The procedure is applicable for use with other alcohols and for acid-catalyzed pretreatment of high FFA feedstocks. Though, molar ratios

**Table 1.2** Homogeneous catalysts and reaction conditions used for alkaline transesterification.

| Catalyst type | Examples | Reaction conditions | Oils and fats | Alcohol | Esters yield | References |
|---|---|---|---|---|---|---|
| Alkali metals (dissolved in alcohol) | $AlCl_3 \cdot 6H_2O$ | Alcohol:oil = 10 : 1, $T = 72\,°C$, $t = 2\,h$, 1.5 wt% catalyst loading | Waste oil | Methanol | 94% | [105] |
| Alkali metal alcoholates and hydroxide | KOH | Alcohol:oil = 9 : 1, $T = 70\,°C$, $t = 1\,h$, catalyst loading = 1.0 wt% | Waste cooking oil | Methanol | 98.2% | [106] |
| | KOH | Alcohol:oil = 20.39 wt%, $T = 57.1\,°C$, $t = 54.1\,min$, catalyst loading = 0.4 wt% | Black mustard | Methanol | 97.3% | [107] |
| | NaOH | Alcohol:oil = 10:1, $T = 65\,°C$, $t = 1.5\,h$, catalyst loading = 1.5 wt% | Waste cooking oil | Methanol | 88.1 | [108] |
| | $CH_3ONa$ | Alcohol:oil = 3.37:1, $T = 60\,°C$, $t = 1\,h$, catalyst loading = 0.5 wt% | Sunflower oil | Methanol | 99.7 | [109] |
| | $CH_3OK$ | Alcohol:oil = 5 : 1, $T = 86\,°C$, $t = 1.5\,h$, catalyst loading = 2 wt% | *Thevetia peruviana* seed oil | Dimethyl carbonate | 97.1 | [110] |

of alcohol:oil and other parameters are affected by the addition of the cosolvents. Here, some extra complexity also occurs due to recovering and recycling the cosolvent. This can be minimized by choosing a cosolvent with a boiling point near that of the alcohol being used. However, there may be some hazards associated with its most common cosolvents, THF and MTBE.

Nevertheless, the traditional homogeneous catalysis offers a series of advantages; its major disadvantage is the fact that homogeneous catalysts cannot be reused. Moreover, catalyst residues have to be removed from the ester product, usually necessitating several washing steps, which increases production costs. Thus, there have been various attempts at simplifying product purification by applying heterogeneous catalysts, which can be recovered by decantation or filtration or are alternatively used in a fixed-bed catalyst arrangement. The most frequently cited heterogeneous alkaline catalysts are alkali metal and alkaline earth metal carbonates and oxides. For the production of biofuels in tropical countries, Vargas et al. [117] recommended utilizing the ashes of oil crop waste (e.g. coconut fibers, shells, and husks) as catalysts. Such natural catalysts are rich in carbonates and potassium oxide and have shown considerable activity in transesterifications of coconut oil with methanol and water-free ethanol. Some studies reveal the use of heterogeneous catalysts for transesterification of vegetable oils [118, 119]. No heterogeneous catalysts are commercially feasible in the 45–65 °C range. Some may be feasible at 100–150 °C; however, reactor residence times are more than 4 h, involving large amounts of catalysts. At temperature higher than 100–150 °C, the high pressures needed to keep the methanol in the liquid phase can significantly increase equipment costs [16].

The application of calcium carbonate may seem particularly promising, as it is a readily available, low-cost substance. Moreover, Ho et al. reported that this catalyst showed no decrease in activity even after several weeks of utilization, and the spent calcium carbonate could easily be disposed of in cement kilns [120]. However, the high reaction temperatures and pressures and the high alcohol volumes required in this technology are likely to prevent its commercial applications. The alkali and alkaline earth metals as a catalyst are also in practice for transesterification of vegetable oils. Arzamendi et al. [121] investigated the methanolysis of refined sunflower oil with a series of catalysts consisting of alkaline and alkaline earth metals. Abdelhady et al. studied the activity of activated CaO as a heterogeneous catalyst in the production of BD by transesterification of sunflower oil with methanol [122]. In another study, Riso et al. investigated the performance of calcium methoxide as a solid base catalyst, and it was observed that 98% BD yields within 2 h [94]. However, drawbacks as associated with heterogeneous catalyst are reported for alkali metal or alkaline earth metal salts of carboxylic acids. The use of strong basic ion-exchange resins as catalysts, on the other hand, is limited by their low stability at temperatures higher than 40 °C and by the fact that FFAs in the feedstock neutralize the catalysts even in low concentrations. Finally, glycerol released during the transesterification process has a strong affinity to polymeric resin material, which can result in complete impermeability of the catalysts [9].

Other possibilities for accelerating the transesterification are microwave [123] or ultrasonic [28] irradiation. Further fundamental materials, such as alkylguanidines, which were anchored to or entrapped in various supporting materials such as polystyrene and zeolite [124], also catalyze transesterification. Such schemes may provide for easier catalyst recovery and reuse. A review article on various transesterification strategies [125] suggested replacing conventional sodium and potassium compounds by guanidines, such as TBD (1,5,7-triazabicyclo[4.4.0]dec-5-ene). These compounds enable high conversion under comparatively mild reaction conditions like conventional alkaline catalysts, while they will not cause the formation of soaps. Moreover, it was found that guanidines can be fixed on organic polymers, such as modified polystyrene, or can be entrapped in a $SiO_2$ sol–gel matrix, which facilitates heterogeneous catalysis and thus enables the repeated use of the catalyst preparation. However, guanidines tend to leach from the carrier, so that the activity of the fixed catalysts markedly decreases in repeated use.

## 1.7 Acid-Catalyzed Transesterification

Acid-catalyzed transesterification offers the advantage of esterifying FFAs contained in the fats and oils and is therefore especially suited for the transesterification of highly acidic fatty materials, such as palm oil or waste edible oils. Used cooking oils typically contain 2–7% FFA, and animal fats contain 5–30% FFA. A few very low quality feedstocks, such as trap grease, can approach 100% FFA level. Also, acid-catalyzed transesterification enables the production of long- or branched-chain esters, which pose considerable difficulty in alkaline transesterification because the FFA react with the catalyst to form soap and water [126] as shown:

$$\text{R-COOH} + \text{KOH} \rightarrow \text{R-COOK} + \text{H}_2\text{O}$$

Up to 5% FFA, the reaction can still be catalyzed with an alkali catalyst, but additional catalyst must be added to compensate for that lost to soap. The soap produced during the reaction is either removed with the glycerol or washed out during the water wash. When the FFA level is >5%, the soap inhibits separation of the glycerol from the methyl esters and contributes to emulsion formation during the water wash. Intended for these cases, an acid catalyst such as sulfuric acid can be used to esterify the FFA to methyl esters as shown in the following reaction:

$$\text{R-COOH} + \text{CH}_3\text{OH} \rightarrow \text{R-COOCH}_3 + \text{H}_2\text{O}$$

This process can be used as a pretreatment to convert the FFA to methyl esters, thereby reducing the FFA level. In that case, the low FFA pretreated oil can be transesterified with an alkali catalyst to convert the triglycerides to methyl esters [127]. As depicted in the reaction, water is produced and, if it accumulates, it can stop the reaction well before completion. It was projected to allow the alcohol to separate from the pretreated oil or fat after the reaction. Exclusion of this alcohol also removes the water formed by the esterification reaction and permits for a second step of esterification; alternatively, one may proceed directly to alkali-catalyzed transesterification. It is important to note that the methanol–water mixture will also contain some dissolved oil and FFA that should be recovered and reprocessed. The pretreatment by an acidic ion-exchange resin has also been described [128]. It was revealed [129, 130] that acid-catalyzed esterification can be used to produce BD from low-grade by-products of the oil refining industry such as soap stock.

Acid catalysts provide high yields, but the transesterification reaction is slow than that by alkali catalysis and requires higher temperatures. The most common acids used are phosphoric, hydrochloric, sulfonic, and sulfuric acids. They are also employed in pretreatment steps, to esterify the FFA, prior to basic catalyzed reaction. Nevertheless, acid catalysis is also affected by the presence of water that inhibits the reaction. Transesterification happens at a faster rate in the presence of an alkaline catalyst than in the presence of the same amount of acid catalyst [131].

The typical reaction conditions for homogeneous acid-catalyzed methanolysis are temperatures of up to 100 °C and pressures of up to five bars in order to keep the alcohol liquid [132]. A further disadvantage of acid catalysis – probably prompted by the higher reaction temperatures – is an increased formation of unwanted secondary products, such as dialkyl ethers or glycerol ethers [76]. Finally, in contrast to alkaline reactions, the presence of water in the reaction mixture proves absolutely detrimental for acid catalysis. Fonseca et al. reported that the addition of 0.5% water to a mixture comprising soybean oil, methanol, and sulfuric acid reduced ester conversion from 95% to below 90% [127]. At a water content of 5%, ester conversion decreased to only 5.6%. It should also be noted that water released during esterification of FFA might inhibit further reaction, so that very acidic raw materials might give moderate conversion even in acid-catalyzed alcoholysis. Shu et al. investigated the effect of reaction variables such as feed composition, temperature, and rate of mixing on the kinetics of the acid-catalyzed transesterification of waste frying oils. The optimal yield of 99% was achieved after a 4 h reaction [133].

For acid-catalyzed transesterification, the concentrated sulfuric acid is the most frequently used catalytic substance. Its advantages are its low price and its hygroscopicity, which is important for the esterification of FFAs, removing released water from the reaction mixture. Drawbacks include its corrosiveness, its tendency to attack double bonds in unsaturated FAs, and the fact that concentrated $H_2SO_4$ may cause dark coloring in the ester product [65]. Besides, also the use of various sulfonic acids as homogeneous catalysts is reported. These substances have lower catalytic activity than mineral acids. However, they pose fewer problems in handling and do not attack double bonds within the starting material.

## 1.8 Enzymatic-Catalyzed Transesterification

Although at present BD is successfully produced chemically, there are several associated problems such as glycerol recovery and the need to use refined oils and fats as primary feedstocks [127]. The use of lipases from various microorganisms is becoming important in BD production. Lipases are enzymes that catalyze both the hydrolytic cleavage and the synthesis of ester bonds in glycerol esters. The disadvantages of using chemical catalysts can be overcome by using lipases as the catalysts for ester synthesis [134]. Advantages mentioned for lipase catalysis over chemical methods in the production of simple alkyl esters include the ability to esterify both acylglycerol linked and FFA in one step, the production of a glycerol side stream with minimal water content and with little or no inorganic material, and catalyst reuse. Other advantages include the occurrence of transesterification under mild temperature, pressure, and pH conditions; neither the ester product nor the glycerol phase has to be purified from basic catalyst residues or soaps. This means that phase separation is easier, high quality glycerol can be obtained as a by-product, and environmental problems due to alkaline wastewater are eliminated [135]. Moreover, both the transesterification of triglycerides and the esterification of FFAs occur in one process step. Consequently, also highly acidic fatty materials, such as palm oil or waste oils, can be used without pretreatment [136]. Finally, many lipases show considerable activity in catalyzing transesterifications with long- or branched-chain alcohols, which can hardly be converted to FA esters in the presence of conventional alkaline catalysts.

Early work on the application of enzymes for BD synthesis was conducted using sunflower oil as the feedstock [137] and various lipases to perform alcoholysis reactions in petroleum ether. From the tested lipases, only three were found to catalyze alcoholysis with an immobilized lipase preparation of a *Pseudomonas* sp. offering the maximum ester yields. Maximum conversion (99%) was obtained with ethane, and when the reaction was repeated without solvent, only 3% product was produced with methanol as alcohol, whereas with absolute ethanol and 96% ethanol and 1-butanol, the ester yields were ranged between 70 and 82%, respectively. Reactions by a progression of homologous alcohols showed that reaction rates, with or without the addition of water, increased with increasing chain length of the alcohol. For methanol, the highest conversion was obtained without the addition of water, but for other alcohols the addition of water increased the esterification rate two to five times.

Pedro et al. reported the lipase-catalyzed alcoholysis of low erucic acid rapeseed oil without organic solvent in a stirred batch reactor. The best results were obtained with a *Candida rugosa* lipase, and under optimal conditions nearly complete conversion of oil to ester was obtained [138]. Other studies [139] reported the ethanolysis of sunflower oil with lipozyme in a medium totally composed of sunflower oil and ethanol. In this case the factors studied for the conversion of the oil to esters included substrate molar ratio, reaction temperature and time, and enzyme load. Ethyl ester yields, however, did not exceed 85% even under the optimized reaction conditions. These authors also reported that the ester yields could be improved by adding silica to the medium. The positive effect of silica on yield was attributed to the adsorption of the polar glycerol coproduct onto the silica, which reduced glycerol deactivation of the enzyme. The reuse of the enzyme was also investigated, but ester yields decreased significantly with enzyme recycle, even in the presence of added silica.

In other studies [140, 141], mixtures of soybean and rapeseed oils were treated with various immobilized lipase preparations in the presence of methanol. Lipase from *Candida antarctica* was found to be the most effective in methyl ester formation. To attain high levels of conversion of oil to methyl ester, three equivalents of methanol were needed because this level of methanol resulted in lipase deactivation. It was necessary to add methanol in three separate additions. Under these conditions, >97% conversion of oil to methyl ester was achieved. In another study [142], it was reported that the lipase of *Rhizopus oryzae* catalyzed the methanolysis of soybean oil in the presence of 4–30% water in the starting materials but was inactive in the absence of water. Methyl ester yields of >90% could be obtained with stepwise additions of methanol to the reaction mixture. Lately, the conversion of soy oil to BD in a continuous batch operation catalyzed by an immobilized lipase of *Thermomyces lanuginosus* was reported [143]. These instigators also used a stepwise addition of methanol to the reaction, and in this manner complete conversion of oil to ester was achieved. Replicates recycle of the lipase was made possible by removing the bound glycerol by washing with isopropanol. When crude soy oil was used as substrate, a much lower yield of methyl ester was obtained compared with that using refined oil [144]. The reduction in ester yields was directly related to the phospholipids content of the oil, which apparently deactivated the lipase. Maximum esterification activity could be attained by pre-immersion of the lipase in the crude oil before methanolysis.

During the transesterification of tallow with secondary alcohols, the lipases from *C. antarctica* (trade name SP435) and *Pseudomonas cepacia* (PS30) offered the best oil conversions to esters [145]. Reactions, run without the addition of water, were sluggish for both lipases, and conversions of only 60–84% were obtained overnight (16 h). The accumulation of small amounts of water improved the yields. The converse effect was observed in the case of methanolysis, which was extremely sensitive to the presence of water. For the branched-chain alcohols, isopropanol and 2-butanol, better ester yields were obtained when the reactions were run without solvent [146]. Reduced yields when using the normal alcohols methanol and ethanol, in solvent-free reactions were attributed to enzyme deactivation by these more polar alcohols. Similar effects were observed for both the methanolysis and iso-propanolysis of soybean and rapeseed oils [147]. The enzymatic conversion of lard to methyl and ethyl esters was reported [148] using a three-step addition of alcohol to the substrate in solvent-free medium [149]. The conversion of Nigerian palm oil and the lauric oils, palm kernel and coconut, to simple alkyl esters for use as BD fuels was also reported [150]. The best ester yields (>95%) were of ethyl esters.

Low-cost lipids, such as waste deep fat fryer grease, usually have relatively high levels of FFA (>8%). The lipases are of particular interest as catalysts to produce fatty esters from such feedstocks because they accept both free and glyceride-linked FAs as substrates for ester synthesis. On the other hand, BD production from such mixed feedstocks (e.g. spent rapeseed oil) using inorganic catalysts requires multistep processing [141]. To develop these attractive features of lipase catalysis, studies were conducted using a lipase from *P. cepacia* and recycled restaurant grease with 95% ethanol in batch reactions [151]. Subsequent work showed that methyl and ethyl esters of lard could be obtained by lipase-catalyzed alcoholysis [152]. The restaurant greases using a series of immobilized lipases from *T. lanuginosus, C. antarctica,* and *P. cepacia* in solvent-free medium utilizing a one-step addition of alcohol to the reaction system for methanolysis and ethanolysis were reported [153]. The continuous production of ethyl esters of grease using a phyllosilicate sol–gel immobilized lipase from *Burkholderia cepacia* (IM BS-30) as catalyst was investigated [154]. Enzymatic transesterification was carried out in a recirculating packed column reactor using 1M BS-30 as the stationary phase and ethanol and restaurant grease as the substrates, without solvent addition. The bioreactor was operated at temperatures (40–60 °C), flow rates (5–50 ml $min^{-1}$), and times (8–48 h) to optimize ester production. Under optimum operating conditions (flow rate, 30 ml $min^{-1}$; temperature, 50 °C; mole ratio of substrates, 4 : 1, ethanol:grease; reaction time, 48 h), the ester yields were >96%.

Nasaruddin et al. devised a two-step enzymatic protocol for the conversion of acid oils, a mixture of FFA and partial glycerides obtained after acid dilution of soap stock, to fatty esters. In the first step, the lipids in the acid oil were hydrolyzed using *Caulerpa cylindracea* lipase. In the second step, the high acid oils were esterified to short- and long-chain esters using an immobilized *Mucor miehei* lipase [155].

Another important aspect of lipase-catalyzed transesterifications is whether or not to use an organic lipophilic solvent. In general, alcoholysis with long-chain or branched alcohols proceeds efficiently even in a solvent-free medium, whereas solvent-free methanolysis tends to give low ester yields. This may be attributed to the poor solubility of methanol in fats and oils [152]. Depending on the type of lipase used, various solvents for fatty material and methanol have been suggested, including petroleum ether [156], hexane [147], isooctane [142], commercial fossil diesel fuel [157], 1,4-dioxan [158], and supercritical carbon dioxide [159]. From an economic viewpoint, however, the use of organic solvents is hardly useful [137], even more so as these have to be removed from the ester product by evaporation. Moreover, the toxicity and inflammability of organic solvents is also another issue to be considered. As a consequence, considerable effort has been directed toward conducting lipase-catalyzed methanolysis reactions in a solvent-free medium.

Zhang et al. reported regenerating enzyme preparations by using them with 2-butanol or tert-butanol [154], which proved successful for mobilized *C. antarctica* lipase. A recommendation for further treatment with 1-propanol for immobilized *Thermomyces iamgmosa* lipase [160]. If the enzyme chosen for transesterification turns out to be particularly sensitive to glycerol released by ester formation, it might make sense to use methyl acetate instead of methanol [161]. The authors claim that triacetylglycerol, which is produced instead of glycerol in this process, has no negative effects on the enzyme activity of immobilized *C. antarctica* lipase and does not affect the quality of the resulting fuel either.

Finally, BD producers can choose between several methods of preventing enzyme inactivation, which is a phenomenon frequently reported for lipase-catalyzed methanolysis. Enzymes are easily inactivated by compounds contained in the oil or fat. Quayson et al. found that phospholipids present in crude soybean oil efficiently inhibit methanolysis, as these bind to the immobilized enzyme and interfere in the interaction of lipase and substrate [162]. They concluded that for enzymatic methanolysis, vegetable oils have to be degummed. The enzyme-catalyzed reactions have the following disadvantages: (i) lose some initial activity due to volume of the oil molecule; (ii) number of support enzyme is not uniform; (iii) biocatalyst is more expensive than the natural enzyme; (iv) inactivation by acyl acceptors, such as methanol, and inactivation by minor components in the crude oil and waste oils; and (v) desorption from immobilization support and fouling in packed bed bioreactors. Due to said disadvantages, the enzymatic catalyzed transesterification reactions are not in common practice for commercial scale BD production.

## 1.9 Fuel Properties and Quality Specifications for Biodiesel

While BD is produced in reasonably differently scaled plants from vegetable oils of varying origin and quality, it was essential to install a standardization of fuel quality for assurance of engine performance exclusive of any difficulties. Austria was the first country in the world to define and approve the standards for rapeseed oil methyl esters as diesel fuel. As standardization is a prerequisite for successful market introduction and penetration of BD, standards or guidelines for the quality of BD have also been defined in other countries like Germany, Italy, France, the Czech Republic, and in the United States.

Quality standards are prerequisites for the commercial use of any fuel product. They serve as guidelines for the production process, guarantee customers that they are buying high quality fuels, and provide authorities with approved tools for the assessment of safety risks and environmental pollution [163]. Moreover, engine and automobile manufacturers rely on fuel standards for releasing warranties for their vehicles to be operated on BD.

In 1997, the European Committee for Standardization was mandated to develop a uniform standard for FAME fuels and come up with respective measurement procedures. The resulting standard EN 14214, which has come into picture in 2004, is valid for all member states of the European Union and thus replaces the respective national specifications. Apart from Australia and the United States of America, which have already drawn up BD quality norms, a number of countries worldwide (e.g. Brazil, Canada, Japan) are currently working on their introduction, having released drafts or preliminary specifications. Then, ASTM D 6751 and EN 14214 conditions as well as their analysis methods for BD are illustrated in Table 1.3.

The constraints that are utilized to describe the quality of BD can be divided in two groups [164]. One of them is also used for mineral diesel, and the second describes the composition and purity of fatty esters. The former includes, for example, density, viscosity, flash point, sulfur percentage, Conradson carbon residue, sulfate ash percentage, cetane number, and acid number. The latter comprises, for example, methanol, free glycerol, total glycerol, phosphorus contents, water and esters content, and other properties described in Table 1.3. Thus, ASTM D 6751 and EN 14214 specifications methods for BD are illustrated in Table 1.3.

## 1.10 Conclusion

Currently, the uses of BD as an eco-friendly alternative to petrodiesel are gaining much recognition. The production of BD from nonconventional oils may simultaneously reduce dependence on imported fossil fuels and help alleviate the food versus fuel dilemma that plagues rapeseed, soybean, palm, and other oilseed crops that are also traditional oil sources. As a result of development of local BD industry and market, opportunities would be raised for the farmers to grow new oilseed crops and increase production of traditional and nonconventional oils, generating profit and income for all the stakeholders. Moreover, the establishment of local BD industry not only will generate opportunities for employment and personnel training but also might help reduce the dependence on imported petroleum and fuel derived from it, which continues to decrease in availability and affordability. More research and advancements in BD technology coupled with large-scale cultivation of oilseed crops, especially the nontraditional crops, additional subsidies, and the relevant technological sector, may lead to further reduction of the cost of this renewable fuel. Furthermore, there is a real need to appraise the environmental benefits of producing BD and to consider such attributes while determining the cost incurred in the production of such green fuels.

**Table 1.3** Biodiesel specifications according to ASTM D6751 and EN 14214 standards.

| | ASTM D 6751 | | EN 14214 | |
|---|---|---|---|---|
| **Property** | **Test method** | **Limits** | **Test method** | **Limits** |
| Density (15 °C) | — | — | EN ISO 3675 | 860–900 kg $m^{-3}$ |
| Kinematic viscosity (40 °C) | ASTM D 445 | 1.9–6.0 $mm^2 s^{-1}$ | EN ISO 3104 | 3.5–5.0 $mm^2 s^{-1}$ |
| Flash point | ASTM D 93 | 130 °C, min | EN ISO 3679 | 120 °C, min |
| Cloud point | ASTM 2500 | Not specified | — | — |
| Sulfur content | ASTM 5453 | 0.05% (w/w), max | EN ISO 20864 | 10.0 mg $kg^{-1}$, max |
| Carbon residue | ASTM D 4530 | 0.050% (w/w), max | EN ISO 10370 | 0.30% ($molmol^{-1}$) |
| Cetane number | ASTM D 613 | 47, min | EN ISO 5165 | 51, min |
| Sulfated ash | ASTM 874 | 0.020% (w/w), max | ISO 3987 | 0.02% ($molmol^{-1}$) |
| Distillation temperature | ASTM D 1160 | 360 °C, max | — | — |
| Copper strip corrosion (3 h, 50 °C) | ASTM D 130 | No. 3, max | EN ISO 2160 | 1 (degree of corrosion) |
| Acid number or acid value | ASTM 664 | 0.50 mg KOH $g^{-1}$, max | EN 14104 | 0.50 mg KOH $g^{-1}$, max |
| Iodine value | — | — | EN 14111 | 120 $gI_2 \cdot 100 g^{-1}$, max |
| P content | ASTM D 4951 | 0.001% (w/w), max | EN 14107 | 10.0 mg $kg^{-1}$, max |
| Water content | ASTM D 2709 | 0.050% (v/v), max | EN ISO 12937 | 500 mg $kg^{-1}$, max |
| Oxidative stability | — | — | EN 14112 | 6 h, min |
| Methanol content | — | — | EN 14110 | 0.20% ($molmol^{-1}$) |
| Free glycerine | ASTM D 6584 | 0.020% (w/w), max | EN 14105 | 0.020% ($molmol^{-1}$), max |
| Total glycerine | ASTM D 6584 | 0.240% (w/w), max | EN 14105 | 0.25% ($molmol^{-1}$), max |
| Ester content | — | — | EN 14103 | 96.5% ($molmol^{-1}$) |

## References


**1** Stephen, J.L. and Periyasamy, B. (2018). Innovative developments in biofuels production from organic waste materials: a review. *Fuel* 214: 623–633.

**2** Nayak, A., Pulidindi, I.N., and Rao, C.S. (2020). Novel strategies for glucose production from biomass using heteropoly acid catalyst. *Renew. Energy* 159: 215–220.

**3** Sovtić, N., Predrag, K.S., Bera, O.J. et al. (2020). A review of environmentally friendly rubber production using different vegetable oils. *Polym. Eng. Sci.* 60 (60): 1097–1117.

4 Wang, A., Sudarsanam, P., Xu, Y. et al. (2020). Functionalized magnetic nanosized materials for efficient biodiesel synthesis: via acid-base/enzyme catalysis. *Green Chem.* 22 (10): 2977–3012.

5 Demirbas, A., Ak, N., Aslan, A., and Sen, N. (2018). Calculation of higher heating values of hydrocarbon compounds and fatty acids. *Pet. Sci. Technol.* 36 (11): 712–717.

6 Thushari, I., Babel, S., and Samart, C. (2018). Biodiesel production in an autoclave reactor using waste palm oil and coconut coir husk derived catalyst. *Renew. Energy* 134: 125–134.

7 Mahmudul, H.M., Hagos, F.Y., Mamat, R. et al. (2017). Production, characterization and performance of biodiesel as an alternative fuel in diesel engines – a review. *Renew. Sustain. Energy Rev.* 72: 497–509.

8 Singh, D., Sharma, D., Soni, S.L. et al. (2021). A comprehensive review of biodiesel production from waste cooking oil and its use as fuel in compression ignition engines: 3rd generation cleaner feedstock. *J. Clean. Prod.* 307: 127299.

9 Abdullah, S.H.Y.S., Hanapi, N.H.M., Azid, A. et al. (2017). A review of biomass-derived heterogeneous catalyst for a sustainable biodiesel production. *Renew. Sustain. Energy Rev.* 70: 1040–1051.

10 Lin, L., Cunshan, Z., Vittayapadung, S. et al. (2011). Opportunities and challenges for biodiesel fuel. *Appl. Energy* 88 (4): 1020–1031.

11 Ellabban, O., Abu-Rub, H., and Blaabjerg, F. (2014). Renewable energy resources: current status, future prospects and their enabling technology. *Renew. Sustain. Energy Rev.* 39: 748–764.

12 Boonyuen, S., Smith, S.M., Malaithong, M. et al. (2018). Biodiesel production by a renewable catalyst from calcined Turbo jourdani (Gastropoda: Turbinidae) shells. *J. Clean. Prod.* 177: 925–939.

13 Knothe, G. and Razon, L.F. (2017). Biodiesel fuels. *Prog. Energy Combust. Sci.* 58: 36–59.

14 Takase, M., Zhao, T., Zhang, M. et al. (2015). An expatiate review of neem, jatropha, rubber and karanja as multipurpose non-edible biodiesel resources and comparison of their fuel, engine and emission properties. *Renew. Sustain. Energy Rev.* 43: 495–520.

15 Gebremariam, S.N. and Marchetti, J.M. (2018). Economics of biodiesel production: review. *Energy Convers. Manag.* 168: 74–84.

16 Chua, S.Y., Periasamy, L.A., Goh, C.M.H. et al. (2020). Biodiesel synthesis using natural solid catalyst derived from biomass waste – a review. *J. Ind. Eng. Chem.* 81: 41–60.

17 Zhang, H., Li, H., Hu, Y. et al. (2019). Advances in production of bio-based ester fuels with heterogeneous bifunctional catalysts. *Renew. Sustain. Energy Rev.* 114: 109296.

18 Konwar, L.J., Boro, J., and Deka, D. (2014). Review on latest developments in biodiesel production using carbon-based catalysts. *Renew. Sustain. Energy Rev.* 29: 546–564.

19 Yaşar, F. (2020). Comparision of fuel properties of biodiesel fuels produced from different oils to determine the most suitable feedstock type. *Fuel* 264: 116817.

20 Dhawane, S.H., Kumar, T., and Halder, G. (2018). Recent advancement and prospective of heterogeneous carbonaceous catalysts in chemical and enzymatic transformation of biodiesel. *Energy Convers. Manag.* 167: 176–202.

21 D'Souza, R., Vats, T., Chattree, A., and Siril, P.F. (2018). Graphene supported magnetically separable solid acid catalyst for the single step conversion of waste cooking oil to biodiesel. *Renew. Energy* 126: 1064–1073.

22 Kataria, J., Mohapatra, S.K., and Kundu, K. (2019). Biodiesel production from waste cooking oil using heterogeneous catalysts and its operational characteristics on variable compression ratio CI engine. *J. Energy Inst.* 92 (2): 275–287.

23 Joshi, G., Rawat, D.S., Lamba, B.Y. et al. (2015). Transesterification of Jatropha and Karanja oils by using waste egg shell derived calcium based mixed metal oxides. *Energy Convers. Manag.* 96: 258–267.

24 Adepoju, T.F., Ibeh, M.A., and Asuquo, A.J. (2021). Elucidate three novel catalysts synthesized from animal bones for the production of biodiesel from ternary non-edible and edible oil blend: a case of *Jatropha curcus*, *Hevea brasiliensis*, and *Elaeis guineensis* oil. *S. Afr. J. Chem. Eng.* 36: 58–73.

25 Mansir, N., Teo, S.H., Rashid, U. et al. (2018). Modified waste egg shell derived bifunctional catalyst for biodiesel production from high FFA waste cooking oil. A review. *Renew. Sustain. Energy Rev.* 82: 3645–3655.

26 Eze, V.C., Phan, A.N., and Harvey, A.P. (2018). Intensified one-step biodiesel production from high water and free fatty acid waste cooking oils. *Fuel* 220: 567–574.

27 Mohamed, M.M., Bayoumy, W.A., El-Faramawy, H. et al. (2020). A novel $\alpha$-$Fe_2O_3$/ AlOOH($\gamma$-$Al_2O_3$) nanocatalyst for efficient biodiesel production from waste oil: kinetic and thermal studies. *Renew. Energy* 160: 450–464.

28 Shahraki, H., Entezari, M.H., and Goharshadi, E.K. (2015). Sono-synthesis of biodiesel from soybean oil by KF/$\gamma$-$Al_2O_3$ as a nano-solid-base catalyst. *Ultrason. Sonochem.* 23: 266–274.

29 Mazumdar, P., Borugadda, V.B., Goud, V.V., and Sahoo, L. (2012). Physico-chemical characteristics of *Jatropha curcas* L. of North East India for exploration of biodiesel. *Biomass Bioenergy* 46: 546–554.

30 Abomohra, A.E.F., Elsayed, M., Esakkimuthu, S. et al. (2020). Potential of fat, oil and grease (FOG) for biodiesel production: a critical review on the recent progress and future perspectives. *Prog. Energy Combust. Sci.* 81: 100868.

31 Maroa, S. and Inambao, F. (2020). *Biodiesel, Combustion, Performance and Emissions Characteristics*. Springer Publishers.

32 Tabatabaei, M. and Aghbashlo, M. (2019). *Biodiesel. From Production to Combustion*, 239. Springer Publishers.

33 Farid, M.A.A., Roslan, A.M., Hassan, M.A. et al. (2020). Net energy and techno-economic assessment of biodiesel production from waste cooking oil using a semi-industrial plant: a Malaysia perspective. *Sustain. Energy Technol. Assess.* 39: 100700.

34 Lani, N.S., Ngadi, N., Yahya, N.Y., and Rahman, R.A. (2017). Synthesis, characterization and performance of silica impregnated calcium oxide as heterogeneous catalyst in biodiesel production. *J. Clean. Prod.* 146: 116–124.

35 Talebian-Kiakalaieh, A., Amin, N.A.S., and Mazaheri, H. (2013). A review on novel processes of biodiesel production from waste cooking oil. *Appl. Energy* 104: 683–710.

36 Long, Y.D., Fang, Z., Su, T.C., and Yang, Q. (2014). Co-production of biodiesel and hydrogen from rapeseed and Jatropha oils with sodium silicate and Ni catalysts. *Appl. Energy* 113: 1819–1825.

37 Martínez, S.L., Romero, R., Natividad, R., and González, J. (2014). Optimization of biodiesel production from sunflower oil by transesterification using $Na_2O$/NaX and methanol. *Catal. Today* 220–222: 12–20.

**38** Tamjidi, S., Esmaeili, H., and Moghadas, B.K. (2021). Performance of functionalized magnetic nanocatalysts and feedstocks on biodiesel production: a review study. *J. Clean. Prod.* 305: 127200.

**39** Shan, R., Lu, L., Shi, Y. et al. (2018). Catalysts from renewable resources for biodiesel production. *Energy Convers. Manag.* 178: 277–289.

**40** Pinto, B.F., Garcia, M.A.S., Costa, J.C.S. et al. (2019). Effect of calcination temperature on the application of molybdenum trioxide acid catalyst: screening of substrates for biodiesel production. *Fuel* 239: 290–296.

**41** Murray, R., King, G., and Wyse-Mason, R. (2019). Micro-emulsification vs. transesterification: an investigation of the efficacy of methanol use in improving vegetable oil engine performance. *Biofuels* https://doi.org/10.1080/17597269.2019.1598316.

**42** Marwaha, A., Rosha, P., Mohapatra, S.K. et al. (2018). Waste materials as potential catalysts for biodiesel production: current state and future scope. *Fuel Process. Technol.* 181: 175–186.

**43** Rashid, U., Anwar, F., Yunus, R., and Al-Muhtaseb, A.H. (2015). Transesterification for biodiesel production using thespesia populnea seed oil: an optimization study. *Int. J. Green Energy* 12: 479–484.

**44** Thiyagarajan, S., Sonthalia, A., Edwin Geo, V. et al. (2020). Effect of manifold injection of methanol/n-pentanol in safflower biodiesel fuelled CI engine. *Fuel* 261: 116378.

**45** Lertpanyapornchai, B. and Ngamcharussrivichai, C. (2015). Mesostructured Sr and Ti mixed oxides as heterogeneous base catalysts for transesterification of palm kernel oil with methanol. *Chem. Eng. J.* 264: 789–796.

**46** Rashid, U., Anwar, F., Ashraf, M. et al. (2011). Application of response surface methodology for optimizing transesterification of *Moringa oleifera* oil: biodiesel production. *Energy Convers. Manag.* 52 (8–9): 3034–3042.

**47** Rashid, U., Ibrahim, M., Yasin, S. et al. (2013). Biodiesel from *Citrus reticulata* (mandarin orange) seed oil, a potential non-food feedstock. *Ind. Crop. Prod.* 45: 355–359.

**48** Buasri, A., Lukkanasiri, M., Nernrimnong, R. et al. (2016). Rapid transesterification of *Jatropha curcas* oil to biodiesel using novel catalyst with a microwave heating system. *Korean J. Chem. Eng.* 33: 3388–3400.

**49** Koutsouki, A.A., Tegou, E., Kontakos, S. et al. (2015). In situ transesterification of *Cynara cardunculus* L. seed oil via direct ultrasonication for the production of biodiesel. *Fuel Process. Technol.* 134: 122–129.

**50** Anwar, F., Rashid, U., Ashraf, M., and Nadeem, M. (2010). Okra (*Hibiscus esculentus*) seed oil for biodiesel production. *Appl. Energy* 87 (3): 779–785.

**51** Rashid, U., Ibrahim, M., Ali, S. et al. (2012). Comparative study of the methanolysis and ethanolysis of maize oils using alkaline catalysts. *Grasas Aceites* 63 (1): 35–43.

**52** Barminas, J.T., Maina, H.M., Tahir, S. et al. (2001). A preliminary investigation into the biofuel characteristics of tigernut (*Cyperus esculentus*). *Bioresource Technology* 79 (1): 87–89.

**53** Sbihi, H.M., Nehdi, I.A., Blidi, L.E. et al. (2015). Lipase/enzyme catalyzed biodiesel production from *Prunus mahaleb*: a comparative study with base catalyzed biodiesel production. *Ind. Crop. Prod.* 76: 1049–1054.

**54** Rashid, U., Knothe, G., Yunus, R., and Evangelista, R.L. (2014). Kapok oil methyl esters. *Biomass Bioenerg.* 66: 419–425.

**55** Sharma, S., Saxena, V., Baranwal, A. et al. (2018). Engineered nanoporous materials mediated heterogeneous catalysts and their implications in biodiesel production. *Mater. Sci. Energ. Technol.* 1 (1): 11–21.

**56** Mokbli, S., Nehdi, I.A., Sbihi, H.M. et al. (2018). *Yucca aloifolia* seed oil: a new source of bioactive compounds. *Waste Biomass Valori.* 9 (7): 1087–1093.

**57** Syam, A.M., Rashid, U., Yunus, R. et al. (2016). Conversion of *Oleum papaveris seminis* oil into methyl esters via esterification process: optimization and kinetic study. *Grasas Aceites* 67 (1): e115.

**58** Karmakar, B., Samanta, S., and Halder, G. (2020). Delonix regia heterogeneous catalyzed two-step biodiesel production from *Pongamia pinnata* oil using methanol and 2-propanol. *J. Clean. Prod.* 255: 120313.

**59** Angin, D. and Şensöz, S. (2014). Effect of pyrolysis temperature on chemical and surface properties of biochar of rapeseed (*Brassica napus* L.). *Int. J. Phytoremed.* 16 (7–8): 684–693.

**60** Kumar, S. and Dinesha, P. (2019). Use of alternative fuels in compression ignition engines: a review. *Biofuels* 10 (4): 525–535.

**61** Lokman, I.M., Rashid, U., and Taufiq-Yap, Y.H. (2016). Meso- and macroporous sulfonated starch solid acid catalyst for esterification of palm fatty acid distillate. *Arab. J. Chem.* 9 (2): 179–189.

**62** Jayakumar, M., Karmegam, N., Gundupalli, M.P. et al. (2021). Heterogeneous base catalysts: synthesis and application for biodiesel production – a review. *Bioresour. Technol.* 331: 125054.

**63** Ajala, E.O., Ajala, M.A., Ajao, A.O. et al. (2020). Calcium-carbide residue: a precursor for the synthesis of $CaO–Al_2O_3–SiO_2–CaSO_4$ solid acid catalyst for biodiesel production using waste lard. *Chem. Eng. J. Adv.* 4: 100033.

**64** Hajjari, M., Tabatabaei, M., Aghbashlo, M., and Ghanavati, H. (2017). A review on the prospects of sustainable biodiesel production: a global scenario with an emphasis on waste-oil biodiesel utilization. *Renew. Sustain. Energy Rev.* 72: 445–464.

**65** Dhawane, S.H., Karmakar, B., Ghosh, S., and Halder, G. (2018). Parametric optimisation of biodiesel synthesis from waste cooking oil via Taguchi approach. *J. Environ. Chem. Eng.* 6 (4): 3971–3980.

**66** Lokman, I.M., Rashid, U., Yunus, R., and Taufiq-Yap, Y.H. (2014). Carbohydrate-derived solid acid catalysts for biodiesel production from low-cost feedstocks: a review. *Catal. Rev. Sci. Eng.* 56 (2): 187–219.

**67** Kächele, R., Nurkowski, D., Martin, J. et al. (2019). An assessment of the viability of alternatives to biodiesel transport fuels. *Appl. Energy* 251: 113363.

**68** Musa, I.A. (2016). The effects of alcohol to oil molar ratios and the type of alcohol on biodiesel production using transesterification process. *Egypt. J. Pet.* 25 (1): 21–31.

**69** Jain, S. (2019). The production of biodiesel using Karanja (*Pongamia pinnata*) and Jatropha (*Jatropha curcas*) oil. *Biomass, Biopolymer-Based Materials, and Bioenergy* 17: 397–408.

**70** Alsharifi, M., Znad, H., Hena, S., and Ang, M. (2017). Biodiesel production from canola oil using novel $Li/TiO_2$ as a heterogeneous catalyst prepared via impregnation method. *Renew. Energy* 114: 1077–1089.

**71** Ye, W., Gao, Y., Ding, H. et al. (2016). Kinetics of transesterification of palm oil under conventional heating and microwave irradiation, using CaO as heterogeneous catalyst. *Fuel* 180: 574–579.

**72** Boz, N., Degirmenbasi, N., and Kalyon, D.M. (2013). Transesterification of canola oil to biodiesel using calcium bentonite functionalized with K compounds. *Appl. Catal. B Environ.* 138–139: 236–242.

**73** Ramli, A., Farooq, M., Naeem, A. et al. (2017). Bifunctional heterogeneous catalysts for biodiesel production using low cost feedstocks: a future perspective. In: *Front Bioenergy Biofuels* (ed. E. Jacob-Lopes and L.Q. Zepka). IntechOpen.

**74** Dehghan, L., Golmakani, M.T., and Hosseini, S.M.H. (2019). Optimization of microwave-assisted accelerated transesterification of inedible olive oil for biodiesel production. *Renew. Energy* 138: 915–922.

**75** Banković-Ilić, I.B., Stojković, I.J., Stamenković, O.S. et al. (2014). Waste animal fats as feedstocks for biodiesel production. *Renew. Sustain. Energy Rev.* 32: 238–254.

**76** Ambat, I., Srivastava, V., and Sillanpää, M. (2018). Recent advancement in biodiesel production methodologies using various feedstock: a review. *Renew. Sustain. Energy Rev.* 90: 356–369.

**77** Anto, S., Karpagam, R., Renukadevi, P. et al. (2019). Biomass enhancement and bioconversion of brown marine microalgal lipid using heterogeneous catalysts mediated transesterification from biowaste derived biochar and bionanoparticle. *Fuel* 255: 115789.

**78** Awogbemi, O., Von Kallon, D.V., and Aigbodion, V.S. (2021). Trends in the development and utilization of agricultural wastes as heterogeneous catalyst for biodiesel production. *J. Energy Inst.* 98: 244–258.

**79** Zailan, Z., Tahir, M., Jusoh, M., and Zakaria, Z.Y. (2021). A review of sulfonic group bearing porous carbon catalyst for biodiesel production. *Renew. Energy* 175: 430–452.

**80** Dahman, Y., Dignan, C., Fiayaz, A., and Chaudhry, A. (2019). An introduction to biofuels, foods, livestock, and the environment. *Biomass, Biopolymer-Based Materials, Bioenergy* 241–276.

**81** Kumar, D. and Singh, B. (2020). Effect of winterization and plant phenolic-additives on the cold-flow properties and oxidative stability of Karanja biodiesel. *Fuel* 262: 116631.

**82** Mohiddin, M.N.B., Tan, Y.H., Seow, Y.X. et al. (2021). Evaluation on feedstock, technologies, catalyst and reactor for sustainable biodiesel production: a review. *J. Ind. Eng. Chem.* 98: 60–81.

**83** Chong, C.C., Aqsha, A., Ayoub, M. et al. (2020). A review over the role of catalysts for selective short-chain polyglycerol production from biodiesel derived waste glycerol. *Environ. Technol. Innov.* 19: 100859.

**84** Yusuf, N.N.A.N., Kamarudin, S.K., and Yaakub, Z. (2011). Overview on the current trends in biodiesel production. *Energy Convers. Manag.* 52 (7): 2741–2751.

**85** Bora, P., Boro, J., Konwar, L.J., and Deka, D. (2016). Formulation of microemulsion based hybrid biofuel from waste cooking oil – a comparative study with biodiesel. *J. Energy Inst.* 89 (4): 560–568.

**86** Mohd Noor, C.W., Noor, M.M., and Mamat, R. (2018). Biodiesel as alternative fuel for marine diesel engine applications: a review. *Renew. Sustain. Energy Rev.* 94: 127–142.

**87** Jitputti, J., Kitiyanan, B., Rangsunvigit, P. et al. (2006). Transesterification of crude palm kernel oil and crude coconut oil by different solid catalysts. *Chem. Eng. J.* 116 (1): 61–66.

**88** Abdul Kapor, N.Z., Maniam, G.P., Rahim, M.H.A., and Yusoff, M.M. (2017). Palm fatty acid distillate as a potential source for biodiesel production-a review. *J. Clean. Prod.* 143: 1–9.

**89** Li, H., Niu, S., Lu, C. et al. (2014). Transesterification catalyzed by industrial waste – lime mud doped with potassium fluoride and the kinetic calculation. *Energy Convers. Manag.* 86: 1110–1117.

**90** de Luna, M.D.G., Cuasay, J.L., Tolosa, N.C., and Chung, T.W. (2017). Transesterification of soybean oil using a novel heterogeneous base catalyst: synthesis and characterization of Na-pumice catalyst, optimization of transesterification conditions, studies on reaction kinetics and catalyst reusability. *Fuel* 209: 246–253.

**91** Bayat, A., Baghdadi, M., and Bidhendi, G.N. (2018). Tailored magnetic nano-alumina as an efficient catalyst for transesterification of waste cooking oil: optimization of biodiesel production using response surface methodology. *Energy Convers. Manag.* 177: 395–405.

**92** Olutoye, M.A., Wong, S.W., Chin, L.H. et al. (2016). Synthesis of fatty acid methyl esters via the transesterification of waste cooking oil by methanol with a barium-modified montmorillonite K10 catalyst. *Renew. Energy* 86: 392–398.

**93** Wong, Y.C., Tan, Y.P., Taufiq-Yap, Y.H. et al. (2015). Biodiesel production via transesterification of palm oil by using $CaO–CeO_2$ mixed oxide catalysts. *Fuel* 162: 288–293.

**94** Risso, R., Ferraz, P., Meireles, S. et al. (2018). Highly active Cao catalysts from waste shells of egg, oyster and clam for biodiesel production. *Appl. Catal. A Gen.* 567: 56–64.

**95** Mansir, N., Teo, S.H., Rabiu, I., and Taufiq-Yap, Y.H. (2018). Effective biodiesel synthesis from waste cooking oil and biomass residue solid green catalyst. *Chem. Eng. J.* 347: 137–144.

**96** Hazmi, B., Rashid, U., Taufiq-yap, Y.H., and Ibrahim, M.L. (2020). Supermagnetic nano-bifunctional catalyst from rice husk: synthesis, characterization and application for conversion of used cooking oil to biodiesel. *Catalysts* 10 (2): 225.

**97** Yu, J., Tang, L., Pang, Y. et al. (2019). Magnetic nitrogen-doped sludge-derived biochar catalysts for persulfate activation: internal electron transfer mechanism. *Chem. Eng. J.* 364: 146–159.

**98** Tan, Y.H., Abdullah, M.O., Nolasco-Hipolito, C., and Ahmad Zauzi, N.S. (2017). Application of RSM and Taguchi methods for optimizing the transesterification of waste cooking oil catalyzed by solid ostrich and chicken-eggshell derived CaO. *Renew. Energy* 114: 437–447.

**99** Ayoob, A.K. and Fadhil, A.B. (2020). Valorization of waste tires in the synthesis of an effective carbon based catalyst for biodiesel production from a mixture of non-edible oils. *Fuel* 264: 116754.

**100** Essamlali, Y., Amadine, O., Fihri, A., and Zahouily, M. (2019). Sodium modified fluorapatite as a sustainable solid bi-functional catalyst for biodiesel production from rapeseed oil. *Renew. Energy* 133: 1295–1307. https://doi.org/10.1016/j.renene.2018.08.103.

**101** Nisar, J., Razaq, R., Farooq, M. et al. (2017). Enhanced biodiesel production from Jatropha oil using calcined waste animal bones as catalyst. *Renew. Energy* 101: 111–119.

**102** Muciño, G.G., Romero, R., Ramírez, A. et al. (2014). Biodiesel production from used cooking oil and sea sand as heterogeneous catalyst. *Fuel* 138: 143–148.

**103** Uzun, B.B., Kiliç, M., Özbay, N. et al. (2012). Biodiesel production from waste frying oils: optimization of reaction parameters and determination of fuel properties. *Energy* 44: 347–351.

**104** Tan, Y.H., Abdullah, M.O., and Nolasco Hipolito, C. (2016). Comparison of biodiesel production between homogeneous and heterogeneous base catalysts. *Appl. Mech. Mater.* 833: 71–77.

**105** Pastore, C., Barca, E., Del Moro, G. et al. (2015). Recoverable and reusable aluminium solvated species used as a homogeneous catalyst for biodiesel production from brown grease. *Appl. Catal. A Gen.* 501: 48–55.

**106** Agarwal, M., Chauhan, G., Chaurasia, S.P., and Singh, K. (2012). Study of catalytic behavior of KOH as homogeneous and heterogeneous catalyst for biodiesel production. *J. Taiwan Inst. Chem. Eng.* 43 (1): 89–94.

**107** Aslan, V. and Eryilmaz, T. (2020). Polynomial regression method for optimization of biodiesel production from black mustard (*Brassica nigra* L.) seed oil using methanol, ethanol, NaOH, and KOH. *Energy* 209: 118386.

**108** Vaishnavi Sree, J., Akhil Chowdary, B., Santosh Kumar, K. et al. (2021). Optimization of the biodiesel production from waste cooking oil using homogeneous catalyst and heterogeneous catalysts. *Mater. Today Proc.* 46: 4900–4908.

**109** Koohikamali, S., Tan, C.P., and Ling, T.C. (2012). Optimization of sunflower oil transesterification process using sodium methoxide. *Sci. World J.* 2012: 475027.

**110** Panchal, B.M., Deshmukh, S.A., and Sharma, M.R. (2016). Biodiesel from *Thevetia peruviana* seed oil with dimethyl carbonate using as an active catalyst potassium-methoxide. *Sains Malaysiana* 45 (10): 1461–1468.

**111** Sai, B.A.V.S.L., Subramaniapillai, N., Khadhar Mohamed, M.S.B., and Narayanan, A. (2020). Optimization of continuous biodiesel production from rubber seed oil (RSO) using calcined eggshells as heterogeneous catalyst. *J. Environ. Chem. Eng.* 8: 103603.

**112** Wei, Z., Wang, Z., Tait, W.R.T. et al. (2018). Synthesis of green phosphors from highly active amorphous silica derived from rice husks. *J. Mater. Sci.* 53 (3): 1824–1832.

**113** Atadashi, I.M., Aroua, M.K., Abdul Aziz, A.R., and Sulaiman, N.M.N. (2013). The effects of catalysts in biodiesel production: a review. *J. Ind. Eng. Chem.* 19 (1): 14–26.

**114** Firouzjaee, M.H. and Taghizadeh, M. (2017). Optimization of process variables for biodiesel production using the nanomagnetic catalyst CaO/NaY-$Fe_3O_4$. *Chem. Eng. Technol.* 40 (6): 1140–1148.

**115** Roschat, W., Siritanon, T., Kaewpuang, T., and Yoosuk, B. (2016). Economical and green biodiesel production process using river snail shells-derived heterogeneous catalyst and co-solvent method. *Bioresour. Technol.* 209: 343–350.

**116** Singh, V., Yadav, M., and Sharma, Y.C. (2017). Effect of co-solvent on biodiesel production using calcium aluminium oxide as a reusable catalyst and waste vegetable oil. *Fuel* 203: 360–369.

**117** Vargas, E.M., Neves, M.C., Tarelho, L.A.C., and Nunes, M.I. (2019). Solid catalysts obtained from wastes for FAME production using mixtures of refined palm oil and waste cooking oils. *Renew. Energy* 136: 873–883.

**118** Zhao, X., Wei, L., Cheng, S., and Julson, J. (2017). Review of heterogeneous catalysts for catalytically upgrading vegetable oils into hydrocarbon biofuels. *Catalysts* 7 (3): 83.

**119** Saba, T., Estephane, J., El Khoury, B. et al. (2016). Biodiesel production from refined sunflower vegetable oil over KOH/ZSM5 catalysts. *Renew. Energy* 90: 301–306.

**120** Ho, W.W.S., Ng, H.K., Gan, S., and Tan, S.H. (2014). Evaluation of palm oil mill fly ash supported calcium oxide as a heterogeneous base catalyst in biodiesel synthesis from crude palm oil. *Energy Convers. Manag.* 88: 1167–1178.

**121** Arzamendi, G., Campo, I., Arguiñarena, E. et al. (2008). Synthesis of biodiesel from sunflower oil with silica-supported NaOH catalysts. *Journal of Chemical Technology and Biotechnology*. 83: 862–870. https://doi.org/10.1002/jctb.1881.

**122** Abdelhady, H.H., Elazab, H.A., Ewais, E.M. et al. (2020). Efficient catalytic production of biodiesel using nano-sized sugar beet agro-industrial waste. *Fuel* 261: 116481.

**123** Rajasekhar Reddy, B. and Vinu, R. (2018). Microwave-assisted co-pyrolysis of high ash Indian coal and rice husk: product characterization and evidence of interactions. *Fuel Process. Technol.* 178: 41–52.

**124** Fattahi, N., Triantafyllidis, K., Luque, R., and Ramazani, A. (2019). Zeolite-based catalysts: a valuable approach toward ester bond formation. *Catalysts* 9 (9): 758.

**125** Rosazley, R., Shazana, M., Izzati, M. et al. (2016). Characterization of nanofibrillated cellulose produced from oil palm empty fruit bunch fiber (OPEFB) using ultrasound. *J. Tempor. Issue Thoughts* 6: 28–35.

**126** Bala, D.D., Misra, M., and Chidambaram, D. (2017). Solid-acid catalyzed biodiesel production, part I: biodiesel synthesis from low quality feedstock. *J. Clean. Prod.* 142: 4169–4177.

**127** Fonseca, J.M., Teleken, J.G., de Cinque Almeida, V., and da Silva, C. (2019). Biodiesel from waste frying oils: methods of production and purification. *Energy Convers. Manag.* 184: 205–218.

**128** Zeng, D., Zhang, Q., Chen, S. et al. (2016). Synthesis porous carbon-based solid acid from rice husk for esterification of fatty acids. *Microporous Mesoporous Mater.* 219: 54–58.

**129** Wang, Y.T., Fang, Z., Yang, X.X. et al. (2018). One-step production of biodiesel from Jatropha oils with high acid value at low temperature by magnetic acid-base amphoteric nanoparticles. *Chem. Eng. J.* 348: 929–939.

**130** Liu, R., Sarker, M., Rahman, M.M. et al. (2020). Multi-scale complexities of solid acid catalysts in the catalytic fast pyrolysis of biomass for bio-oil production – a review. *Prog. Energy Combust. Sci.* 80: 450–464.

**131** Hindryawati, N., Pragas, G., Karim, R., and Feng, K. (2014). Transesterification of used cooking oil over alkali metal (Li, Na, K) supported rice husk silica as potential solid base catalyst. *Eng. Sci. Technol. Int. J.* 17 (2): 95–103.

**132** Zhang, F., Wu, X.H., Yao, M. et al. (2016). Production of biodiesel and hydrogen from plant oil catalyzed by magnetic carbon-supported nickel and sodium silicate. *Green Chem.* 18 (11): 3302–3314.

**133** Shu, Q., Nawaz, Z., Gao, J. et al. (2010). Synthesis of biodiesel from a model waste oil feedstock using a carbon-based solid acid catalyst: reaction and separation. *Bioresour. Technol.* 101 (14): 5374–5384.

**134** Sangaletti-Gerhard, N., Cea, M., Risco, V., and Navia, R. (2015). In situ biodiesel production from greasy sewage sludge using acid and enzymatic catalysts. *Bioresour. Technol.* 179: 63–70.

**135** Sirisomboonchai, S., Abuduwayiti, M., Guan, G. et al. (2015). Biodiesel production from waste cooking oil using calcined scallop shell as catalyst. *Energy Convers. Manag.* 95: 242–247.

**136** Zulqarnain, Yusoff, M.H.M., Ayoub, M. et al. (2021). Solvent extraction and performance analysis of residual palm oil for biodiesel production: experimental and simulation study. *J. Environ. Chem. Eng.* 9 (4): 105519.

**137** Sagiroglu, A. (2008). Conversion of sunflower oil to biodiesel by alcoholysis using immobilized lipase. *Artif Cells, Blood Substitutes, Biotechnol.* 36 (2): 138–149.

**138** Pedro, K.C.N.R., Parreira, J.M., Correia, I.N. et al. (2018). Enzymatic biodiesel synthesis from acid oil using a lipase mixture. *Quim Nova* 41 (3): 284–291.

**139** Santaraite, M., Sendzikiene, E., Makareviciene, V., and Kazancev, K. (2020). Biodiesel production by lipase-catalyzed in situ transesterification of rapeseed oil containing a high free fatty acid content with ethanol in diesel fuel media. *Energies* 13 (10): 2588.

**140** Peñarrubia Fernandez, I.A., Liu, D.H., and Zhao, J. (2017). LCA studies comparing alkaline and immobilized enzyme catalyst processes for biodiesel production under Brazilian conditions. *Resour. Conserv. Recycl.* 119: 117–127.

**141** Zhong, L., Jiao, X., Hu, H. et al. (2021). Activated magnetic lipase-inorganic hybrid nanoflowers: a highly active and recyclable nanobiocatalyst for biodiesel production. *Renew. Energy* 171: 825–832.

**142** Rachmadona, N., Amoah, J., Quayson, E. et al. (2020). Lipase-catalyzed ethanolysis for biodiesel production of untreated palm oil mill effluent. *Sustain. Energy Fuels* 4 (3): 1105–1111.

**143** Sarno, M. and Iuliano, M. (2018). Active biocatalyst for biodiesel production from spent coffee ground. *Bioresour. Technol.* 266: 431–438.

**144** Petronikolou, N. and Nair, S.K. (2015). Biochemical studies of mycobacterial fatty acid methyltransferase: a catalyst for the enzymatic production of biodiesel. *Chem. Biol.* 22 (11): 1480–1490.

**145** Guo, J., Sun, S., and Liu, J. (2020). Conversion of waste frying palm oil into biodiesel using free lipase A from Candida antarctica as a novel catalyst. *Fuel* 267: 117323.

**146** Shalini, P., Deepanraj, B., Vijayalakshmi, S., and Ranjitha, J. (2021). Synthesis and characterisation of lipase immobilised magnetic nanoparticles and its role as a catalyst in biodiesel production. *Mater. Today Proc.* https://doi.org/10.1016/j.matpr.2021.07.027.

**147** Lai, J.Q., Hu, Z.L., Sheldon, R.A., and Yang, Z. (2012). Catalytic performance of cross-linked enzyme aggregates of Penicillium expansum lipase and their use as catalyst for biodiesel production. *Process Biochem.* 47 (12): 2058–2063.

**148** Adewale, P., Dumont, M.J., and Ngadi, M. (2016). Enzyme-catalyzed synthesis and kinetics of ultrasonic assisted methanolysis of waste lard for biodiesel production. *Chem. Eng. J.* 284: 158–165.

**149** Firdaus, M.Y., Guo, Z., and Fedosov, S.N. (2016). Development of kinetic model for biodiesel production using liquid lipase as a biocatalyst, esterification step. *Biochem. Eng. J.* 105: 52–61.

**150** Ribeiro, L.M.O., da Santos, B.C., S., and Almeida, R.M.R.G. (2012). Studies on reaction parameters influence on ethanolic production of coconut oil biodiesel using immobilized lipase as a catalyst. *Biomass Bioenerg.* 47: 498–503.

**151** Toldrá-Reig, F., Mora, L., and Toldrá, F. (2020). Developments in the use of lipase transesterification for biodiesel production from animal fat waste. *Appl. Sci.* 10 (15): 5085.

**152** João, J.H., Tres, M.V., Jahn, S.L., and de Oliveira, J.V. (2020). Lipases in liquid formulation for biodiesel production: current status and challenges. *Biotechnol. Appl. Biochem.* 67: 648–667.

**153** Yan, J., Zheng, X., Du, L., and Li, S. (2014). Integrated lipase production and in situ biodiesel synthesis in a recombinant Pichia pastoris yeast: an efficient dual biocatalytic system composed of cell free enzymes and whole cell catalysts. *Biotechnol. Biofuels* 7 (1): 1–8.

**154** Zhang, H., Liu, T., Zhu, Y. et al. (2020). Lipases immobilized on the modified polyporous magnetic cellulose support as an efficient and recyclable catalyst for biodiesel production from Yellow horn seed oil. *Renew. Energy* 145: 1246–1254.

**155** Nasaruddin, R.R., Alam, M.Z., Jami, M.S., and Salihu, A. (2016). Statistical optimization of ethanol-based biodiesel production from sludge palm oil using locally produced *Candida cylindracea* lipase. *Waste Biomass Valor.* 7 (1): 87–95.

**156** Yan, W., Li, F., Wang, L. et al. (2017). Discovery and characterizaton of a novel lipase with transesterification activity from hot spring metagenomic library. *Biotechnol. Rep.* 14: 27–33.

**157** Jayaraman, J., Alagu, K., Appavu, P. et al. (2020). Enzymatic production of biodiesel using lipase catalyst and testing of an unmodified compression ignition engine using its blends with diesel. *Renew. Energy* 145: 399–407.

**158** Garlapati, V., Kant, R., Kumari, A. et al. (2013). Lipase mediated transesterification of Simarouba glauca oil: a new feedstock for biodiesel production. *Sustain. Chem. Process.* 1 (1): 11.

**159** Pollardo, A.A., Lee, H.S., Lee, D. et al. (2017). Effect of supercritical carbon dioxide on the enzymatic production of biodiesel from waste animal fat using immobilized *Candida antarctica* lipase B variant. *BMC Biotechnol.* 17 (1): 70.

**160** Binhayeeding, N., Klomklao, S., Prasertsan, P., and Sangkharak, K. (2020). Improvement of biodiesel production using waste cooking oil and applying single and mixed immobilised lipases on polyhydroxyalkanoate. *Renew. Energy* 162: 1819–1827.

**161** Nguyen, H.C., Liang, S.H., Chen, S.S. et al. (2018). Enzymatic production of biodiesel from insect fat using methyl acetate as an acyl acceptor: optimization by using response surface methodology. *Energy Convers. Manag.* 158: 168–175.

**162** Quayson, E., Amoah, J., Hama, S. et al. (2020). Immobilized lipases for biodiesel production: current and future greening opportunities. *Renew. Sust. Energ. Rev.* 134: 110355.

**163** Allami, H.A. and Nayebzadeh, H. (2021). The assessment of the engine performance and emissions of a diesel engine fueled by biodiesel produced using different types of catalyst. *Fuel* 305: 121525.

**164** Gupta, J., Agarwal, M., and Dalai, A.K. (2020). An overview on the recent advancements of sustainable heterogeneous catalysts and prominent continuous reactor for biodiesel production. *J. Ind. Eng. Chem.* 88: 58–77.

# 2

# Green Technologies in Valorization of Waste Cooking Oil to Biodiesel

*Bisheswar Karmakar and Gopinath Halder*

*Department of Chemical Engineering, National Institute of Technology, Durgapur, India*

## 2.1 Introduction

### 2.1.1 The Necessity for Biodiesel

The current energy scenario around the world can be summed up as an ever-increasing demand for energy per capita, while fossilized energy reserves being limited are projected to last less than half a century [1]. Since energy demands cannot be reduced without impacting the advancements in our daily life, the most obvious solution would be to harness renewable sources of energy for consumption and in fuel development. Additionally, the emissions from petrofuels comprise mostly of nitrogen oxides ($NO_X$), sulphur oxides ($SO_X$), $CO_2$, and other potent greenhouse gases (GHGs), which contribute massively in global warming and acid rain, causing climate changes and soil acidification [2]. One of the most promising fuels that can meet the exorbitant demands in the transport sector is biodiesel, emitting mostly $CO_2$. Biodiesel also has zero hydrocarbon, particulate, and smoke release with minimal NOX and SOX emissions.

### 2.1.2 Sourcing the Correct Precursor

Biodiesel can be produced from various feedstock, which are broadly classified according to type such as edible oils, nonedible oils, animal fats, and other miscellaneous sources as shown in Table 2.1. The use of edible oils has been discontinued since the growing population needs oil crops for sustenance, and this helps avoid the food vs. fuel dispute [2]. The use of nonedible oil crops is being thoroughly investigated by researchers worldwide, with quite a few crops such as jatropha, castor, karanja, etc. being used for small-scale commercial production [3–6]. However, most of these crops are available in the tropical and subtropical belt, and, as such, these belts exhibit quite the

*Biodiesel Production: Feedstocks, Catalysts, and Technologies*, First Edition. Edited by Samuel Lalthazuala Rokhum, Gopinath Halder, Suttichai Assabumrungrat, and Kanokwan Ngaosuwan.

**Table 2.1** Feedstock for biodiesel classified according to type.

| Edible oils | Nonedible oils | Animal fats | Other sources |
|---|---|---|---|
| Sunflower | Tobacco seed | Poultry fat | Switchgrass |
| Soybean | Rubber seed | Pork lard | Spent coffee grounds |
| Sesame | Neem | Fish oil | Poplar |
| Safflower | Nag champa | Chicken fat | Olive stones |
| Rice bran | Mahua | Beef tallow | Miscanthus |
| Rapeseed | Krating oil | | Microalgae |
| Peanut | Karanja | | Fungi |
| Palm | Jojoba | | Cyanobacteria |
| Mustard | *Jatropha curcas* | | *Calophyllum inophyllum* |
| Groundnut | *Croton megalocarpus* | | |
| Corn | Cotton seed | | |
| Coconut | Castor | | |
| Canola | *Camelina sativa* | | |
| Barley | | | |

diversity in flora. They can also be classified according to different generations described as follows:

- The earliest oil crops used for fuel synthesis have been designated as the first-generation feedstock. Including mostly edible oils such as palm, rapeseed, sunflower, soybean, etc. [7–10], these are easy to convert due to their simple fatty acid composition. However, growing concerns over food security inhibit the use of these oils in fuel synthesis.
- With advantages ranging from being nonedible to having the potential of growing in harsh and arid climates as well as being useful as decorative plants, feedstock designated as second generation includes both these nonedible oil crops and waste oils or fats such as waste cooking oil (WCO), brown and yellow grease, and animal tallow. The fuel yield is analogous to that obtained from edible oilseeds and hence is a much better alternative, provided that the chosen feedstock is sufficiently available to sustain commercial scale production.
- Using plant-based oils usually means that the oil produced per kilogram biomass is far less lucrative in terms of net energy benefits. Also plants tend to have long growth cycles with typically one harvest per year, which is a setback in terms of annual oil yield and fuel production. This is where lipid accumulating microalgae, designated as third-generation feedstock, are beneficial since (i) they do not need arable land [11]; (ii) algae show rapid growth rates and can accumulate biomass indefinitely without nutrient constraints [2]; (iii) they have high oil to biomass ratio; (iv) genetic manipulation on individual strains is easy; and (v) they can be engineered for high oil yield and for improved carbon capture and biomass accumulation. However, genetic modifications are often associated with very low success rates, and extensive screening is required. Also the culture medium must be properly aerated for $CO_2$ solubilization, as well as to eliminate local accumulation of formed by-products.

## 2.2 Importance of Valorization

Conversion of oils into esters and glycerol (or other glycerol-free products) must be energy and cost efficient while being fairly controllable and adaptable with different oils. Transesterification (occasionally coupled with esterification) is the most widely sought approach because of its simplicity combined with versatility [6]. Mostly in conventional transesterification approaches, the use of catalysts is also an important factor both chemically (saving energy) and economically (reusability of heterogeneous catalysts compared with high reagent use in homogeneous systems). Catalyst preparation and/or use comes with added costs. It had been earlier established by Karmakar et al. [3] that feedstock procurement and refining can comprise up to 70% of the fuel price, and this goes up when uncommon nonedible oils are used, due to collection and processing. A simple way to overcome this is through the use of WCO, which are commonly available and can drastically cut down production costs [12]. Using waste biomass for catalyst synthesis is another step toward cost efficiency; however, that lies beyond the scope of the presented discussion.

In frying, the oils are used for multiple rounds (usually 8–10), and the process of reheating the oils past its smoke point results in thermal degradation of the glycerides and fatty acids, burning as well as inducing rancidity. The oils thus turn bitter and become unfit for further use, at which point they are discarded. Although small in quantity for each food outlet, the global scenario is quite insurmountable. According to the American Petroleum Institute, about 63.5 million barrels of WCO are discarded annually, whereas the average household needs only 0.063 barrels of this oil to produce enough electricity for a day [13]. Animal fats also contain very high amounts of free fatty acids (FFAs) and water, and thus cannot be transesterified without proper pretreatment [14].

## 2.3 Purification and Characterization

Purification to remove water, suspended solids, and FFAs is needed for optimal biodiesel synthesis from WCO. The glycerides in the oil are vulnerable to hydrolysis at elevated temperatures, which break these molecules into FFAs and glycerol [2]. Hence, the most common process of water removal is through drying in a hot air oven at 100 °C. During frying, polymerized oil and food residue are generated, which need filtration for removal. The repeated heating of the oil also results in breakdown of glycerides into FFAs, which can be esterified using acids.

The residual FFA in the product, which also affects its total acid number (TAN) must also be low to reduce its corrosiveness [15]. In order to determine TAN, $m$ quantity of oil is dissolved in 2-propanol along with phenolphthalein. Titration against KOH solution of $N$ normality gives titer value $t$ shown in Eqs. (2.1) and (2.2), where 56.1 = molar weight of KOH, 28.2 weight for 0.1 M oleic acid.

$$\mathrm{TAN} = \frac{56.1 \times N \times t}{m} \tag{2.1}$$

$$\mathrm{FFA}(\%) = \frac{28.2 \times N \times t}{m} \tag{2.2}$$

The oil is also tested for calorific value using a bomb calorimeter, which determines the energy content of a sample per mole. The kinematic viscosity of the sample is also measured using a standard viscometer (capillary glass, Saybolt, or Redwood), which gives us an idea about the flow properties of the oil [16].

## 2.4 Transesterification: A Comprehensive Look

Alcoholysis is another term used for the process of transesterification, which involves the molecular breakdown of a glyceride (mono-, di-, or tri-) using an alcohol, which after multiple rounds produces esters and glycerol as a by-product [17, 18]. Stepwise conversion occurring in a typical triglyceride transesterification with methanol is depicted in Figure 2.1.

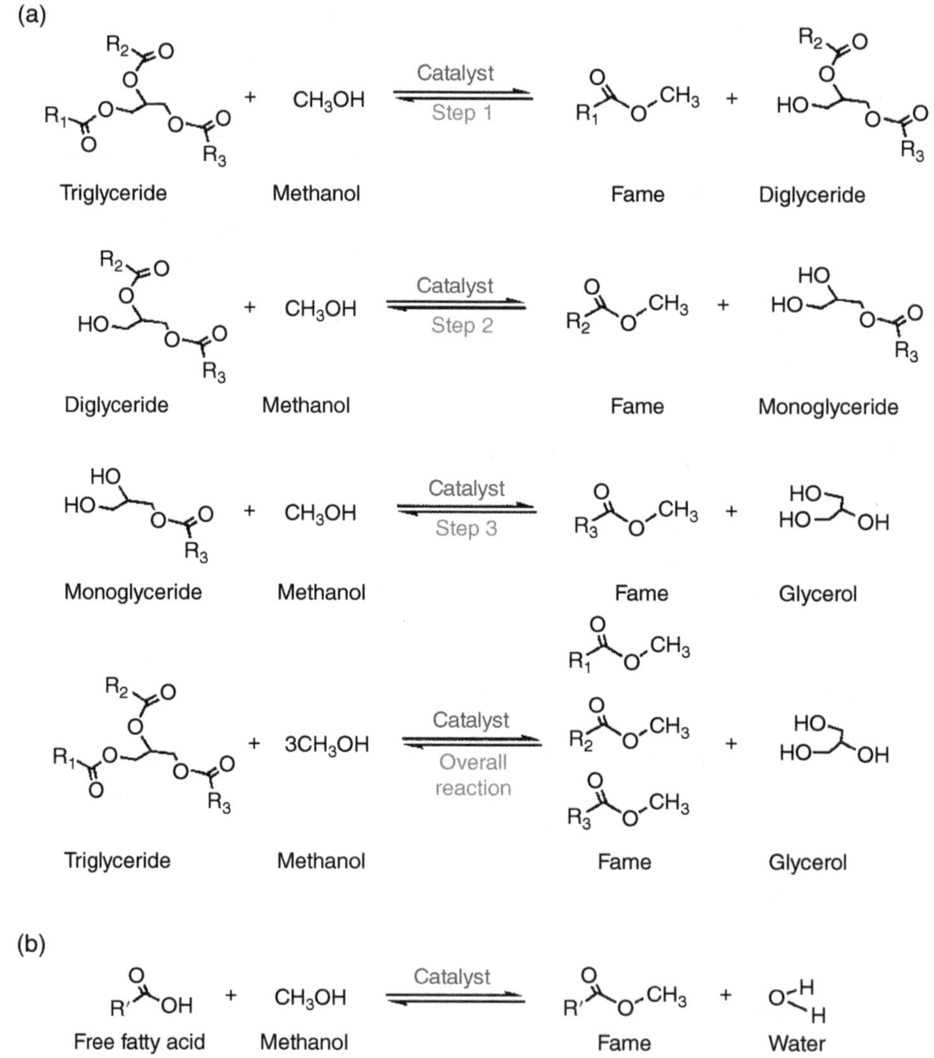


**Figure 2.1** Catalyzed conversion of triglyceride and FFA into esters using methanol.

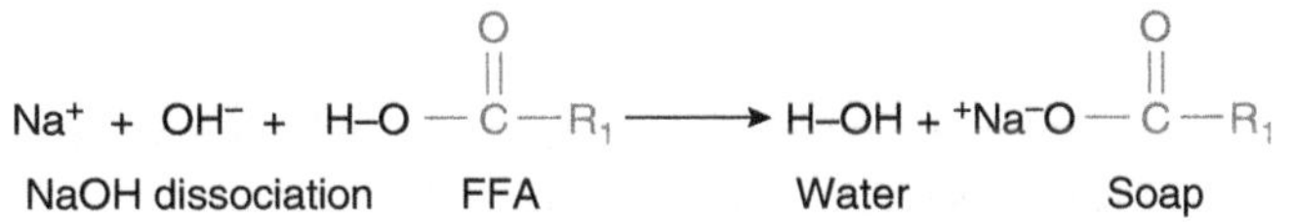


**Figure 2.2** Dissociation of alkali and saponification with FFA.

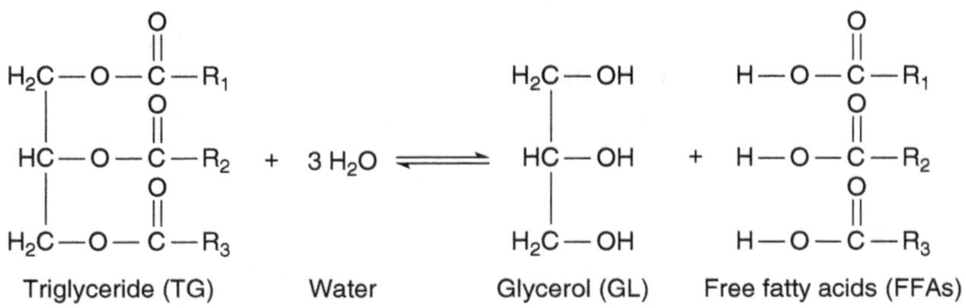


**Figure 2.3** Hydrolysis of triglyceride leading to FFA formation.

Alcohols used can be polar (such as ethanol and methanol) or nonpolar (such as 2-propanol or butanol). Although polar alcohols are capable of providing good fuel yields during transesterification, they are not miscible with the oils, and it has been reported by Karmakar et al. in their works with castor–karanja oil blend that 2-propanol provides better conversion compared to methanol when used simultaneously, due to being miscible [1]. This can lead to the speculation that while activation energy for the nonpolar system would be lower (from reduced diffusion resistance), the reaction rate would be lower since nonpolar compounds are less reactive. Conversions of this nature can be facilitated with the use of alkali or enzyme catalysts, and while alkalis are used more frequently, the system is left vulnerable to saponification due to the presence of FFAs in most feedstock as depicted in Figure 2.2. Triglycerides are also degraded into FFAs at elevated temperatures in the presence of water, through a process termed as "hydrolysis," depicted in Figure 2.3.

Other modes of transesterification include the use of supercritical fluids heated to 300–400 °C under 10–30 MPa; superheated alcohols, which are heated to above 150 °C and injected to hot oil (above 250 °C); and enzyme catalysts under slightly elevated temperatures [19]. However, these processes are not without drawbacks, discussed in later sections in detail. The formed fatty acid alkyl esters (FAAE) have a much lower viscosity compared to the parent oil due to their smaller molecular sizes, which gives the fuel its ability to be compatible in diesel engines. This process is applicable to all feedstock in which glycerides exist and thus can be used for a wide variety of sources [2].

## 2.5 Conversion Techniques

When it comes to conversion of reactive components of an oil into fuel-grade products, there are a surprisingly large number of approaches available. Here, we discuss only the most commonly used techniques that are still in use for research as well as in commercial biodiesel synthesis.

### 2.5.1 Traditional Conversion Approaches

These approaches include acids, bases, enzymes, or other novel catalysts, which have been successfully used in batch or semicontinuous studies by researchers to obtain biodiesel or for the pretreatment of feedstock prior to actual conversion from a wide variety of edible and nonedible oils including WCO [3]. Usually, the use of homogeneous catalysts can provide greater conversion in a single round, but for process economics it is always better to dope the necessary functional groups derived from acid, base, enzyme, or transition metals (among other novel catalysts) and then use them as they can typically be reused for a few times before being discarded or recharged. An in-depth discussion about this is presented in the following sections.

#### 2.5.1.1 Acid Catalysis

The acid-catalyzed conversion is termed as "esterification" and involves the conversion of FFAs present in the oil to esters and water using polar alcohols such as methanol or ethanol [6]. Along with esterification, transesterification of glycerides also takes place albeit to a smaller degree. Also the effect of nonpolar alcohols such as 2-propanol had been studied, which revealed that while 2-propanol is able to enhance the acid-catalyzed transesterification, it has absolutely no impact in FFA conversion, which resulted in the formed esters becoming rancid with 72 h upon storage [6]. Usually, mineral acids (such as $HNO_3$, $H_2SO_4$, or HCl) are used as they provide greater reactivity compared with other weaker acids.

The esterification process involves adding $H^+$ to the O atom in a carboxyl group, which is followed by the alcohol performing a nucleophilic attack on the fatty acid aided by the protonated acid catalyst, which results in breakage of a $H_2O$ molecule and a proton (adds back to the acid catalyst), resulting in ester formation [2]. Heterogeneous doped catalysts are preferred for reusability (verified up to four uses in our studies) before needing to be recharged with fresh acid. It was also shown in our other reported works with both MFL and *Delonix regia* char that the inert carbon supports are able to efficiently go through multiple rounds of redoping since the supports can withstand mechanical stress from agitation during reactions [4, 5]. Various other researchers have performed extensive studies on a wide variety of oils including WCO through the use of both homogeneous acids as well as acid-doped inert supports; a select few of which are summarized in Table 2.2. An interesting point to note is that this process is not sensitive to the presence of small amounts of water, since hydrolysis produces FFAs, which are readily esterified [2].

#### 2.5.1.2 Alkali Catalysis

The use of bases in transesterification is probably the most common practice of conversion of compatible feedstock (low in FFA and moisture) into esters. The conversion involves formation of an alkoxide ion from the base and the alcohol, which then targets the triglyceride (for example), attacking the carbonyl carbon, and forming a tetrahedral shaped intermediate. These compounds then undergo reaction with an alcohol molecule, undergoing a structural rearrangement during the process to give off an ester molecule and leading to the formation of a diglyceride [2]. This process will now repeat itself twice to finally yield glycerol as a by-product along with 2 mol of alkyl ester [28], as shown in Figure 2.1. The process

**Table 2.2** Traditional catalyzed conversions using acids, bases, enzymes, or other catalysts. *Source:* Modified from Ref. [2].

| | | Reaction conditions | | | | | | |
|---|---|---|---|---|---|---|---|---|
| **Feedstock** | **Catalyst used** | **Reaction temperature (°C)** | **Reaction time (h)** | **Catalyst concentration (% w/w)** | **Alcohol:oil (ratio or wt%)** | **Agitation speed (rpm)** | **Yield/ conversion (%)** | **References** |
| **Acid-catalyzed conversion** | | | | | | | | |
| Castor oil | $H_2SO_4$ | 50 | 1 | 1 | 20 : 1 | 700 | 90.83 | [4] |
| Waste cooking oil | $H_2SO_4$ | 60 | 3 | 5 | 12 : 1 | 800 | 95.376 | [3] |
| Mahua oil | Sulfonated *Delonix regia* char | 50 | 1 | 4 | 6 : 1 | 1000 | 97.04 | [5] |
| Castor oil | Sulfonated MFL char | 60 | 1.5 | 7 | 70 | 950 | 92 | [16] |
| **Base-catalyzed conversion** | | | | | | | | |
| Waste cooking oil | KOH | 50 | 1.67 | 0.75 | 9 : 1 | — | 90 | [20] |
| Waste cooking oil | CaO | 75 | 1 | 5 | 9.8 : 1 | 450 | 96.6 | [21] |
| Waste cooking oil | CaO | 65 | 3 | 7.5 | 15 : 1 | 1200 | 90 | [22] |
| Waste cooking oil | Calcined egg shells | 65 | 5.5 | 3.5 | 22.5 : 1 | 600 | 91 | [23] |
| **Enzyme-catalyzed conversion** | | | | | | | | |
| Waste cooking oil | Lipase | 50 | 10 | 6 | 20 | — | 94 | [24] |
| Jatropha oil | Lipase | 40 | 8 | 0.5 | 4 : 1 | — | 71 | [25] |
| **Other novel catalysts for conversion** | | | | | | | | |
| Waste cooking oil | Fe(II)-doped anthill | 60 | 1.5 | 1.2 | 6 : 1 | — | 99.73 | [15] |
| Rubber seed oil | Fe(II)-doped *Delonix regia* char | 40 | 15 | 5 | 3 : 1 | 500 | 96.31 | [26] |
| **Two-step catalyzed conversion** | | | | | | | | |
| Waste cooking oil | $H_2SO_4$ | 65 | 3 | 1 | 3 : 7 | 400 | 21.5 | [27] |
| | NaOH | 50 | 3 | 1 | 3 : 7 | 400 | 90.6 | |
| Karanja oil | Sulfonated *Delonix regia* char | 50 | 0.75 | 3 | 12 : 1 | 900 | 99.86 | [6] |
| | KOH-doped *Delonix regia* char | 60 | 1.5 | 4 | 6 : 1 | 700 | 99.39 | |

cannot tolerate even traces of FFA and moisture except for KOH, which is why it is favored over other bases. The process can also include homogeneous and heterogeneous catalysts, with doped basic groups being more stable compared to acidic groups [1, 6]. Many edible and nonedible oils (including waste oils) have been converted into biodiesel through this process; a select few of which are summarily presented in Table 2.2.

#### 2.5.1.3 Enzyme Catalysis

The wastewater generated from washing of fuels using acid or base catalysts is not pH neutral, and thus can hamper water bodies connected to the discharge stream if the generated effluent is left untreated before drainage. Enzymes can also be beneficial in conversion of high FFA feedstock, since saponification cannot take place as both glycerides and FFAs are converted to esters [29]. They also offer the following advantages:

- A low alcohol-to-oil ratio is needed, with deemulsification conferring reusability of enzymes.
- Low product inhibition and reacting temperature, with easy separation in case of heterogeneous immobilized enzymes.
- Single-step conversion with appreciable yields and are insensitive to moisture exposure.

However, they do require far longer durations to complete conversion, and since enzymes are very temperature sensitive, the reaction must be closely monitored. Another problem associated with them is their high costs as well as limited reusability due to structural denaturation and moderate conversion efficiencies compared to acid- or base-catalyzed systems [30, 31]. Lipase is the most common enzyme used and is obtained from animals, plants, or microbes, and must not be stereospecific for maximum conversion efficiency. Bacterial and fungal lipases (example being Novozym 435 obtained from *Candida antarctica* or other enzymes extracted from sources such as *Penicillium* spp., *Rhizopus* spp., and *Aspergillus niger*) used can show maximum yield up to 90%, when operated between 30 and 50 °C for anywhere between 8 and 90 h depending on feedstock [2]. The variety of studies reported by researchers are numerous; a select few of which have been summarily presented in Table 2.2. However, Nelson et al. reported that polar alcohols tend to inactivate enzymes much faster than nonpolar alcohols [28].

#### 2.5.1.4 Other Novel Heterogeneous Catalysts

There exist quite a number of reports by researchers on successful conversion of oil into biodiesel using catalysts that do not fall under the general spectrum of acids, bases, or enzymes. Mostly heterogeneous in nature, they are usually insensitive to the presence of FFAs and can convert them as well into esters (Table 2.2). The preparation strategies for each catalyst, therefore, vary greatly as they can be the source material itself (albeit modified to a certain extent) [32], a chemical compound that exists naturally as a salt [33], or other inert supports (carbonaceous or siliceous) that have been doped with transition metals, which are able to catalyze the transition much more efficiently [15, 26]. The form of doping in the last category is usually by the use of analytical grade salts containing the metal ion, which gets impregnated, leaving the anion to be washed off. Natural waste materials containing such elements can also be processed and used as a cost-efficient alternative (such as cow bones for calcium doping).

#### 2.5.1.5 Two-Step Catalyzed Process

Many researchers opt for this method, in which acid esterification is used for pretreating the oil in order to make it suitable for base-catalyzed conversion before performing alkali-catalyzed transesterification, which can completely convert the glycerides into esters, since bases are sensitive to high FFAs (owing to saponification) as well as moisture (owing to hydrolysis) [27]. The process can comprise of either esterification–transesterification steps or hydrolysis and esterification steps (Table 2.2) [34]. Hydrolysis combined with esterification is comparatively more wasteful as generation of FFAs is an energy-intensive process since high temperatures (exceeding 300 °C) and pressure (exceeding 10 MPa) are required. In the two-step catalyzed process involving esterification and transesterification, acid catalysts remove almost all of the FFAs through conversion to esters and water, which can be then purified and dried prior to using base catalysts, which convert the glycerides into esters and glycerol [22]. The glycerol and excess alcohol can be removed through washing or by ultracentrifugation before being tested for suitability as fuel. As mentioned in Section 2.4, nonpolar alcohols result in better biodiesel yield compared with polar alcohols, and, thus, they hold great potential for use in biodiesel production [1, 6].

### 2.5.2 Modern Conversion Approaches

#### 2.5.2.1 Supercritical Fluids

Critical point of a fluid designates the temperature and pressure at which compounds exit the liquid–vapor phase equilibrium. Beyond this stage (under supercritical conditions), the formed vapor cannot return to liquid state even under high pressure. For pressure-driven chemical conversions, this is a huge benefit that ensures spontaneous product removal after reaction (for continuous production). Hence the use of supercritical fluids is another PI approach, which has been successfully used for complete conversion of various feedstock (fed-batch or continuous). First proposed by Saka and Dadan [35], supercritical transesterification uses alcohols (polar or nonpolar) that have been preheated at 250–400 °C under 10–65 MPa pressure, during which the alcohol shows an increase in viscosity, resulting in decreased dielectric constant. This ensures that even polar alcohols become completely miscible, forming a single phase with superheated oil. The mixture is then pumped into the supercritical reactor (small capacity stainless steel reactor in a bath-type heater). After the desired duration, the reactor is withdrawn from the bath, depressurized, and cooled for product collection. The supercritical state ensures that diffusional resistances among the reactants are drastically reduced; with $E_a$ no longer a hindrance, the reaction occurs spontaneously with complete conversion in a few minutes. However, the supercritical reactor must have high durability against such extreme pressure and temperature while requiring frequent maintenance, which are setbacks regarding process economics and energy efficiency. Formed glycerol is also not usable due to high impurity and exposure to such extreme conditions [36]. Consequently, few modifications were proposed and tested by researchers to minimize these issues, making the process more lucrative for small-scale commercial production (Table 2.3).

Using cosolvents such as $CO_2$, n-hexane, tetrahydrofuran, propane, etc. facilitates solubilization, achieving homogeneous phase under much benign conditions, thereby increasing

**Table 2.3** Modern approaches in biodiesel production from nonedible/waste feedstock.

**Supercritical fluid-assisted biodiesel production**

| Feedstock | Solvent (+ catalyst/cosolvent) | Reaction temperature (°C) | Reaction pressure (MPa) | Alcohol:oil ratio | Residence time (min) | Yield (%) | References |
|---|---|---|---|---|---|---|---|
| Waste frying oil | Methanol | 300 | 20 | 40 : 1 | Continuous | 81.7 | [37] |
| Castor oil | Methanol<br>Ethanol | 350 | 20 | 40 : 1 | ~ 40 | >99 | [38] |
| Linseed oil | Methanol<br>Ethanol | 350 | 20 | 40 : 1 | ~ 40 | >99 | |

**Ultrasound-assisted biodiesel production**

| Feedstock | Solvent + catalyst | Reaction temperature (°C) | Reaction time (h) | Catalyst concentration (%w/w) | Alcohol:oil (ratio or %w/w) | Sonication (kHz) | Conversion/ yield (%) | References |
|---|---|---|---|---|---|---|---|---|
| Waste cooking oil | Dimethyl carbonate + Novozym 435 | 60 | 4 | 10 | 6 : 1 | 25 | 86.61 | [39] |
| Waste lard | *Candida antarctica* lipase B | 50 | 0.33 | 6 | 4 : 1 | 5 | 96.8 | [40] |
| Waste tallow | *Candida antarctica* lipase B | 27 | 0.33 | 6 | 4 : 1 | 5 | 85.6 | [41] |

**Microwave-assisted biodiesel production**

| Feedstock | Catalyst | Reaction temperature (°C) | Reaction time (h) | Catalyst concentration (%w/w) | Alcohol:oil (ratio or %w/w) | Power (W) | Yield (%) | References |
|---|---|---|---|---|---|---|---|---|
| Waste cooking oil | Sodium methoxide | 27 | 0.05 | 0.75 | 6 : 1 | 750 | 97.9 | [42] |
| Waste cooking oil | $SrO/SiO_2$ | 65 | 0.41 | 0.75 | 12 : 1 | 242 | 99.2 | [43] |
| Waste cooking oil – *Calophyllum inophyllum* oil blend | KOH | 27 | 0.12 | 0.774 | 59.6 | 850 | 97.4 | [44] |
| Waste cooking oil | $H_2SO_4$ | 27 | 6 | 0.5 | 9 : 1 | 800 | 92 | [45] |

energy efficiency [46]. Metal oxides (ZnO, $SrO_2$, $TiO_2$, etc.) were successfully used for facilitating conversions under lowered pressure and temperature; however, catalyst separation is a hurdle that impedes smooth operation [47]. As a better alternative, uncatalyzed subcritical hydrolysis followed by supercritical esterification was proposed and tested by Kusdiana and Saka [48]. Also glycerol-free processes (alcohol-free) have been developed that use compounds such as methyl acetate to yield triacetin (used in leavening of bread and flavor enhancing of beverages), dimethyl carbonate to yield glycerol carbonate (used as fuel additive) and citramalic acid (used in cosmetics for skin toning), and methyl tert-butyl ether, which yields glycerol tert-butyl ether, used as biodiesel additive to lower cloud point and enhance cetane number.

#### 2.5.2.2 Microwave Irradiation

The dipole rotation and ionic conduction achieved by means of microwave irradiation is useful for rapid heating in a system since heat is generated from inside out, thus generated heat and mass transfer gradients travel in the same direction, compared with being opposite during conventional heating. In a microwave, the electricity is converted to heat according to Eq. (2.3), where $P$ = power dispersion per unit volume of reaction mixture, $K$ = constant, $f$ = applied frequency, $\varepsilon$ = dielectric constant of compound (average for mixtures), $E$ = electric field strength, and tan $\delta$ = dielectric loss tangent [2]. From this, the power dispersion is visible dependent on reactant polarity and $E$ as selected from (irradiation strength) during the conversion process.

$$P = Kf\varepsilon E^2\left(\tan\delta\right) \tag{2.3}$$

Compared with conventional processes, the dispersion of generated heat is thus rapid, facilitating very high conversion rates and drastically reducing required time and energy [49]. However, slight disadvantages exist since presence of solids greatly disrupts the microwave penetration, resulting in irregular diffusion and reactivity. Also volatile compounds pose greater risks when exposed to microwaves compared to uniform heating due to rapid heat generation [50]. This process has been successfully applied in both fed-batch and continuous processes, as listed in Table 2.3.

#### 2.5.2.3 Ultrasonication

Vibrations ranging between 18 kHz and 100 MHz can be classified as ultrasound, and the process of ultrasonication involves the use of sound waves generated from a probe or in a bath for enhancement of chemical conversions such as transesterification [51]. In the bulk liquid, ultrasound generates acoustic pressure ($P_a$), which adds to the already existing hydrostatic pressure ($P_h$). $P_a$ depends on exposure time $t$, amplitude pressure of wave $P_A$, and acoustic cycle sin $2\pi$ as depicted in Eq. (2.4):

$$P_a = P_A \sin 2\pi\, ft \tag{2.4}$$

The generated sound waves have regions of compression during one half of the acoustic cycle where the liquid is compacted, creating a high pressure ($P_a + P_h$) along with heat generation, and regions of rarefaction during the other half when the liquid experiences a local vacuity ($P_a - P_h$), experiencing sudden cooling owing to decreasing local pressure [52]. The wave intensity $I$ is related to $P_A$ as shown in Eq. (2.5), where ρ is liquid density, while $c$ is the speed of sound in the liquid:

$$I = P_A^2 (2\rho c)^{-1} \tag{2.5}$$

The intensity of the wave is distributed as energy (generating heat) and thus gradually decreases over increasing distance as shown in Eq. (2.6), where $I_0$ = sound intensity at probe, α = absorption coefficient, and $d$ = distance of molecule from probe:

$$I = I_0 e^{-2\alpha d} \tag{2.6}$$

Regions in the rarefaction region often experience a fall in pressure exceed the critical distance when $P_a - P_h$ is quite large, and this results in molecules being stretched beyond the critical distance $R$, which causes cavitation bubbles to form, subsequently when the compression wave hits this region. Pressure increase causes temperature and pressure inside the cavitation bubble to increase abnormally (up to 4000 K and 90 MPa) before it implodes, spreading the heat afterward, which greatly enhance chemical conversion such as transesterification. The investment costs in this process are low, and the process itself is very energy efficient; however, exposing the oil to such extreme changes in temperature and pressure at a molecular level may have damaging effects on the glycerides and FFAs, hampering fuel yield. Nevertheless, it is still one of the most sought PI approaches in commercial biodiesel production [46], as seen in Table 2.3.

## 2.6 Economics and Environmental Impact

Despite extensive research data reported every year and the successful utilization of biodiesel blends in a few countries, the feasibility of utilizing biodiesel globally is still debated since the process is complex, involving costly equipment and high reactant losses, while the feedstock used are not sufficiently available. However, the cultivation and/or utilization of these nonedible and waste feedstock for commercial fuel synthesis provides monetary incentives to farmers, which can bolster economy [53]. The distribution of potential sources of WCO collection also plays an important function in determining the best site for establishing a production plant, since transportation must be minimized. Highly populated areas remain a lucrative choice due to increased number of sources; however, the plant must be strategically placed to avoid local contamination from industrial discharge and emissions. Unfortunately, no extensive studies have been reported so far on this aspect.

Life cycle assessment (LCA) studies provide an insight into the impact of biodiesel synthesis on the market and environment [54]. Lee et al. reported an estimated cost reduction by 10–20% on manufacturing and 42% feedstock cost when using WCO in supercritical systems [55]. A notable point is that waste oil collected from cafes, eateries, and restaurants generally consists of WCO mixed with grease trap waste, which cannot be utilized for fuel synthesis without incurring heavy energy and other purification losses during intensive pretreatment. This also adds to $CO_2$ emissions (estimated at 92%) during biodiesel synthesis life cycle, with 25% of it generated from transportation of this unusable waste. Anaerobic digester employing methanogens, for example, can be used to solve this [55]. Thus WCO is physically separated before transport and use. Another aspect that can generate more revenue is the processing and sale of glycerol or other products generated from glycerol-free processes, since they boast widespread use [53]. Carbon footprint remains uninfluenced in a worst-case scenario when using WCO, since WCO generation can be equated to our dietary activities, and its utilization expectedly resulted in negligible changes for global $CO_2$ uptake, while nonedible oil usage showed positive impact.

## 2.7 Conclusion and Perspectives

Not without challenges, WCO is remarkably lucrative for biodiesel synthesis. The sourcing and utilization of WCO benefits both economy and the environment, especially due to its easy availability. The removal of impurities and pretreatment prior to production is a hassle, which is eclipsed by its applicability to various production approaches with polar or nonpolar alcohols. While catalyzed processes boast smoother operation and low capital costs at the cost of increased time and labor, continuous PI approaches boast rapid conversion and operation ease while being high on investments, maintenance, and alcohol consumption. However, with high variance in composition each time it is procured, the efficiency under any given set of reaction conditions is still unpredictable. The separation, collection, and purification of by-products for sale can further help bolster competitiveness of biodiesel with petrodiesel, making the inevitable transition easier.

## References


1. Karmakar, B., Hossain, A., Jha, B. et al. (2021). Factorial optimization of biodiesel synthesis from castor-karanja oil blend with methanol-isopropanol mixture through acid/base doped *Delonix regia* heterogeneous catalysis. *Fuel* 285: 119197.
2. Karmakar, B. and Halder, G. (2019). Progress and future of biodiesel synthesis: advancements in oil extraction and conversion technologies. *Energy Convers. Manag.* 189: 309–337.
3. Dhawane, S.H., Karmakar, B., Ghosh, S., and Halder, G. (2018). Parametric optimisation of biodiesel synthesis from waste cooking oil via Taguchi approach. *J. Environ. Chem. Eng.* 6: 3971–3980.
4. Karmakar, B., Dhawane, S.H., and Halder, G. (2018). Optimization of biodiesel production from castor oil by Taguchi design. *J. Environ. Chem. Eng.* 6: 2684–2695.
5. Karmakar, B., Ghosh, B., Samanta, S., and Halder, G. (2020). Sulfonated catalytic esterification of *Madhuca indica* oil using waste *Delonix regia*: L16 Taguchi optimization and kinetics. *Sustain. Energy Tech. Assess.* 37: 100568.
6. Karmakar, B., Samanta, S., and Halder, G. (2020). Delonix regia heterogeneous catalyzed two-step biodiesel production from Pongamia pinnata oil using methanol and 2-propanol. *J. Clean. Prod.* 255: 120313.
7. Lang, X., Dalai, A.K., Bakhshi, N.N. et al. (2001). Preparation and characterization of bio-diesels from various bio-oils. *Bioresour. Technol.* 80: 53–62.
8. Rattanaphra, D. and Srinophakun, P. (2010). Biodiesel production from crude sunflower oil and crude jatropha oil using immobilized lipase. *J. Chem. Eng. Japan* 43 (1): 104–108.
9. Haas, M.J. (2005). Improving the economics of biodiesel production through the use of low value lipids as feedstocks: vegetable oil soapstock. *Fuel Process. Technol.* 86: 1087–1096.

**10** Kalam, M.A. and Masjuki, H.H. (2002). Biodiesel from palm oil – an analysis of its properties and potential. *Biomass Bioenergy* 23: 471–479.

**11** Mata, T.M., Martins, A.A., and Caetano, N.S. (2010). Microalgae for biodiesel production and other applications: a review. *Renew. Sustain. Energy Rev.* 14 (1): 217–232.

**12** Atabani, A.E., Silitonga, A.S., Badruddin, I.A. et al. (2012). A comprehensive review on biodiesel as an alternative energy resource and its characteristics. *Renew. Sustain. Energy Rev.* 16: 2070–2093.

**13** Wastes – Resource Conservation – Common Wastes & Materials. US Environmental Protection Agency. https://archive.epa.gov/wastes/conserve/materials/usedoil/web/html/oil.html.

**14** Ma, F. and Hanna, M.A. (1999). Biodiesel production: a review. *Bioresour. Technol.* 70: 1–15.

**15** Olubunmi, B.E., Karmakar, B., Aderemi, O.M. et al. (2020). Parametric optimization by Taguchi L9 approach towards biodiesel production from restaurant waste oil using Fe-supported anthill catalyst. *J. Environ. Chem. Eng.* 8: 104288.

**16** Karmakar, B., Ghosh, B., and Halder, G. (2020). Sulfonated *Mesua ferrea linn* seed shell catalyzed biodiesel synthesis from castor oil – response surface optimization. *Front. Energy Res.* 8: 576792. https://doi.org/10.3389/fenrg.2020.576792.

**17** Srivastava, A. and Prasad, R. (2000). Triglycerides-based diesel fuels. *Renew. Sustain. Energy Rev.* 4 (2): 111–133.

**18** Freedman, B., Pryde, E.H., and Mounts, T.L. (1984). Variables affecting the yields of fatty esters from transesterified vegetable oils. *J. Am. Oil Chem. Soc.* 61 (10): 1638–1643.

**19** Farobie, O., Leow, Z.Y.M., Samanmulya, T., and Matsumura, Y. (2016). New insights in biodiesel production using supercritical 1-propanol. *Energy Convers. Manag.* 124: 212–218.

**20** Phan, A.N. and Phan, T.M. (2008). Biodiesel production from waste cooking oils. *Fuel* 87: 3490–3496.

**21** Singh, T.S. and Verma, T.N. (2019). Taguchi design approach for extraction of methyl ester from waste cooking oil using synthesized CaO as heterogeneous catalyst: response surface methodology optimization. *Energy Convers. Manag.* 182: 383–397.

**22** Maneerung, T., Kawi, S., Dai, Y., and Wang, C.H. (2016). Sustainable biodiesel production via transesterification of waste cooking oil by using CaO catalysts prepared from chicken manure. *Energy Convers. Manag.* 123: 487–497.

**23** Ngoya, T., Aransiola, E.F., and Oyekola, O. (2017). Optimization of biodiesel production from waste vegetable oil and eggshell ash. *S. Afr. J. Chem. Eng.* 23: 145–156. https://doi.org/10.1016/j.sajce.2017.05.003.

**24** Fan, Y., Wu, G., Su, F. et al. (2016). Lipase oriented-immobilized on dendrimer-coated magnetic multi-walled carbon nanotubes toward catalyzing biodiesel production from waste vegetable oil. *Fuel* 178: 172–178.

**25** Shah, S., Sharma, S., and Gupta, M.N. (2004). Biodiesel preparation by lipase catalyzed transesterification of Jatropha oil. *Energy Fuels* 18: 154–159.

**26** Dhawane, S.H., Kumar, T., and Halder, G. (2016). Parametric effects and optimization on synthesis of iron (II) doped carbonaceous catalyst for the production of biodiesel. *Energy Convers. Manag.* 122: 310–320.

**27** Jain, S., Sharma, M.P., and Rajvanshi, S. (2011). Acid base catalyzed transesterification kinetics of waste cooking oil. *Fuel Process. Technol.* 92: 32–38.

**28** Nelson, L.A., Foglia, T.A., and Marmer, W.N. (1996). Lipase-catalyzed production of biodiesel. *J. Am. Oil Chem. Soc.* 73: 1191–1195.

**29** Lee, S., Posarac, D., and Ellis, N. (2011). Process simulation and economic analysis of biodiesel production processes using fresh and waste vegetable oil and supercritical methanol. *Chem. Eng. Res. Des.* 89: 2626–2642.

**30** Shimada, Y., Watanabe, Y., Samukawa, T., and Sugihara, A. (1999). Conversion of vegetable oil to biodiesel using immobilized *Candida antarctica* lipase. *J. Am. Oil Chem. Soc.* 76 (7): 789–793.

**31** Jaeger, K.E. and Eggert, T. (2002). Lipases for biotechnology. *Curr. Opin. Biotech.* 3 (4): 390–397.

**32** Bora, D.K., Saha, R., Saikia, M. et al. (2014). Biofuel production from *Mesua ferrea* L. seed oil. *Int. J. Eng. Tech. Res.* 2 (9): 397–402. ISSN: 2321-0869.

**33** Lin, Y.C., Chen, S.C., Chen, C.E. et al. (2014). Rapid Jatropha-biodiesel production assisted by a microwave system and a sodium amide catalyst. *Fuel* 135: 435–442.

**34** Phuenduang, S., Chatsirisook, P., Simasatitkul, L. et al. (2012). Heat-integrated reactive distillation for biodiesel production from Jatropha oil. *Proceedings 11th International Symposium on Process Systems Engineering*, Singapore (15–19 July 2012).

**35** Saka, S. and Kusdiana, D. (2001). Biodiesel fuel from rapeseed oil as prepared in supercritical methanol. *Fuel* 80: 225–231.

**36** Farobie, O. and Matsumura, Y. (2017). State of the art of biodiesel production under supercritical conditions. *Prog. Energy Combust. Sci.* 63: 173–203.

**37** Gonzalez, S.L., Sychoski, M.M., Navarro-Díaz, H.J. et al. (2013). Continuous catalyst-free production of biodiesel through transesterification of soybean fried oil in supercritical methanol and ethanol. *Energy Fuels* 27: 5253–5259. https://doi.org/10.1021/ef400869y.

**38** Varma, M.N. and Madras, G. (2007). Synthesis of biodiesel from castor oil and linseed oil in supercritical fluids. *Ind. Eng. Chem. Res.* 46: 1–6.

**39** Gharat, N. and Rathod, V.K. (2013). Ultrasound assisted enzyme catalyzed transesterification of waste cooking oil with dimethyl carbonate. *Ultrason. Sonochem.* 20 (3): 900–905.

**40** Adewale, P., Dumont, M.J., and Ngadi, M. (2016). Enzyme-catalyzed synthesis and kinetics of ultrasonic-assisted methanolysis of waste lard for biodiesel production. *Chem. Eng. J.* 284: 158–165.

**41** Adewale, P., Dumont, M.J., and Ngadi, M. (2015). Enzyme catalyzed synthesis and kinetics of ultrasonic-assisted biodiesel production from waste tallow. *Ultrason. Sonochem.* 27: 1–9.

**42** Chen, K.S., Lin, Y.C., Hsu, K.H., and Wang, H.K. (2012). Improving biodiesel yields from waste cooking oil by using sodium methoxide and a microwave heating system. *Energy* 38: 151–156.

**43** Tangy, A., Pulidindi, I.N., Perkas, N., and Gedanken, A. (2017). Continuous flow through a microwave oven for the large-scale production of biodiesel from waste cooking oil. *Biores. Technol.* 224: 333–341.

**44** Milano, J., Ong, H.C., Masjuki, H.H. et al. (2018). Optimization of biodiesel production by microwave irradiation-assisted transesterification for waste cooking oil – *Calophyllum inophyllum* oil via response surface methodology. *Energy Convers. Manag.* 158: 400–415.

**45** Patil, P.D., Gude, V.G., Reddy, H.K. et al. (2012). Biodiesel production from waste cooking oil using sulfuric acid and microwave irradiation processes. *J. Environ. Prot.* 3: 107–113.

**46** Imahara, H., Xin, J., and Saka, S. (2009). Effect of $CO_2/N_2$ addition to supercritical methanol on reactivities and fuel qualities in biodiesel production. *Fuel* 88: 1329–1332.

**47** Demirbas, A. (2007). Biodiesel from sunflower oil in supercritical methanol with calcium oxide. *Energy Convers. Manag.* 48: 937–941.

**48** Kusdiana, D. and Saka, S. (2004). Two-step preparation for catalyst-free biodiesel fuel production: hydrolysis and methyl esterification. *Appl. Biochem. Biotech.* 113 (116): 781–791.

**49** Mandal, V., Mohan, Y., and Hemalatha, S. (2007). Microwave-assisted extraction-an innovative and promising extraction tool for medicinal plant research. *Pharmaco. Rev.* 1 (1): 7–18.

**50** Farhat, A., Ginies, C., Romdhane, M., and Chemat, F. (2009). Eco-friendly and cleaner process for isolation of essential oil using microwave energy: experimental and theoretical study. *J. Chromato. A* 1216 (26): 5077–5085.

**51** Karim Assami, D.P. (2012). Ultrasound-induced intensification and selective extraction of essential oil from *Carum carvi* L. seeds. *Chem. Eng. Process. Process Intensif.* 62: 99–105.

**52** Bhaskaracharya, R.K., Kentish, S., and Ashokkumar, M. (2009). Selected applications of ultrasonics in food processing. *Food Eng. Rev.* 1: 31–49.

**53** Glisic, S.B., Pajnik, J.M., and Orlovic, A.M. (2016). Process and techno-economic analysis of green diesel production from waste vegetable oil and the comparison with ester type biodiesel production. *Appl. Energy* 170: 176–185.

**54** Tran, N.P.L., Ong, L.K., and Ju, Y.H. (2019). Biodiesel production from waste oils. In: *Biofuels: Alternative Feedstocks and Conversion Processes for the Production of Liquid and Gaseous Biofuels* (ed. A. Pandey, C. Larroche, C.-G. Dussap, et al.). Academic Press.

**55** Tu, Q. and McDonnell, B.E. (2016). Monte Carlo analysis of life cycle energy consumption and greenhouse gas (GHG) emission for biodiesel production from trap grease. *J. Clean. Prod.* 112: 2674–2683.

# 3

# Non-edible Oils for Biodiesel Production

## State of the Art and Future Perspectives

*Valeria D'Ambrosio, Enrico Scelsi, and Carlo Pastore*

*Istituto di Ricerca Sulle Acque, Consiglio Nazionale delle Ricerche (IRSA-CNR), Bari, Italy*

## 3.1 Introduction

The depletion of fossil feedstock, together with the increasing need to reduce the environmental impact related to the consumption of liquid fuel, has led to a greater effort being made to research alternative sources and energy vectors. In addition, the demand for and consumption of liquid fuel is continually increasing, meaning that new alternatives are required, which satisfy technical, economic, and environmental issues. Biodiesel, namely, fatty acid methyl esters (FAMEs), is an attractive alternative to diesel, since it can be produced from renewable resources and involves lower carbon dioxide emissions than fossil fuel.

Vegetable feedstocks are normally the main source for biodiesel production, but there are a number of economic, ethic, and environmental concerns regarding their use. These are raw materials that are both expensive and in a high demand, meaning that final products are not economically competitive with diesel. In addition, they raise the *food vs. fuel* ethical debate, as oil exploited for biodiesel production can be alternatively used to alleviate starvation and malnutrition. Indirectly, they can even cause deforestation and loss of biodiversity, due to the intensive cultivation of crops to meet the high demand of oils for fuel.

Non-edible oils are not subject to these concerns and can be considered as a promising feedstock for biodiesel production, with the potential to substitute conventional edible vegetable oils.

They may contain toxic compounds, making them unsuitable for human consumption, and consequently they have a lower demand and a lower cost [1].

In this chapter, a review of the main non-edible oils is carried out: cottonseed oil (*Gossypium hirsutum*) [1, 2], *Crambe abyssinica* [3, 4], croton oil (*Croton megalocarpus*) [5, 6], desert date (*Balanites aegyptiaca*) [7, 8], jatropha oil (*Jatropha curcas*) [9–12], jojoba oil (*Simmondsia chinensis*) [13–15], karanja oil (*Pongamia pinnata*) [16–18], kemiri sunan oil (*Reutealis trisperma*) [19–21], linseed oil (*Linum usitatissimum*) [22, 23], mahua oil

*Biodiesel Production: Feedstocks, Catalysts, and Technologies*, First Edition. Edited by Samuel Lalthazuala Rokhum, Gopinath Halder, Suttichai Assabumrungrat, and Kanokwan Ngaosuwan.

(*Madhuca indica*) [24], moringa oil (*Moringa oleifera*) [25, 26], neem oil (*Azadirachta indica*) [27], polanga oil (*Calophyllum inophyllum*) [28, 29], rubber seed oil (*Hevea brasiliensis*) [30, 31], sal oil (*Shorea robusta*) [32, 33], sea mango (*Cerbera odollam*) [34, 35], soap nut oil (*Sapindus mukorossi*) [36, 37], stillingia oil (*Sapium sebiferum*) [38, 39], taramira oil (*Eruca sativa*) [40, 41], and tobacco seed oil (*Nicotiana tabacum*) [42–45] were examined. Particular attention was given to plants distribution, growth conditions, oil content in seeds (or in kernels), and yield per hectare, all factors that allow us to assess the relevant availability, cost of production, and an eventual economic competition with other agricultural crops used in the production of biodiesels. Following this, the fatty acids composition of non-edible oils is reported and the properties of the relevant biodiesel critically discussed. A brief summary of the main oil extraction methods and reaction conditions for producing biodiesel is reported. Criticisms related to the use of these feedstocks and future alternatives are then reported and discussed.

## 3.2 Vegetable Non-edible Oils

### 3.2.1 General Cultivation Data

Non-edible oils represent a promising feedstock for biodiesel production. Some advantages in their use are associated to their non-edible nature, which eliminates problems that affect edible oil (e.g. the food vs. fuel debate, high cost). Most of them are actually produced by cultivating crops which (Table 3.1) have the advantage of being resistant to unfavorable conditions (high salinity, drought and frost, presence of insects and pests) and can grow in soils unsuitable for other species. For this reason, the indirect improper use of land can also be avoided.

Distribution and growth conditions provide an idea as to where different crops can be cultivated, whereas oil yields per hectare directly gives a measure of the profitability, in a particular region, for a specific non-edible oil plant. The definition of the sustainability of a productive chain is an extremely complex matter to be evaluated and includes many factors. In any case, the use of marginal land for cultivating non-edible oils crops generally has a particularly positive effect in terms of green house gases (GHG) emissions [50]. In several cases reported in Table 3.1, non-edible oils can be reasonably obtained from crops which can also be cultivated under extreme conditions. In some cases, even very high productivity can also be achieved, with over 1500 kg $ha^{-1}$ per year being produced, without requiring too much water and using arid lands.

### 3.2.2 Composition and Chemical–Physical Properties of Biodiesel Obtained from Non-edible Vegetable Oils

The Fatty acids composition, along with information on distribution, growing conditions, and oil yield per hectare, allow for a more accurate evaluation of biodiesel production from these alternative feedstocks and the relative market. Indeed, the composition of fatty acids determines the quality of the final product and strongly indicates how (and where) they should be best employed.

**Table 3.1** Non-edible oils sources: distribution, growth conditions, oil content, and yield for hectare.

| Non-edible oil source | Distribution | Growth conditions | Oil content (% wt) | | Yield for hectare (kg $ha^{-1}$) | References |
|---|---|---|---|---|---|---|
| | | | Seed | Kernel | | |
| *Gossypium hirsutum* (Cottonseed) | Native to Southern Mexico and Guatemala. Widespread in India, China, Europe, and the United States. | | 17–25% | | | [2] |
| *Crambe abyssinica* | Native of Mediterranean region, from Ethiopia to Tanzania. Widespread in Russia and the United States. It can be grown also in Europe as a spring crop. | It can grow in sites with rainfall in the range of 350–1200 mm, an annual average temperature in the range of 5.7–16.2 °C and soils with a pH range of 5.0–7.8. Great tolerance to drought and frost. | 36–43% | | 450–4000[a] | [3] |
| *Croton megalocarpus* (croton oil) | | | 32% | | 50[b] | [6] |
| *Balanites aegyptiaca* (desert date) | Arid regions in Africa and Asia | It can grow in adverse arid desert environments. | | 39–55% | 1600[a] | [7, 8] |
| *Jatropha curcas* (jatropha) | Native to tropical America and spread all over the tropics and subtropics of Asia and Africa. | It can grow on diverse wasteland without any agricultural impute (irrigation and fertilization). It has an easy propagation, rapid growth, drought tolerance, and pest resistance. | 30–60% | | 1900–2500[a] | [9, 11, 46] |
| *Simmondsia chinensis* (jojoba) | Growing naturally in the Sonora desert (Mexico) and in the southwest of the United States. Now cultivated in some countries: South America, Israel, the United States, and some Mediterranean and African lands | It can tolerate extremely high temperature up to 54 °C and low temperature up to 5 °C. It is moderately drought tolerant and can produce regular crops with 420 mm of precipitation annually. It can grow in soils of marginal fertility, and it is a relatively pest and disease resistant plant. | 45–55% | | 1125–2250[a] | [13] [14, 46] |

(*Continued*)

**Table 3.1** (Continued)

| Non-edible oil source | Distribution | Growth conditions | Oil content (% wt) | | Yield for hectare (kg ha$^{-1}$) | References |
|---|---|---|---|---|---|---|
| | | | Seed | Kernel | | |
| *Pongamia pinnata* (karanja) | Growing in Indian subcontinent and Southeast Asia. Successfully introduced to humid tropical regions of the world as well as parts of Australia, New Zealand, China, and the United States. | It is fast growing, drought resistant, moderately frost hardy, and highly tolerant of salinity. | | 30–40% | 225–2250[a] | [18, 47] |
| *Reutealis trisperma (kemiri sunan)* | Native to the Philippines and Southeast Asia. | | 50–52% | | — | [20] [21] |
| *Linum usitatissimum (linseed)* | Distributed to the regions extending from the Eastern Mediterranean to India. There is wide cultivation of this crop in Europe. | | 35–45% | | 300[a] | [2, 46] |
| *Madhuca indica (mahua)* | Mainly found in India. | It is adapted to arid environments. | | 50% | | [48] |
| *Moringa oleifera (moringa)* | Native to the Northwest India, Africa, Arabia, Southeast Asia, and South America. Now distributed in the Philippines, Cambodia, and Central and North America. | It thrives best in a tropical insular climate and is plentiful near the sandy beds of rivers and streams.<br>It is fast growing, drought tolerant and it can grow in poor soils with a pH from 5.0 to 9.0, adapting to a wide rainfall range (25–300 cm yr$^{-1}$). | 33–41% | | 900[a] | [25] |
| *Azadirachta indica (neem)* | Native to India, Pakistan, Sri Lanka, Burma, Malaysia, Indonesia, Japan, and the tropical regions of Australia. | It can grow in almost all kinds of soil. It thrives well in arid and semiarid climate with a maximum temperature of 49 °C and rainfall as low as 250 mm. | | 45% | 2670[a] | [27, 28, 48] |
| *Calophyllum inophyllum (polanga)* | Native to East Africa, India, Southeast Asia, and Australia. | It grows best in deep soils or on exposed sea sands.<br>The rainfall requirement is 750–5000 mm yr$^{-1}$. | 40% | 60–65% | 4680[a] | [28, 29, 46] |

| | | | | | | |
|---|---|---|---|---|---|---|
| *Hevea brasiliensis (rubber)* | Native to the Amazon rainforest (Brazil). Distributed mainly in Indonesia, Malaysia, Liberia, India, Sri Lanka, Sarawak, and Thailand. | It requires heavy rainfall. | | 40–50% | 40–50[a] | [46, 48] |
| *Shorea robusta (sal)* | Native to Southern Asia (India, Myanmar, Nepal, and Bangladesh) and widely distributed in tropical regions of India. | | 19–20% | | — | [32] |
| *Cerbera odollam (sea mango)* | Native to India and other parts of Southern Asia. | It grows well in coastal salt swamps and creeks. | 40–54% | | — | [48] |
| *Sapindus mukorossi (soap nut)* | Found in tropical and subtropical climate areas and in various parts of the world including Asia, America, and Europe. | It grows very well in deep loamy soils and leached soils, in altitudes from 200 to 1500 m, in regions where precipitation varies from 150 to 200 cm $yr^{-1}$. | | 51% | — | [48, 49] |
| *Sapium sebiferum (stillingia)* | Native to Eastern Asia (China, Japan, and India) and widespread in the Southern coastal United States. | It can grow on marginal land and is adapted to alkaline, saline, droughty, and acidic soils. | | 13–32% | | [39] |
| *Eruca sativa* (taramira) | Widespread in Middle East, India, and Pakistan. | It can grow in areas where rainfall and soil fertility are too low to cultivate food crops. Although vulnerable to sawfly attack at the seedling stage, mature plants are highly resistant to insect. | 28–35% | | 420–590[a] | [41, 46] |
| *Nicotiana tabacum (tobacco)* | Widespread in more than 100 countries such as Macedonia, Turkey, South Serbia, and North and South America. | | 35–49% | | 590–1480[a] | [44, 45] |

[a] Yield of oil (kg $ha^{-1}$).
[b] Yield of seeds (kg $ha^{-1}$).

The first parameter that should be considered is the calorific value (or heating value). This is a key indicator of the fuel energy content. Typically, for biodiesel mainly composed (>96%) of fatty acid (C14–C18) methyl esters, the calorific value is similar to that one of petrol diesel (38–40 MJ $kg^{-1}$) and for this reason suitable for use as a substitute for fossil fuel. The fatty acids composition of non-edible oils (reported in Table 3.2), greatly influences biodiesel properties, such as cloud point, pour point, flash point, kinematic viscosity, cetane number, and iodine value. Cloud point and pour point are defined as the temperature (°C) at which crystals become visible in the fuel and the temperature at which the amount of crystals is sufficient to gel the fuel, respectively. These values increase with the length of chains of methyl esters and decrease with the increase of unsaturation. Data reported in Table 3.2 shows this trend: biodiesel from *cottonseed* and *linseed* oils, which mostly consist of unsaturated fatty acids, has very low pour point values; biodiesel from *sal* and *moringa* oils, which mostly consist of saturated or monounsaturated oils, respectively, has high values. The flash point, defined as the temperature (°C) at which the fuel will ignite when exposed to a flame or a spark, varies inversely with regard to volatility, having, therefore, the same trend of cloud and pour points. For the investigated non-edible oils, flash point values are in the range of 108–243 °C (Table 3.2). Kinematic viscosity has almost the same trend; in general, biodiesel viscosity is greater than that of fossil diesel, due to the presence of oxygenated moieties [57]. The cetane number is a measure of the ignition quality of diesel fuel during combustion ignition, and it provides information regarding the ignition delay time of a diesel fuel upon injection into the combustion chamber [56]. The longer the fatty acid carbon chains and the more saturated the molecules, the higher the cetane number, which implies a short ignition delay. For the investigated non-edible oils biodiesel cetane number values vary from 41 to 64 (Table 3.2).

Lastly, the iodine value (g $I_2$/100 g) is related to the unsaturation degree of biodiesel, which greatly influences oxidation stability and the ability to polymerize. This value increases with the increasing of unsaturation degree and varies from 28 for *kemiri sunan* to 184 for *linseed* oil biodiesel, the last mostly constituted by linolenic acid (C18 : 3) (Table 3.2). In order to assess the suitability of a feedstock for producing biodiesel, it is important to check if the previously reported parameters suit the values specified in ASTM D6751 or EN 14214 [56]. Some of the specifications are also reported in Table 3.2.

### 3.2.3 Biodiesel Production from Non-edible Vegetable Oil

Biodiesel production from non-edible oil seeds or kernels commonly requires pretreatment steps (e.g. grinding, drying [38, 58]), which depend on the seeds variety and on the following steps to be taken, and then a two-step process: extraction of the oil (mechanical, solvent, or enzymatic extraction) and methanolysis of the oil catalyzed by an acidic or alkaline catalyst, homogeneous or heterogeneous.

#### 3.2.3.1 Extraction Methods

The extraction of the oil from seeds or kernels can be carried out mechanically, by using either a manual ram press or an engine-driven screw press, which can extract most of the oil available in the seeds [56]. In the case of *J. curcas* seeds, the amount of extracted oil was 60–65% and 75–80% of the contained oil, respectively [59]. However, mechanical

**Table 3.2** Fatty acids composition of non-edible oils and relative biodiesel properties.

| | 14 : 0 | 16 : 1 | 16 : 0 | 18 : 3 | 18 : 2 | 18 : 1 | 18 : 0 | 20 : 4 | 20 : 1 | 20 : 0 | 22 : 4 | 22 : 1 | 22 : 0 | 24 : 0 | CP[a] | PP[b] | FP[c] | KV[d] | CN[e] | IV[f] | References |
|---|---|---|---|---|---|---|---|---|---|---|---|---|---|---|---|---|---|---|---|---|---|
| Cottonseed (*Gossypium hirsutum*) | 0.7 | | 11.7–20.1 | | 55.2–55.5 | 19.2–23.6 | 2.6–3.2 | | | 0.4 | | | | | 2 | −10 to −15 | 210–243 | 4.0–4.9 | 41–60 | 65.48 | [2, 51] |
| *Crambe abyssinica* | | | 2.1 | 6.9 | 9.0 | 18.9 | 0.7 | | | 2.1 | | 58.5 | | | – | – | 179 | 5.8 | – | 89 | [4] |
| Croton (*Croton megalocarpus*) | | | 7.4 | 3.4 | 71.2 | 12.2 | 4.1 | | 0.9 | | | | | | – | −5 | 190 | 4.1 | 42 | – | [5] |
| Desert date (*Balanites aegyptiaca*) | | | 16.7 | | 47.9 | 22.2 | 11.7 | | | 0.3 | | | | | 3–7 | – | 122–131 | 3.7–4.2 | 54 | 97–100 | [7] |
| Jatropha (*Jatropha curcas*) | | | 13.6–15.1 | | 31.4–43.2 | 34.3–44.7 | 7.1–7.4 | | | | | | | | −4 to 1 | −8 to 2 | 159 | 4.6 | 58 | 109 | [10] |
| Jojoba (*Simmondsia chinensis*) | | | 1.2 | | | 10.7 | | 9.1 | 59.5 | | 3.7 | 12.3 | 0.2 | | – | −4.5 | 180 | 3.1 | 63 | – | [15, 52, 53] |
| Karanja (*Pongamia pinnata*) | | | 11.6 | 2.6 | 16.5 | 51.6 | 7.5 | | | | | | | | 19 | 17 | 154 | 3.8 | 56 | 89 | [16, 54] |
| Kemiri sunan (*Reutealis trisperma*) | 0.1 | 0.6 | 16.2–22.9 | 0.3–0.6 | 13.6–38.2 | 30.2–32.1 | 22.0–6.6 | | 0.2 | 0.2 | | 4.2 | 1.1 | | −13 | −15 | 148 | 6.7 | 64 | 28 | [19, 20] |
| Linseed (*Linum usitatissimum*) | | 0.3 | 6.0 | 64.3 | 13.7 | 12.5 | 3.0 | | | | | | | | 1.7 | −18 to −4 | 108–148 | 3.4–4.2 | 28–48 | 184 | [2, 22, 23] |
| Mahua (*Madhuca indica*) | | | 21.5 | 0.2 | 19.6 | 39.1 | 19.0 | | | 0.6 | | | | | 13 | 6 | 170 | 4.2 | 57 | 71 | [24] |
| Moringa (*Moringa oleifera*) | | 0–2.5 | 6.5–7.9 | 1.1 | 1.0 | 67.3–72.2 | 4.5–6.0 | | 0–2.0 | 4.0–5.5 | | | 4.1–7.1 | | 18 | 17 | 103–162 | 3.7–4.8 | 55–67 | – | [25, 26] |

(*Continued*)

**Table 3.2** (Continued)

| | 14 : 0 | 16 : 1 | 16 : 0 | 18 : 3 | 18 : 2 | 18 : 1 | 18 : 0 | 20 : 4 | 20 : 1 | 20 : 0 | 22 : 4 | 22 : 1 | 22 : 0 | 24 : 0 | CP[a] | PP[b] | FP[c] | KV[d] | CN[e] | IV[f] | References |
|---|---|---|---|---|---|---|---|---|---|---|---|---|---|---|---|---|---|---|---|---|---|
| Neem (*Azadirachta indica*) | 0.2–2.5 | | 14.1–16.3 | | 2.4–14.8 | 48.9–62.7 | 15.2–23.7 | | | 0.9–3.3 | | | | | 9 | 5 | 184 | 4.6 | 58 | 75 | [27] |
| Polanga (*Calophyllum inophyllum*) | | 2.5 | 14.2 | 0.8 | 28.1 | 38.9 | 15.9 | | | 1.0 | | | | | 4 | 2 | 127 | 5.2 | 57 | 80 | [29] |
| Rubber (*Hevea brasiliensis*) | | | 9.1 | 14.2 | 46.2 | 24.0 | 5.6 | | | | | | | | 3 | −5 | 120–187 | 4.5–4.8 | – | – | [30, 31] |
| Sal (*Shorea robusta*) | | | 5.1 | | 1.8 | 43.1 | 45.2 | | | 4.8 | | | | | – | 18 | 160 | 4.8–5.0 | 52–53 | 40–47 | [32, 33] |
| Sea mango (*Cerbera odollam*) | | 0.8 | 24.7–31.5 | | 13.7–17.0 | 47.9–52.8 | 3.6–5.8 | | | | | | | | – | – | 119–138 | 2.9–4.5 | 53–56 | - | [34, 35] |
| Soap nut (*Sapindus mukorossi*) | | | 3.8 | 2.4 | 8.3 | 55.2 | 1.1 | | 23.9 | 5.3 | | | | | 14 | 5 | 177 | 4.9 | 58 | 84 | [37, 55] |
| Stillingia (*Sapium sebiferum*) | | | 7.7 | 39.5 | 29.1 | 15.9 | 2.5 | | | | | | | | −13 | – | 137 | 4.8 | 50 | – | [39] |
| Taramira oil (*Eruca sativa*) | | 0.4 | 28.3 | 12.4 | 22.7 | 28.3 | 2.5 | | | | | | | | – | −10 | 185 | 4.2 | 48 | – | [40] |
| Tobacco (*Nicotiana tabacum*) | 0.1–0.2 | | 8.9–11.9 | 0.8–4.3 | 69.5–73.8 | 12.2–14.3 | 3.3–3.6 | | | | | | | | | <15 | 174 | | 35 | 110 | [42, 43] |
| ASTM D6751 | | | | | | | | | | | | | | | −3 to −12 | −15 to 16 | >130 | 1.9–6.0 | >47 | | [56] |
| EN 14214 | | | | | | | | | | | | | | | – | – | >101 | 3.5–5.0 | >51 | <120 | [56] |

[a] Cloud point (°C).
[b] Pour point (°C).
[c] Flash point (°C).
[d] Kinematic viscosity ($mm^2\ s^{-1}$).
[e] Cetane number.
[f] Iodine value ($g\ I_2/100\ g$).

extraction has some problems: mechanical presses do not work well for all types of seed with the extracted oils requiring further treatment (filtration and degumming) [60]. Solvent extraction, which is more economically convenient on an industrial scale [61], has been used for different types of seeds. Solvent extractions are commonly carried out by using hexane [25, 60, 62], which provides high yields; other solvents that have been evaluated for oil extraction from seeds include diethyl ether, ethanol, n-heptane, isopropanol, acetone, chloroform, methanol, 1-butanol, and ethyl acetate [63]. Another extractive method, which is more environmentally friendly than solvent extraction, is aqueous enzymatic oil extraction, which involves an enzyme (e.g. cellulase, protozyme) that disrupts the structure of the seed, releasing the oil [64]. This method makes it possible to obtain high yields when ultrasonic pretreatment is carried out (e.g. ultrasonication for 5 min followed by aqueous enzymatic oil extraction using an alkaline protease at pH 9.0 allows to obtain a maximum yield of 74% of oil in *J. curcas* kernels [64]). However, this method requires a lengthy process time [60].

#### 3.2.3.2 Biodiesel Production

Biodiesel is produced by the transesterification reaction of oils with methanol commonly catalyzed by homogeneous alkaline catalysts (NaOH, KOH), which are 4000 times faster than acidic catalysts [65]. Reaction conditions differ and depend mostly on the starting composition of the oil; reaction times in the range 24–180 min, temperature in the range 50–70 °C, methanol : oil molar ratio (4 : 1–9 : 1), and catalyst concentration 0.55–1.5% (% wt of the oil) give yields higher than 90% wt for different types of oils [47, 66–68]. For oils with a high content of free fatty acids, preliminary esterification with methanol, by using acidic catalysts (e.g. HCl, $H_2SO_4$, $AlCl_3{\cdot}6H_2O$), is required [67, 69, 70], in order to increase FAMEs yield and to avoid the formation of soaps, which hinder the recovery of reaction products. Heterogeneous catalysts ($SO_4^{2-}/SnO_2–SiO_2$; $SO_4^{2-}/TiO_2$; sulfated zirconia alumina [6, 62, 71]) and enzymatic catalysts (Novozym 435, Lipozyme TLIM, and Lipozyme RMIM [39]) were also tested, obtaining high yields. However, these require drastic reaction conditions in terms of methanol : oil ratio, time, and reaction temperature.

### 3.2.4 Criticisms Related to Non-edible Oils

The use of non-edible oils as alternative biodiesel feedstock has been the subject of some criticism. Paradoxically, the main aspect that negatively affects the use of non-edible oils is strictly related to their principal strong point: the cultivation of crops in marginal areas leads to negative aspects in terms of economic and environmental impacts. The use of scattered locations for cultivating crops to obtain non-edible oils generates difficulties in picking and harvesting, as well as collecting seeds or kernels to be treated for the extraction of oil. Low-quality planting material or seed, a limited period of availability, unreliable and improper marketing channels, the possible lack of postharvesting technologies, and their processing negatively complete the present scenario. The resulting price of non-edible oil is often prohibitive. In addition, the unpredictability of the yield per hectare (Table 3.1) plays a negative role in planning a reliable economic projection and appropriate business plan.

Besides the economic factors involved, the environment might also be negatively affected by cultivating these crops in a marginal area, especially if it is located too far from the place where they will be used. Observing Figure 3.1, it is clear that the cultivation of non-edible oil crops is mainly focused in a limited area, often far from more populated territories, where the consumption of biodiesel is actually higher. In addition, the principal chemical–physical properties of biodiesel play a pivotal role in defining where they can be used without incurring problems. Too much saturated biodiesel needs to be consumed in temperate places, to avoid clogging and fillings of engines. While in locations that are too warm, biodiesel with a high level of unsaturation might incur polymerization or rancidity problems, which negatively affect their final use. All these points concur in highlighting another important issue related to the use of non-edible oil-derived biofuel – that of transportation. Indeed, the transportation of extracted oils or biodiesel obtained from such non-edible oils could significantly and negatively affect the balance of final economic and GHG emissions. The use of derived biodiesel should be limited to geographical areas located not far from where they are produced (proximity constraints). Where not possible, the use of additives and appropriate blends represents the best way to solve these technical difficulties.

## 3.3 Future Perspectives of Non-edible Oils: Oils from Waste

The full exploitation of waste to produce biodiesel could represent a strategic way to substitute fossil feedstock with renewable forms. Indeed, several residual biomasses can have promising potential in this regard. More specifically, the lipid component of several types of waste could in some cases be profitably recovered and converted into biodiesel. Differently from virgin non-edible oils, this category of lipids does not require the use of land and harvesting, and collection expenses do not need to be considered since the network of collection is already existent. In addition, they easily satisfy the previously mentioned proximity requirements. Finally, waste utilization would lead to the relevant minimization, with a consequent decrease of environmental impact and financial losses due to the disposal. Generating value from waste is the key concept here, upon which the "circular economy" has been founded on. Therefore, efforts are underway to develop integrated strategies to promote the generation of new networks and value chains based on the use of these residues.

For example, agro-industrial processing produces considerable waste, particularly residue coming from seeds, pods, and shells, which can cause environmental problems and financial losses due to the relevant high cost of disposal. A challenging point related to this kind of biomass is the diversity and the complexity requiring specifically designed solutions, which can completely change depending on latitude, climatic area, and country considered.

By way of example, there are a multitude of agro-industrial wastes whose exploitation would not subtract resources (inappropriate consumption of land included) for food consumption. Avocado seeds [72], olive oil residues [73], bran rice [46], and tobacco [56] are only a few examples of possible feedstocks, which already have their own chain of production, which could be implemented to achieve an oil to be converted into biodiesel.

Analogously, food waste might represent a possible valid alternative [74].

As reported for agro-industrial waste, such "renewable" biomass is already collected and widely available. In Europe and North America, the production per capita of food

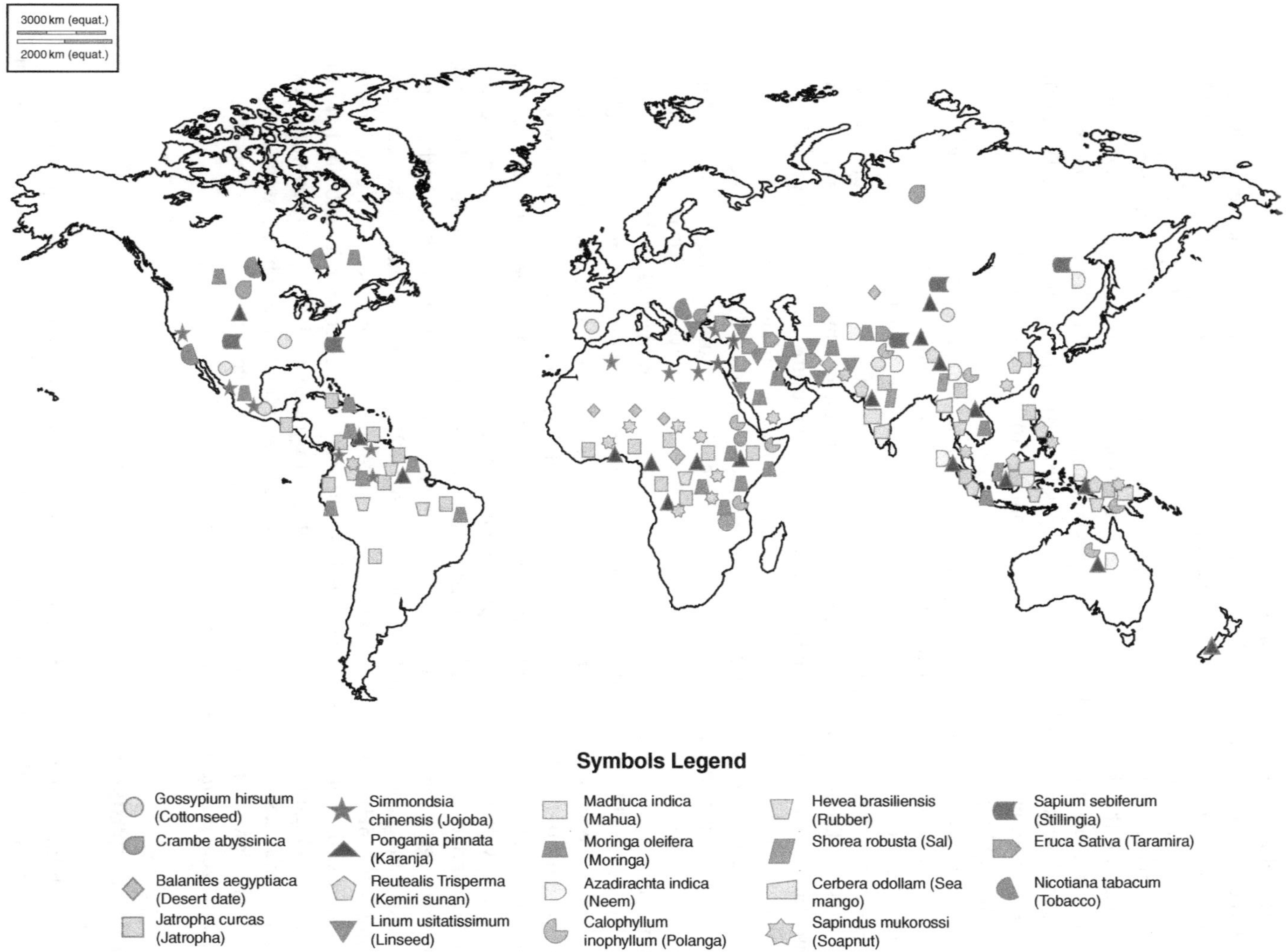


**Figure 3.1** Locations of potential cultivation areas for non-edible oils.

waste is 95–115 kg $yr^{-1}$ [73]. In developing Asian countries, the production of food waste is soon expected to increase up to 416 million tons per year [75]. As a whole, 1.3 billion tons per year of food are presently wasted.

Typically, food waste is mainly composed of simple and relatively complex (polymeric) sugars, such as starches and pectins, whereas cellulose content is very limited [76]. However, lipids are also present in the range of 4–8% by weight, to the point of making food waste a most proficient feedstock [77]. Such huge potential is today mainly exploited by combusting the biogas generated from anaerobic digestion, through cogeneration. However, more recently, direct alcoholysis was found to be effective to produce biodiesel from food waste. Ethyl levulinate (a new possible candidate as liquid biofuel) and fine chemicals (ethyl esters of amino acids) coming from alcoholysis of protein were also coproduced [78].

Another valuable feedstock satisfying the proximity constraint, which is continuously available and maintains a constant composition independently of where its place of production, is sewage sludge. Urban sewage sludge generated in wastewater treatment plants, thanks to its relative abundant lipid content, might meet the requirements of sustainable lipids feedstock since it is a low-cost starting material, available in large amounts, which is also expected to increase in the near future along with the growth of the global population and increasing of urbanization and industrialization.

In a wastewater treatment plant sewage grease could be easily collected and profitably converted into biodiesel. Most of the lipids are recoverable from sewage scum and primary sludge. The full exploitation of these biomasses could have a major impact in terms of volumes recovered: according to data found in the literature [79], over 5 Mton $yr^{-1}$ of biodiesel could be potentially obtained annually in Europe from sewage sludge alone. This massive amount would meet half of the present European demand. Very recently, it was demonstrated that direct alcoholysis of wet sludge into biofuels is possible [80]. This procedure does not need preliminary separation of lipids from sludge. However, the usual scheme of valorization, based on a preliminary recovery of lipids to be subsequently reacted and converted into biodiesel, is also positively investigated. More specifically, while lipids from sewage scum can be recovered through a solvent-free treatment [81], in the case of primary sludge, hexane could be proficiently used [82]. These lipids could be then converted using $AlCl_3 \cdot 6H_2O$, which has been very effective in promoting the direct esterification reaction with methanol [83]. $AlCl_3 \cdot 6H_2O$ not only satisfies economic viability but also guarantees high robustness in reactivity of brown grease, which is highly contaminated by salts, water, waxes, etc. After its use, it could be eventually added into sewage sludge in the final dewatering process before centrifugation as a coagulant [84]. In this way, an almost "zero discharge" system could be achieved.

## 3.4 Conclusion

Research into new feedstocks will continue to play a key role in biodiesel production in the future. Non-edible oils have great potential to contribute to overcoming a series of drawbacks related to conventional vegetable feedstocks. Unfortunately, however, uncertainty regarding the actual profitability of their cultivation will not help in making this transition. In addition, the low availability of noncultivable lands, especially in developed and populated areas, represents a fundamental limitation to the development of this solution. In this context it might be extremely profitable to include as marginal areas the contaminated sites, which cannot be used for food production. In this way, the recovery of depressed lands would be promoted in the production of biomasses for energetical purposes. On the other hand, the exploitation of the potential of different waste products might significantly influence biodiesel production. Food and agro-industrial waste and sewage sludge may represent the feedstocks of the future of lipids to be used for the production of fuel. In any case, to promote the development of these alternatives, political support can accelerate and contribute to overcoming the abovementioned obstacles. Regarding the topic of renewable energies, it will not be possible to meet the worldwide demand for energy through one solution alone. On the contrary, several well-balanced options represent the only sustainable and viable answer.

## Acknowledgments

This research was supported by FESR"PON Ricerca e Innovazione 2014–2020. Progetto: Energie per l'Ambiente TARANTO—Cod. ARS01_00637" and IProPBio "Integrated Process and Product Design for Sustainable Biorefineries (MSCA – RISE 2017: Research and Innovation Staff Exchange," Project ID: 778168.

## References

**1** Dhamodaran, G., Krishnan, R., Pochareddy, Y.K. et al. (2017). A comparative study of combustion, emission, and performance characteristics of rice-bran-, neem-, and cottonseed-oil biodiesels with varying degree of unsaturation. *Fuel* 187: 296–305.

**2** Ashraful, A.M., Masjuki, H.H., Kalam, M.A. et al. (2014). Production and comparison of fuel properties, engine performance, and emission characteristics of biodiesel from various non-edible vegetable oils: a review. *Energy Convers. Manag.* 80: 202–228.

**3** Falasca, S.L., Flores, N., Lamas, M.C. et al. (2015). *Crambe abyssinica*: an almost unknown crop with a promissory future to produce biodiesel in Argentina. *Int. J. Hydrogen Energy* 35 (11): 5808–5812.

**4** Costa, E., Almeida, M.F., Alvim-Ferraz, C., and Dias, J.M. (2019). The cycle of biodiesel production from *Crambe abyssinica* in Portugal. *Ind. Crops Prod.* 129 (November 2018): 51–58.

**5** Ruhul, A.M., Kalam, M.A., Masjuki, H.H. et al. (2017). Evaluating combustion, performance and emission characteristics of *Millettia pinnata* and *Croton megalocarpus* biodiesel blends in a diesel engine. *Energy* 141: 2362–2376.

**6** Kafuku, G., Kee, M., Kansedo, J. et al. (2010). Bioresource technology croton megalocarpus oil: a feasible non-edible oil source for biodiesel production. *Bioresour. Technol.* 101 (18): 7000–7004.

**7** Chapagain, B.P., Yehoshua, Y., and Wiesman, Z. (2009). Bioresource technology desert date (*Balanites aegyptiaca*) as an arid lands sustainable bioresource for biodiesel. *Bioresour. Technol.* 100 (3): 1221–1226.

8 Jamil, F., Al-Riyami, M., Al-Haj, L. et al. (2020). Waste *Balanites aegyptiaca* seed oil as a potential source for biodiesel production in the presence of a novel mixed metallic oxide catalyst. *Int. J. Energy Res.* 45: 1–14.

9 Kamel, D.A., Farag, H.A., Amin, N.K. et al. (2018). Smart utilization of jatropha (*Jatropha curcas* Linnaeus) seeds for biodiesel production: optimization and mechanism. *Ind. Crops Prod.* 111 (October 2017): 407–413.

10 Silitonga, A.S., Hassan, M.H., Ong, H.C., and Kusumo, F. (2017). Analysis of the performance, emission and combustion characteristics of a turbocharged diesel engine fuelled with *Jatropha curcas* biodiesel-diesel blends using kernel-based extreme learning machine. *Environ. Sci. Pollut. Res.* 24 (32): 25383–25405.

11 Ewunie, G.A., Morken, J., Lekang, O.I., and Yigezu, Z.D. (2020). Factors affecting the potential of *Jatropha curcas* for sustainable biodiesel production: a critical review. *Renew. Sustain. Energy Rev.* 137: 110500.

12 Moniruzzaman, M., Yaakob, Z., Shahinuzzaman, M. et al. (2017). Jatropha biofuel industry: the challenges. In: *Front Bioenergy Biofuels* (ed. E. Jacob-Lopes and L.Q. Zepka). InTech.

13 Canoira, L., Alcántara, R., Jesús García-Martínez, M., and Carrasco, J. (2006). Biodiesel from Jojoba oil-wax: transesterification with methanol and properties as a fuel. *Biomass Bioenergy* 30 (1): 76–81.

14 Sandouqa, A. and Al-Hamamre, Z. (2019). Energy analysis of biodiesel production from jojoba seed oil. *Renew. Energy* 130: 831–842.

15 Agarwal, S., Kumari, S., Mudgal, A., and Khan, S. (2020). Green synthesized nanoadditives in jojoba biodiesel-diesel blends: an improvement of engine performance and emission. *Renew. Energy* 147: 1836–1844.

16 Shrivastava, P., Salam, S., Verma, T.N., and Samuel, O.D. (2020). Experimental and empirical analysis of an IC engine operating with ternary blends of diesel, karanja and roselle biodiesel. *Fuel* 262 (X): 116608.

17 Sahoo, P.K., Das, L.M., Babu, M.K.G. et al. (2009). Comparative evaluation of performance and emission characteristics of jatropha, karanja and polanga based biodiesel as fuel in a tractor engine. *Fuel* 88 (9): 1698–1707.

18 Scott, P.T., Pregelj, L., Chen, N. et al. (2008). *Pongamia pinnata*: an untapped resource for the biofuels industry of the future. *BioEnergy Res.* 1 (1): 2–11.

19 Holilah, H., Prasetyoko, D., Oetami, T.P. et al. (2015). The potential of *Reutealis trisperma* seed as a new non-edible source for biodiesel production. *Biomass Convers. Biorefin.* 5 (4): 347–353.

20 Silitonga, A.S., Mahlia, T.M.I., Ong, H.C. et al. (2017). A comparative study of biodiesel production methods for *Reutealis trisperma* biodiesel. *Energy Sources Part A Recover. Util. Environ. Eff.* 39 (20): 2006–2014.

21 Riayatsyah, T.M.I., Ong, H.C., Chong, W.T. et al. (2017). Life cycle cost and sensitivity analysis of reutealis trisperma as non-edible feedstock for future biodiesel production. *Energies* 10 (7): 877.

22 Veinblat, M., Baibikov, V., Katoshevski, D. et al. (2018). Impact of various blends of linseed oil-derived biodiesel on combustion and particle emissions of a compression ignition engine – a comparison with diesel and soybean fuels. *Energy Convers. Manag.* 178 (August): 178–189.

**23** Puhan, S., Jegan, R., Balasubbramanian, K., and Nagarajan, G. (2009). Effect of injection pressure on performance, emission and combustion characteristics of high linolenic linseed oil methyl ester in a DI diesel engine. *Renew. Energy* 34 (5): 1227–1233.

**24** Acharya, N., Nanda, P., Panda, S., and Acharya, S. (2017). Analysis of properties and estimation of optimum blending ratio of blended mahua biodiesel. *Eng. Sci. Technol. Int. J.* 20 (2): 511–517.

**25** Rashid, U., Anwar, F., Moser, B.R., and Knothe, G. (2008). *Moringa oleifera* oil: a possible source of biodiesel. *Bioresour. Technol.* 99: 8175–8179.

**26** Omonhinmin, C., Olomukoro, E., Ayoola, A., and Egwim, E. (2020). Utilization of *Moringa oleifera* oil for biodiesel production: a systematic review. *AIMS Energy* 8 (1): 102–121.

**27** Oni, B.A. and Oluwatosin, D. (2020). Emission characteristics and performance of neem seed (*Azadirachta indica*) and Camelina (*Camelina sativa*) based biodiesel in diesel engine. *Renew. Energy* 149: 725–734.

**28** Azam, M.M., Waris, A., and Nahar, N.M. (2005). Prospects and potential of fatty acid methyl esters of some non-traditional seed oils for use as biodiesel in India. *Biomass Bioenergy* 29 (4): 293–302.

**29** Silitonga, A.S., Masjuki, H.H., Chyuan, H. et al. (2016). Pilot-scale production and the physicochemical properties of palm and Calophyllum inophyllum biodiesels and their blends. *J. Clean. Prod.* 126: 654–666.

**30** Roschat, W., Siritanon, T., Yoosuk, B. et al. (2017). Rubber seed oil as potential non-edible feedstock for biodiesel production using heterogeneous catalyst in Thailand. *Renew. Energy* 101: 937–944.

**31** Morshed, M., Ferdous, K., Khan, M.R. et al. (2011). Rubber seed oil as a potential source for biodiesel production in Bangladesh. *Fuel* 90 (10): 2981–2986.

**32** Vedaraman, N., Puhan, S., Nagarajan, G. et al. (2012). Methyl ester of Sal oil (Shorea robusta) as a substitute to diesel fuel – a study on its preparation, performance and emissions in direct injection diesel engine. *Ind. Crops Prod.* 36 (1): 282–288.

**33** Hajra, B. and Pathak, A.K. (2014). Optimal synthesis of methyl ester of Sal oil (*Shorea robusta*) using ion-exchange resin catalyst. *Int. J. Ind. Chem.* 5: 95–106.

**34** Lie, J., Bangun, M., Felycia, R. et al. (2018). Production of biodiesel from sea mango (*Cerbera odollam*) seed using in situ subcritical methanol – water under a non – catalytic process. *Int. J. Ind. Chem.* 9: 53–59.

**35** Tin, G., Nee, S., Tat, K. et al. (2015). Optimization and kinetic studies of sea mango (*Cerbera odollam*) oil for biodiesel production via supercritical reaction. *Energy Convers. Manag.* 99: 242–251.

**36** Misra, R.D. and Murthy, M.S. (2011). Performance, emission and combustion evaluation of soapnut oil – diesel blends in a compression ignition engine. *Fuel* 90 (7): 2514–2518.

**37** Chen, Y., Chiang, T., and Chen, J. (2013). Properties of soapnut (*Sapindus mukorossi*) oil biodiesel and its blends with diesel. *Biomass Bioenergy* 52: 15–21.

**38** Wang, R., Hanna, M.A., Zhou, W. et al. (2011). Bioresource technology production and selected fuel properties of biodiesel from promising non-edible oils: *Euphorbia lathyris* L., *Sapium sebiferum* L. and *Jatropha curcas* L. *Bioresour. Technol.* 102 (2): 1194–1199.

**39** Liu, Y., Xin, H., and Yan, Y. (2009). Physicochemical properties of stillingia oil: feasibility for biodiesel production by enzyme transesterification. *Ind. Crops Prod.* 30: 431–436.

40 Tayari, S., Abedi, R., and Tahvildari, K. (2020). Experimental investigation on fuel properties and engine characteristics of biodiesel produced from *Eruca sativa*. *SN Appl. Sci.* 2 (1): 1–13.

41 Li, S., Wang, Y., Dong, S. et al. (2009). Biodiesel production from *Eruca Sativa* Gars vegetable oil and motor, emissions properties. *Renew. Energy* 34 (7): 1871–1876.

42 García-Martínez, N., Andreo-Martínez, P., Quesada-Medina, J. et al. (2017). Optimization of non-catalytic transesterification of tobacco (*Nicotiana tabacum*) seed oil using supercritical methanol to biodiesel production. *Energy Convers. Manag.* 131: 99–108.

43 Guntur, R. and Prasanthi, G. (2018). Effect of compression ratio on thermal characteristics of vcr diesel engine using *Nicotiana tabacum* L. seed oil methyl ester. *J. Eng. Sci. Technol.* 13 (3): 558–572.

44 Usta, N., Aydoğan, B., On, A.H. et al. (2011). Properties and quality verification of biodiesel produced from tobacco seed oil. *Energy Convers. Manag.* 52 (5): 2031–2039.

45 Grisan, S., Polizzotto, R., Raiola, P. et al. (2016). Alternative use of tobacco as a sustainable crop for seed oil, biofuel, and biomass. *Agron. Sustain. Dev.* 36 (4): 55.

46 Razon, L.F. (2009). Alternative crops for biodiesel feedstock. *CAB Rev. Perspect. Agric. Vet. Sci. Nutr. Nat. Resour.* 4 (56): 1–15.

47 Moser, B.R. (2009). Biodiesel properties, and feedstocks. *Soc. Vitr. Biol.* 45 (3): 229–266.

48 Kumar, A. and Sharma, S. (2011). Potential non-edible oil resources as biodiesel feedstock: an Indian perspective. *Renew. Sustain. Energy Rev.* 15 (4): 1791–1800.

49 Asif, S., Ahmad, M., Zafar, M., and Ali, N. (2017). Prospects and potential of fatty acid methyl esters of some non-edible seed oils for use as biodiesel in Pakistan. *Renew. Sustain. Energy Rev.* 74 (February): 687–702.

50 Atta, P.A., N'Guessan, Y., Morin, C. et al. (2017). Calculation of greenhouse gas emissions of jatropha oil and jatropha biodiesel as alternative fuels for electricity production in Côte d'Ivoire. *AIP Conf. Proc.* 1814: 020068.

51 Thapa, S., Indrawan, N., and Bhoi, P.R. (2018). An overview on fuel properties and prospects of Jatropha biodiesel as fuel for engines. *Environ. Technol. Innov.* 9: 210–219.

52 Matsumoto, Y., Ma, S., Tominaga, T. et al. (2019). Acute effects of transdermal administration of jojoba oil on lipid metabolism in mice. *Medicina* 55 (9): 594.

53 Shah, S.N., Sharma, B.K., Moser, B.R., and Erhan, S.Z. (2010). Preparation and evaluation of jojoba oil methyl esters as biodiesel and as a blend component in ultra-low sulfur diesel fuel. *Bioenergy Res.* 3 (2): 214–223.

54 Verma, P., Dwivedi, G., Behura, A.K. et al. (2020). Experimental investigation of diesel engine fuelled with different alkyl esters of Karanja oil. *Fuel* 275 (May): 117920.

55 Tiwari, M. and Singh, A.P. (2017). A comparative evaluation of biodiesel blends of soapnut, Palm, and Karanja for usage in CI engine. *Int. J. Appl. Eng. Res.* 12 (17): 6440–6446.

56 Atabani, A.E., Silitonga, A.S., Ong, H.C. et al. (2013). Non-edible vegetable oils: a critical evaluation of oil extraction, fatty acid compositions, biodiesel production, characteristics, engine performance and emissions production. *Renew. Sustain. Energy Rev. [Internet]* 18: 211–245. http://dx.doi.org/10.1016/j.rser.2012.10.013.

57 Knothe, G. and Steidley, K.R. (2005). Kinematic viscosity of biodiesel fuel components and related compounds. Influence of compound structure and comparison to petrodiesel fuel components. *Fuel* 84 (9): 1059–1065.

**58** Kibazohi, O. and Sangwan, R.S. (2011). Vegetable oil production potential from *Jatropha curcas, Croton megalocarpus, Aleurites moluccana, Moringa oleifera* and *Pachira glabra*: assessment of renewable energy resources for bio-energy production in Africa. *Biomass Bioenergy* 35 (3): 1352–1356.

**59** Achten, W.M.J., Verchot, L., Franken, Y.J. et al. (2008). Jatropha bio-diesel production and use. *Biomass Bioenergy* 32 (12): 1063–1084.

**60** Mahanta, P. and Shrivastava, A. (2004). Technology development of bio-diesel as an energy alternative. In: *Challenges and Strategies for Sustainable Energy Efficiency and Environment* (ed. S. Onkar), 1–19. New Age International.

**61** Adriaans, T. (2006). Sustainability of solvent extraction for *Jatropha curcas*. FACT Found Eindhoven, Netherlands.

**62** Kansedo, J., Lee, K.T., and Bhatia, S. (2009). *Cerbera odollam* (sea mango) oil as a promising non-edible feedstock for biodiesel production. *Fuel* 88 (6): 1148–1150.

**63** Keneni, Y.G., Bahiru, L.A., and Marchetti, J.M. (2020). Effects of different extraction solvents on oil extracted from Jatropha seeds and the potential of seed residues as a heat provider. *Bioenergy Res.* 14: 1207–1222.

**64** Shah, S., Sharma, A., and Gupta, M.N. (2005). Extraction of oil from *Jatropha curcas* L. seed kernels by combination of ultrasonication and aqueous enzymatic oil extraction. *Bioresour. Technol.* 96 (1): 121–123.

**65** Atabani, A.E., Silitonga, A.S., Badruddin, I.A. et al. (2012). A comprehensive review on biodiesel as an alternative energy resource and its characteristics. *Renew. Sustain. Energy Rev.* 16 (4): 2070–2093.

**66** Berchmans, H.J. and Hirata, S. (2008). Biodiesel production from crude *Jatropha curcas* L. seed oil with a high content of free fatty acids. *Bioresour. Technol.* 99 (6): 1716–1721.

**67** Pinzi, S., Garcia, I.L., Lopez-Gimenez, F.J. et al. (2009). The ideal vegetable oil-based biodiesel composition: a review of social, economical and technical implications. *Energy Fuels* 23 (5): 2325–2341.

**68** Satyanarayana, M. and Muraleedharan, C. (2011). A comparative study of vegetable oil methyl esters (biodiesels). *Energy* 36 (4): 2129–2137.

**69** Jain, S. and Sharma, M.P. (2010). Biodiesel production from *Jatropha curcas* oil. *Renew. Sustain. Energy Rev.* 14 (9): 3140.

**70** di Bitonto, L. and Pastore, C. (2019). Metal hydrated-salts as efficient and reusable catalysts for pre-treating waste cooking oils and animal fats for an effective production of biodiesel. *Renew. Energy* 143: 1193–1200. https://doi.org/10.1016/j.renene.2019.05.100.

**71** Lu, H., Liu, Y., Zhou, H. et al. (2009). Production of biodiesel from *Jatropha curcas* L. oil. *Comput. Chem. Eng.* 33 (5): 1091–1096.

**72** di Bitonto, L., Reynel-Ávila, H.E., Mendoza-Castillo, D.I. et al. (2020). Residual Mexican biomasses for bioenergy and fine chemical production: correlation between composition and specific applications. *Biomass Convers. Biorefin.* 11: 619–631.

**73** Berbel, J. and Posadillo, A. (2018). Review and analysis of alternatives for the valorisation of agro-industrial olive oil by-products. *Sustainability* 10 (1): 1–9.

**74** Karmee, S.K. (2016). Liquid biofuels from food waste: current trends, prospect and limitation. *Renew. Sustain. Energy Rev.* 53: 945–953. http://dx.doi.org/10.1016/j.rser.2015.09.041.

**75** Melikoglu, M., Lin, C.S.K., and Webb, C. (2013). Analysing global food waste problem: pinpointing the facts and estimating the energy content. *Cent. Eur. J. Eng.* 3 (2): 157–164.

**76** Pagliaccia, P., Gallipoli, A., Gianico, A. et al. (2019). Variability of food waste chemical composition: impact of thermal pre-treatment on lignocellulosic matrix and anaerobic biodegradability. *J. Environ. Manag.* 236: 100–107.

**77** Uçkun Kiran, E., Trzcinski, A.P., Ng, W.J., and Liu, Y. (2014). Bioconversion of food waste to energy: a review. *Fuel* 134: 389–399.

**78** di Bitonto, L., Antonopoulou, G., Braguglia, C. et al. (2018). Lewis-Brønsted acid catalysed ethanolysis of the organic fraction of municipal solid waste for efficient production of biofuels. *Bioresour. Technol.* 266: 297–305.

**79** di Bitonto, L., Todisco, S., Gallo, V., and Pastore, C. (2020). Urban sewage scum and primary sludge as profitable sources of biodiesel and biolubricants of new generation. *Bioresour. Technol. Rep. [Internet]* 9 (January): 100382. https://doi.org/10.1016/j.biteb.2020.100382.

**80** di Bitonto, L., Locaputo, V., D'Ambrosio, V., and Pastore, C. (2020). Direct Lewis-Brønsted acid ethanolysis of sewage sludge for production of liquid fuels. *Appl. Energy* 259 (July 2019): 114163. https://doi.org/10.1016/j.apenergy.2019.114163.

**81** di Bitonto, L., Lopez, A., Mascolo, G. et al. (2016). Efficient solvent-less separation of lipids from municipal wet sewage scum and their sustainable conversion into biodiesel. *Renew. Energy* 90: 55–61.

**82** Pastore, C., Lopez, A., Lotito, V., and Mascolo, G. (2013). Biodiesel from dewatered wastewater sludge: a two-step process for a more advantageous production. *Chemosphere* 92 (6): 667–673.

**83** Pastore, C., Barca, E., Del Moro, G. et al. (2016). Recoverable and reusable aluminium solvated species used as a homogeneous catalyst for biodiesel production from brown grease. *Appl. Catal. A Gen.* 501: 48–55.

**84** Pastore, C., Lopez, A., and Mascolo, G. (2014). Efficient conversion of brown grease produced by municipal wastewater treatment plant into biofuel using aluminium chloride hexahydrate under very mild conditions. *Bioresour. Technol.* 155: 91–97.

# 4

# Algal Oil as a Low-Cost Feedstock for Biodiesel Production

*Michael Van Lal Chhandama*[1], *Kumudini Belur Satyan*[1], *and Samuel Lalthazuala Rokhum*[2,3]

[1] *Department of Biotechnology, School of Sciences (Block-I), JAIN (Deemed-to-be University), Bengaluru, Karnataka, India*
[2] *Hamid Yusuf Department of Chemistry, University of Cambridge, Cambridge, UK*
[3] *Department of Chemistry, National Institute of Technology, Silchar, Assam, India*

## 4.1 Introduction

The overexploitation of fossil fuels for the production of plastics, textiles, fertilizers, and in automotive industry has led to global energy and environmental crisis [1, 2]. The global demand for these exhaustible sources of energy sources can be attributed to rapid population growth, economic development, and urbanization, which was further predicted to increase by 40% by 2025 [3, 4]. A study has shown that the world's capacity to produce oil almost reaches its geological limit. A lowering trend was shown in the value of energy return per energy investment for the extraction of natural oil and gas [4, 5]. In addition to the alarming decline in the availability, combustion of fossil fuels led to emission of greenhouse gases like carbon dioxide ($CO_2$), methane ($CH_4$), nitrous oxide ($N_2O$), carbon monoxide (CO), and air pollutants like ammonia ($NH_3$), volatile organic compounds (VOC), particulate matter (PM), nitrogen oxides (NOx), and sulfur oxides (SOx), which have harmful effects on biotic lives [6]. Anthropogenic $CO_2$ emissions account for 80% of all green house gases (GHG) produced and regarded as the chief cause of rising global temperatures [7].

Research has been carried out for the past few decades to find an alternative source of energy that could replace or reduce the use of fossil fuels. Biofuel, a fuel derived from organic biomass, including plants, lignocellulosic waste, microorganism, and metabolic by-products of living organisms, posed to be a promising substitute for conventional fuels [8, 9]. Biofuels are inexhaustible, and the emission of GHG from biofuel combustion is reported to be much lower than those of fossil fuels. As a result, biofuel can be used to promote climate change mitigation [10]. The most commonly produced biofuel includes bioethanol and biodiesel. Based on the feedstock, biofuels are divided into different generations. First-generation biofuels are derived from sugars, starch, vegetable oils, and animal fats. Biofuels derived from

*Biodiesel Production: Feedstocks, Catalysts, and Technologies*, First Edition. Edited by Samuel Lalthazuala Rokhum, Gopinath Halder, Suttichai Assabumrungrat, and Kanokwan Ngaosuwan.

lignocellulosic biomass are referred to as second-generation biofuels. Biofuels derived from microorganism like algae (specifically microalgae) are regarded as third-generation biofuels. Fourth-generation biofuels are the fuels derived through genetic engineering and the redesign of existing, natural biological systems like microalgae [11, 12]. Microalgae were regarded as the most promising renewable feedstock for biodiesel production credited to their annual productivity and oil content, which is significantly higher than any oil seed crop. For instance, in a high-yielding oil crop, oil palm requires 24% of total cultivable land to meet 50% of transport fuel demands in the United States, whereas microalgae require only 1–3% of total land area to meet the exact same demand. Cultivating microalgae in the same acreage for oil crops would have easily sated the demands for diesel, and considering sustainability and renewability into account, microalgae cultivation posed a more suitable and realistic approach toward future biodiesel production [13–15].

### 4.1.1 Microalgae for Biodiesel Production

Microalgae are photosynthetic organisms of diverse species differing in physiology and growth requirements. They may be prokaryotic or eukaryotic or unicellular or multicellular, and depending on their mode of nutrition, they may be autotrophs, heterotrophs, mixotrophs, and photoheterotrophs [16, 17]. The average size of the cell varies between 5 and 50 μm, and they have a large surface-to-volume ratio, allowing them to have efficient growth rate [18]. The cell surface is negatively charged, which averts them from aggregation; as a result of which, the cells are able to get suspended in the media. The ultrastructure of microalgae was given in Figure 4.1. Microalgae have the ability to thrive in both

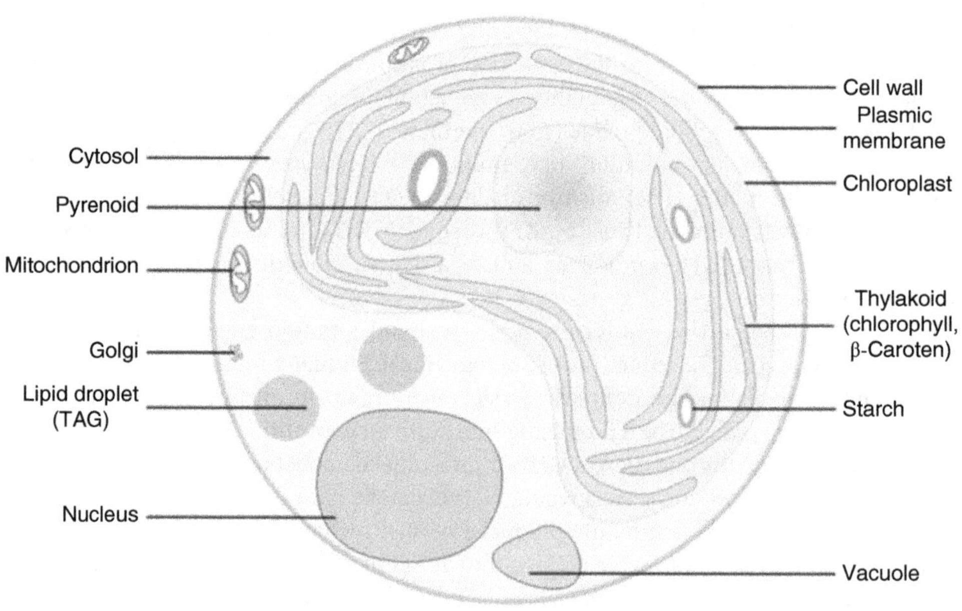


**Figure 4.1** Ultrastructure of microalgae. *Source:* Reproduced with authorised permission from Ref. [101].

**Table 4.1** Comparison of oil yield between microalgae and oil crops.
*Source:* Based on Ref. [13].

| Sl. no. | Source | Oil yield (L/acre) |
|---|---|---|
| 1 | Palm oil | 2407 |
| 2 | Jatropha | 764 |
| 3 | Coconut | 1086 |
| 4 | Corn | 68 |
| 5 | **Microalgae** | **23781–55416** |

aquatic (fresh, saline, or brackish water) and terrestrial ecosystem. They are able to endure varying environmental stress like nutrient deficiency, salinity, drought, light, osmotic pressure, temperature, and ultraviolet radiation [19, 20], which conveniently leads to optimization of oil production [10].

Microalgae were regarded as a feasible feedstock for biodiesel production mainly due to their high lipid content and biomass production [13]. Microalgae also have high growth rate, efficiently converting solar energy and carbon dioxide into different forms of useful metabolites such as oils, carbohydrates, and proteins. Microalgae were reported to produce ~30 times more oil per acre than food crops like corn and soybean. Oil productivity of microalgae was significantly higher than that of oil crops as shown in Table 4.1. Some of the advantages of using microalgae as a source of biodiesel include the following [21]:

- Microalgae have efficient photosynthetic system, which resulted in rapid growth rate, high biomass, and lipid yield.
- Microalgae represent nonedible feedstock, thereby posing no competition or lower risk of competition with feeds and foods. Microalgal cultivation do not require fertile and agriculturally productive land. The organism showed a great potential to replace terrestrial biomass for producing bioenergy.
- Microalgae are highly adaptive to wide range of ecosystem and environment. They have the ability to flourish in industrial, municipal, and agricultural wastewaters, contaminated ponds, and on wastelands, which further leads to bioremediation of degraded and contaminated areas and wastewaters. The cost of production can be reduced by making use of these wastewaters for cultivation.
- Microalgae are efficient sequester of $CO_2$ while generating $O_2$. Combustion of microalgal biodiesel resulted in reduction of emission of harmful greenhouse gases, harmful trace elements, and particulates in the atmosphere.
- Biodiesel derived from microalgae represents diversification of fuel supply and energy security. They can regulate the negative effects of fluctuating oil prices and reduce the dependency on foreign oil imports. Microalgae are sustainable, renewable, competent, and eco-friendly feedstock for biodiesel with the potential to meet the global energy demand.
- Cultivation of microalgae creates rural economy revitalization and convey social benefits with creation of new jobs and income.
- Modification of engine is not required while using biodiesel. Biodiesel derived from microalgae has high reactivity with low initial ignition and combustion temperatures.

## 4.2 Lipid and Biosynthesis of Lipid in Microalgae

Lipids constitute the main biomass for production of biodiesel from microalgae [22]. Lipids are biomolecules that are soluble in organic solvents but insoluble in water. The total lipid content of some of the commonly studied microalgae was given in Table 4.2. Based on the chemical structure and polarity, lipids can be classified as polar and nonpolar [23]. Polar lipids like phospholipids and glycolipids mostly constitute the cell membranes, while nonpolar lipids (or neutral lipids) are used as energy source and comprise acylglycerols (mono-, di-, and tri-) and free fatty acids [24]. The main constituents of both neutral and polar lipid molecules are the fatty acids (containing 12–22 carbons in length), which may be of saturated (no double bonds) or unsaturated types (at least one double bond). Not all lipids can be converted into biodiesel. The composition of fatty acids determined the quality and production of biodiesel [25]. The biodiesel yield greatly depends on the lipid content of individual cells [26], which varies considerably for different microalgal species. Among the various classes of lipids, triacylglycerols (TAGs) are the lipids required for biodiesel production, based on their lower degree of unsaturation than the other lipid fractions [27]. Therefore, microalgae having relatively low amounts of neutral lipids (TAG) cannot be considered as potential candidates for biodiesel [25]. Microalgae, such as *Chlorella* sp., *Scenedesmus* sp., and *Nannochloropsis* sp., have been regarded as the most potential species for biodiesel production and have been studied for biofuel applications [28]. Ideal microalgal candidates for biodiesel production require not only high TAG accumulation but also suitable fatty acid composition [29, 30]. Previous investigations [29, 31, 32] showed that microalgae enriched with C16–C18 fatty acids, including C16 : 0 (palmitic acid), C18 : 0 (stearic acid), C18 : 1 (oleic acid), C18 : 2 (linoleic acid), and C18 : 3 (linolenic acid), have high potential for producing quality biodiesel.

**Table 4.2** Different species of microalgae with their total lipid content [102].

| Sl. no. | Microalgae | Lipid content (% dcw) |
|---|---|---|
| 1 | *Schizochytrium* sp. | 50–77 |
| 2 | *Botryococcus braunii* | 25–75 |
| 3 | *Nannochloropsis* sp. | 31–68 |
| 4 | *Chlorella emersonii* | 25–63 |
| 5 | *Chlorella protothecoides* | 15–58 |
| 6 | *Pleurochrysis carterae* | 30–50 |
| 7 | *Nannochloris* sp. | 20–50 |
| 8 | *Neochloris oleoabundans* | 45–47 |
| 9 | *Chlorella pyrenoidosa* | 46.7 |

### 4.2.1 Lipid Biosynthesis

Different lipids are synthesized through different metabolic pathways in microalgae, and one of the most detailed pathways studied is the de novo synthesis of lipids as shown in Figure 4.2 [7, 33, 34]:

- The first step in de novo pathway is the reaction of $CO_2$, adenosine triphosphate (ATP), and acetylCoA to form malonyl-CoA, which is catalyzed by acetyl-CoA-carboxylase (ACCase). This is the first irreversible step in biosynthesis of lipids. The product formed in this reaction is further used as the central carbon donor for the extension of the acyl group.
- The second step involves transacylation between a protein cofactor, acyl carrier protein (ACP), and the two acyl-CoA molecules. The enzyme complex malonyl-CoA-ACP has high specificity, while the specificity of acetyl-CoA-ACP is low; as a result of which, it easily reacts with other low molecular weight acyl-CoA compounds, which produce an uneven carbon number and branching in fatty acid.
- In the next step, the fatty acids produced get accumulated either in the chloroplast or in the endoplasmic reticulum.

De novo synthesis of lipid utilizes glycerol-3-phosphate for the serial addition of fatty acids provided by acyl-CoA. The production and fate of acetyl-CoA is an important

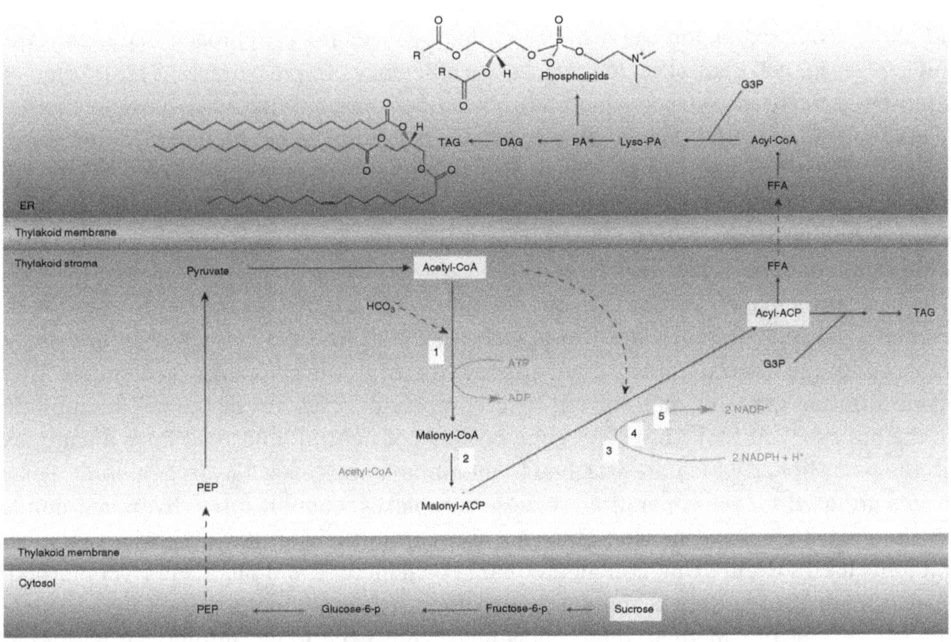


**Figure 4.2** Lipid biosynthesis in microalgae. *Source:* Based on Ref. [7].

regulatory step in de novo pathway. Acetyl-CoA comes from either glycolysis or Calvin cycle. During cytosolic glycolysis in plants, fructose-6-phosphate, glucose-6-phosphate, and eventually phosphoenolpyruvate (PEP) are produced from the metabolism of sucrose, which are transported into the chloroplast. Inside the chloroplast, PEP enters the de novo lipid biosynthesis. Another metabolic pathway for the biosynthesis of lipid involves recycling of existing fatty acids from other cell components whose regulation highly depends on the metabolic status of the cell [35].

Majority of the lipid classes in microalgae are synthesized de novo. Optimization of lipid production usually occurs during stress conditions, and during those conditions half of the increased lipids are synthesized de novo, whereas the rest of the accumulated lipids are derived from recycling of existing fatty acids. Biosynthesis of lipid in microalgae is quite complex and is efficiently regulated based on the signals reflecting nutritional and environmental stress. For example, the activity acyl hydrolase, which is specific to a membrane lipid (galactolipid), increased; *Dunaliella salina* was grown in nitrogen-deficient medium [7, 35].

### 4.2.2 Lipid Extraction

Lipid extraction is one of the most important steps in microalgal biodiesel production, which is carried out after harvesting and drying. Harvesting technique includes sedimentation, centrifugation, and flotation [36], and drying is the moisture removal process by providing heat energy or vapor transport [37]. Organic solvents are commonly used for extraction because they can be directly used in wet feedstock and show less reactivity with lipid [20]. Solvent extraction can also be assisted by mechanical approach like high press homogenizer for cell disruption to increase the efficiency of lipid extraction [28]. Some of the most effectively used solvents include acetone, benzene, chloroform, n-hexane, methanol, ethanol, butanol, and cyclohexane that can dissolve lipids, but methanol, chloroform, and hexane have been considered as the most potential solvents for lipid extraction [38, 39]. An ideal solvent for lipid extraction should have high level of specificity for intracellular lipids, low boiling point, infiltrate through the cell matrix, insoluble in water, cost effective, volatile, and nontoxic in nature [40].

The basic principle of solvent extraction is that when a nonpolar solvent is mixed with the microalgal cells, it penetrates through the cell wall into the cytoplasm and forms a complex with the neutral lipids, which already existed in the cytoplasm as globules. Due to concentration gradient, this solvent-lipid complex diffuses out of the cell membrane into the bulk solvent [28]. However, some portions of neutral lipids still form a complex with the polar lipids, which are attached to membrane proteins via hydrogen bond. Polar solvents are used for the separation of these complexes where it forms hydrogen bonds with the polar lipids, and neutral lipids are freed from the complex [41]. The two most commonly used solvent extraction methods are Folch method and Bligh and Dyer method. In Folch method of solvent extraction, chloroform and methanol are used in a volume ratio of 2 : 1. For this method, the homogenized cells need to be equilibrated with one-fourth volume of saline solution [42]. On the other hand, Bligh and Dyer method used 1 : 2 (v/v) chloroform and methanol [43], and this method has been used in large-scale processes [37, 44].

## 4.3 Optimization of Lipid Production in Microalgae

Biodiesel production from microalgae has attained numerous advances in laboratory scale. The key factor to surge the economic feasibility of large-scale production is minimizing cost of production and optimizing lipid production [45]. Nutrient deprivation and environmental stresses such as light, temperature, and pH induce higher lipid synthesis and accumulation in microalgae [46].

### 4.3.1 Nitrogen Stress

Nitrogen plays a significant role in the biosynthesis of lipid in microalgae. Microalgae usually assimilate nitrogen in the form of ammonia through glutamine synthetase and glutamine-2-oxoglutarate amidotransferase pathway (GS/GOGAT pathway). Ammonia, mostly as glutamine, diffuse through the cell membrane and integrate into amino acids for protein synthesis [47]. Higher lipid accumulation under nitrogen starvation in several microalgae is well documented. For example, under nitrogen starvation, lipid production increased twofold in *Neochloris oleoabundans* and onefold in *Nannochloropsis* sp. [48, 49]. *Auxenochlorella protothecoides* also showed optimized lipid productivity with decreasing nitrate concentration in the growth medium [50]. The total lipid content of *Chlorella pyrenoidosa* increased from 18 to 26% dcw when the concentration of $KNO_3$ in the medium was reduced [51]. A study reported that *Chlorella vulgaris* showed an elevated lipid accumulation (20–53% dcw) under nitrogen starvation [52]. *Chlorella* sp. undergo metabolic changes under nitrogen deficiency, and neutral lipid content increased significantly from 15.48 to 50.34% [53]. Another report has shown that the total lipid content of *Chlorella sorokiniana* increased when grown under low concentration of $NaNO_3$ concentration [18]. Chakravarty and Mallick [54] observed an elevated lipid accumulation (16.12–48.6% dcw) in *Selenastrum* sp. under nitrogen starvation [54]. Similarly, *Isochrysis galbana* also showed increased total lipid content (17.2–30.6% dcw) under nitrogen starvation [55]. A recent study also discovered that total lipid content of *Eustigmatos vischeri* increased from 49.5 to 52.3% when cultivated under limited nitrogen supply [56]. *Schizochytrium* sp. showed efficient lipid accumulation under nitrogen starvation [57]. When *Spirulina platensis* was grown at three nitrogen concentrations (N-limited, N-optimal, and N-rich medium), the highest total lipid content was observed at N-limited medium [58]. An increasing trend in the total lipid yield was observed under a nitrogen-starved culture condition in *Chlorolobion* sp. and *Chlorella* sp. [59].

### 4.3.2 Phosphorous Stress

Phosphorus accounts for less than 1% of total algal biomass (~0.03–0.06%) but a crucial component in the growth medium because it has a significant impact on the growth and metabolism of microalgae [60–62]. Phosphorus is one of the essential nutrients for the synthesis of important molecules in microalgae such as nucleic acids and ATP, which is the primary product of photosynthesis and is essential for cell development. Microalgae utilize phosphorus in two ways, for construction of organic cellular components such as phospholipids, and the remaining excess is used to produce inorganic polyphosphate granules [63].

Different studies observed higher lipid production in microalgae under phosphorus stress. Induction of phosphorus deficiency resulted in increased lipid accumulation (30% dcw) in *Tetradesmus obliquus* [64]. Both *Scenedesmus* sp. and *Botryococcus braunii* showed optimized lipid productivity when grown in a medium with limited phosphorus supply [65, 66]. Similarly, reduction of phosphate concentration induced elevation in the total lipid content of *Phormidium* sp. from 17 to 27.0% dcw [67]. Chakravarty and Mallick [54] have shown an elevated lipid production (~ 31% dcw) under phosphate deprivation in *Selenastrum* sp. [54]. Recent study also showed that deprivation of phosphorus in the medium increases the accumulation of cellular lipids by ~ 53% in *Scenedesmus* sp. [68]. Severe phosphorus reduction increased oil content from 5.88 to 24.86% in *D. salina* [69]. Reduction of phosphorus supply by 10-fold also resulted in threefold increase in the total lipid content [70].

### 4.3.3 pH Stress

Alteration in the pH of the culture medium changes the lipid accumulation in microalgae. For example, optimization of TAG production was induced by lowering the pH in *Chlorella*, and was not dependent on nutrient levels [71]. A study has shown that alkaline pH inhibited the growth of microalgae and divert the energy to form TAG [72]. The effects of pH on the lipid and fatty acid composition of a *Chlamydomonas* sp. and *Chlorella reinhardtii* was estimated, and the concentration of fatty acids in *Chlamydomonas* sp. was more than that in *C. reinhardtii*, and increasing the pH higher than the normal condition resulted in the relative increase in the accumulation of TAG in *Chlamydomonas sp.* [73]. Three species of microalgae, *Nitzschia* sp., *Porphyridium* sp., and *Tetraselmis* sp., showed increased lipid production at pH varied from neutral pH where the highest lipid accumulation was found in *Tetraselmis* sp. cultivated at pH 9.5 [74]. Similarly, *Chlamydomonas reinhardtii* exhibited higher lipid accumulation at pH 7.8 when compared with the lipid content at controlled pH 7.0. [75]. Another study has shown that alkaline pH (7.8–8.0) increased lipid content by 60.6% in *C. pyrenoidosa* [76]. The effect of pH in the accumulation of lipid was also studied by comparing the total lipid content of *Nannochloropsis* sp. at different pH. The total lipid content was found to increase from 20.58% at control pH to 36.83% at altered pH condition [77].

### 4.3.4 Temperature Stress

Temperature was observed to have a major effect on the composition of fatty acid in microalgae [78, 79]. A common pattern where decreasing temperature increased the unsaturation of fatty acids and increasing saturation with increasing temperature has been observed in many microalgae [72, 76, 80]. In an earlier study, changes in the lipid profile *of D. salina* was analyzed where a decrease in culture temperature from 30 to 12 °C increased the concentration of unsaturated lipids significantly by 20% [81]. In *C. vulgaris* and *B. braunii*, raising the culture temperature resulted in declining level of unsaturated intracellular fatty acids [82]. A maximum yield of polyunsaturated fatty acids was observed in *Phaeodactylum tricornutum* when the temperature changes from 25 to 10 °C [83]. A study has shown that reducing the temperature from 25 to 20 °C resulted in optimization of total lipid content by 14.92% in *Nannochloropsis oculata* [84]. *Porphyridium purpureum*, *Fistulifera* sp., and *C. vulgaris* showed elevated lipid content at 20 °C when compared with their normal controlled temperature at 10 °C [85]. In a recent study, cold-water marine diatom *Porosira glacialis* was cultivated four different temperatures lower than the control temperature (12 °C); the highest lipid content was measured 2 °C (33.4%) [86]. *Cylindrotheca closterium* total lipid content was also optimized from cultivating at 20 °C (control) to a lower temperature (11 °C) [87]. In *Chlamydomonas* sp. the cell growth increased under high temperature, but fatty acid accumulations were enhanced at low temperature [88].

### 4.3.5 Light

Phototrophic microalgae have the ability to grow on various light intensities and alter their photosynthetic activity, chemical composition, and production of metabolites [72]. Different light intensities and wavelengths modify biosynthesis of lipid, thereby changing the lipid profile [72, 89, 90]. In an earlier study, exposure to high light intensity reduced phospholipid content but increased the concentration of TAG in *Cladophora* sp. [91]. Similarly, low light intensity resulted in increased levels of membrane lipids, and higher light intensities resulted in the elevation of TAG concentration in *Tichocarpus crinitus* [92]. Under higher light intensity the electron acceptors required by the photosystem might be depleted, and microalgae increased TAG production under these conditions as a protective mechanism where TAG potentially assists the cell to reproduce its electron acceptor pool [72]. Increasing light intensity of microalgal culture from 250 to 400 μmol photons $m^{-2}$ $s^{-1}$ was reported to enhance the accumulation of total lipid and neutral lipid in *Scenedesmus* sp. [93]. Light intensity of 400 μmol photons $m^{-2}$ $s^{-1}$ induced optimization of neutral lipid by allocation of carbon for lipid biosynthesis and decreased carbohydrate content, in *Chlorella* sp. and *Monoraphidium* sp. [94]. The total lipid content of *B. braunii* differed when cultured at varied light intensity and highest was observed at $48.6 \pm 2.7$ μmol photons $m^{-2}$ $s^{-1}$ [95].

The wavelength of the light also plays an important role in the accumulation of lipid in microalgae. An experiment has shown that growing *Cyclotella cryptica* at yellow wavelengths (~580 nm) doubled the lipid content, while blue wavelengths (~450 nm) resulted in a fourfold optimization [96]. Another study has revealed that growth and lipid production in *Chlorella* sp. varies when yellow, red, blue, and green fluorescent paint solutions were incorporated into the culture. Maximum cell biomass (1.7 g $l^{-1}$) was produced with red paint, while maximum lipid content (30% dcw) was estimated using a blue paint solution [97]. Lipid productivity of *C. vulgaris* has shown 2.3-fold increase when grown in medium with these organic dyes (rhodamine101 and 9,10-diphenylanthracene) [98]. Exposure of growing cells to blue (450–475 nm) and red (630–675 nm) light induced maximum light absorption by chlorophyll pigments and, thus, improves the working of photosystems I and II [99], and exposure to light-emitting diode (LED) of red wavelengths optimized lipid production in *Chlorella* sp. [100].

## 4.4 Conclusion

Microalgae emerged as a potent feedstock for the production of biodiesel. Application of nonconventional and eco-friendly fuel in transportation sector is crucial to combat the rising global warming and energy crisis. Microalgae are efficient photosynthetic organisms capable of yielding high amount of lipid as one of their biomasses while sequestrating atmospheric $CO_2$ for their growth and development. They produce different classes of lipid out of which triacylglycerides (TAG) can be converted into their respective alkyl esters (biodiesel) by reacting with alcohol using acid or base catalyst. The possibility of further optimizing the lipid accumulation in microalgae gives an added advantage on the economic feasibility of large-scale production of biodiesel from microalgae. Biodiesel production from microalgae offers a potential sustainable approach for generation of clean fuel without causing harm to the engine of the vehicle and the environment.

## References


1 Omer, A.M. (2012). The energy crisis, the role of renewable and global warming. *Greener J. Environ. Manag. Publ. Saf.* 1 (1): 38–70.

2 Erdiwansyah, Mamat, R., Sani, M.S. et al. (2019). An overview of higher alcohol and biodiesel as alternative fuels in engines. *Energy Rep.* 5 (1): 467–479.

3 Tariyal, K., Bartwal, D.M., and Bartwal, S. (2013). Algal biofuel: a symbol of sustainability in the developmental era. *Sch. Acad. J. Biosci.* 1 (5): 192–197.

4 Kumar, D. and Singh, B. (2019). Algal biorefinery: an integrated approach for sustainable biodiesel production. *Biomass Bioenerg.* 131 (1): 1–16.

5 Rogers, Z., Kelly, T.G., Rogers, D.S., and Carter, C.R. (2007). Alternative fuels: are they achievable? *Int. J. Logist. Res. Appl.* 10 (3): 269–282.

6 Correa, D.F., Beyer, H.F., Fargione, J.E. et al. (2019). Towards the implementation of sustainable biofuel production systems. *Renew. Sustain. Energy Rev.* 107 (1): 250–263.

7 Greenwell, H.C., Laurens, L.M.L., Shields, R.J. et al. (2009). Placing microalgae on the biofuels priority list: a review of the technological challenges. *J. R. Soc. Interface* 7 (1): 703–726.

8 Demirbas, A. (2009). Political, economic and environmental impacts of biofuels: a review. *Appl. Energy* 86 (1): 108–117.

9 Khan, M.I., Shin, J.H., and Kim, J.D. (2018). The promising future of microalgae: current status, challenges, and optimization of a sustainable and renewable industry for biofuels, feed, and other products. *Microb. Cell Fact.* 17 (1): 1–21.

10 Abdullah, B., Muhammad, S.A., Shokravi, Z. et al. (2019). Fourth generation biofuel: a review on risks and mitigation strategies. *Renew. Sustain. Energy Rev.* 107 (1): 37–50.

11 Aro, E.M. (2016). From first generation biofuels to advanced solar biofuels. *Ambio* 45 (1): 24–31.

12 Zabed, H., Suely, A., Yun, J. et al. (2019). Recent advances in biological pretreatment of microalgae and lignocellulosic biomass for biofuel production. *Renew. Sustain. Energy Rev.* 105 (1): 105–128.

13 Chisti, Y. (2007). Biodiesel from microalgae. *Biotechnol. Adv.* 25 (3): 294–306.

14 Duffy, J.E., Canuel, E.A., Adey, W., and Swaddle, J.P. (2009). Biofuels: algae. *Science* 326 (1): 1345.

15 Leong, W.H., Lim, J.W., Lam, M.K. et al. (2018). Third generation biofuels: a nutritional perspective in enhancing microbial lipid production. *Renew. Sustain. Energy Rev.* 91 (1): 950–961.

**16** Mohan, V.S., Rohit, M.V., Chiranjeevi, P. et al. (2015). Heterotrophic microalgae cultivation to synergize biodiesel production with waste remediation: progress and perspectives. *Bioresour. Technol.* 184 (1): 169–178.

**17** Arif, M., Wang, L., Salama, E.S. et al. (2020). Microalgae isolation for nutrient removal assessment and biodiesel production. *Bioenerg. Res.* 13 (1): 1247–1259.

**18** Zhu, Y. and Huang, Y. (2017). Use of flux balance analysis to promote lipid productivity in *Chlorella sorokiniana*. *J. Appl. Phycol.* 29 (1): 889–902.

**19** Abou-Shanab, R.A., Matter, I.A., Kim, S.N. et al. (2011). Characterization and identification of lipid-producing microalgae species isolated from a freshwater lake. *Biomass Bioenerg.* 37 (1): 3079–3085.

**20** Yadav, M., Chavan, S.B., Singh, R. et al. (2019). Experimental study on emissions of algal biodiesel and its blends on a diesel engine. *J. Taiwan Inst. Chem. Eng.* 96 (1): 160–168.

**21** Vassilev, S.V. and Vassileva, C.G. (2016). Composition, properties and challenges of algae biomass for biofuel application: an overview. *Fuel* 181 (1): 1–33.

**22** Upadhyay, A.K., Singh, R., Singh, J.S., and Singh, D.P. (2019). Microalgae-assisted phyco-remediation and energy crisis solution: challenges and opportunity. In: *New and Future Developments in Microbial Biotechnology and Bioengineering* (ed. J.S. Singh and D.P. Singh), 295–307. India: Elsevier.

**23** Yin, Z., Zhu, L., Li, S. et al. (2020). A comprehensive review on cultivation and harvesting of microalgae for biodiesel production: environmental pollution control and future directions. *Bioresour. Technol.* 301 (1): 1–19.

**24** Subhash, G.V., Rajvanshi, M., Kumar, B.N. et al. (2017). Carbon streaming in microalgae: extraction and analysis methods for high value compounds. *Bioresour. Technol.* 244 (2): 1304–1316.

**25** D'Alessandro, E.B. and Filho, N.R. (2016). Concepts and studies on lipid and pigments of microalgae: a review. *Renew. Sustain. Energy Rev.* 58 (1): 832–841.

**26** Scott, S.A., Davey, M.P., Dennis, J.S. et al. (2010). Biodiesel from algae: challenges and prospects. *Curr. Opin. Biotechnol.* 21 (3): 277–286.

**27** Halim, R., Danquah, M.K., and Webley, P.A. (2012). Extraction of oil from microalgae for biodiesel production: a review. *Biotechnol. Adv.* 30 (3): 709–732.

**28** Sati, H., Mitra, M., Mishra, S., and Baredar, P. (2019). Microalgal lipid extraction strategies for biodiesel production: a review. *Algal Res.* 38 (1): 1–12.

**29** Ma, Y., Wang, Z., Yu, C. et al. (2014). Evaluation of the potential of 9 *Nannochloropsis* strains for biodiesel production. *Bioresour. Technol.* 167 (1): 503–509.

**30** Selvarajan, R., Felföldi, T., Tauber, T. et al. (2015). Screening and evaluation of some green algal strains (chlorophyceae) isolated from freshwater and soda lakes for biofuel production. *Energies* 8 (7): 7502–7521.

**31** Song, M., Pei, H., Hu, W., and Ma, G. (2013). Evaluation of the potential of 10 microalgal strains for biodiesel production. *Bioresour. Technol.* 141 (1): 245–251.

**32** Chowdhury, R., Keen, P.L., and Tao, W. (2019). Fatty acid profile and energy efficiency of biodiesel production from an alkaliphilic algae grown in the photobioreactor. *Bioresour. Technol. Rep.* 6 (1): 229–236.

**33** Chung, Y.S., Lee, J.W., and Chung, C.H. (2017). Molecular challenges in microalgae towards cost-effective production of quality biodiesel. *Renew. Sustain. Energy Rev.* 74 (1): 139–144.

**34** Wang, H., Zhang, Y., Zhou, W. et al. (2018). Mechanism and enhancement of lipid accumulation in filamentous oleaginous microalgae *Tribonema minus* under heterotrophic condition. *Biotechnol. Biofuels* 11 (1): 1–14.

**35** Ran, W., Wang, H., Liu, Y. et al. (2019). Storage of starch and lipids in microalgae: biosynthesis and manipulation by nutrients. *Bioresour. Technol.* 291 (1): 1–12.

**36** Duran, S.K., Kumar, P., and Sandhu, S.S. (2019). Prospect of algal biodiesel as a fuel in engine. *Int. J. Eng. Adv. Technol.* 8 (6): 2739–2744.

**37** Yew, G.Y., Lee, S.Y., Show, P.L. et al. (2019). Recent advances in algae biodiesel production: from upstream cultivation to downstream processing. *Bioresour. Technol. Rep.* 7 (1): 1–10.

**38** Wahidin, S., Idris, A., and Shaleh, S.R. (2014). Rapid biodiesel production using wet microalgae via microwave irradiation. *Energy Convers. Manag.* 84 (1): 227–233.

**39** Bundhoo, Z.M. (2018). Microwave-assisted conversion of biomass and waste materials to biofuels. *Renew. Sustain. Energy Rev.* 82 (1): 1149–1177.

**40** Garoma, T. and Janda, D. (2016). Investigation of the effects of microalgal cell concentration and electroporation, microwave and ultrasonication on lipid extraction efficiency. *Renew. Energy* 86 (1): 117–123.

**41** DosSantos, R.R., Moreira, D.M., Kunigami, C.N. et al. (2015). Comparison between several methods of total lipid extraction from *Chlorella vulgaris* biomass. *Ultrason. Sonochem.* 22 (1): 95–99.

**42** Folch, J., Lees, M., and Stanley, G.H. (1957). A simple method for the isolation and purification of total lipids from animal tissues. *J. Biol. Chem.* 226 (1): 497–509.

**43** Bligh, E.G. and Dyer, W.J. (1959). A rapid method of total lipid extraction and purification. *Can. J. Biochem. Physiol.* 37 (1): 911–917.

**44** Breil, C., Abert, V.M., Zemb, T. et al. (2017). "Bligh and Dyer" and Folch methods for solid–liquid–liquid extraction of lipids from microorganisms. Comprehension of solvatation mechanisms and towards substitution with alternative solvents. *Int. J. Mol. Sci.* 18 (1): 1–21.

**45** Miranda, C.T., de Lima, D.V., Atella, G.C. et al. (2016). Optimization of nitrogen, phosphorus and salt for lipid accumulation of microalgae: towards the viability of microalgae biodiesel. *Nat. Sci.* 8 (12): 557–573.

**46** Park, S., Nguyen, T.H., and Jin, E. (2019). Improving lipid production by strain development in microalgae: strategies, challenges and perspectives. *Bioresour. Technol.* 292 (1): 1–11.

**47** Cai, T., Park, S.Y., and Li, Y. (2013). Nutrient recovery from wastewater streams by microalgae: status and prospects. *Renew. Sustain. Energy Rev.* 19 (1): 360–369.

**48** Li, Y., Horsman, M., Wang, B. et al. (2008). Effects of nitrogen sources on cell growth and lipid accumulation of green alga *Neochloris oleoabundans*. *Appl. Microbiol. Biotechnol.* 8 (1): 629–636.

**49** Rodolfi, L., Chini, Z.G., Bassi, N. et al. (2009). Microalgae for oil: strain selection, induction of lipid synthesis and outdoor mass cultivation in a low-cost photobioreactor. *Biotechnol. Bioeng.* 102 (1): 100–112.

**50** Shen, Y., Pei, Z., Yuan, W., and Mao, E. (2009). Effect of nitrogen and extraction method on algae lipid yield. *Int. J. Agric. Biol. Eng.* 2 (1): 51–57.

**51** Nigam, S., Prakash, M., and Sharma, R. (2011). Effect of nitrogen on growth and lipid content of *Chlorella pyrenoidosa*. *Am. J. Biochem. Biotechnol.* 7 (3): 124–129.

**52** Yeh, K.L. and Chang, J.S. (2012). Effects of cultivation conditions and media composition on cell growth and lipid productivity of indigenous microalga *Chlorella vulgaris* ESP-31. *Bioresour. Technol.* 105 (1): 120–127.

**53** Rai, V., Muthuraj, M., Gandhi, M.N. et al. (2017). Real-time iTRAQ-based proteome profiling revealed the central metabolism involved in nitrogen starvation induced lipid accumulation in microalgae. *Sci. Rep.* 7 (1): 1–16.

**54** Chakravarty, S. and Mallick, N. (2019). Optimization of lipid accumulation in an aboriginal green microalga *Selenastrum* sp. GA66 for biodiesel production. *Biomass Bioenerg.* 126 (1): 1–13.

**55** Zarrinmehr, M.J., Farhadian, O., Heyrati, F.P. et al. (2020). Effect of nitrogen concentration on the growth rate and biochemical composition of the microalga, *Isochrysis galbana*. *Egypt. J. Aquat. Res.* 46 (2): 153–158.

**56** Xu, J., Li, T., Li, C.L. et al. (2020). Lipid accumulation and eicosapentaenoic acid distribution in response to nitrogen limitation in microalga *Eustigmatos vischeri* JHsu-01 (Eustigmatophyceae). *Algal. Res.* 48 (1): 1–11.

**57** Ju, J.H., Ko, D.H., Heo, S.Y. et al. (2020). Regulation of lipid accumulation using nitrogen for microalgae lipid production in *Schizochytrium* sp. ABC101. *Renew. Energy* 153 (1): 580–587.

**58** Baky, H.H., Baroty, G.S., and Mostafa, E.M. (2018). Optimization growth of *Spirulina* (Arthrospira) platensis in photobioreactor under varied nitrogen concentration for maximized biomass, carotenoids and lipid contents. *Recent Pat. Food Nutr. Agric.* 10 (1): 40–48.

**59** Arguelles, E.D. and Martinez-Goss, M.R. (2020). Lipid accumulation and profiling in microalgae *Chlorolobion sp.* (BIOTECH 4031) and *Chlorella sp.* (BIOTECH 4026) during nitrogen starvation for biodiesel production. *J. Appl. Phycol.* https://doi.org/10.1007/s10811-020-02126-z.

**60** Hu, Q. (2004). Environmental effects on cell composition. In: *Handbook of Microalgal Culture* (ed. A. Richmond), 83–93. Oxford: Blackwell.

**61** Hannon, M., Gimpel, J., Tran, M. et al. (2010). Biofuels from algae: challenges and potential. *Biofuels* 1 (5): 763–784.

**62** Fan, J., Cui, Y., Wan, M. et al. (2014). Lipid accumulation and biosynthesis genes response of the oleaginous *Chlorella pyrenoidosa* under three nutrition stressors. *Biotechnol. Biofuels* 7 (1): 1–14.

**63** Goh, B.H., Ong, H.C., Cheah, M.Y. et al. (2019). Sustainability of direct biodiesel synthesis from microalgae biomass: a critical review. *Renew. Sustain. Energy Rev.* 107 (1): 59–74.

**64** Mandal, S. and Mallick, N. (2009). Microalga *Scenedesmus obliquus* as a potential source for biodiesel production. *Appl. Microbiol. Biotechnol.* 84 (1): 281–291.

**65** Xin, L., Hu, H.Y., Ke, G., and Sun, Y.X. (2010). Effects of different nitrogen and phosphorus concentrations on the growth, nutrient uptake, and lipid accumulation of a freshwater microalga *Scenedesmus* sp. *Bioresour. Technol.* 101 (14): 5494–5500.

**66** Venkatesan, S., Swamy, M.S., Jayavel, D., and Senthil, C. (2013). Effects of nitrate and phosphate on total lipid content and pigment production in *Botryococcus braunii* Kutzing KM-104. *J. Acad. Indus. Res.* 1 (1): 820–825.

**67** Salama el, S., Kim, H.C., Abou-Shanab, R.I. et al. (2013). Biomass, lipid content, and fatty acid composition of freshwater *Chlamydomonas mexicana* and *Scenedesmus obliquus* grown under salt stress. *Bioprocess. Biosyst. Eng.* 36 (6): 827–833.

**68** Jaiswal, K.K., Banerjee, I., Singh, D. et al. (2020). Ecological stress stimulus to improve microalgae biofuel generation: a review. *Octa J. Biosci.* 8 (1): 48–54.

**69** Almutairi, A.W. (2020). Effects of nitrogen and phosphorus limitations on fatty acid methyl esters and fuel properties of *Dunaliella salina*. *Environ. Sci. Pollut. Res.* 27 (1): 32296–32303.

**70** Shrestha, N., Dandinpet, K.K., and Schneegurt, M.A. (2020). Effects of nitrogen and phosphorus limitation on lipid accumulation by *Chlorella kessleri* str. UTEX 263 grown in darkness. *J. Appl. Phycol.* 32 (1): 2795–2805.

**71** Guckert, J.B. and Cooksey, K.E. (1990). Triglyceride accumulation and fatty acid profile changes in *Chlorella* (chlorophyta) during high pH-induced cell cycle inhibition. *J. Phycol.* 26 (1): 72–79.

**72** Sharma, K., Schuhmann, H., and Schenk, P.M. (2012). High lipid induction in microalgae for biodiesel production. *Energies* 5 (5): 1532–1553.

**73** Tatsuzawa, H., Takizawa, E., Wada, M., and Yamamoto, Y. (1996). Fatty acid and lipid composition of the acidophilic green alga *Chlamydomonas* sp. *J. Phycol.* 32 (1): 598–601.

**74** Fadhilah, I., Annisa, T., Widiastuti, E.L., and Maharani, H.W. (2019). Lipid contain of three microalgae on culture with different pH and salinity. In: *Proceedings of the International Conference on Marine and Coastal Engineering and Sciences [conference proceedings on the internet]*, Indonesia (23–24 August 2019). Indonesia: ICMACES; [cited 2020 Nov 13]. p. 1–12. http://repository.lppm.unila.ac.id/15612.

**75** Ochoa-Alfaro, A.E., Gaytán-Luna, D.E., González-Ortega, O. et al. (2019). pH effects on the lipid and fatty acids accumulation in *Chlamydomonas reinhardtii*. *Biotechnol. Prog.* 35 (6): 1–8.

**76** Zhao, X.C., Tan, X.B., Yang, L.B. et al. (2019). Cultivation of *Chlorella pyrenoidosa* in anaerobic wastewater: the coupled effects of ammonium, temperature and pH conditions on lipids compositions. *Bioresour. Technol.* 284 (1): 90–97.

**77** Xiaoling, P., Fanping, M., Yuejie, W. et al. (2020). Effect of pH, temperature, and $CO_2$ concentration on growth and lipid accumulation of *Nannochloropsis* sp. MASCC 11. *J. Ocean Univ. China* 19 (1): 1183–1192.

**78** Guschina, I.A. and Harwood, J.L. (2006). Lipids and lipid metabolism in eukaryotic algae. *Prog. Lipid Res.* 45 (2): 160–186.

**79** Morgan-Kiss, R.M., Priscu, J.C., Pocock, T. et al. (2006). Adaptation and acclimation of photosynthetic microorganisms to permanently cold environments. *Microbiol. Mol. Biol. Rev.* 70 (1): 222–252.

**80** Renaud, S.M., Thinh, L.V., Lambrinidis, G., and Parry, D.L. (2002). Effect of temperature on growth, chemical composition and fatty acid composition of tropical Australian microalgae grown in batch cultures. *Aquaculture* 211 (4): 195–214.

**81** Thompson, G.A. (1996). Lipids and membrane function in green algae. *Biochim. Biophys. Acta* 1302 (1): 17–45.

**82** Sushchik, N.N., Kalacheva, G.S., Zhila, N.O. et al. (2003). A temperature dependence of the intra- and extracellular fatty-acid composition of green algae and cyanobacterium. *Russ. J. Plant Physiol.* 50 (1): 374–380.

**83** Jiang, H. and Gao, K. (2004). Effects of lowering temperature during culture on the production of polyunsaturated fatty acids in the marine diatom *Phaeodactylum tricornutum* (bacillariophyceae). *J. Phycol.* 40 (1): 651–654.

**84** Converti, A., Casazza, A.A., Ortiz, E.Y. et al. (2009). Effect of temperature and nitrogen concentration on the growth and lipid content of *Nannochloropsis oculata* and *Chlorella vulgaris* for biodiesel production. *Chem. Eng. Process.* 48 (6): 1146–1151.

**85** Felpeto, A.B., Sliwinska-Wilczewska, S., Klin, M. et al. (2019). Temperature-dependent impacts of allelopathy on growth, pigment, and lipid content between a subpolar strain of *Synechocystis* sp. CCBA MA-01 and coexisting microalgae. *Hydrobiologia* 835 (1): 117–128.

**86** Svenning, J.B., Dalheim, L., Eilertsen, H.C., and Vasskog, T. (2019). Temperature dependent growth rate, lipid content and fatty acid composition of the marine cold-water diatom *Porosira glacialis*. *Algal Res.* 37 (1): 11–16.

**87** Almeyda, M.D., Bilbao, P.G., Popovich, C.A. et al. (2020). Enhancement of polyunsaturated fatty acid production under low-temperature stress in *Cylindrotheca Closterium*. *J. Appl. Phycol.* 32 (1): 989–1001.

**88** Ma, R., Zhao, X., Ho, S.H. et al. (2020). Co-production of lutein and fatty acid in microalga *Chlamydomonas* sp. JSC4 in response to different temperatures with gene expression profiles. *Algal Res.* 47 (1): 1–9.

**89** Harwood, J.L. (1998). Membrane lipids in algae. In: *Lipids in Photosynthesis: Structure, Function and Genetics* (ed. P.A. Siegenthaler and N. Murata), 53–64. The Netherlands: Kluwer Academic Publishers.

**90** Chu, W.L. (2017). Strategies to enhance production of microalgal biomass and lipids for biofuel feedstock. *Eur. J. Phycol.* 52 (4): 419–437.

**91** Napolitano, G.E. (1994). The relationship of lipids with light and chlorophyll measurements in freshwater algae and periphyton. *J. Phycol.* 30 (1): 943–950.

**92** Khotimchenko, S.V. and Yakovleva, I.M. (2005). Lipid composition of the red alga *Tichocarpus crinitus* exposed to different levels of photon irradiance. *Phytochemistry* 66 (1): 73–79.

**93** Liu, J., Yuan, C., Hu, G., and Li, G. (2012). Effects of light intensity on the growth and lipid accumulation of microalga *Scenedesmus* sp. 11-1 under nitrogen limitation. *Appl. Biochem. Biotechnol.* 166 (8): 2127–2137.

**94** He, Q., Yang, H., Xu, L. et al. (2015). Sufficient utilization of natural fluctuating light intensity is an effective approach of promoting lipid productivity in oleaginous microalgal cultivation outdoors. *Bioresour. Technol.* 180 (1): 79–87.

**95** Abdel-Hamid, M.I., Azab, Y.A., Hussain, M.H. et al. (2019). Effect of incubation temperature, light intensity and salinity on growth, lipid content and composition of two Egyptian race a of *Botryococcus Braunii*. *J. Earth Environ. Sci. Res.* 1 (2): 1–9.

**96** Shih, S.C., Mufti, N.S., Chamberlain, M.D. et al. (2014). A droplet-based screen for wavelength-dependent lipid production in algae. *Energy Environ. Sci.* 7 (1): 1–10.

**97** Seo, Y.H., Cho, C., Lee, J.Y., and Han, J.I. (2014). Enhancement of growth and lipid production from microalgae using fluorescent paint under the solar radiation. *Bioresour. Technol.* 173 (1): 193–197.

**98** Seo, Y.H., Lee, Y., Jeon, D.Y., and Han, J.I. (2015). Enhancing the light utilization efficiency of microalgae using organic dyes. *Bioresour. Technol.* 181 (1): 355–359.

**99** Choi, Y.K., Kumaran, R.S., Jeon, H.J. et al. (2015). LED light stress induced biomass and fatty acid production in microalgal biosystem, *Acutodesmus obliquus*. *Spectrochim. Acta, Part A* 145 (1): 245–253.

**100** Severes, A., Hegde, S., D'Souza, L., and Hegde, S. (2017). Use of light emitting diodes (LEDs) for enhanced lipid production in micro-algae based biofuels. *J. Photochem. Photobiol., B.* 170 (1): 235–240.

**101** Pignolet, O., Jubeau, S., Vaca-Garcia, C., and Michaud, P. (2013). Highly valuable microalgae: biochemical and topological aspects. *J. Ind. Microbiol. Biotechnol.* 40: 781–796.

**102** Ashmita, G., Saumyakanti, K., Madhumanti, M. et al. (2016). Progress toward isolation of strains and genetically engineered strains of microalgae for production of biofuel and other value added chemicals: a review. *Energy Convers. Manag.* 113 (1): 104–118.

# Part 2

# Different Catalysts Used in Biodiesel Production

# 5

# Homogeneous Catalysts Used in Biodiesel Production

*Bidangshri Basumatary[1], Biswajit Nath[1,2], and Sanjay Basumatary[1]*

[1] *Department of Chemistry, Bodoland University, Kokrajhar, Assam, India*
[2] *Department of Chemistry, Science College, Kokrajhar, Assam, India*

## 5.1 Introduction

Energy is an important aspect of today's human civilization, and its global demand is increasing drastically. Intense consumption of fossil fuel is not only diminishing the fossil fuel reservoirs and creating a situation of energy crisis but also degrading the environment by releasing greenhouse gases such as $SO_2$, $CO_2$, and $N_2O$ as well as droplets of tar, shoot, ashes, etc., causing pollution and global warming [1]. Observing the scenario of rapid climate change as well as the huge energy demand, researchers have drawn their attention toward developing alternative and renewable energy to substitute fossil fuel. Biodiesel is one of the promising biofuels due to its renewability and emission of less pollutant to the environment, and it can mitigate the future energy demand across the globe [1, 2]. Biodiesel with the emission of less greenhouse gases on its combustion, high cetane number, high flash point, low viscosity, good lubricity, and biodegradability are some of its advantageous points compared to petrodiesel [3, 4]. Besides, due to comparable chemical, flow, and combustion properties with diesel, biodiesel can be used in a diesel engine by blending with diesel directly or with a slight modification of the engine [5, 6]. The concept of biofuel was originated from the proposal of Dr. Rudolf Diesel in 1911 to introduce peanut oil and various other oil sources for use as energy in the diesel engine. Accordingly, in 1925, Henry Ford successfully used bioethanol to run his model T ford engine and hailed the term biofuel as "fuel for the future" [7]. Consequently, in 1940, vegetable oil methyl and ethyl esters were used as fuel to run the buses in France and Belgium. Following that, the use and development of vegetable oil as a biofuel is being studied since World War II [8].

As a viable and sustainable alternative to fossil fuel, biodiesel stands as promising hope for the future generation. The major factors behind quality and sustainability are the availability of feedstock oil resources. Algae biomass, microbial oil, edible, and nonedible vegetable oils, animal fat/oil, etc. are being used as feedstocks for biodiesel production. Coconut [9], sunflower [10], soybean [10], palm [11], mustard [11], etc. are some of the

*Biodiesel Production: Feedstocks, Catalysts, and Technologies*, First Edition. Edited by Samuel Lalthazuala Rokhum, Gopinath Halder, Suttichai Assabumrungrat, and Kanokwan Ngaosuwan.

edible oil sources used as feedstock for biodiesel production. Due to probable future and present food crisis, the use of edible oils as biodiesel feedstocks created controversies, and, accordingly, the use of nonedible oil sources is preferred over edible oil sources in biodiesel production. Because of that, the synthesis of biodiesel from nonedible oil sources like *Jatropha curcas* [9, 12], cotton seed [9], karanja [9], kusum [9], mahua [11], etc. is investigated and being developed. *J. curcas* is considered the best due to its high oil content (27–40%) as well as easy growing nature in all climates with minimum effort [9, 13]. In this context, the United States identified and planted 165 varieties of plants to produce ~7100 million liters of biodiesel annually, and, similarly, European Union (EU) has 120 plant varieties to produce ~6900 million liters of biodiesel per annum [14]. Countries like Brazil, Spain, France, Italy, Ireland, Indonesia, Malaysia, Australia, China, Germany, Canada, India, and Ghana are also contributing to the production of biodiesel [15]. The world recorded the production of 36 billion liters of biodiesel in 2017 in which the United States and EU were the major contributors [16].

Among the four different biodiesel processing methods such as direct use of vegetable oil, microemulsion, thermal cracking, and transesterification, the most widely used and well-accepted method for biodiesel production is the transesterification [17]. In the presence of a catalyst, transesterification successfully converts vegetable oil to mixtures of fatty acid alkyl esters mainly methyl and ethyl esters, which is termed as biodiesel and are being used as compatible biofuel [18, 19]. For the production of cost-effective biodiesel, the optimization of reaction is important and is influenced by the factors like suitable ratio of methanol to oil, catalyst loading, and temperature, among which catalyst plays a vital role in transesterification, and various types of catalysts like homogeneous, heterogeneous, and enzyme catalysts are being used. Acids such as sulfuric acid, hydrochloric acid, sulfonic acid, etc. are utilized as the homogeneous acid catalyst in biodiesel synthesis [20]. Sodium hydroxide, potassium hydroxide, sodium methoxide, etc. are some of the reported homogeneous base catalysts used for biodiesel synthesis [21]. Vegetable oil usually contains few percentages of free fatty acids (FFAs). Oils with FFAs >1% undergo saponification with the base used as a catalyst in homogeneous base-catalyzed transesterification [8]. In such cases, the FFAs are to be reduced by converting them to fatty acid esters by esterification followed by the transesterification to overcome the problems related to saponification [11]. In this chapter, the performances of various homogeneous acids and bases used as catalysts in biodiesel production are reviewed.

## 5.2 Transesterification in Biodiesel Synthesis

Transesterification also known as alcoholysis is one of the easiest and most effective methods for biodiesel production in which oils are transesterified with methanol preferably in the presence of a catalyst to afford fatty acid methyl ester (FAME) and glycerol (Scheme 5.1) [22–25]. Production of biodiesel through transesterification can be catalyzed using acid, base, and enzyme catalysts. The reaction mechanisms for the transesterification of vegetable oil (triglyceride) with methanol under acid and base-catalyzed reactions are depicted in Schemes 5.2 and 5.3, respectively. During the reaction, the triglyceride is converted to diglyceride; diglyceride is converted to monoglyceride, and finally monoglyceride to glycerol yielding one molecule of FAME (biodiesel) at each step [26–28]. In addition to being corrosive, acid-catalyzed reactions require high alcohol to oil molar ratio, high

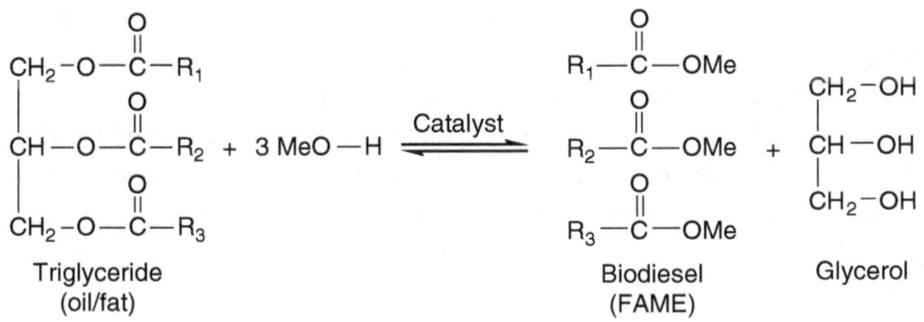


**Scheme 5.1** Transesterification of vegetable oil (triglyceride) to biodiesel (FAME). *Source:* Based on Refs. [22–25].

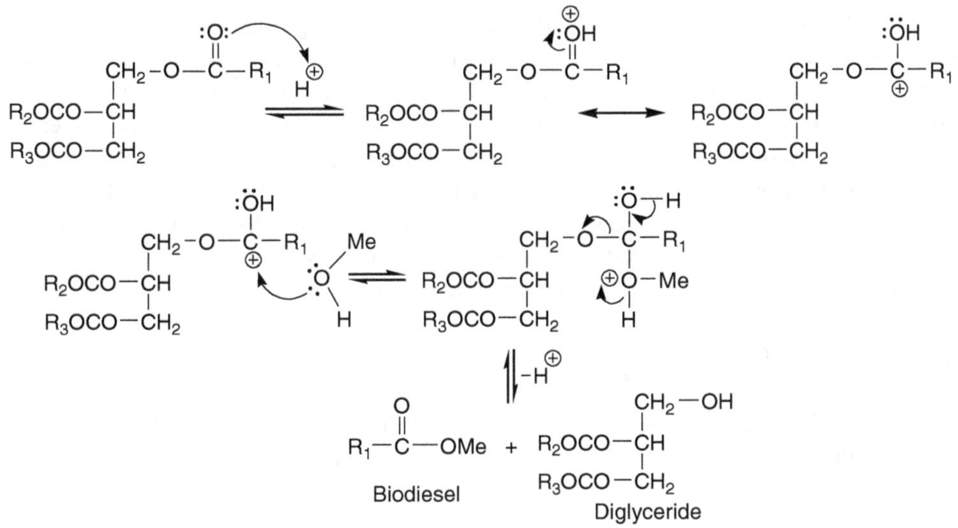


**Scheme 5.2** Mechanism of acid-catalyzed reaction of triglyceride.

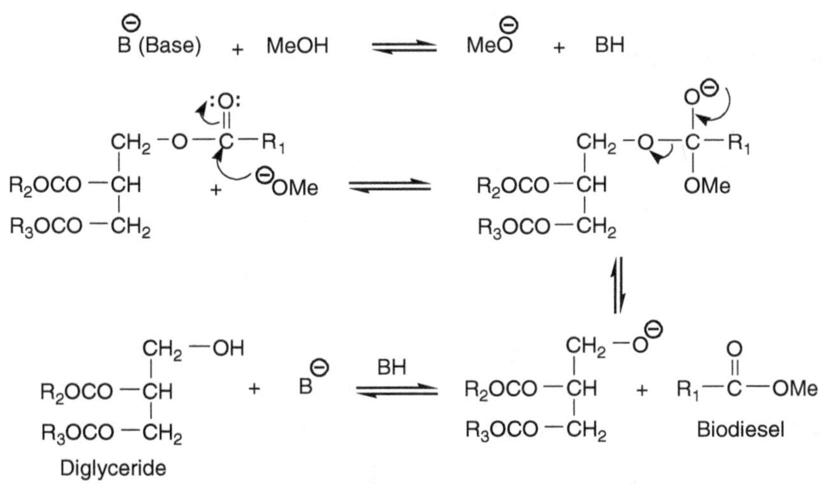


**Scheme 5.3** Mechanism of base-catalyzed reaction of triglyceride.

temperature, and pressure as well as longer reaction times for which the process has drawbacks in many aspects [29, 30]. Although homogeneous base-catalyzed transesterification reactions of oil require milder conditions and shorter reaction times with high conversion, they have several drawbacks like soap formation in presence of FFAs in the oil and complications in the separation of catalysts from the product mixtures, for which catalysts cannot be recycled or reused as they are consumed during the reaction [31, 32]. The cost of enzyme catalyst is very high, and the rate of oil transesterification with this catalyst is also very slow and hence hinders wider application [33, 34].

## 5.3 Homogeneous Catalyst in Biodiesel Synthesis

Various acids and bases are used as homogeneous catalysts in biodiesel synthesis. Both homogeneous acid and base-catalyzed transesterification reactions possess few limitations. The presence of FFA in oils greater than 2 wt% creates the problem of soap formation in base-catalyzed reaction, which makes the biodiesel production process difficult and creates difficulty in the separation process of glycerol from biodiesel mixture [35, 36]. In such cases, the amount of FFA is being reduced first via esterification (Schemes 5.4 and 5.5) before the transesterification process to form biodiesel. The oils with less than 2 wt% FFA can directly be transesterified with a base catalyst [36]. A schematic representation of biodiesel production using homogeneous catalyst is depicted in Figure 5.1.

### 5.3.1 Homogeneous Acid Catalyst

Some raw oil feedstock contains high FFAs. The base-catalyzed transesterification of the oils with high FFA is not feasible due to the formation of soaps, leading to problems in the

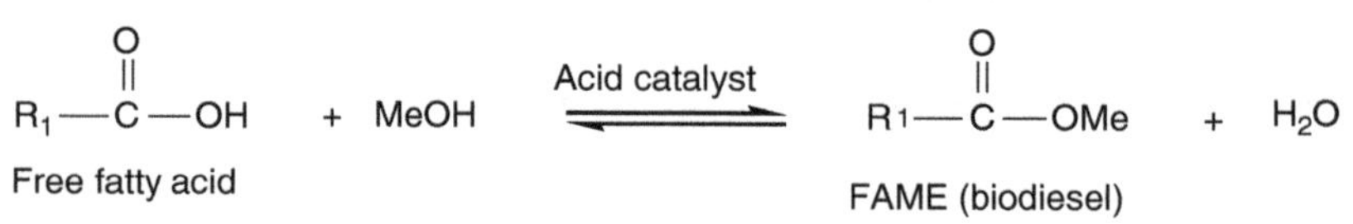


**Scheme 5.4** Esterification of free fatty acid.

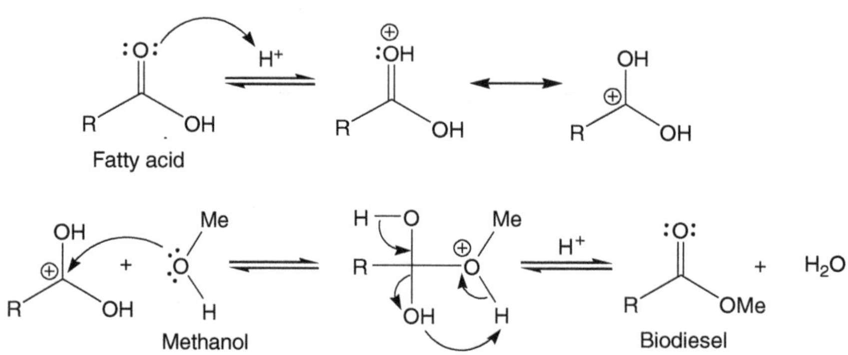


**Scheme 5.5** Mechanism of acid-catalyzed esterification of free fatty acid to FAME.

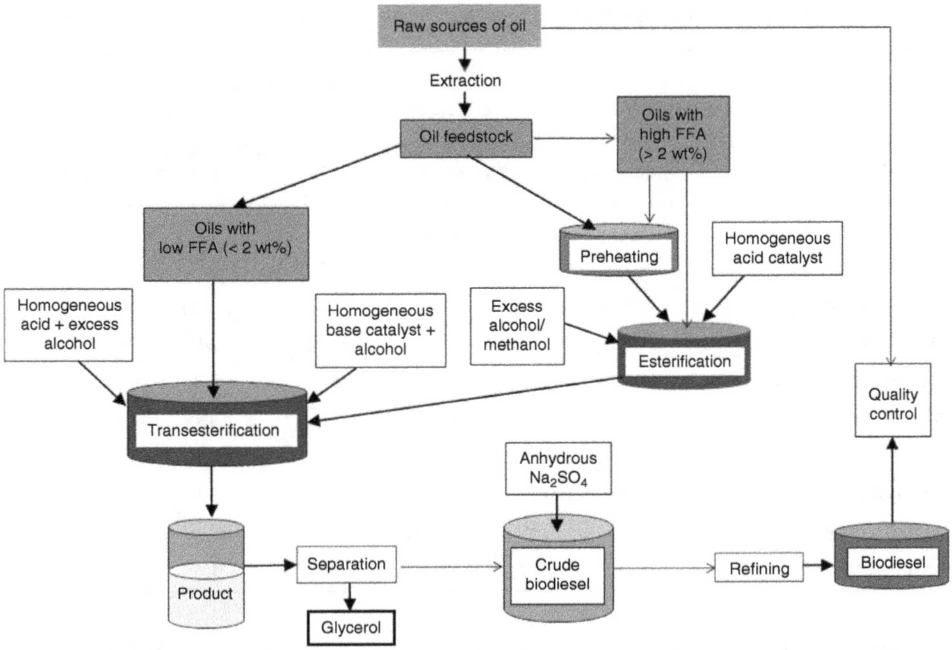


**Figure 5.1** Representation of biodiesel production process using homogeneous catalyst.

purification and separation of fatty acid esters and glycerols. Therefore, it is suggested to use an acid catalyst for transesterification reaction in the case of oils containing high FFA contents to avoid problems of saponification reaction. Among the various acid catalysts, hydrochloric acid and sulfuric acid are the most favorable. Comparative data of transesterification reaction with the homogeneous acid catalyst is depicted in Table 5.1. Javidialesaadi et al. [37] carried out their study to produce biodiesel from oil containing 44.5% of FFA using 98% of sulfuric acid catalyst in esterification reaction at 50 °C, resulting in a maximum yield less than 90% in 6 : 1 methanol-to-oil ratio at the fixed reaction time of 1 h. They reported that increasing the catalyst loading of $H_2SO_4$ by more than 3% has no significant effect on the rate of transesterification reaction and product yield. Cao et al. [38] used 0.5 M $H_2SO_4$ as the catalyst in the production of biodiesel from *Chlorella pyrenoidosa*, a type of green microalgae, by reacting with 4 ml methanol and 8 ml *n*-hexane at the reaction temperature of 120 °C for 3 h via transesterification reaction and converted 92.5% of it to biodiesel. Zhang et al. [39] reported esterification of crude *Zanthoxylum bungeanum* oil using 24 : 1 methanol-to-oil ratio with 2% of $H_2SO_4$ in the 1.33 h. Similarly, Bhatti et al. [40] reported the applicability of $H_2SO_4$ catalyst in transesterification to convert chicken/mutton tallow into biodiesel under catalyst load of 1.2/2.5% with 30 : 1 methanol-to-oil ratio at 50/60 °C producing 99.01/93.21% yield of biodiesel in 24 h. Veljkovic et al. [41] esterified tobacco seed oil using $H_2SO_4$ catalyst loaded in 1% v/v on 18 : 1 methanol to oil molar ratio at 60 °C in 0.41 h to yield 91% of FAME. They reported that the acid-catalyzed esterification reaction reduced the FFA content. Canakci et al. [30] studied the production of biodiesel from soybean oil via acid-catalyzed transesterification

**Table 5.1** Homogeneous acid-catalyzed biodiesel synthesis. *Source:* Based on Refs. [30, 37–45].

| | | Parameters | | | | |
|---|---|---|---|---|---|---|
| **Biodiesel feedstock** | **Catalyst** | **Catalyst load (%)** | **Methanol/ oil** | **Temperature (°C)** | **Reaction time (h)** | **Yield (%)** |
| Vegetable oil [37] | $H_2SO_4$ | 3 | 6 : 1 | 50 | 1 | >90 |
| *Chlorella pyrenoidosa* [38] | $H_2SO_4$ | 0.5 | 40 : 1 | 120 | 3 | 92.5 |
| *Zanthoxylum bungeanum* [39] | $H_2SO_4$ | 2 | 24 : 1 | 60 | 1.33 | 98 |
| Chicken/mutton tallow [40] | $H_2SO_4$ | 1.2/2.5 | 30 : 1 | 50/60 | 24 | 99.01/93.21 |
| Tobacco seed oil [41] | $H_2SO_4$ | 1 | 18 : 1 | 60 | 0.41 | 91 |
| Soybean oil [30] | $H_2SO_4$ | 3 | 6 : 1 | 60 | 48 | 98 |
| Waste cooking oil [42] | $H_2SO_4$ | 4 | 20 : 1 | 95 | 10 | 90 |
| Used frying oil [43] | $H_2SO_4$ | 0.1 | 3.6 : 1 | 65 | 40 | 79.3 |
| Soybean oil [44] | $C_2HF_3O_2$ | 2 M | 20 : 1 | 120 | 5 | 98.4 |
| Sunflower oil [45] | HCl | 1.85 | 25 : 1 | 100 | 1 | 95.2 |

and reported that utilization of 3% loading of $H_2SO_4$ catalyst and 6 : 1 methanol-to-oil ratio at the reaction temperature of 60 °C resulted in a much slower reaction and took 48 h to convert 98% of oil to biodiesel. Accordingly, Wang et al. [42] studied waste cooking oil (WCO) for its conversion into biodiesel. They utilized 4% sulfuric acid as the catalyst in transesterification reaction with 20 : 1 methanol to triglyceride ratio at a reaction temperature of 95 °C, and 90% yield of product could be achieved in 10 h. They found that the rate of acid-catalyzed transesterification reaction is quite slow. Nye et al. [43] studied the alcoholysis of used frying oil using 0.1% concentrated sulfuric acid under 3.6 : 1 ratio of methanol to oil at a reaction temperature of 65 °C. It was stated that conversion is slow and took a longer reaction time of 40 h yielding only 79.3% biodiesel, which is very low compared to other results. Miao et al. [44] reported the use of trifluoroacetic acid ($C_2HF_3O_2$)-catalyzed transesterification reaction to convert soybean oil into biodiesel. They achieved 98.4% yield of biodiesel under the optimized conditions of 2.0 M $C_2HF_3O_2$ concentration with 20 : 1 molar ratio of methanol to soybean oil at 120 °C of the reaction temperature in a reaction time of 5 h. They also reported the reduction of specific gravity of the product biodiesel on applying acid catalyst. Sagirogluet al. [45] reported that sunflower oil, soybean oil, safflower oil, canola oil, corn oil, olive oil, and hazelnut oil can be converted into biodiesel through transesterification reaction by employing hydrochloric acid as catalyst. From their experiments, a maximum of 95.2% biodiesel was yielded in 1 h of reaction time with 2 g of sunflower oil and 50 ml of cool-dry methanol in the presence of 1.85% HCl (37%, density = 1.19 g $mol^{-1}$) at 100 °C.

### 5.3.2 Homogeneous Base Catalyst

Homogeneous base catalyst is extensively utilized in the transesterification reaction for some advantages against homogeneous acid catalyst. Homogenous base catalysts are cheap and easily accessible. The reaction is also relatively faster than acid-catalyzed reaction and can be operated at room temperature and atmospheric pressure. The most commonly used homogeneous base catalysts are KOH, NaOH, $NaOCH_3$, and $NaOCH_2CH_3$ [8]. Various homogeneous base catalysts were studied in the production of biodiesel, and their results are summarized in Table 5.2. Alamu et al. [46] reported the use of 1% KOH solution as a

**Table 5.2** Homogeneous base-catalyzed biodiesel synthesis. *Source:* Based on Refs. [46–72].

| | | Parameters | | | | |
|---|---|---|---|---|---|---|
| **Biodiesel feedstock** | **Catalyst** | **Catalyst load (%)** | **Methanol/oil** | **Temperature (°C)** | **Reaction time (h)** | **Yield (%)** |
| Palm kernel oil [46] | KOH | 1 | Ethanol | 60 | 1.5 | 96 |
| *Pongamia pinnata* oil [47] | KOH | 1 | 10 : 1 | 60 | 1.5 | 92 |
| Rapeseed oil [48] | KOH | 1 | 6 : 1 | 65 | 2 | 95–96 |
| Waste vegetable oil [49] | KOH | 1 | 6 : 1 | 65 | 1 | 96.15 |
| Vegetable oil [50] | KOH | 1 | 6 : 1 | 25 | 0.66 | 51–87 |
| Crude rubber/palm oil [51] | KOH | 2 | 8 : 1 | 55 | 5 | 98 |
| Roselle oil [52] | KOH | 1.5 | 8 : 1 | 60 | 1 | 99.4 |
| Frying oil [53] | KOH | 1 | 12 : 1 | 60 | 2 | 72.5 |
| Used olive oil [54] | KOH | 1.26 | 12 : 1 | 25 | 1.5 | 94 |
| Duck tallow [55] | KOH | 1 | 6 : 1 | 65 | 3 | 83.6 |
| *Pongamia pinnata* oil [56] | KOH | 0.1 M | 4 : 1 | 60 | 1 | 98.51 |
| Jatropha oil [57] | KOH | 0.075 | 6 : 1 | 50 | 5 | 87 |
| *Calophyllum inophyllum* oil [58] | KOH | 1 | 9 : 1 | 55 | 1 | 98.53 |
| Waste cooking oil [59] | KOH | 1.16 | 9.4 : 1 | 62.4 | 0.33 | 98.26 |
| Soybean oil [60] | KOH | 1 | 6 : 1 | 60 | 1 | ~96 |
| Soybean oil [60] | $NaOCH_3$ | 0.6 | 6 : 1 | 60 | 1 | 97 |
| Waste cooking oil [61] | $NaOCH_3$ | 0.75 | 6 : 1 | 65 | 1.5 | 96.6 |
| Rice bran oil [62] | $NaOCH_3$ | 0.88 | 7.5 : 1 | 55 | 1 | 83.3 |
| *Sesamum indicum* oil [63] | $NaOCH_3$ | 0.75 | 6 : 1 | 50 | 0.5 | 87.8 |
| Soybean oil [64] | NaOH | 1 | 6 : 1 | 60 | 1 | 98.5 |
| Cotton seed oil [64] | NaOH | 1 | 6 : 1 | 60 | 1 | 97 |
| Waste cooking oil [65] | NaOH | 1 | 6 : 1 | 50 | 1.5 | 89.8 |
| Canola oil [66] | NaOH | 1 | 6 : 1 | 45 | 0.25 | 93.5 |
| Waste frying oil [67] | NaOH | 0.5 | 7.5 : 1 | 0.5 | 0.5 | 96 |
| Sunflower oil [68] | NaOH | 1 | 6 : 1 | 60 | 2 | 97 |
| Refined palm oil [69] | NaOH | 1 | 6 : 1 | 60 | 0.5 | 95 |
| Waste frying oil [70] | NaOH | 0.6 | 4.8 : 1 | 65 | 1 | 98 |
| Rapeseed oil [71] | NaOH | 0.1 | 475 : 1 | 60 | 1 | 68.5 |
| Soybean oil [72] | NaOH | 1.3 | 9 : 1 | 40 | 1.33 | 95 |

catalyst in the transesterification of palm kernel oil to biodiesel. In the study, 100 g of palm kernel oil was reacted with 20 g of ethanol at 60 °C for 1.5 h and achieved an optimum biodiesel yield of 96.0%. Similarly, Karmee et al. [47] achieved 92% of biodiesel in 1.5 h from crude pongamia oil using 1% KOH as a catalyst in transesterification reaction with 10 : 1 methanol to oil molar ratio at the reaction temperature of 60 °C. Rashid et al. [48] investigated the conversion of rapeseed oil to biodiesel via transesterification reaction. The catalysts used were 1% of KOH, NaOH, $CH_3ONa$, and $CH_3OK$. They reported that the use of 1% KOH as catalyst could produce 95–96% yield of biodiesel under 6 : 1 methanol to oil molar ratio with a mixing speed of 600 rpm at 65 °C in 2 h. Refaat et al. [49] reported that a maximum of 96.15% biodiesel was produced from waste vegetable oil catalyzed by 1% of KOH at the reaction temperature of 65 °C in 1 h, and the fuel properties of biodiesel are in agreement with properties of standard biodiesel. Dmytryshyn et al. [50] utilized 1% of KOH as a catalyst in the transesterification reaction of vegetable oil to biodiesel and reported the conversion of 51–87% oil to biodiesel under 6 : 1 methanol-to-oil ratio at 25 °C in 0.66 h. Yusup et al. [51] investigated the conversion of palm oil to biodiesel by employing 2% KOH as the catalyst and achieved 98% of biodiesel by setting the reaction at 55 °C with 8 : 1 methanol to oil molar in a reaction time of 5 h. Nakpong et al. [52] successfully transesterified roselle oil in the presence of KOH catalyst producing 99.4% of biodiesel in the optimum reaction conditions of 8 : 1 methanol-to-oil ratio and 1.5% of catalyst at 60 °C in 1 h. Frying oil containing 308 and 882 g $mol^{-1}$ of fatty acid and oil, respectively, was successfully transesterified by Encinar et al. [53] using KOH, NaOH, $CH_3ONa$, and $CH_3OK$ catalysts in optimum reaction conditions of 12 : 1 methanol-to-oil ratio, 1% of catalyst load, and 60 °C of reaction temperature. The reaction yielded 72.5% of biodiesel in a reaction time of 2 h. Dorado et al. [54] studied the comparison of NaOH and KOH as base catalysts in the transesterification of waste olive oil collected from hospitals using ethanol and methanol. They reported that the best result was obtained with methanol in the optimum reaction conditions of 12 : 1 methanol-to-oil ratio, 1.26% KOH of catalyst at room temperature (25 °C) yielding 94% of biodiesel in 1.5 h. Similarly, Chung et al. [55] compared the use of NaOH, $CH_3ONa$, and KOH as base catalysts. They reported that KOH showed better catalytic activity in the reaction of duck tallow oil with methanol in the ratio of 6 : 1 (methanol to oil) and 1% of KOH catalyst at 65 °C in 3 h producing 83.6% yield of biodiesel. Porwal et al. [56] directly transesterified pongamia seeds and successfully transformed them into biodiesel by utilizing 0.1 M solution of KOH catalyst at 60 °C yielding 98.51% biodiesel in 1 h with 4 : 1 solvent/seed (w/w). Kartika et al. [57] utilized KOH catalyst (0.075 mol $l^{-1}$) in transesterification reaction that was conducted at 50 °C with 6 : 1 methanol to jatropha oil ratio. The reaction yielded 87% of FAME in 5 h of reaction time. *Calophyllum inophyllum* oil containing high FFA was transesterified successfully by Silitonga et al. [58] using 1% KOH catalyst in 9 : 1 methanol to oil molar ratio in 1 h at the reaction temperature of 55 °C. They reported a maximum yield of 98.53% biodiesel. Mohadesi et al. [59] examined the efficiency of KOH catalyst for the production of biodiesel from WCO. The optimum conditions for biodiesel production reported were methanol to waste oil ratio of 9.4 : 1 with 1.16 wt% of KOH as the catalyst at 62.4 °C. They achieved 98.26% of biodiesel only in 0.033 h. Dias et al. [60] also studied the use of KOH, NaOH, and $CH_3ONa$ catalysts in transesterification reaction to convert sunflower oil, waste frying oil, and soybean oil to biodiesel. Under the optimum conditions of 6 : 1 methanol-to-oil ratio and 1% of catalyst loading at 60 °C in 1 h of reaction time, KOH could produce 96% of yield whereas $NaOCH_3$ yielded 97% biodiesel. Chen et al. [61] reported that the biodiesel production through transesterification with $NaOCH_3$ catalyst produces higher biodiesel yield (96.6%) in optimum conditions of 0.75% catalyst loading and 6 : 1 methanol-to-oil ratio at 65 °C in 1.5 h. Similarly, $NaOCH_3$ catalyst was studied by Rashid et al. [62] in the conversion of rice bran oil taking 7.5 : 1 molar ratio of methanol to oil and 0.88% of $NaOCH_3$ catalyst loaded at 55 °C, and a yield of 83.3% biodiesel was achieved in 1 h. Dawadu et al. [63] also used $NaOCH_3$ catalyst to convert *Sesamum indicum* oil to biodiesel. They reported 87.8% conversion of oil to biodiesel at optimum methanol-to-oil ratio of 6 : 1 and 0.5% of catalyst amount in 30 min at 50 °C. Keera et al. [64] investigated the effect in yield and characteristics of biodiesel by varying catalyst load of 0.5–1.5 of NaOH and methanol-to-oil ratio of 3 : 1–9 : 1. They reported that 1% NaOH could produce the highest yield of 98.5% (soybean biodiesel) and 97% (cotton oil biodiesel) using methanol-to-oil ratio of 6 : 1 at the reaction temperature of 60 °C in 1 h of reaction time. Meng et al. [65] reported the transesterification of WCO using 1% NaOH catalyst and 6 : 1 methanol-to-oil ratio at 50 °C in 1.5 h that yielded 89.8% of biodiesel. Under optimal reaction conditions of 6 : 1 methanol-to-oil ratio, 1% of catalyst load at 45 °C, Leung et al. [66] could produce 93.5% of canola biodiesel in 0.25 h of reaction time. Similarly, the utilization of NaOH catalyst was also reported by Uzun et al.[67], Rashid et al.[68], Lubes et al. [69], Felizardo et al. [70], Zakaria et al. [71], and Silva et al. [72] in the transesterification of different oils such as waste frying oil, sunflower oil, refined palm oil, waste frying oil, rapeseed oil, and soybean oil, respectively. Ghadge et al. [73] studied the production of biodiesel from crude mahua oil having high FFA contents of 19%. The mahua oil was first esterified in the presence of 1% v/v $H_2SO_4$ acid catalyst under (0.30–0.35 v/v) methanol-to-oil ratio at 60 °C for 1 h and reduced the FFA to less than 1%. Thereafter, the product mixture having an acid value of less than 2 mg KOH $g^{-1}$ was transesterified using 0.7% w/v of KOH catalyst with 0.25 v/v of methanol and reported 98% yield of biodiesel. Similarly, Ramadhas et al. [74] reported the lowering of high FFA (<2%) of rubber seed oil by acid-catalyzed (0.2–2% of sulfuric acid) esterification. The esterified product was then transesterified using NaOH as the catalyst. Usta et al. [75] investigated the production of biodiesel from tobacco seed oil by transesterification process using NaOH catalyst (7.5 g $l^{-1}$ of the oil) by heating at 55 °C that yielded 86% of biodiesel.

## 5.4 Properties of Biodiesel Produced by Homogeneous Acid and Base-Catalyzed Reactions

Impurities in biodiesel affect its quality and engine performance. The quality of produced biodiesel is very important. The properties of biodiesel should meet the criteria prescribed in the international biodiesel standards before it is used as fuel in a diesel engine either in its pure form or blended with diesel fuel. Comparison of properties of biodiesel produced using homogeneous acid and base catalysts along with the specified limits prescribed in ASTM D6751 and EN 14214 standards is presented in Table 5.3. Various biodiesel properties such as density, kinematic viscosity (KV), cetane number (CN), pour point (PP), flash point (FP), cloud point (CP), acid value (AV), iodine value (IV), saponification number (SN), calorific value (CV), etc. are prescribed in the international standards. It can be seen

**Table 5.3** Properties of biodiesel produced from homogeneous acid and base catalysts.

| Biodiesel feedstock | Density (g $cm^{-3}$) | KV (40 °C, $mm^2$ $s^{-1}$) | CN | PP (°C) | FP (°C) | CP (°C) | AV (mg KOH $g^{-1}$) | IV (g $I_2$/100 g) | CV (MJ $kg^{-1}$) |
|---|---|---|---|---|---|---|---|---|---|
| Rice bran oil [62] | 0.876 | 5.18 | 63 | −2.00 | — | 8.00 | 0.33 | — | 43.10 |
| Canola oil [50] | 0.875 | 4.23 | — | −8.0 to −10.0 | — | 1.0–2.0 | 0.154 | 127 | — |
| Crude palm/crude rubber oil [51] | 0.874 | 4.22 | 50.19 | — | 150 | — | 0.46 | — | — |
| Roselle oil [52] | 0.880 | 4.588 | — | −1 | >130 | — | 0.43 | 62 | — |
| Vegetable oil [49] | 0.882 | 5.07 | — | −6 | 174 | 3 | — | — | — |
| Waste vegetable oil [49] | 0.886 | 5.64 | — | −6 | 176 | 3 | — | — | — |
| Waste vegetable oil [49] | 0.879 | 6.40 | — | 9 | 168 | 15 | — | — | — |
| Used olive oil [54] | 0.882–0.887 | 5.29–6.46 | — | −6 | 169 | −1 to −2 | 0.1 | 78 | — |
| Palm kernel oil [46] | 0.883 | 4.839 | — | 2 | 167 | 6 | — | — | — |
| Pongamia oil [47] | 0.48 | — | — | — | 150 | — | 0.62 | — | — |
| Waste cooking oil [65] | 0.890 | 4.23 | 54.5 | — | 171 | — | 0.48 | — | 32.9 |
| Waste frying oil [67] | 0.890 | 4.36 | 53.4 | −3 | 175.4 | — | — | — | 45.34 |
| Sunflower oil [68] | — | 4.90 | — | 4.0 | 170.0 | 1.0 | 0.24 | — | 45.3 |
| Soybean oil [64] | 0.881 | 4.16 | 57.6 | 0 | 171 | 3 | — | — | — |
| Cottonseed oil [64] | 0.886 | 4.89 | 57.1 | 3 | 178 | 6 | — | — | — |
| Jatropha oil [57] | 0.885 | 3.5 | 47 | 0 | 107 | 11 | 0.35 | 107 | 40 |
| Jatropha oil [58] | 0.881 | 4.48 | — | −2 | 160.5 | −3 | 0.39 | — | 40.224 |
| *Calophyllum inophyllum* oil [58] | 0.896 | 4.57 | — | −1 | 158.5 | −1 | 0.45 | — | 40.104 |
| *Ceiba pentandra* oil [58] | 0.876 | 4.61 | — | −2 | 156.5 | −2 | 0.40 | — | 40.493 |
| Waste cooking oil [59] | 0.880 | 4.6 | — | 5 | — | 6 | — | — | — |

| | | | | | | | | | |
|---|---|---|---|---|---|---|---|---|---|
| Tobacco seed oil [41] | 0.882 | 5.2 | — | — | — | — | 0.66 | 109.8 | — |
| *Madhuca indica* oil [73] | 0.880 | 3.98 | — | 6 | 208 | — | 0.41 | — | 37 |
| Rubber seed oil [74] | 0.874 | 5.81 | — | −8 | 130 | 4 | 0.118 | — | 36.50 |
| Tobacco seed oil [75] | 0.886 | 3.5 | 51 | — | — | — | 0.45 | 112 | 39.811 |
| ASTM–D6751 standard | 0.860–0.894 | 1.9–6.0 | 47 (min) | NS | 130 (min) | NS | 0.5 max | NS | NS |
| EN–14214 standard | 0.86–0.90 | 3.5–5.0 | 51 (min) | NS | 120 (min) | NS | 0.5 max | 120 (max) | NS |

KV–Kinematic viscosity; CN–cetane number; PP–pour point; FP–flash point; CP–cloud point; AV–acid value; IV–iodine value; CV–calorific value; NS–not specified; min–minimum; max–maximum.

from Table 5.3 that the fuel property of the biodiesels produced from various oil sources via transesterification using different homogeneous acid and base catalysts is comparable, and most of them are found within the specified limits prescribed in the standards.

## 5.5 Relevance of Homogeneous Acid and Base Catalysts in Biodiesel Synthesis

Comparative analysis of Tables 5.1 and 5.2 shows that the methanol-to-oil ratios in transesterification catalyzed by homogeneous acid catalysts are higher than that of homogeneous base-catalyzed reactions. Excessive amounts of alcohol (methanol) in biodiesel production create difficulty in recovering glycerol during biodiesel separation process [8]. The temperature reported in the homogeneous acid-catalyzed transesterification is also higher (>65 °C) compared with homogeneous base-catalyzed transesterification (≤65 °C) confirming the consumption of higher energy in biodiesel production using a homogeneous acid catalyst. In addition to that homogeneous acid-catalyzed transesterifications are recorded to be slower than the base-catalyzed transesterification reactions. Thus, homogeneous base-catalyzed transesterifications are the best for biodiesel production in terms of methanol requirement, reaction temperature, and time. Conclusively, these have higher catalytic activities and are cheap. However, in the case of oils with the high FFA content, the FFA may react with the employed base catalyst, leading to the formation of soap by saponification that may lead to lower activity and yield of biodiesel. The soap formation also creates the problem in purification of the product requiring more water, resulting in the waste of water. The oil feedstock having low FFA content (<2 wt%) can be directly utilized in transesterification with alkaline catalyst. Oil feedstock with high FFA (>2 wt%) is first esterified with an acid catalyst to lower the FFA followed by the transesterification with the base catalyst to produce biodiesel [36]. Biodiesel produced by acid-catalyzed transesterification possesses high acid value, causing equipment corrosion as well as difficulty in separation of catalyst from the product. Homogeneous acid and base catalysts are nonrecyclable and are the major demerit in using as a catalyst in biodiesel production. Considering that, researchers are working on developing other good, recyclable, green catalyst for biodiesel synthesis.

## 5.6 Conclusion

Biodiesel as a green, renewable, and sustainable alternative to fossil fuel is being experimented, and various edible and nonedible oil sources along with animal fat, microalgae, etc. are recommended as feedstocks. Due to scarcities in food, nonedible oils are preferred over edible oils. The most preferable method for the production of biodiesel is the transesterification of oils with alcohol in presence of a catalyst. Various homogeneous, heterogeneous, and enzyme catalysts are utilized in biodiesel synthesis. With the growing demand for biodiesel, a cost-effective production process is in search, and, in this regard, catalyst plays a vital role. Enzyme-catalyzed transesterification is less feasible in biodiesel production as the reaction is slow and the cost of enzyme is high. Homogeneous acid and base-catalyzed transesterification reactions are used in industrial-scale production with some demerits. Homogeneous acid-catalyzed transesterification is slow and requires high temperature and pressure. Homogenous base-catalyzed reactions are fast with good yield but form soap during the reaction. Due to this, additional precaution is needed to be taken, which makes the production process costlier. Heterogeneous base catalyst is recently being investigated and resulted in good efficiency in terms of biodiesel conversion, yield, product quality, and the economic point of view. It has the property to overcome the demerits of homogeneous acid, base, and enzyme-catalyzed reactions. In this regard, the heterogeneous base catalyst may stand with the expectations. Investigation on the development of catalysts with all these prospects is mostly needed for the commercial-scale production of biodiesel.

## References


1 Pathak, G., Das, D., Rajkumari, K., and Rokhum, L. (2018). Exploiting waste: towards a sustainable production of biodiesel using *Musa acuminata* peel ash as a heterogeneous catalyst. *Green Chem.* 20 (10): 2365–2373.

2 Nath, B., Kalita, P., Das, B., and Basumatary, S. (2020). Highly efficient renewable heterogeneous base catalyst derived from waste *Sesamum indicum* plant for synthesis of biodiesel. *Renew. Energy* 151: 295–310.

3 Mendonça, I.M., Paes, O.A., Maia, P.J. et al. (2019). New heterogeneous catalyst biodiesel production from waste tucumã peels (*Astrocaryum aculeatum* Meyer): parameters optimization study. *Renew. Energy* 130: 103–110.

4 Deka, D.C. and Basumatary, S. (2011). High quality biodiesel from yellow oleander (*Thevetia peruviana*) seed oil. *Biomass Bioenergy* 35 (5): 1797–1803.

5 Abdullah, S.H.Y.S., Hanapi, N.H.M., Azid, A. et al. (2017). A review of biomass-derived heterogeneous catalyst for a sustainable biodiesel production. *Renew. Sustain. Energy Rev.* 70: 1040–1051.

6 Muniyappa, P.R., Brammer, S.C., and Noureddini, H. (1996). Improved conversion of plant oils and animal fats into biodiesel and co-product. *Bioresour. Technol.* 56: 19–24.

7 Dash, S.K. and Lingfa, P. (2017). A review on production of biodiesel using catalyzed transesterification. *AIP Conf. Proc.* 1859 (1): 020100. AIP Publishing LLC.

8 Narasimharao, K., Lee, A., and Wilson, K. (2007). Catalysts in production of biodiesel: a review. *J. Biobased Mater. Bioenergy* 1 (1): 19–30.

9 Bhatia, S.K., Bhatia, R.K., Jeon, J.M. et al. (2021). An overview on advancements in biobased transesterification methods for biodiesel production: oil resources, extraction, biocatalysts, and process intensification technologies. *Fuel* 285: 119117.

10 Ambat, I., Srivastava, V., and Sillanpää, M. (2018). Recent advancement in biodiesel production methodologies using various feedstock: a review. *Renew. Sustain. Energy Rev.* 90: 356–369.

11 Verma, P. and Sharma, M.P. (2016). Review of process parameters for biodiesel production from different feedstocks. *Renew. Sustain. Energy Rev.* 62: 1063–1071.

12 Basumatary, S. (2013). Non-conventional seed oils as potential feedstocks for future biodiesel industries: a brief review. *Res. J. Chem. Sci.* 3 (5): 99–103.

**13** ANR, R., Saleh, A.A., Islam, M.S. et al. (2016). Biodiesel production from crude Jatropha oil using a highly active heterogeneous nanocatalyst by optimizing transesterification reaction parameters. *Energy Fuels* 30 (1): 334–343.

**14** Alptekin, E. and Canakci, M. (2008). Determination of the density and the viscosities of biodiesel–diesel fuel blends. *Renew. Energy* 33 (12): 2623–2630.

**15** Owolabi, R.U., Adejumo, A.L., and Aderibigbe, A.F. (2012). Biodiesel: fuel for the future (a brief review). *Int. J. Energy Eng.* 2 (5): 223–231.

**16** Ewunie, G.A., Morken, J., Lekang, O.I., and Yigezu, Z.D. (2021). Factors affecting the potential of *Jatropha curcas* for sustainable biodiesel production: a critical review. *Renew. Sustain. Energy Rev.* 137: 110500.

**17** Yusuf, N.N., Kamarudin, S.K., and Yaakob, Z. (2012). Overview on the production of biodiesel from *Jatropha curcas* L. by using heterogenous catalysts. *Biofuels Bioprod. Biorefin.* 6 (3): 319–334.

**18** Zhang, H., Ding, J., Qiu, Y., and Zhao, Z. (2012). Kinetics of esterification of acidified oil with different alcohols by a cation ion-exchange resin/polyether sulfone hybrid catalytic membrane. *Bioresour. Technol.* 112: 28–33.

**19** Odude, V.O., Adesina, A.J., Oyetunde, O.O. et al. (2019). Application of agricultural waste-based catalysts to transesterification of esterified palm kernel oil into biodiesel: a case of banana fruit peel versus cocoa pod husk. *Waste Biomass Valor.* 10 (4): 877–888.

**20** Atadashi, I.M., Aroua, M.K., Aziz, A.A., and Sulaiman, N.M. (2013). The effects of catalysts in biodiesel production: a review. *J. Ind. Eng. Chem.* 19 (1): 14–26.

**21** Talha, N.S. and Sulaiman, S. (2016). Overview of catalysts in biodiesel production. *ARPN J. Eng. Appl. Sci.* 11 (1): 439–448.

**22** Ma, F. and Hanna, M.A. (1999). Biodiesel production: a review. *Bioresour. Technol.* 70: 1–15.

**23** Singh, S.P. and Singh, D. (2010). Biodiesel production through the use of different sources and characterization of oils and their esters as the substitute of diesel: a review. *Renew. Sustain. Energy Rev.* 14: 200–216.

**24** Tariq, M., Ali, S., and Khalid, N. (2012). Activity of homogeneous and heterogeneous catalysts, spectroscopic and chromatographic characterization of biodiesel: a review. *Renew. Sustain. Energy Rev.* 16: 6303–6316.

**25** Kouzu, M., Kasumo, T., Tajika, M. et al. (2008). Calcium oxide as a solid base catalyst for transesterification of soybean oil and its application to biodiesel production. *Fuel* 87: 2798–2806.

**26** Basumatary, S. (2014–2015). Yellow oleander (*Thevetia peruviana*) seed oil biodiesel as an alternative and renewable fuel for diesel engines: a review. *Int. J. ChemTech. Res.* 7 (6): 2823–2840.

**27** Schuchardt, U., Sercheli, R., and Vargas, R.M. (1998). Transesterification of vegetable oils: a review. *J. Braz. Chem. Soc.* 9 (1): 199–210.

**28** Meher, L.C., Sagar, D.V., and Naik, S.N. (2006). Technical aspects of biodiesel production by transesterification – a review. *Renew. Sustain. Energy Rev.* 10: 248–268.

**29** Freedman, B., Pryde, E.H., and Mounts, T.L. (1984). Variables affecting the yields of fatty esters from transesterified vegetable oils. *J. Am. Oil Chem. Soc.* 61: 1638–1643.

**30** Canacki, M. and Gerpen, J.V. (1999). Biodiesel production via acid catalysis. *Trans. ASAE* 42: 1203–1210.

**31** Boro, J., Deka, D., and Thakur, A.J. (2012). A review on solid oxide derived from waste shells as catalyst for biodiesel production. *Renew. Sustain. Energy Rev.* 16: 904–910.

**32** Dube, M.A., Tremblay, A.Y., and Liu, J. (2007). Biodiesel production using a membrane reactor. *Bioresour. Technol.* 98 (3): 639–647.

**33** Marchetti, J.M., Miguel, V.U., and Errazu, A.F. (2007). Possible methods for biodiesel production. *Renew. Sustain. Energy Rev.* 11: 1300–1311.

**34** Leung, D.Y.C., Wu, X., and Leung, M.K.H. (2010). A review on biodiesel production using catalyzed transesterification. *Appl. Energy* 87: 1083–1095.

**35** Jacobson, K., Gopinath, R., Meher, L.C., and Dalai, A.K. (2008). Solid acid catalyzed biodiesel production from waste cooking oil. *Appl. Catal. B-Environ.* 85 (1–2): 86–91.

**36** Changmai, B., Vanlalveni, C., Ingle, A.P. et al. (2020). Widely used catalysts in biodiesel production: a review. *RSC Adv.* 10 (68): 41625–41679.

**37** Javidialesaadi, A. and Raeissi, S. (2013). Biodiesel production from high free fatty acid-content oils: experimental investigation of the pretreatment step. *APCBEE Procedia* 5: 474–478.

**38** Cao, H., Zhang, Z., Wu, X., and Miao, X. (2013). Direct biodiesel production from wet microalgae biomass of *Chlorella pyrenoidosa* through *in situ* transesterification. *Biomed. Res. Int.* 930686: 1–6.

**39** Zhang, J. and Jiang, L. (2008). Acid-catalyzed esterification of *Zanthoxylum bungeanum* seed oil with high free fatty acids for biodiesel production. *Bioresour. Technol.* 99 (18): 8995–8998.

**40** Bhatti, H.N., Hanif, M.A., and Qasim, M. (2008). Biodiesel production from waste tallow. *Fuel* 87 (13–14): 2961–2966.

**41** Veljković, V.B., Lakićević, S.H., Stamenković, O.S. et al. (2006). Biodiesel production from tobacco (*Nicotiana tabacum* L.) seed oil with a high content of free fatty acids. *Fuel* 85 (17–18): 2671–2675.

**42** Wang, Y., Ou, S., Liu, P. et al. (2006). Comparison of two different processes to synthesize biodiesel by waste cooking oil. *J. Mol. Catal. A Chem.* 252 (1–2): 107–112.

**43** Nye, M.J., Williamson, T.W., Deshpande, W. et al. (1983). Conversion of used frying oil to diesel fuel by transesterification: preliminary tests. *J. Am. Oil Chem. Soc.* 60 (8): 1598–1601.

**44** Miao, X., Li, R., and Yao, H. (2009). Effective acid-catalyzed transesterification for biodiesel production. *Energy Convers. Manag.* 50 (10): 2680–2684.

**45** Sagiroglu, A., Isbilir, S.Ş., Ozcan, M.H. et al. (2011). Comparison of biodiesel productivities of different vegetable oils by acidic catalysis. *Chem. Ind. Chem. Eng.* 17 (1): 53–58.

**46** Alamu, O.J., Waheed, M.A., Jekayinfa, S.O., and Akintola, T.A. (2007). Optimal transesterification duration for biodiesel production from Nigerian palm kernel oil. *Agric. Eng. Int.* IX: EE07018.

**47** Karmee, S.K. and Chadha, A. (2005). Preparation of biodiesel from crude oil of *Pongamia pinnata*. *Bioresour. Technol.* 96 (13): 1425–1429.

**48** Rashid, U. and Anwar, F. (2008). Production of biodiesel through optimized alkaline-catalyzed transesterification of rapeseed oil. *Fuel* 87 (3): 265–273.

**49** Refaat, A.A., Attia, N.K., Sibak, H.A. et al. (2008). Production optimization and quality assessment of biodiesel from waste vegetable oil. *Int. J. Environ. Sci. Technol.* 5 (1): 75–82.

**50** Dmytryshyn, S.L., Dalai, A.K., Chaudhari, S.T. et al. (2004). Synthesis andcharacterization of vegetable oil derived esters: evaluation for their diesel properties. *Bioresour. Technol.* 92 (1): 55–64.

51 Yusup, S. and Khan, M.A. (2010). Base catalyzed transesterification of acid treated vegetable oil blend for biodiesel production. *Biomass Bioenergy* 34 (10): 1500–1504.

52 Nakpong, P. and Wootthikanokkhan, S. (2010). Roselle (*Hibiscus sabdariffa* L.) oil as alternative feedstock for biodiesel production in Thailand. *Fuel* 89 (8): 1806–1811.

53 Encinar, J.M., González, J.F., and Rodríguez-Reinares, A. (2007). Ethanolysis of used frying oil. Biodiesel preparation and characterization. *Fuel Process. Technol.* 88 (5): 513–522.

54 Dorado, M.P., Ballesteros, E., Mittelbach, M., and López, F.J. (2004). Kinetic parameters affecting the alkali-catalyzed transesterification process of used olive oil. *Energy Fuels* 18 (5): 1457–1462.

55 Chung, K.H., Kim, J., and Lee, K.Y. (2009). Biodiesel production by transesterification of duck tallow with methanol on alkali catalysts. *Biomass Bioenergy* 33 (1): 155–158.

56 Porwal, J., Bangwal, D., Garg, M.O., and Kaul, S. (2012). Reactive-extraction of pongamia seeds for biodiesel production. *J. Sci. Ind. Res.* 72: 822–828.

57 Kartika, I.A., Yani, M., Ariono, D. et al. (2013). Biodiesel production from Jatropha seeds: solvent extraction and in situ transesterification in a single step. *Fuel* 106: 111–117.

58 Silitonga, A.S., Ong, H.C., Mahlia, T.M. et al. (2014). Biodiesel conversion from high FFA crude *Jatropha curcas*, *Calophyllum inophyllum* and *Ceiba pentandra* oil. *Energy Procedia* 61: 480–483.

59 Mohadesi, M., Aghel, B., Maleki, M., and Ansari, A. (2019). Production of biodiesel from waste cooking oil using a homogeneous catalyst: study of semi-industrial pilot of microreactor. *Renew. Energy* 136: 677–682.

60 Dias, J.M., Alvim-Ferraz, M.C., and Almeida, M.F. (2008). Comparison of the performance of different homogeneous alkali catalysts during transesterification of waste and virgin oils and evaluation of biodiesel quality. *Fuel* 87 (17–18): 3572–3578.

61 Chen, K.S., Lin, Y.C., Hsu, K.H., and Wang, H.K. (2012). Improving biodiesel yields from waste cooking oil by using sodium methoxide and a microwave heating system. *Energy* 38 (1): 151–156.

62 Rashid, U., Anwar, F., Ansari, T.M. et al. (2009). Optimization of alkaline transesterification of rice bran oil for biodiesel production using response surface methodology. *J. Chem. Technol. Biotechnol.* 84 (9): 1364–1370.

63 Dawodu, F.A., Ayodele, O.O., and Bolanle-Ojo, T. (2014). Biodiesel production from *Sesamum indicum* L. seed oil: an optimization study. *Egypt. J. Pet.* 23 (2): 191–199.

64 Keera, S.T., El Sabagh, S.M., and Taman, A.R. (2011). Transesterification of vegetable oil to biodiesel fuel using alkaline catalyst. *Fuel* 90 (1): 42–47.

65 Meng, X., Chen, G., and Wang, Y. (2008). Biodiesel production from waste cooking oil via alkali catalyst and its engine test. *Fuel Process. Technol.* 89 (9): 851–857.

66 Leung, D.Y. and Guo, Y. (2006). Transesterification of neat and used frying oil: optimization for biodiesel production. *Fuel Process. Technol.* 87 (10): 883–890.

67 Uzun, B.B., Kılıç, M., Özbay, N. et al. (2012). Biodiesel production from waste frying oils: optimization of reaction parameters and determination of fuel properties. *Energy* 44 (1): 347–351.

68 Rashid, U., Anwar, F., Moser, B.R., and Ashraf, S. (2008). Production of sunflower oil methyl esters by optimized alkali-catalyzed methanolysis. *Biomass Bioenergy* 32 (12): 1202–1205.

**69** Ilham, Z. (2009). Analysis of parameters for fatty acid methyl esters production from refined palm oil for use as biodiesel in the single-and two-stage processes. *Malaysian J. Biochem. Mol. Biol.* 17 (1): 5–9.

**70** Felizardo, P., Correia, M.J., Raposo, I. et al. (2006). Production of biodiesel from waste frying oils. *Waste Manage.* 26 (5): 487–494.

**71** Zakaria, R. and Harvey, A.P. (2012). Direct production of biodiesel from rapeseed by reactive extraction/in situ transesterification. *Fuel Process. Technol.* 102: 53–60.

**72** Silva, G.F., Camargo, F.L., and Ferreira, A.L. (2011). Application of response surface methodology for optimization of biodiesel production by transesterification of soybean oil with ethanol. *Fuel Process. Technol.* 92 (3): 407–413.

**73** Ghadge, S.V. and Raheman, H. (2005). Biodiesel production from mahua (*Madhuca indica*) oil having high free fatty acids. *Biomass Bioenergy* 28 (6): 601–605.

**74** Ramadhas, A.S., Jayaraj, S., and Muraleedharan, C. (2005). Biodiesel production from high FFA rubber seed oil. *Fuel* 84 (4): 335–340.

**75** Usta, N. (2005). Use of tobacco seed oil methyl ester in a turbocharged indirect injection diesel engine. *Biomass Bioenergy* 28 (1): 77–86.

# 6

# Application of Metal Oxides Catalyst in Production of Biodiesel

*Hui Li*

*School of Thermal Engineering, Shandong Jianzhu University, Jinan, PR China*

Biodiesel, a biomass-derived fuel, is primarily composed of fatty acid methyl ester (FAME), which is normally synthesized by transesterification of triglyceride with methanol [1]. As illustrated in Figure 6.1, catalyst is distinctly important to drive the transesterification forward to biodiesel production, where the metal oxides, zeolites, ion exchange resin, etc. are employed. Thereinto, metal oxides are gaining extensive attention for the favorable catalytic activity, simple preparation, and wide sources. Metal oxide is composed of oxygen and metal element ($M_xO_y$), such as CaO, $Fe_2O_3$, ZnO, $ZrO_2$, and so on. As far as active site is concerned, metal oxide can be divided into two types of basic and acid.

## 6.1 Basic Metal Oxide

In general, basic metal oxides could be divided into monometal oxide, multimetal oxide, and active site doped metal oxide.

### 6.1.1 Monobasic Metal Oxide

#### 6.1.1.1 Alkaline Earth Metal Oxide

Alkaline earth metals refer to the Group II of the periodic table of elements, including magnesium (Mg), calcium (Ca), strontium (Sr), and barium (Ba). The basic center is linked to the presence of $M^{2+}$ and $O^{2-}$ ion pair (M is alkaline earth metal) and surface hydroxyl group in a different coordination environment [2]. Alkaline earth metals oxides are verified to be a kind of efficient basic metal oxide for transesterification [3]. The order of catalytic activity among alkaline earth oxide is BaO>SrO>CaO>MgO, which is in accordance with the metallicity sequence [4].

Margellou et al. [5] synthesized MgO in the presence of polyvinyl alcohol (PVA), and the specific surface area was increased from $18\,m^2\,g^{-1}$ to $26\,m^2\,g^{-1}$ in comparison with that of MgO prepared without PVA. As a result, the biodiesel conversion of 98.4% was achieved under the

*Biodiesel Production: Feedstocks, Catalysts, and Technologies*, First Edition. Edited by Samuel Lalthazuala Rokhum, Gopinath Halder, Suttichai Assabumrungrat, and Kanokwan Ngaosuwan.

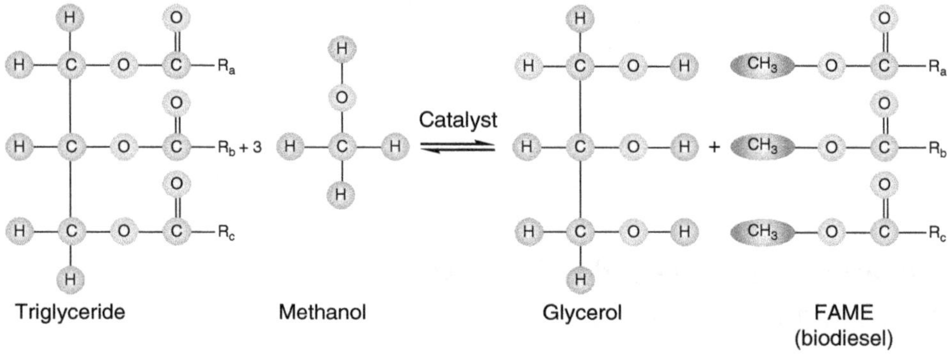


**Figure 6.1** Transesterification process for biodiesel production. *Source:* Based on Ref. [1].

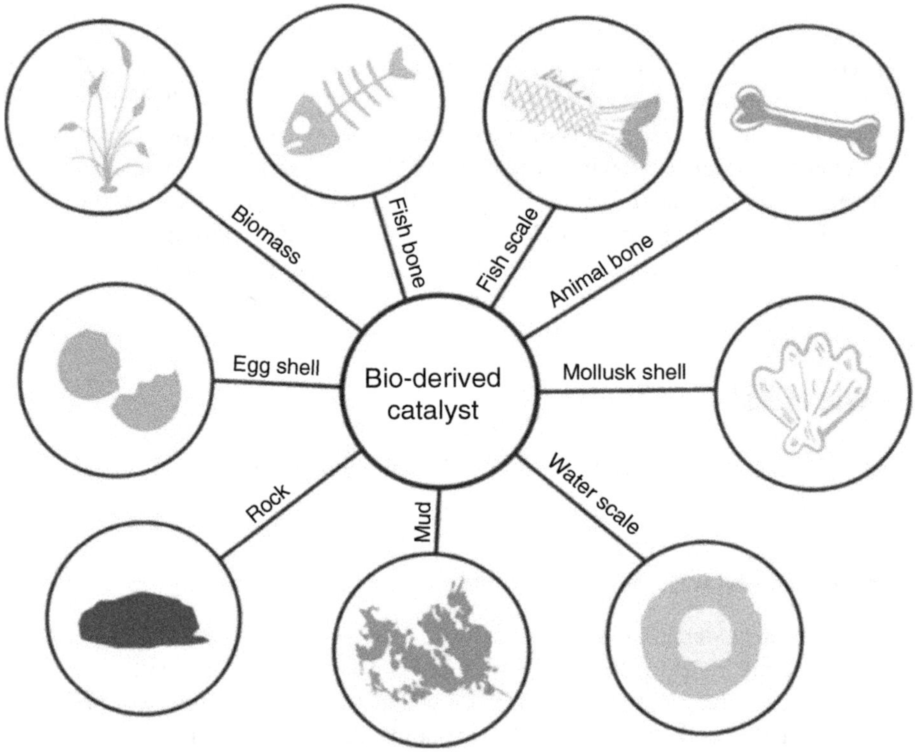


**Figure 6.2** Potential materials for alkaline earth metal oxide synthesis. *Source:* Ref. [7].

optimum conditions. Du et al. [6] prepared MgO by urea combustion and used it to catalyze the transesterification of castor oil and ethanol for biodiesel production. The catalytic activity showed that the biodiesel conversion only decreased from 96.5 to 80.2% after five cycles with catalyst amount of 6 wt% and molar ratio of methanol/oil of 12 : 1 at 75 °C within 1 h.

Calcium is largely abundant in nature, which exists in natural materials such as limestone, mollusk, eggs shells, and waste animal bone as catalyst, as described in Figure 6.2 and Table 6.1. Calcium oxide (CaO) is a promising basic catalyst by the reason of its

**Table 6.1** Catalytic performance of CaO.

| Resource | Catalyst | Reactants | Catalyst amount (wt%) | Molar ratio of methanol to oil | Reaction temperature (°C) | Reaction time (h) | Conversion (%) |
|---|---|---|---|---|---|---|---|
| Mollusk shells | CaO | Cotton seed oil/ methanol | 5 | 12 : 1 | 65 | 4.5 | 95 [8] |
| Waste animal bones | CaO | Jatropha oil/ methanol | 6 | 9 : 1 | 70±3 | 3 | 96.1 [9] |
| Chicken and fish bones | CaO | WCO/ methanol | 1.98 | 10 : 1 | 65 | 1.54 | 89.5 [10] |
| Sea sand | CaO | Cooking oil/ methanol | 7.5 | 12 : 1 | 60 | 6 | 95.4 [11] |

cost-effectiveness, high efficiency, superior basicity, and easy availability [12, 13]. Boonyuen et al. [14] prepared CaO by calcining turbo jourdani shells and used it to produce biodiesel via transesterification of palm oil. The conversion of 99.33% was obtained with catalyst amount of 10 wt% and molar ratio of methanol/oil of 3 : 1 at 80 °C for 7 h. Farooq et al. [15] synthesized CaO by calcining egg shell and treated it via hydration. It was found that the surface area of CaO increased significantly from 1.342 $m^2 g^{-1}$ to 66.7 $m^2 g^{-1}$ after hydration, and the conversion was 93.5% with catalyst amount of 5 wt% and molar ratio of methanol/oil of 12 : 1 at 65 °C just within 1.5 h.

By contrast to CaO and MgO, SrO with strong basic sites is insoluble in methanol. Liu et al. [4] reported that the transesterification of soybean oil and ethanol was catalyzed by SrO, and the conversion of 96% was obtained with catalyst amount of 3 wt% and molar ratio of methanol/oil of 6 : 1 at 65 °C within 0.5 h. In addition, the conversion only declined by 2% even after 10 cycles. Compared with CaO, MgO, and SrO, BaO is rarely used in biodiesel production because of the toxicity of barium.

#### 6.1.1.2 Transition Metal Oxide

Zinc oxide (ZnO) is a transition metal oxide with high chemical and mechanical stability [16, 17]. Justine et al. [16] synthesized ZnO nanocomposite and used it to produce biodiesel via transesterification of waste cooking oil (WCO). The conversion of 81.57% was obtained with catalyst amount of 1 wt% and molar ratio of methanol/oil of 6 : 1 at 110 °C within 2 h. A study was conducted by Zhou et al. [18] to investigate the activity of nano $La_2O_3$ prepared from sonochemical methods. Results revealed that the activity of nano $La_2O_3$ was significantly higher than that of conventional $La_2O_3$ on account of the high basic strength, small particle size, and large specific surface area. Under the optimal conditions of 180 °C, catalyst amount of 10 wt%, molar ratio of methanol/oil of 28 : 1 and 2 h, the biodiesel conversion was 90.3%.

### 6.1.2 Multibasic Metal Oxide

Transesterification is occurred on the catalyst surface, where the available active site would trigger the reaction between feedstock oil and alcohol. However, monometal oxide suffers from relatively low specific surface area and small pore size, resulting in poor catalytic

**Table 6.2** Catalytic performance of multimetal oxides.

| Catalyst | Reactants | Catalyst amount (wt%) | Molar ratio of methanol/oil | Reaction temperature (°C) | Reaction time (h) | Conversion (%) |
|---|---|---|---|---|---|---|
| Zn–CaO | Eucalyptus oil/ methanol | 5 | 6 : 1 | 65 | 1.5 | 93.2 [19] |
| Cu–CaO | Eucalyptus oil/ methanol | 5 | 6 : 1 | 65 | 2.5 | 90.6 [19] |
| $CaO–SiO_2$ | WCO/ methanol | 6 | 15 : 1 | 60 | 2 | 94 [20] |
| MgO/ZSM-5 | *Spirulina platensis*/ methanol | 3 | 15 : 1 | 75 | 1 | 92.1 [21] |
| $CaO/ZrO_2$ | Rapeseed oil/ methanol | 8 | 72 : 1 | 120 | 6 | 92.60 [22] |
| $MgO@Na_2O$ | Chicken fat/ methanol | 2 | 24 : 1 | 65 | 2 | 95.22 [23] |
| $CaO–MgO/Al_2O_3$ | Cottonseed oil/ methanol | 12.5 | 8.5 : 1 | 95 | 3 | 97.62 [24] |
| CaO@MgO | Waste edible oil /methanol | 4.571 | 16.7 : 1 | 69.37 | 7.08 | 98.37 [25] |
| $La_2O_3–CaO$ | Palm oil/ methanol | 7 | 18 : 1 | 150 | 2 | 97.5 [26] |

activity and weak reusability. To address these defects, by loading active basic metal oxide on porous support like $Al_2O_3$, $SiO_2$, and $ZrO_2$, activated carbon and metal organic framework (MOF) is a feasible route to prepare multibasic metal oxide (Table 6.2).

#### 6.1.2.1 Supported on Metal Oxide

Marinković et al. [27] synthesized spherically shaped $CaO/\gamma\text{-}Al_2O_3$ catalyst; the maximum biodiesel conversion of 94.3% was obtained with the following reaction conditions: catalyst amount of 0.5 wt% and molar ratio of methanol/oil of 12 : 1 at 60 °C within 5 h. In comparison to the powdery $CaO/Al_2O_3$, spherical-shaped $CaO/\gamma\text{-}Al_2O_3$ possessed higher specific surface area and catalytic activity. Wu et al. [28] prepared CaO/NaZSM-5 by a microwave irradiation and used it directly for transesterification of soybean oil with methanol. Biodiesel conversion of 92.1% was achieved with catalyst amount of 3 wt% and molar ratio of methanol/oil of 9 : 1 at 65 °C within 3 h. Besides, basic catalyst of $ZrO_2/CaO$ was successfully synthesized through hydrothermal method by Sakti La Ore et al. [29], in which the stable $CaZrO_3$ was generated, resulting in biodiesel conversion of 69.52%.

#### 6.1.2.2 Supported on Activated Carbon

Activated carbon is regarded as an ideal support in light of its low cost, high surface area, and porous structure. Faria et al. [30] used activated carbon to support Ca, and the specific surface area of the obtained catalyst was 869 $m^2\ g^{-1}$, which was much larger than that of commercial CaO. Tabah et al. [31] loaded SrO on activated carbon by using sonochemically deposition; the biodiesel conversion was 98.5% when the catalyst amount was 7.1 wt%, the molar ratio of methanol/oil was 6 : 1, and the reaction time was 1 h at 46 °C. Moreover, the conversion was only reduced by 3% after four cycles, thus confirming the high stability of the catalyst.

#### 6.1.2.3 Supported on Metal Organic Framework

MOF is a novel class of porous material assembled with metal ions and organic linker, which has the outstanding characteristics of large specific surface area, adjustable porosity, orderly pore structure, and diverse topology structure [32, 33]. By means of such merits, Li group [32] supported calcium acetate on MIL-100(Fe) to prepare the highly effective basic metal oxide catalyst. The active site of $Ca_2Fe_2O_5$ and $CaFe_3O_5$ was obtained by calcination in inert atmosphere, in which the active site of CaO was stabilized. In comparison to commercial CaO (10.09 $m^2\ g^{-1}$), the specific surface area of this catalyst was high as 90.01 $m^2\ g^{-1}$. The palm oil conversion of 95.09% was obtained with catalyst amount of 4 wt% and methanol to palm oil of 9 : 1 at 65 °C for 2 h. In addition, MOF could be first carbonized and used to support the active site. Li group [34] further carbonized MIL-100(Fe) and used it to support SrO. The attained catalyst (Fe@C/Sr) could achieve the biodiesel conversion of 98.12% with catalyst amount of 4 wt% and molar ratio of methanol/oil of nine at 65 °C just within 30 min. This is almost equal to the catalytic activity of homogeneous basic catalyst. Most importantly, the two catalysts synthesized from MIL-100(Fe) have strong magnetic properties that can be easily separated from the reaction system by magnet.

### 6.1.3 Active Site-Doped Basic Metal Oxide

In order to improve the catalytic activity of monometal oxide, doping active site is an ideal strategy, where the alkali metal and active metal oxide are adopted.

#### 6.1.3.1 Alkali Metal Doped

Alkali metals like Li, Na, and K possess strong metallic property, and the corresponding alkali metal salt is commonly employed to modify the metal oxide. Khatibi et al. [35] doped Na/K on CaO by impregnation method. The Na–K/CaO catalyst showed that the highest biodiesel conversion of 97.6% was achieved with only catalyst amount of 3 wt% and molar ratio of methanol/oil of 9 : 1 at 50 °C within 3 h. Malhotra et al. [36] prepared Na/ZnO–SBA15 by impregnation and used it directly for transesterification of cotton seed oil with methanol. Biodiesel conversion merely decreased from 98 to 74% after five cycles with catalyst amount of 12 wt% and molar ratio of methanol/oil of 24 : 1 at 65 °C within 4 h.

#### 6.1.3.2 Active Metal Oxide Doped

Abukhadra et al. [37] prepared MgO/CaO nanorod catalysts, which was recognized with specific surface area of 112.8 $m^2\ g^{-1}$ and used to catalyze transesterification of castor oil and ethanol. It was found that MgO/CaO possessed higher specific surface area and stability than that of CaO, where the conversion only decreased from 98.8 to 91.2% after

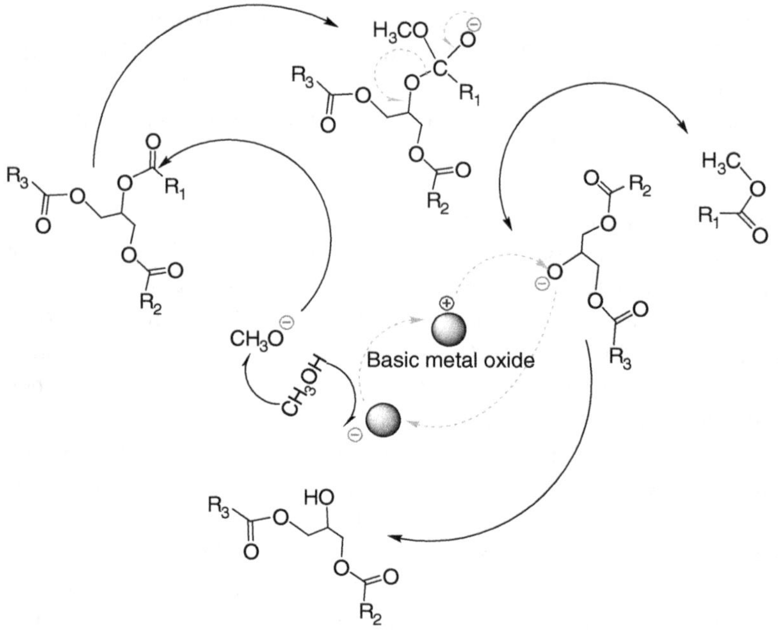


**Figure 6.3** Transesterification mechanism catalyzed by basic metal oxide. *Source:* Based on Ref. [1].

five cycles. Sudsakorn et al. [38] obtained a catalyst with large specific surface area of 24.67 $m^2\ g^{-1}$ by doping Sr into CaO/MgO, and the conversion remained above 90% after repeated four cycles.

### 6.1.4 Mechanism of Transesterification Catalyzed by Basic Metal Oxide

As depicted in Figure 6.3, transesterification begins with sufficient adsorption of methanol on active sites, and then a proton ($H^+$) would be extracted from methanol by using basic metal oxide, which is transformed into methoxy ions ($CH_3O^-$). $CH_3O^-$ will attack the $sp^2$ carbonyl carbon with formation of tetrahedral intermediate. Owing to instability of the intermediate, it would be rearranged to generate ester and diglyceride. With that, diglyceride is proceeded with the previously mentioned process, and the formed monoglyceride would repeat the process until the glycerol is finally obtained [1, 3].

## 6.2 Acid Metal Oxide

Although basic metal oxide could achieve the high catalytic activity with mild condition, it is easily disabled as it confronts with water and free fatty acid [39, 40]. In comparison, acid metal oxide is insensitive to free fatty acids and water, which could realize the esterification of free fatty acids and the transesterification of triglycerides with alcohols at the same time [41]. Even better, acid metal oxide can avoid equipment corrosion caused by homogeneous acid [42]. Acid metal oxide contains two acid site type, namely, Brønsted and Lewis. Brønsted acid site is a highly polarized hydroxyl group used as $H^+$–donor, while the Lewis acid site is a coordination unsaturated cation center, which makes exposed metal ion interact with guest molecules and acts as acceptor of electron pairs [43, 44].

### 6.2.1 Monoacid Metal Oxide

Monometal oxide has been widely studied in biodiesel production via catalyzing esterification/transesterification. Yoo et al. [45] reported that the surface area of $ZrO_2$ was only $4.3\,m^2\,g^{-1}$, which has failed to give full catalytic effect due to the low surface area. As a result, the rapeseed oil conversion was only 68% with catalyst amount of 1 wt% and molar ratio of methanol/oil of 40 : 1 at 270 °C within 0.17 h. It is reported that the acidic sites and acidity are directly proportional to the catalytic activity [46]. Hino et al. [47] proved the acidity and activity of $Fe_2O_3$ could be remarkably enhanced by sulfate ion. Later, they paid close attention to the action of such ions and successfully developed sulfated metal oxides, although the immersion of sulfate groups led to reducing surface areas. The pore size and pore volume of the catalyst were enlarged [48]. This exerted positive effect on biodiesel production, because catalyst with larger pore reduces the diffusion limit of molecules, especially for transesterification and esterification. Jitputti et al. [49] utilized impregnation method to load sulfuric acid groups on zirconia. With the introduction of sulfate, the agglomeration of metal oxide crystals during calcination was avoided, and more tetragonal zirconia were produced; the conversion of palm kernel oil catalyzed by the catalyst was 90.3% with catalyst amount of 10 wt% and molar ratio of methanol/oil of 6 : 1 at 240 °C within 4 h [50].

In theory, the active sites could be evenly distributed in virtue of the porous support. For this, Chen et al. [51] successfully supported the sulfated zirconia on mesoporous silica SBA-15 by direct calcination at high temperature under strong acidic circumstance. Catalyst characterizations showed that the high surface area of SBA-15 provided better dispersion of $SO_4^{2-}/ZrO_2$, thus offering more available acid sites. The conversion of palmitic acid was 89.2% with catalyst amount of 20% and the molar ratio of methanol /acid of 10 at 68 °C within 6 h. Senson et al. [52] loaded tungsten oxide on zirconia in inert atmosphere. The specific surface area and acid density of zirconia was increased to $90.2\,m^2\,g^{-1}$, and the acid density was $200\,\mu mol\,g^{-1}$, respectively. As expected, it exhibited strong catalytic activity in transesterification.

### 6.2.2 Multiacid Metal Oxide

Monometal oxide is always limited by the structure properties such as low specific surface area and small pore size [53, 54]. Researches illustrated that the multiacid metal oxide would overcome such defects. Meanwhile, the synergistic effect among different metal oxides would stabilize the active sites, thus promoting the catalytic activity (Table 6.3). Lam et al. [48] synthesized $SO_4^{2-}/SnO_2$ solid acid and used it to catalyze the transesterification of waste oil and methanol for biodiesel production. With introducing $Al_2O_3$ and the molar ratio of $SnO_2$ to $Al_2O_3$ was set as 3, the maximum of surface acidic site was thus achieved. It exhibited an exceptional high activity with an optimum conversion of 92.3%

**Table 6.3** Catalytic performance of metal oxides.

| Catalyst | Reactants | Catalyst amount (wt%) | Molar ratio of methanol/ oil(acid) | Reaction temperature (°C) | Reaction time (h) | Conversion (%) |
|---|---|---|---|---|---|---|
| $ZrO_2$ | Rapeseed oil/ methanol | 1 | 40 | 270 | 10 min | 68.0 [45] |
| $SO_4^{2-}/ZrO_2$ | Crude coconut oil/ methanol | 3 | 6 | 200 | 4 | 86.3 [49] |
| $SO_4^{2-}/SnO_2$ | Crude coconut oil/ methanol | 3 | 6 | 200 | 4 | 80.6 [49] |
| $SO_4^{2-}/ZrO_2/$ SBA–15 | Palmitic acid/ methanol | 20 | 80 | 68 | 6 | 89.2 [51] |
| $SO_4^{2-}$/Fe–Al–$TiO_2$ | Waste cooking oil/ methanol | 3 | 10 | 90 | 2.5 | 95.6 [57] |
| ZnO–$TiO_2$–ICG | Palm oil/methanol | 1 | 6 | 100 | 1.25 | 96.1 [59] |
| UiO–66/SFN | Oleic acid/ methanol | 8 | 8 | 70 | 2 | 96.2 [60] |

with catalyst amount of 3 wt% and molar ratio of methanol/oil of 15 at 150 °C within 3 h. As a result, the bimetallic effect of catalyst significantly reduced the catalyst amount and the molar ratio of alcohol to realize the same biodiesel conversion.

Gardy et al. [58] prepared $Ti(SO_4)O$, and only 73% biodiesel conversion was obtained after being reused for 10 times. Gardy et al. [57] further synthesized the magnetic $SO_4^{2-}$/Fe–Al–$TiO_2$ solid acid by doping alumina and iron oxide on $TiO_2$ (Figure 6.4). The catalyst was used to catalyze the transesterification/esterification of methanol and WCO. The $SO_4^{2-}$/Fe–Al–$TiO_2$ catalyst achieved 95.6% biodiesel conversion with catalyst amount of 3 wt% and molar ratio of methanol/WCO of 10 : 1 at 90 °C within 2.5 h. After being repeated for 10 times, the biodiesel conversion was still above 90%. According to catalyst characterizations, aluminum sulfate was formed in the process of alumina sulfonation, which provided catalytic active sites and improved catalytic activity. The addition of iron oxide made the catalyst magnetic; therefore, it can be easily removed from the products with external magnetic field.

In heterogeneously catalyzed system, the reaction occurs on catalyst surface; therefore, the high specific surface area is preferred for more active sites are available for reactants. For this, Soltani et al. [59] incompletely carbonized glucose (ICG) and used it as support for synthesizing mesoporous catalyst of $SO_3H$@ZnO–$TiO_2$–ICG (Figure 6.5). Then it was used to catalyze the transesterification of waste edible palm oil and methanol. The specific surface area of $SO_3H$@ZnO–$TiO_2$–ICG was up to 324.5 $m^2\ g^{-1}$, which enhanced the effective contact between the active site and the reactants, thus improving catalytic activity. The $SO_3H$@ZnO–$TiO_2$–ICG catalyst showed the highest biodiesel conversion of 96.1% with

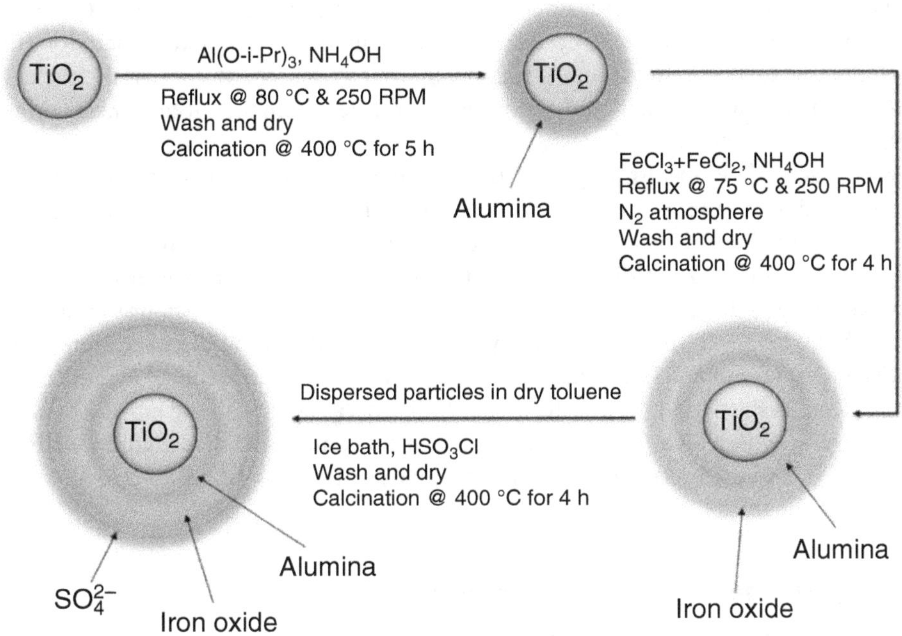


**Figure 6.4** Schematic of $SO_4^{2-}$/Fe–Al–$TiO_2$ catalyst synthesis. *Source:* Ref. [57].

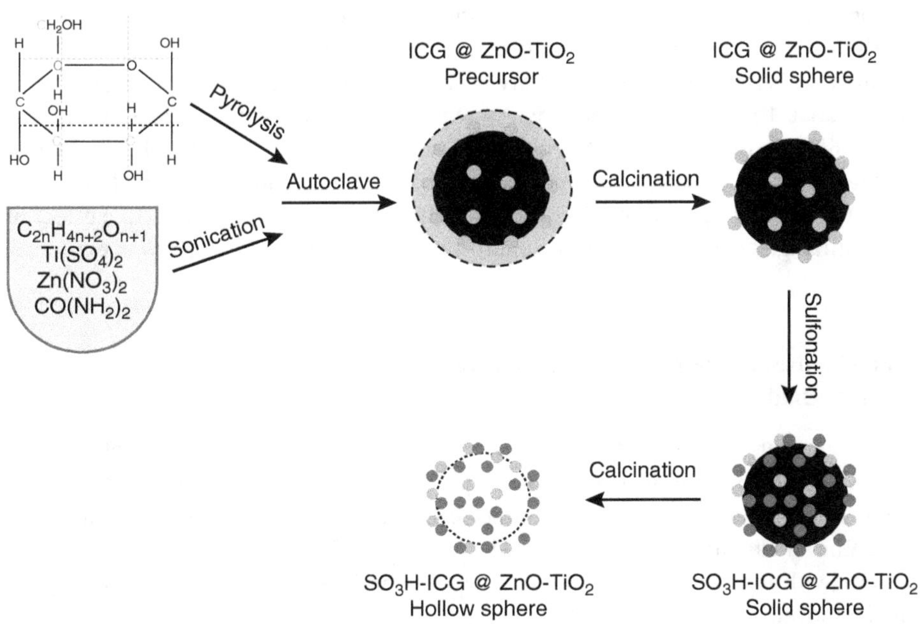


**Figure 6.5** Preparation of $SO_3H$@ZnO–$TiO_2$–ICG. *Source:* Ref. [59].

only catalyst amount of 1 wt% and molar ratio of methanol/oil of 6 : 1 at 100 °C within 1.25 h. Even after being reused in 10 cycles, it still could maintain the catalytic activity. The high stability was related to the special structure, and the active site was not easily leached, which is profited from the good connection between acidic component and mesoporous support.

Heteropoly acids (HPAs) with Keggin structure are widely used as acid catalysts for the strong Brønsted acidity [55]. The activity of per mole HPAs is about 3–100 times higher than that of sulfuric acid. However, the main disadvantages of HPAs are the relatively low surface area (1–10 $m^2 g^{-1}$) and difficult separation from the reaction mixture. To end this, HPAs was uniformly dispersed on porous bentonite (AT–GMB) [56], which was composed of various metal oxides. The AT–GMB catalyst showed the highest biodiesel conversion of 88% with catalyst amount of 3 wt% and molar ratio of methanol/acetic acid of 3 : 1 at 150 °C within 3 h. Results showed that the enhanced activity was attributed to the high dispersion of HPAs on AT–GMB, which provided more surface area and active sites than pure HPAs. Nevertheless, due to the loss of active sites, the activity of the catalyst dropped to 45% as it was reused for the third time.

### 6.2.3 Supported on Metal Organic Framework

Except for the traditional support, MOF is an ideal support for synthesizing acid metal oxide. Li group [60] adopted UiO–66(Zr) and ammonium sulfate to prepare the efficient and stable acid catalyst for biodiesel production. It was found that acid amount of the catalyst calcined under nitrogen (UiO–66/SFN) was higher than that of calcined under air (UiO–66/SFA). The UiO–66/SFN catalyst showed the highest biodiesel conversion of 96.2% with catalyst amount of 8 wt% and molar ratio of methanol/oleic acid of 8 : 1 at 70 °C within 2 h. With that, they creatively employed the second calcination (UiO–66/SSN) to enhance the catalytic stability. As expected, the conversion decrement was reduced by 66.25% of UiO–66/SSN compared with that of UiO–66/SFN within five cycles. Catalyst characterization revealed that either Lewis acid or Brønsted acid was increased after second calcination. This was profited from the strong inductive effect of the sulfate during second calcination, where the electron density of Zr was evidently reduced.

### 6.2.4 Mechanism of Transesterification/Esterification Catalyzed by Acid Metal Oxide

The mechanism of transesterification catalyzed by acid metal oxide is elucidated in Figure 6.6. The carboxyl in triglyceride would be protonated with the assistance of acid metal oxide. Thereafter, the nucleophilic reaction is proceeded with methanol, thus generating the tetrahedral intermediate. The proton is finally broken away from the intermediate with formation of FAME. This process will be repeated three times till the glycerol is removed.

In esterification, the carbonyl in carboxylic acid is initially protonated with acid metal oxide. With more positive charge of carbonyl carbon, the nucleophilic reaction of alcohol

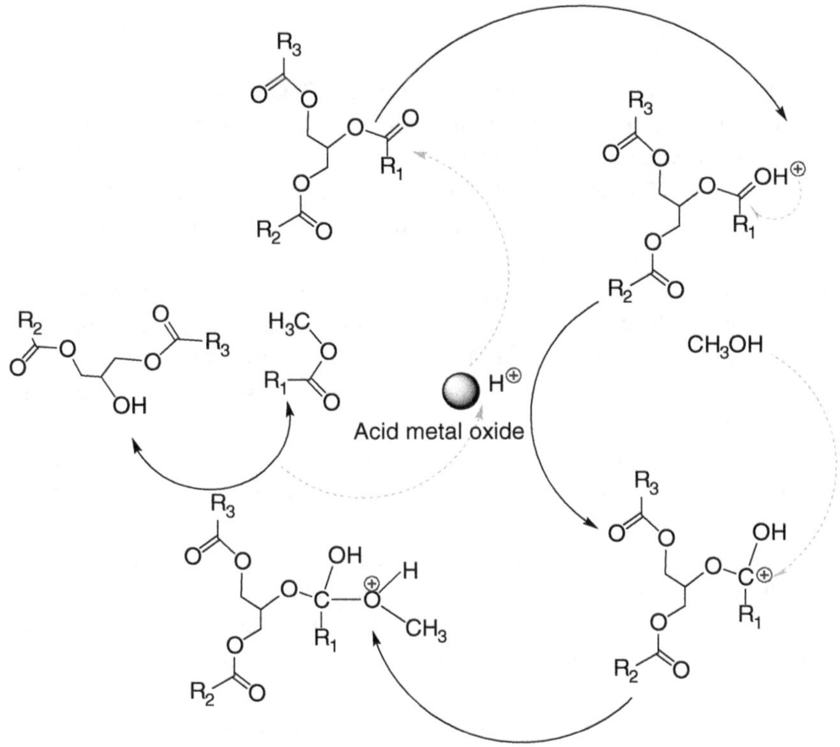


**Figure 6.6** Transesterification mechanism catalyzed by acid catalyst. *Source:* Based on Ref. [1].

is more easily conducted with intermediates generation [1]. Then the proton is transferred with $H_2O$ removal; after that, the proton is finally eliminated with ester formation [61]. This process is similar to that of transesterification catalyzed by acid metal oxide as exhibited in Figure 6.6.

## 6.3 Deactivation of Metal Oxide

Compared with homogeneous catalyst, reusability is the unique advantage of solid catalyst, yet deactivation is still a great challenge during recycle process for metal oxide catalysts. The main reasons of deactivation are aggregation of catalyst particles [18], leaching of active sites [38], and pore blockage [32]. Especially for basic metal oxide, water and $CO_2$ are also able to toxify the active site [62, 63]. To end this, researchers made great efforts, such as loading acidic groups to complex metal oxides to inhibit the leaching of active sites [48, 60], cleaning up the blockage of catalyst channel [56], and so on. Furthermore, studies elucidated that the Lewis acid could improve the stability of solid acid, while Brønsted acid can improve the activity of solid acid [43, 44]. Hence, a balance should be maintained between ensuring the coexistence of activity and stability.

## References


1 Ma, X., Liu, F., Helian, Y. et al. (2021). Current application of MOFs based heterogeneous catalysts in catalyzing transesterification/esterification for biodiesel production: a review. *Energy Conversion and Management* 229: 113760.

2 Mansir, N., Teo, S.H., Rashid, U. et al. (2018). Modified waste egg shell derived bifunctional catalyst for biodiesel production from high FFA waste cooking oil. A review. *Renewable and Sustainable Energy Reviews* 82: 3645–3655.

3 Sahani, S., Roy, T., and Sharma, Y.C. (2020). Smart waste management of waste cooking oil for large scale high quality biodiesel production using Sr–Ti mixed metal oxide as solid catalyst: optimization and E–metrics studies. *Waste Management* 108: 189–201.

4 Liu, X., He, H., Wang, Y., and Zhu, S. (2007). Transesterification of soybean oil to biodiesel using SrO as a solid base catalyst. *Catalysis Communications* 8 (7): 1107–1111.

5 Margellou, A., Koutsouki, A., Petrakis, D. et al. (2018). Enhanced production of biodiesel over MgO catalysts synthesized in the presence of Poly–Vinyl–Alcohol (PVA). *Industrial Crops and Products* 114: 146–153.

6 Du, L., Li, Z., Ding, S. et al. (2019). Synthesis and characterization of carbon–based MgO catalysts for biodiesel production from castor oil. *Fuel* 258: 116122.

7 Alagumalai, A., Mahian, O., Hollmann, F., and Zhang, W. (2021). Environmentally benign solid catalysts for sustainable biodiesel production: a critical review. *Science of The Total Environment* 768: 144856.

8 Thakur, S., Singh, S., and Pal, B. (2021). Superior adsorption removal of dye and high catalytic activity for transesterification reaction displayed by crystalline CaO nanocubes extracted from mollusc shells. *Fuel Processing Technology* 213: 106707.

9 Nisar, J., Razaq, R., Farooq, M. et al. (2017). Enhanced biodiesel production from Jatropha oil using calcined waste animal bones as catalyst. *Renewable Energy* 101: 111–119.

10 Tan, Y.H., Abdullah, M.O., Kansedo, J. et al. (2019). Biodiesel production from used cooking oil using green solid catalyst derived from calcined fusion waste chicken and fish bones. *Renewable Energy* 139: 696–706.

11 Muciño, G.G., Romero, R., Ramírez, A. et al. (2014). Biodiesel production from used cooking oil and sea sand as heterogeneous catalyst. *Fuel* 138: 143–148.

12 Banković–Ilić, I.B., Miladinović, M.R., Stamenković, O.S., and Veljković, V.B. (2017). Application of nano CaO–based catalysts in biodiesel synthesis. *Renewable and Sustainable Energy Reviews* 72: 746–760.

13 Huang, J., Zou, Y., Yaseen, M. et al. (2021). Fabrication of hollow cage–like CaO catalyst for the enhanced biodiesel production via transesterification of soybean oil and methanol. *Fuel* 290: 119799.

14 Boonyuen, S., Smith, S.M., Malaithong, M. et al. (2018). Biodiesel production by a renewable catalyst from calcined *Turbo jourdani* (Gastropoda: Turbinidae) shells. *Journal of Cleaner Production* 177: 925–929.

15 Farooq, M., Ramli, A., Naeem, A. et al. (2018). Biodiesel production from date seed oil (*Phoenix dactylifera* L.) via egg shell derived heterogeneous catalyst. *Chemical Engineering Research and Design* 132: 644–651.

16 Justine, M., Joy Prabu, H., Johnson, I. et al. (2021). Synthesis and characterizations studies of ZnO and ZnO–$SiO_2$ nanocomposite for biodiesel applications. *Materials Today: Proceedings* 36: 440–446.

17 Huang, Y.Q., Meidong, L., Yike, Z. et al. (2001). Preparation and properties of ZnO–based ceramic films for low–voltage varistors by novel sol–gel process. *Materials Science and Engineering: B* 86 (3): 232–236.

18 Zhou, Q., Zhang, H., Chang, F. et al. (2015). Nano $La_2O_3$ as a heterogeneous catalyst for biodiesel synthesis by transesterification of *Jatropha curcas* L. oil. *Journal of Industrial and Engineering Chemistry* 31: 385–392.

19 Rahman, W.U., Fatima, A., Anwer, A.H. et al. (2019). Biodiesel synthesis from eucalyptus oil by utilizing waste egg shell derived calcium based metal oxide catalyst. *Process Safety and Environmental Protection* 122: 313–319.

20 Lani, N.S., Ngadi, N., and Inuwa, I.M. (2020). New route for the synthesis of silica–supported calcium oxide catalyst in biodiesel production. *Renewable Energy* 156: 1266–1277.

21 Qu, S., Chen, C., Guo, M. et al. (2020). Synthesis of MgO/ZSM–5 catalyst and optimization of process parameters for clean production of biodiesel from *Spirulina platensis*. *Journal of Cleaner Production* 276: 123382.

22 Liu, L., Wen, Z., and Cui, G. (2015). Preparation of Ca/Zr mixed oxide catalysts through a birch–templating route for the synthesis of biodiesel via transesterification. *Fuel* 158: 176–182.

23 Bahador, F., Foroutan, R., Nourafkan, E. et al. (2021). Enhancement of biodiesel production from chicken fat using MgO and MgO@$Na_2O$ nanocatalysts. *Chemical Engineering & Technology* 44 (1): 77–84.

24 Mahdavi, V. and Monajemi, A. (2014). Optimization of operational conditions for biodiesel production from cottonseed oil on CaO–MgO/$Al_2O_3$ solid base catalysts. *Journal of the Taiwan Institute of Chemical Engineers* 45 (5): 2286–2292.

25 Foroutan, R., Mohammadi, R., Esmaeili, H. et al. (2020). Transesterification of waste edible oils to biodiesel using calcium oxide@magnesium oxide nanocatalyst. *Waste Management* 105: 373–383.

26 Lin, T., Zhao, S., Niu, S. et al. (2020). Halloysite nanotube functionalized with La–Ca bimetallic oxides as novel transesterification catalyst for biodiesel production with molecular simulation. *Energy Conversion and Management* 220: 113138.

27 Marinković, D.M., Avramović, J.M., Stanković, M.V. et al. (2017). Synthesis and characterization of spherically–shaped CaO/$\gamma$–$Al_2O_3$ catalyst and its application in biodiesel production. *Energy Conversion and Management* 144: 399–413.

28 Wu, H., Zhang, J., Wei, Q. et al. (2013). Transesterification of soybean oil to biodiesel using zeolite supported CaO as strong base catalysts. *Fuel Processing Technology* 109: 13–18.

29 Sakti La Ore, M., Wijaya, K., Trisunaryanti, W. et al. (2020). The synthesis of $SO_4/ZrO_2$ and Zr/CaO catalysts via hydrothermal treatment and their application for conversion of low–grade coconut oil into biodiesel. *Journal of Environmental Chemical Engineering* 8 (5): 104205.

30 Faria, D.N., Cipriano, D.F., Schettino, M.A. et al. (2020). Na,Ca–based catalysts supported on activated carbon for synthesis of biodiesel from soybean oil. *Materials Chemistry and Physics* 249: 123173.

31 Tabah, B., Nagvenkar, A.P., Perkas, N., and Gedanken, A. (2017). Solar–heated sustainable biodiesel production from waste cooking oil using a sonochemically deposited SrO catalyst on microporous activated carbon. *Energy & Fuels* 31 (6): 6228–6239.

32 Li, H., Wang, Y., Ma, X., et al. (2020). A novel magnetic CaO-based catalyst synthesis and characterization: Enhancing the catalytic activity and stability of CaO for biodiesel production, *Chemical Engineering Journal* 391: 123549.

33 Zheng, X., Zhang, L., Fan, Z. et al. (2019). Enhanced catalytic activity over MIL-100(Fe) with coordinatively unsaturated $Fe^{2+}/Fe^{3+}$ sites for selective oxidation of $H_2S$ to sulfur. *Chemical Engineering Journal* 374: 793–801.

34 Li, H., Liu, F., Ma, X. et al. (2020). An efficient basic heterogeneous catalyst synthesis of magnetic mesoporous Fe@C support SrO for transesterification. *Renewable Energy* 149: 816–827.

35 Khatibi, M., Khorasheh, F., and Larimi, A. (2021). Biodiesel production via transesterification of canola oil in the presence of Na–K doped CaO derived from calcined eggshell. *Renewable Energy* 163: 1626–1636.

36 Malhotra, R. and Ali, A. (2019). 5–Na/ZnO doped mesoporous silica as reusable solid catalyst for biodiesel production via transesterification of virgin cottonseed oil. *Renewable Energy* 133: 606–619.

37 Abukhadra, M.R., Mohamed, A.S., El-Sherbeeny, A.M. et al. (2020). Sonication induced transesterification of castor oil into biodiesel in the presence of MgO/CaO nanorods as a novel basic catalyst: characterization and optimization. *Chemical Engineering and Processing – Process Intensification* 154: 108024.

38 Sudsakorn, K., Saiwuttikul, S., Palitsakun, S. et al. (2017). Biodiesel production from *Jatropha Curcas* oil using strontium-doped CaO/MgO catalyst. *Journal of Environmental Chemical Engineering* 5 (3): 2845–2852.

39 Fan, M., Si, Z., Sun, W., and Zhang, P. (2019). Sulfonated $ZrO_2$–$TiO_2$ nanorods as efficient solid acid catalysts for heterogeneous esterification of palmitic acid. *Fuel* 252: 254–261.

40 Shatesh, K., Shamsuddin, M.R., Farabi, M.S.A. et al. (2020). Production of methyl esters from waste cooking oil and chicken fat oil via simultaneous esterification and transesterification using acid catalyst. *Energy Conversion and Management* 226: 113366.

41 Melo Junior, C.A.R., Albuquerque, C.E.R., Carneiro, J.S.A. et al. (2010). Solid-acid-catalyzed esterification of oleic acid assisted by microwave heating. *Industrial & Engineering Chemistry Research* 49 (23): 12135–12139.

42 Chaveanghong, S., Smith, S.M., Smith, C.B. et al. (2018). Simultaneous transesterification and esterification of acidic oil feedstocks catalyzed by heterogeneous tungsten loaded bovine bone under mild conditions. *Renewable Energy* 126: 156–162.

43 Su, F. and Guo, Y. (2014). Advancements in solid acid catalysts for biodiesel production. *Green Chemistry* 16 (6): 2934–2957.

44 Cyraniak, A., Skurski, P., and Czapla, M. (2019). The presence of two different central atoms increases the strength of Lewis–Brønsted superacids. *Chemical Physics Letters* 717: 77–81.

45 Yoo, S.J., Lee, H.-s., Veriansyah, B. et al. (2010). Synthesis of biodiesel from rapeseed oil using supercritical methanol with metal oxide catalysts. *Bioresource Technology* 101 (22): 8686–8689.

46 Jiménez-Morales, I., del Río-Tejero, M.A., Braos-García, P. et al. (2012). Preparation of stable sulfated zirconia by thermal activation from a zirconium doped mesoporous MCM-41 silica: application to the esterification of oleic acid with methanol. *Fuel Processing Technology* 97: 65–70.

47 Hino, M., Kobayashi, S., and Arata, K. (1979). Solid catalyst treated with anion. 2. Reactions of butane and isobutane catalyzed by zirconium oxide treated with sulfate ion. Solid superacid catalyst. *Journal of the American Chemical Society* 101 (21): 6439–6441.

**48** Lam, M.K., Lee, K.T., and Mohamed, A.R. (2009). Sulfated tin oxide as solid superacid catalyst for transesterification of waste cooking oil: an optimization study. *Applied Catalysis B: Environmental* 93: 134–139.

**49** Jitputti, J., Kitiyanan, B., Rangsunvigit, P. et al. (2006). Transesterification of crude palm kernel oil and crude coconut oil by different solid catalysts. *Chemical Engineering Journal* 116: 61–66.

**50** Pratap, S.R., Shamshuddin, S.Z.M., and Shyamprasad, K. (2020). Microwave assisted synthesis of propyl esters over modified versions of zirconia: kinetic study. *Chemical Data Collections* 30: 100579.

**51** Chen, X.-R., Ju, Y.-H., and Mou, C.-Y. (2007). Direct synthesis of mesoporous sulfated silica–zirconia catalysts with high catalytic activity for biodiesel via esterification. *Journal of Physical Chemistry C* 111: 18731–18737.

**52** Senso, N., Jongsomjit, B., and Praserthdam, P. (2011). Effect of calcination treatment of zirconia on $W/ZrO_2$ catalysts for transesterification. *Fuel Processing Technology* 92: 1537–1542.

**53** Busto, M., Shimizu, K., Vera, C.R. et al. (2008). Influence of hydrothermal aging on the catalytic activity of sulfated zirconia. *Applied Catalysis A: General* 348: 173–182.

**54** Polychronopoulou, K., Cabello Galisteo, F., López Granados, M. et al. (2005). Novel Fe–Mn–Zn–Ti–O mixed-metal oxides for the low-temperature removal of $H_2S$ from gas streams in the presence of $H_2$, $CO_2$, and $H_2O$. *Journal of Catalysis* 236: 205–220.

**55** Haber, J., Pamin, K., Matachowski, L., and Mucha, D. (2003). Catalytic performance of the dodecatungstophosphoric acid on different supports. *Applied Catalysis A: General* 256 (1): 141–152.

**56** Bhorodwaj, S.K. and Dutta, D.K. (2011). Activated clay supported heteropoly acid catalysts for esterification of acetic acid with butanol. *Applied Clay Science* 53: 347–352.

**57** Gardy, J., Osatiashtiani, A., Céspedes, O., and Ahmed, M.H. (2018). A magnetically separable $SO_4$/Fe–Al–$TiO_2$ solid acid catalyst for biodiesel production from waste cooking oil. *Applied Catalysis B: Environmental* 234: 268–278.

**58** Gardy, J., Hassanpour, A., Lai, X. et al. (2016). Synthesis of $Ti(SO_4)O$ solid acid nano-catalyst and its application for biodiesel production from used cooking oil. *Applied Catalysis A: General* 527: 81–95.

**59** Soltani, S., Khanian, N., Rashid, U., and TSY, C. (2020). Core–shell ZnO–$TiO_2$ hollow spheres synthesized by in-situ hydrothermal method for ester production application. *Renewable Energy* 151: 1076–1081.

**60** Li, H., Chu, H., Ma, X. et al. (2021). Efficient heterogeneous acid synthesis and stability enhancement of UiO-66 impregnated with ammonium sulfate for biodiesel production. *Chemical Engineering Journal* 408: 127277.

**61** Gan, S., Ng, H.K., Ooi, C.W. et al. (2010). Ferric sulphate catalysed esterification of free fatty acids in waste cooking oil. *Bioresource Technology* 101: 7338–7343.

**62** Dutta, S., Jaiswal, K.K., Verma, R. et al. (2019). Green synthesis of zinc oxide catalyst under microwave irradiation using banana (*Musa* spp.) corm (rhizome) extract for biodiesel synthesis from fish waste lipid. *Biocatalysis and Agricultural Biotechnology* 22: 101390.

**63** Kouzu, M., Fujimori, A., Suzuki, T. et al. (2017). Industrial feasibility of powdery CaO catalyst for production of biodiesel. *Fuel Processing Technology* 165: 94–101.

# 7

# Supported Metal/Metal Oxide Catalysts in Biodiesel Production

An Overview

*Pratibha Agrawal*[1] *and Samuel Lalthazuala Rokhum*[2,3]

[1] *Department of Applied Chemistry, Laxminarayan Institute of Technology, RTM Nagpur University, Nagpur, Maharastra, India*
[2] *Hamid Yusuf Department of Chemistry, University of Cambridge, Cambridge, UK*
[3] *Department of Chemistry, National Institute of Technology, Silchar, Assam, India*

## 7.1 Introduction

As the world entered into a technology-reliant modern civilization, the demand for energy sources increases drastically. At the same time, the sharp decline in petroleum resources and the detrimental effects of the fossil fuels are constantly alarming to hunt for newer and cleaner sources of energy [1–3]. Biodiesel stands as a promising candidate to fossil fuels for having renewable and sustainable resources, less $CO_2$ (greenhouse gases [GHGs]) emissions, no sulfur content, nontoxicity, and biodegradable and ecofriendly nature [4–6]. Synthesis of biodiesel is based mainly on three feedstocks – from edible seeds, which is used on commercial scale in Europe and the United States, from nonedible seeds, and from microalgae. Owing to the environmentally beneficial and renewable nature, as well as relatively cleaner production procedure, studies in biodiesel synthesis and optimization of different reaction conditions and material selections have become a distinguished research arena [7, 8]. Among the different methods that are used to produce biodiesel, the most convenient and useful method is transesterification. Though the reaction is reversible, equilibrium favors more toward the forward reaction, and the biodiesel along with glycerin is formed using appropriate catalyst [9].

These catalysts can be of different kinds or types viz. homogeneous, heterogeneous, or enzymatic one, which increase the efficiency of biodiesel formation to the noticeable extent [10]. Various catalysts have been vigorously studied in biodiesel production during the past few years so as to attain an economically viable industrial route for the synthesis of biodiesel. In comparison with homogeneous catalysts, these heterogeneous catalysts have multiple notable advantages, which primarily include the ease of its separation and also reusability of the catalysts, making it cost effective [11]. Different traits of heterogeneous catalysts have been investigated in this regard. Metals and metal oxide catalysts are considered to be one of the most premier and highly explored heterogeneous catalysts [12].

*Biodiesel Production: Feedstocks, Catalysts, and Technologies*, First Edition. Edited by Samuel Lalthazuala Rokhum, Gopinath Halder, Suttichai Assabumrungrat, and Kanokwan Ngaosuwan.

Taking into account all of the demands and considerations, the current chapter has been focused on usage and catalytic activity of various supported solid catalysts.

Heterogeneous catalysts could empower the strategy of efficient, constant process, and they improve the economy of biodiesel production because they can reduce catalyst usage cost, decrease costly separation steps, and permit the use of continuous processing [13]. However, the reactants "diffuse out" to the active sites of catalyst, thereby decreasing transesterification rate as compared to the one with homogeneous catalyst. These active sites also "percolate out" into the reaction mixture producing soap [14]. The catalysts become active in the vicinity of boiling point of methanol. Metal oxides having porous or mesoporous surface area generate large number of active sites [15]. The metal oxide-based catalysts are useful in different reactions, involving oxidation process, acid base catalysis, photocatalysis, or biochar-based transesterification process, also having useful application in fuel cells or sensors [16]. Catalytic activity can be enhanced and achieved by restructuring its properties.

## 7.2 Supported Catalyst

The properties of the catalyst are to be improved as per the demand of industrial sector. In recent years, much more attention is focused on the use of various catalysts for improving the yield and making the process more beneficial [17–19]. Coarse particles of heterogeneous catalyst when supported on some stable surface can get dispersed and gain extra thermal stability [20, 21] as the surface usually remains chemically and thermally stable. There is no as such prescribed outline of the synthesis and the steps involved during the biodiesel synthesis using such supported catalyst. The loading of metal and metal oxides on various porous surfaces may involve various steps and to be verified from the qualitative and quantitative testing of the product obtained. Such supported catalysts provide large surface area and porosity and are also useful in overcoming the mass transfer limitation as metal or metal oxides are immobilized. Various catalysts supported on different materials for the synthesis of biodiesel are discussed next.

## 7.3 Metals and Metal Oxide Supported on Alumina

Many researchers show concern for the activity and selectivity of alumina supported on some metal surface and metal oxide in biodiesel production. Out of many, $NaOH/-Al_2O_3$ catalysts are less favored. Arzamendi et al. [22] reported that due to adsorption of methanol, methanol/oil ratio increased, resulting in favored reaction toward methoxide formation. Thus, the performance of such reaction is largely dependent on methanol/oil molar ratio. Another possibility is the leaching of sodium deposited on catalyst surface. This percentage is found to be very less (5% even after 9 h of reaction), which may not affect the overall performance of $NaOH/-Al_2O_3$ catalysts, but it surely affect the chemical stability of catalyst. The catalyst has been used in synthesis of biodiesel from sunflower oil using methanol at 323 K. The reaction was studied successfully at different concentration of catalyst, varying methanol-to-oil and catalyst-to-methanol molar ratios. A group of researchers studied loading of alkali metal salts or various potassium compounds on alumina in the process of biodiesel formation [23, 24]. It has been observed that only alumina shows less or no activity, but when loaded with different potassium salts at higher temperature, catalytic activity increased. Addition of water molecule in the reaction does not affect the rate, indicating that the catalyst can also be used even in the presence of water moiety. Same problem of leaching of potassium species also incurred in this case. So the excellent catalytic activity (with 99% yield) that was observed in first run diminishes (to 33%) in the second run [25]. Xie and Li [26] reported the decreased catalytic activity of specifically $K_2CO_3/Al_2O_3$ due to its lower basic nature when used in comparison with other compounds of potassium. On the same line, best catalytic activity was shown by 35 wt% KI supported over $Al_2O_3$, (calcined at 500 °C for 3 h). This may be due to the highest value of basicity of KI. Regarding the comparison in leaching, Noiroj et al. [27] studied that when $KOH/Al_2O_3$ is used, 51.26% of K was leached, whereas when KOH/NaY is used, only 3.18% of K was leached.

Some experiments performed in this area [28] proved that the supported catalyst does not show much activity, but it is the loaded active metal or metal oxide influencing the reaction rate and yield. Transesterification of sunflower oil to fatty acid methyl ester (FAME) was observed to be more than 93% after only 15 min of the reaction. $K_2CO_3$ was the active component loaded on alumina, responsible for the increased yield, as similar results were observed for $K_2CO_3$ loaded over Al–O–Si. These catalysts were calcinated from 300 °C onward before use. The optimum temperature for calcination of catalyst is 600 °C at which it has smaller area, and average pore size was increased by 25% when compared with the catalyst that was calcinated at 300 °C. But still, those pore sizes in structure of catalyst do not appreciably affect the rate and yield of reaction.

When only 6% catalyst amount (35 wt% $KNO_3$ loaded on $Al_2O_3$, which was calcinated at 500 °C for 4 h) in methanolysis of jatropha oil is used, after 6 h it showed the highest conversion of 84%. After the reusability study, it was found that the catalyst can be reused for three times successfully [29]. The catalytic activity of $CaO/Al_2O_3$ base catalyst was studied by earlier workers [30]. It has been found that the amount of catalyst and the reaction temperature have remarkable effect on the reaction yield, whereas alcohol/oil molar ratio does not affect the yield. After studying the second order kinetic model, the optimum conditions in transesterification of palm oil were found as 5.97% wt of catalyst in 12.14 : 1 molar ratio of alcohol/oil at 64.29 °C, producing 98.64% of biodiesel with reusability till two cycles. Moreover, the leaching of the catalyst into the reaction in this case was found to be insignificant.

The plausible reaction mechanisms predicted in the transesterification of the soybean oil using $CaO/Al_2O_3$ catalyst proposed by Pasupulety et al. [31] are as follows: $CaO–Al_2O_3$ catalyst (CAC) binds itself with methanol and triglyceride molecules to give intermediates A and B, respectively, which further interact together through active centers producing diglyceride and FAME and regenerating CAC. This diglyceride can again react with CAC to produce an intermediate C, which acts upon A through active centers producing monoglyceride and FAME and regenerating CAC. Finally, monoglyceride interacts with basic sites form the catalytic surfaces to form the intermediate that reacts with intermediate A to produce FAME (required product), regenerating the catalyst and glycerol. The formation of calcium diglyceroxide is the second step in the process (Scheme 7.1).

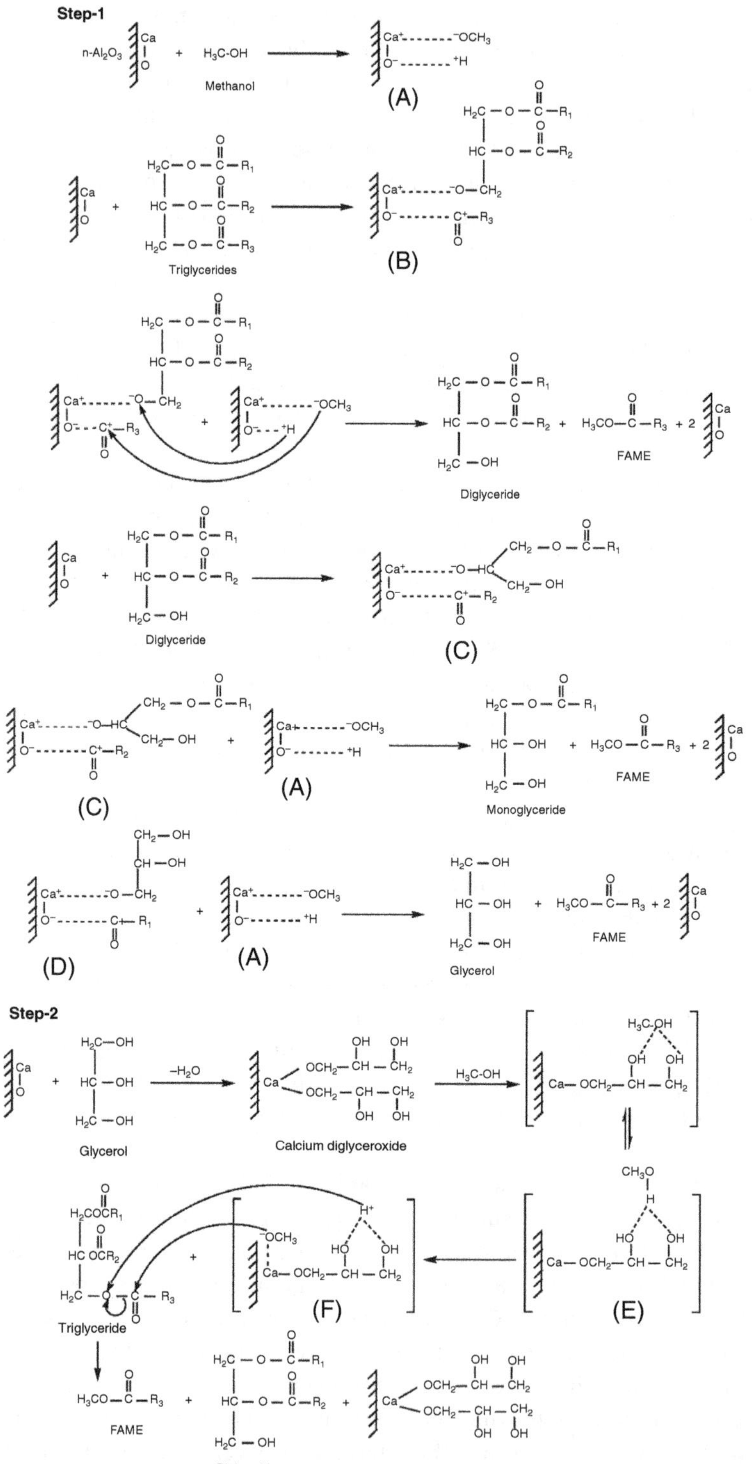


**Scheme 7.1** Proposed mechanism for $CaO/Al_2O_3$-catalyzed synthesis of biodiesel (FAME). *Source:* Ref. [31].

Thus, the process is a self-repeating and continuous one, generating the catalyst again and again and thereby increasing the rate and yield of the transesterification reaction [32]. The main task for using CaO as catalyst will be high free fatty acid (FFA) content feedstocks. So primary emphasis is given by researchers to decrease the amount of FFAs while using CaO catalyst.

The alumina-supported mesoporous metal oxides show well-developed hexagonal symmetry, large pore widths with higher Brunauer–Emmett–Teller (BET) surface area, and more crystalline pore walls. Morris et al. observed larger mesopores and more thermal stability for nickel aluminum oxide when compared with MgO, CaO, $TiO_2$, and $Cr_2O_3$-doped alumina. Those properties prove nickel-doped alumina as a strong candidate for transesterification reaction [33].

The high efficiency of KI-doped alumina was also observed. There might be the formation of $K_2O$ and basic sites and the groups Al–O-K, which may provide active sites for further transesterification process [34, 35]. Figure 7.1 shows insignificant number of particles of the support (up to 100 nm) on the material surface and numerous other particles (0.3–1.9 μm).

These basic sites ($K_2O$) formed increase the basicity of the catalysts, supporting high conversion to methyl esters. The catalyst $KF/Al_2O_3$, which is most active, requires in very small amount. The amounts of biodiesel formed depend on the type of oils, catalyst activity, and the other reaction conditions [34].

## 7.4 Metals and Metal Oxide Supported on Zeolite

Natural zeolites are efficient material producing high yield of triglycerides. The zeolite possesses cubic structure with crystallite size of average 1.4–2 μm. Elements such as Mg, Ca, K, Fe, Si, Na, Al, and O are accumulated on the active site of the catalyst. Metals and metal oxides get dispersed evenly throughout the pores and thus can create additional active centers over zeolites. Metals grafted over zeolite are well known for their potent activity in the process of diesel production. It can be prepared by impregnation method. The zeolite surface shows flakes and granular shaped crystallites with a lumpy surface after successful impregnation of metal and metal oxides.

Powel et al. studied various transesterification reactions of rapeseed oil with methanol using metals like Ru, Pd, Pt, and Ag (0.5 or 2 wt%) supported over zeolites (M/NZ). These catalysts were reduced in a mixture 5%$H_2$–95%Ar at 300 °C for improving the catalytic properties. Increased catalytic activity of Pd/NZ after reduction proved the necessity of that step. The most active catalyst was found to be Pt/NZ maybe due to its easily reducible and high alkaline nature [36].

Ibrahim et al. [37] reported that there is a synergy between the microwave reactor and the calcium oxide loaded zeolite, making the process more effective and sustainable. The process is economical as the catalyst is easily separable and hence reusable for 2–3 times. However, after this there can be some reacting species attached to catalyst surface, which leads to agglomeration, making it difficult to reuse. Among the other metals, when nickel metal is impregnated into zeolite, the chemical potential difference between nickel and aluminum of zeolites makes available excess of an electron free pair, which helps to

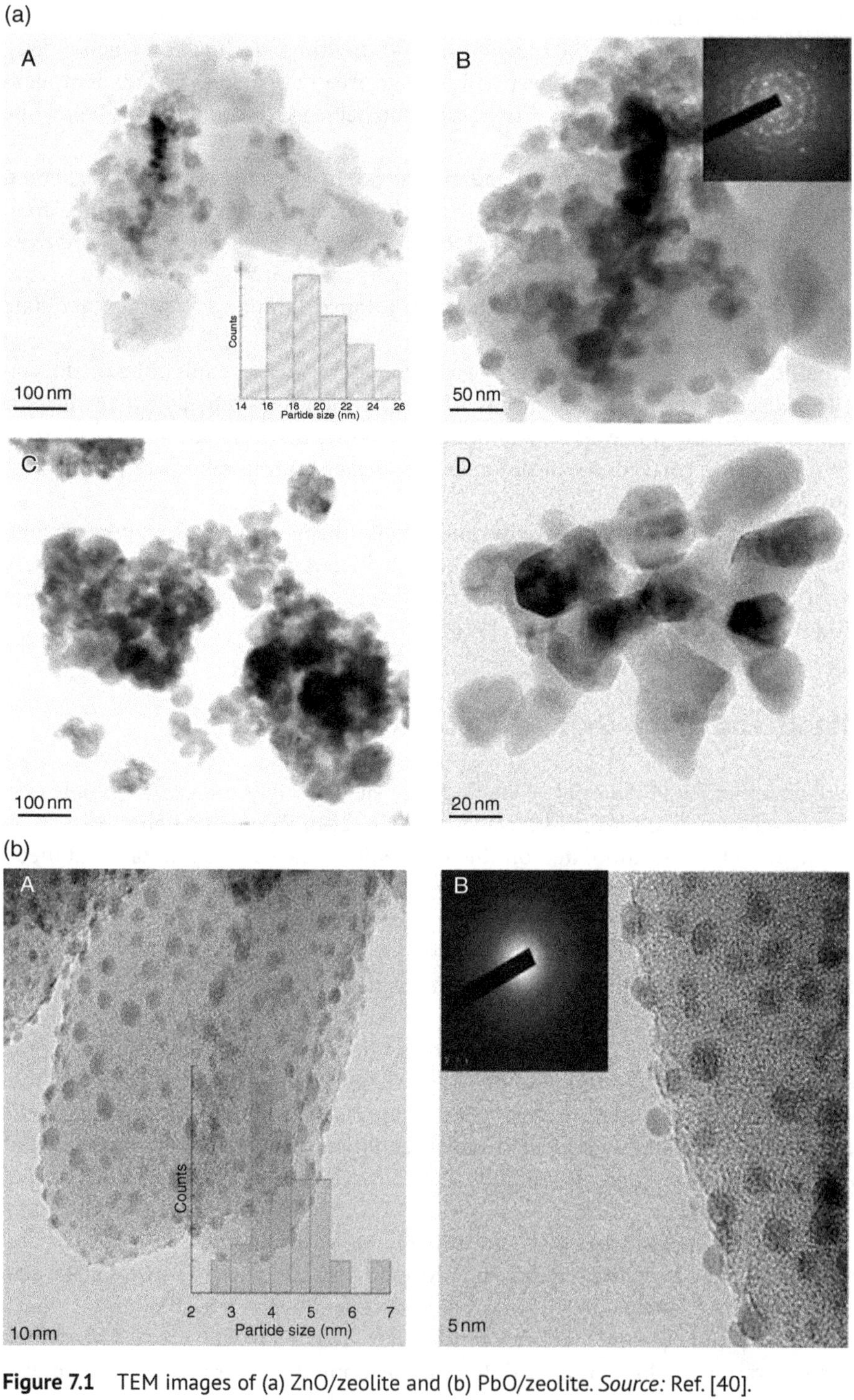


**Figure 7.1** TEM images of (a) ZnO/zeolite and (b) PbO/zeolite. *Source:* Ref. [40].

increase the acidity of such catalyst. It has been found that the highest yield of biodiesel manufacture can be obtained when 5% Ni/zeolite catalyst [38] is used.

The high cost in preparing zeolites restricts its application, and hence comparatively less research work has been carried out on it. Keeping this point in mind, Li et al. [39] prepared the zeolites using fly ash (which has been produced in large amount in power plants and otherwise discharged directly causing pollution) [39]. Using $Li_2CO_3$ and fly ash, recyclable Li/NaY zeolite catalyst was synthesized with large number of alkaline sites and complex channel structure contributing to the increased catalytic property. The components ($Li_4SiO_4$ and $Li_3NaSiO_4$) were also formed at high calcination temperature (650 °C) due to decarbonation and dehydration of catalyst. Those components were thought to be responsible for extra stability with regenerative capacity. It was observed noticeably that even after five cycles of reuse, the catalyst retains complex pore structure. Thus, $LiAlO_2$ acts as active skeleton in the transesterification reaction.

Singh et al. [40] observed 100% efficiency of jatropha oil into the FAME using ZnO supported over zeolite, whereas 93.8% when PbO supported over zeolite was used within 30 min of time. The hydrothermal impregnation precipitation method for catalyst preparation ensured good distribution of the metal oxide ZnO on zeolite support. When supported over zeolite, the particle sizes of ZnO and PbO were confirmed to be 19.5 and 4.2 nm (Figure 7.1), respectively, using transmission electron microscope (TEM) analysis. Besides this, the leaching of Pb and Zn was found to be very less maybe due to strong interactions between ZnO and PbO with zeolite.

The catalytic effects of MgO, CuO, and NiO permeated on zeolite 4A were investigated by Ariffin et al. [41] for transesterification of lipid oil in the production of biodiesel. The highest efficiency of CuO/zeolite may be due to its higher surface area and increased pore volume of the catalysts. Other catalysts have lower surface area restricting the adsorption and transmission of triglyceride in the catalytic framework while reaching to the active sites, thereby reducing the rate of transesterification process. Extra care must be observed in using the excess amount of catalysts to be taken as it might lead to phase separation and accelerate the formation of soap. This will reduce the catalytic efficiency during the productivity of biodiesel.

## 7.5 Metals and Metal Oxide Supported on ZnO

ZnO is thought to be very stable, cheap, easily available, and reusable catalyst of which structural, magnetic, and optical properties can be enhanced by adding different metals and metal oxides either intrinsically or by doping. Chemically, it possesses higher affinity for polar particles due to its wurtzite-type hexagonal structure, and its amphoteric nature (containing both acidic sites due to $Zn^{2+}$ and basic sites due to $O^{2-}$) is responsible for higher affinity for monoglycerides. Modified ZnO stands as a promising candidate supporting biodiesel production due to increased transparency and oxygen vacancy and its structural, optical, and magnetic properties.

Venkatesh et al. [42] observed surface defects (oxygen vacancies as well as oxygen interstitials) in Se-doped ZnO, enhancing basic site reactivity toward biodiesel production. Se doped on the surface of ZnO nanocatalyst is found to be highly stable and active, which might be due to the plausible mechanism [42].

The interstitial oxygen attracts additional electrons from nearby regularly attached oxygen sites, which in turn attract electrons from Se (Se→ $Se^{+2}$ + $2e^{-}$), creating basic sites. Such delocalization of electrons during doping and defect formation is responsible for an increased electronic density on Se doped on the surface of ZnO, increasing their basicity.

Thus, selenium doping on the surface of ZnO increases oxygen interstitials as well as vacancies with an increase in $O^{2-}$ sites, which due to its basic nature would abstract proton from methanol generating methoxide ions, which would initiate the nucleophilic attack at the carbonyl group of triglycerides. Those methoxide ions would be further stabilized by $Zn^{2+}$ counterions.

By coprecipitation method, nickel can be doped in ZnO at calcination temperature of 800 °C. Beyond this, defects in the structure can be produced, thereby decreasing its catalytic activities. Optimization process of transesterification of castor oil using Ni doped on the surface of ZnO nanocatalyst was studied by Aberna et al. [43]. The maximum biodiesel yield of 95.20% (w/w) with Ni-doped ZnO nanocatalyst (in ratio of 11.07 w/w) at 55 °C was reported in 1 h. Moreover, the nanocatalyst was found to be more stable and separable and can be reused for five cycles, making the process more cost effective.

In 2019, Rokhum et al. [44] synthesized zinc oxide-supported silver nanoparticles (ZnO@Ag NPs) in the production of biodiesel from palm oil and observed much higher yields of biodiesel as compared to other catalysts used. ZnO@Ag NPs afford a high FAME yield of 96% with conversion of 97%. The reusability of the catalyst proved to be noteworthy even after five cycles, making it more promising catalyst during such study.

Table 7.1 clearly indicates that on increasing the wt% from 5–10 of Ag in ZnO, biodiesel yield had been increased to 80 and 96% under the same reaction condition (methanol/oil

**Table 7.1** Synthesis of biodiesel using different catalyst and supported catalysts. *Source:* Ref. [44].

| Entry | Catalyst | Conditions[a] | Yield % (wt/wt) |
|---|---|---|---|
| 1 | ZnO | 10, 10 : 1 | 63 |
| 2 | Ag + ZnO | 10, 10 : 1 | 65 |
| 3 | 10 wt% $Al_2O_3$ @Ag (NP-1) | 10, 10 : 1 | 32 |
| 4 | 40 wt% $Al_2O_3$ @Ag (NP-2) | 10, 10 : 1 | 50 |
| 5 | 10 wt% $SiO_2$@Ag (NP-3) | 12, 10 : 1 | 21 |
| 6 | 45 wt% $SiO_2$@Ag (NP-4) | 12, 10 : 1 | 45 |
| 7 | 5 wt% ZnO@Ag (NP-5) | 10, 10 : 1 | 80 |
| 8 | 10 wt% ZnO@Ag (NP-6) | 6, 10 : 1 | 54 |
| 9 | 10 wt% ZnO@Ag (NP-6) | 10, 10 : 1 | 96 |
| 10 | 10 wt% ZnO@Ag (NP-6) | 10, 6 : 1 | 53 |
| 11 | 10 wt% ZnO@Ag (NP-6) | 12, 10 : 1 | 87 |
| 12 | 15 wt% ZnO@Ag (NP-7) | 10, 10 : 1 | 86 |

[a] Catalyst loading (wt%), methanol-to-oil molar ratio.

ratio, 10 : 1; catalyst concentration, 10% [wt/wt]; temperature, 60 °C; reaction time, 60 min). Also further increase in wt% 15 of Ag in ZnO, biodiesel yield had been decreased to 86%, which may be due to increased crystal size of Ag particle that may have blocked the pores of ZnO, thereby reducing its catalytic activity.

Use of Fe (III) doped over the surface of ZnO catalyst was studied for biodiesel production by Saxena et al. [45]. The catalyst formed was found to be soft ferromagnetic substance. It possesses high magnetic and low coercive values. These properties are making its magnetic separation easy, supporting its reusability. Borah et al. [46] observed Co doped on the surface of ZnO is formed by substitution of some Zn atom in the hexagonal lattice that are changing only the microstructural features. It exhibits excellent catalytic property in biodiesel production using *Mesua ferrea* oil. It can be reused for 2–3 cycles, but after that the structure was distorted as the catalyst recovered losses its crystalline structural nature. Such type of distortion in the crystalline structure may be due to the deposition of some organic matters on active sites of catalyst reducing its catalytic use.

The $SO_4^{2}$–ZnO exhibited notable performance during production of biodiesel, which may be due to the Lewis acid nature of active sites on the catalyst surface when sulfate ions are loaded on ZnO surface. It activates the surface strongly by modifying thermodynamically stable monoclinic phase of ZnO to metastable hexagonal or tetragonal phase, promoting its acidity as well as catalytic activity [47].

The zinc and lanthanum mixed oxide catalyst enhanced the catalyst ability in both transesterification and esterification reaction due to increased acidic and basic active sites at the surface. Such catalysts are relatively cheaper due to low manufacturing cost, easy product formation and purification process, and the reusability nature of the catalyst. All those factors contribute in decreasing the cost of biodiesel production [48].

CaO supported on ZnO is effective in the biodiesel production. The process can be made more efficient by synthesizing this catalyst by one-pot method, which is a single step reaction, thereby reducing the reaction time. The main principle beyond this step is that the higher calcination temperature (850 °C) will decompose $CaCO_3$ producing CaO with the evolution of $CO_2$. This in situ produced CaO comes in direct contact with ZnO to produce desired CaO-ZnO catalyst [49]. Such methods are cheap and economically favorable ones as its catalytic properties also help in increasing the yield and rate of biodiesel production.

## 7.6 Metals and Metal Oxide Supported on Silica

Among various catalysts, mesoporous silica is preferred by various researchers due to its high surface area, thermal stability, and large pore structure, reducing its mass transfer and allowing high concentration of active sites per mass of catalyst [50–53]. CaO supported on the surface of silica catalyst was successfully used in biodiesel production. The catalyst showed better catalytic activities for the transesterification reaction. The biodiesel yield reaches to 87.5% at 2 h. This catalyst was found to be thermally more stable, and the interaction between calcium oxide and silica is also strong enough to avoid "the lixiviation" of the active phase in methanol. For the transesterification of castor and sunflower oils using methanol, product yield obtained was 95% after 5 h of reaction.

Various Lewis acids such as aluminum chloride ($AlCl_3$), boron-trifluoride ($BF_3$), and zinc chloride ($ZnCl_2$) are widely used as catalyst. Aluminum chloride ($AlCl_3$) is a very active acidic catalyst, but as it required another reagent as a support to carry out efficient catalysis, it was supported on silica to give better catalyst for transesterification.

During the reaction, $Si^+$ and $O^-$ in silica substance will attack on carboxylic acid group as well as methanol, whereby $H^+$ of acid group will be successfully replaced by $CH_3^+$ (from methanol) to form an intermediate and $H_2O$ molecule. This intermediate further undergoes rearrangement reaction, eliminating Si-O and forms FAME.

Thus silica is capable of adjusting high percentage of water and FFA content in feedstock, making it a good supporting catalyst. Silica also supports to restrain the leaching of calcium species. Due to leaching, CaO is not very much significant in transesterification reaction. But when CaO is supported over SiO, Ca–O–Si bond is formed. This protects Ca from leaching as well as blocks CaO from other feasible side reactions, making it a more stable catalyst. Such catalyst is reported to be reused for eight runs, whereas CaO alone can be reused for four runs only. Such silica-supported CaO catalyst can be effectively used in the transesterification reaction [51].

Likewise, acid sulfated zirconia when dispersed onto mesoporous tetragonal catalytically more active silica shows increased catalytic performance due to increased acid sites. Two species responsible for those increased acid sites are the nanocrystalline sulfated zirconia (IV), and the other is from the mixed silica-zirconia oxides [54]. Due to this, such catalyst shows higher density of acidic sites compared to the unsupported one. Such sites improve the stability and yield of esterification process.

One of the problems associated with this is formation of water during esterification of FFA. This water can hinder the reaction and shift the direction of reaction toward reactant side. At the same time, water molecules formed can block the active surface acidic sites, lowering its acidity and hence the catalytic activity. It is very difficult to recover those lost active sites again during the esterification process. Another problem is if extra amount of the sulfate added, it may segregate extracting zirconia to the surface from the mixed oxides and again stabilize back the tetragonal phase.

## 7.7 Metals and Metal Oxide Supported on Biochar

Biochar is carbon-rich material produced from carbon neutral sources (i.e. biomass). It is also useful for carbon capture and storage with versatile catalytic applications. Carbonized biochar is highly porous and effective catalyst during biodiesel production. It has various pores of different size classified as submicropore, micropores, mesopores, and macropores. Its structure (Figure 7.2) can be modified as per the catalytic role. It was found to be more efficient than non-biochar catalysts only if its stability is improved for better separation and reusability purpose [55].

Activated biochar is used in two ways in preparing the valuable catalysts for biodiesel production – (i) solid acid catalysts and (ii) solid alkali catalysts.

**Figure 7.2** Biochar supporting different metal and metal oxides. *Source:* Modified from Ref. [55].

### 7.7.1 Solid Acid Catalysts

When the oil contains high percentage of acids (cooking oil), then those types of catalyst are used. In this category, acid centers are generated at the surface of biochars usually by sulfonation process (using either $H_2SO_4$ acid solutions or $SO_3$ vapor). As a result of this, $-SO_3H$ groups are deposited on the surface of biochars, making it highly active. The yield of FAME is comparatively lower in those cases, but the conversion achieved is very high [56]. Biochars are highly porous with submicropore, micropores, mesopores, and macropores. Triglyceride molecules are large whereas FFA molecules are smaller. Those molecules access the pores and allow to react with methanol molecules.

However, very few examples of use of solid acid catalyst are seen as compared with solid bases. Solid superacids like sulfated zirconia and tungstated zirconia can be used in transesterification reaction [57]. Sulfated tin oxide has also shown higher acid strength than that of the sulfated zirconia [58–60] in the transesterification of soybean oil with methanol. However, synthesis and use of such catalyst are expensive.

### 7.7.2 Solid Alkali Catalysts

Another type of biochar-supported catalysts that are used for biodiesel production is alkali catalysts supported over biochar [61]. Usually the basic sites are occupied by Ca, K, or Na oxides. Leaching of CaO or other oxides, inside the reaction medium, poses one of the biggest problems in the reactivity. The catalyst can react with glycerol giving calcium diglyceroxide, which is easily soluble in methanol and hydrolyses itself to calcium hydroxide. The active sites of supported catalyst can be poisoned by water molecules and $CO_2$, producing hydroxides and carbonates. To resolve such issues and increase the use of such metal oxide, a cheaper, environment friendly, thermally stable, biodegradable biochar is used as support for such oxides. Bitonot et al. [62] found that 20 wt% of Ca loaded on the biochar showed the best catalytic activity in the studied transesterification process. Scanning electron microscope (SEM) analysis (Figure 7.3) showed significant differences among the surface morphology of the various loaded catalysts. Biochar favors the deposition of (20 wt%) calcium since it is a small ion.

In addition, as the catalyst was easily recovered by the simple centrifugation method, it was regenerated and reused for three cycles of reaction without any significant loss of its catalytic activity. In the IR spectral analysis (Figure 7.4), a strong doublet at 2930 and 2860 $cm^{-1}$ is significant for C-H stretching vibrations present in methyl and methylene groups, and the peak at 1744 $cm^{-1}$ (indicating $-C = O$ stretching frequency of carbonyl groups) was due to the presence of FAME and other organic products (glycerol, mono-, and diglycerides).The organic molecules, which were deposited on the catalyst surface, lead to a reduction of its catalytic activity, resulting in low approachability of the active sites.

Basic active centers are most useful in transesterification reaction. Those sites can easily abstract proton from methanol, producing strong basic methoxide ions easily. In the reaction scheme polar metal oxide of biochar surface ($M^+$-$O^-$) after abstracting proton from methanol gets attached to methoxide ion and comes in the close vicinity of carbocation of ester and forms a tetrahedral intermediate. This intermediate abstracts the $H^+$ ion from either a new methanol molecule or the protonated basic site and finally rearranges itself, giving FAME and alcohol.

The mechanism cycle repeated many times to reach to the formation of biodiesel and glycerol. Such catalyst are very much useful in the biodiesel production process.

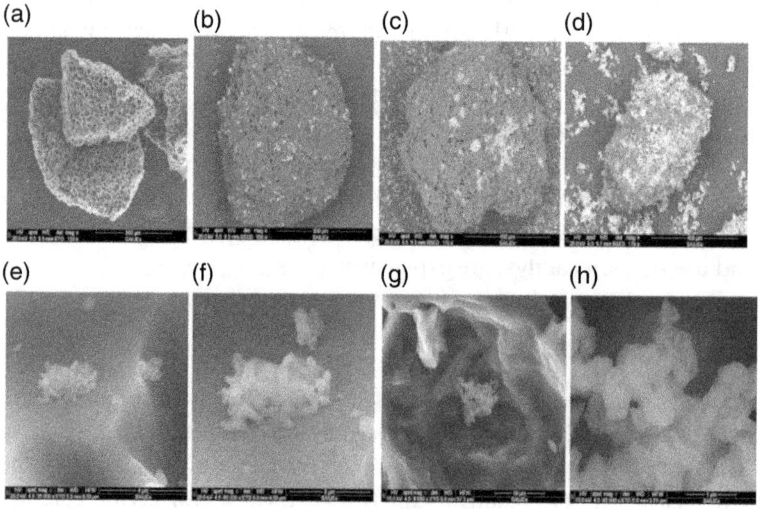


**Figure 7.3** SEM analysis of (a) biochar, (b) 5 % of Ca loaded, (c, e, f) 10% of Ca loaded, (d, g, h) 20% of Ca loaded. *Source:* Ref. [62].

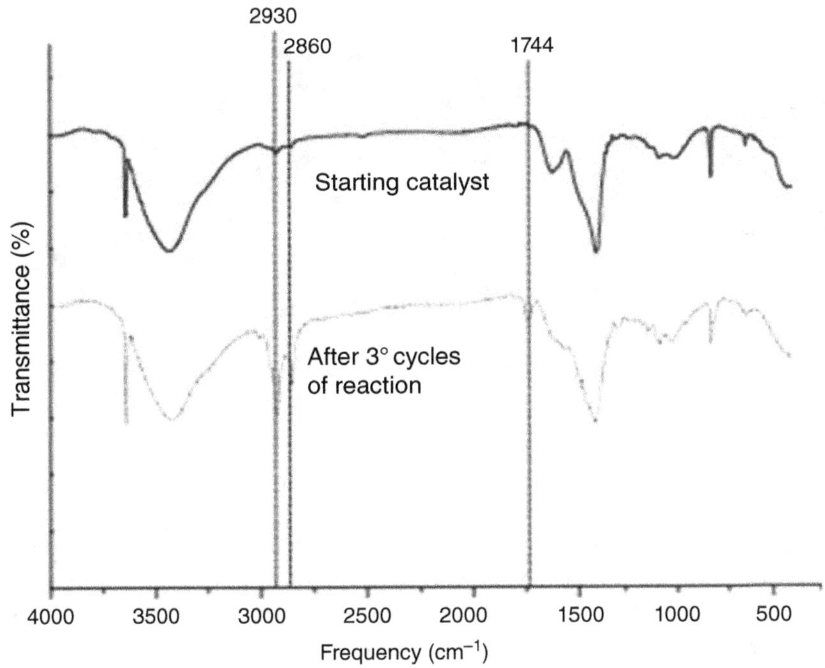


**Figure 7.4** FTIR spectrum of supported catalyst recovered at the end of third cycle of reaction. *Source:* Ref. [62].

## 7.8 Metals and Metal Oxide Supported on Metal Organic Frameworks

Metal organic frameworks (MOFs) are porous crystalline materials containing inorganic–organic hybrid moieties. It can be generated by the self-coagulation of metal clusters or metal ions with multidentate ligands. As compared with other materials (such as "mesoporous materials"), these porous crystalline materials have an improved large surface area and can be used as promising porous materials as heterogeneous catalyst supports application (Figure 7.5).

The rate of acid-catalyzed transesterification is slower than compared with that of base-catalyzed transesterification. Therefore, acid-catalyzed mechanism requires stronger reaction conditions (increased temperature, pressure, and reaction time), disqualifying the process for biodiesel production on industrial scale. Transesterification with catalysts, such as sulfuric acid, provides great conversion capacity and low cost, but water and methanol recovery needs special precautions. Also the corrosive nature of sulfuric acid restricts its use. The steric hindrance occurred by the use of the larger and bulky ligands has lowered the catalytic activity on one side whereas the simple small moieties of salts due to their high solubility and leaching tendency decrease the efficiency of the process. In order to address such issues, heterogeneous catalysts comprising Lewis metal ion coordinated with stable porous ligands that are used for esterification. MOFs are having accessible and

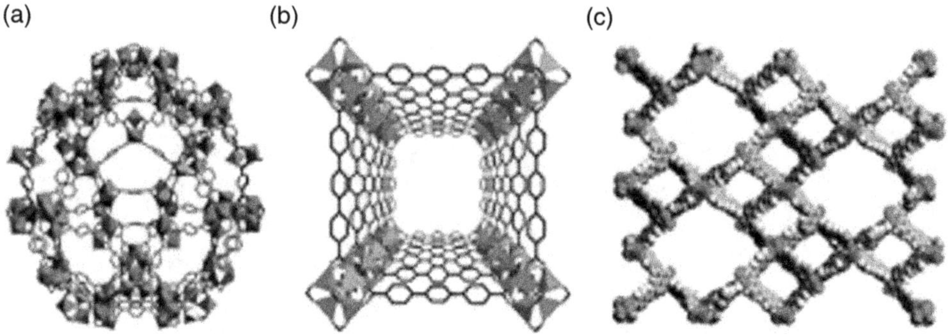


**Figure 7.5** (a) Cage type meso MOFs, (b) channel type meso MFOs, (c) chiral meso MOFs. *Source:* Ref. [63].

adjustable pore size and can be enhanced further for the acid-catalyzed transesterification reaction.

Xie et al. [64] used the polyoxometalate-based sulfonated MOF with a high acidity for transesterification reaction and obtained good conversion efficiency (97.3%). The interaction between the POM acids and the sulfonic acid groups of MOF was so strong that it was able to reduce the leaching of active species into the liquid mixture from the strongly bonded MOF support. However, when the catalyst was separated and reused, the significant loss in the catalytic activity was observed. Therefore, the advantage of catalytic activity for this catalyst was lessened by the disadvantage of severe leaching of active species present, which significantly decreased the catalytic activity of the catalyst in the further cycle. Such application of MOF-supported catalysts for biodiesel production is limited. So the synthesis of calcium-based MOF with carbonate as a versatile precursor is used for optimization of the transesterification process. It efficiently oxidizes alcohols and thus increases its heterogeneous catalytic activity for biodiesel production.

The study carried out by Jamil et al. [65] recommended that among Cu-MOF and Ca-MOF catalysts, Cu-MOF shows the biodiesel yield of 78.3%, while Ca-MOF shows the biodiesel yield of 78%. The two catalysts showed nearly same activities. But the grouping of heterogeneous catalyst Cu-MOF + Ca-MOF produced 85% of biodiesel yield, which was better as compared to the single one, making it suitable for further use. The greater acidic sites can be generated on MOF by hydrated zirconium sulfates. When the water molecule get adsorbed on the surface, hydrogen bond will be formed via Ha (Figure 7.6) with sulfate moiety, which is in turn chelated to the zirconium atom. This Ha acts as the strongest acidic proton, thus providing good and active acidic site for further catalytic applications.

Zr-doped silicotungstic acid supported on MOFs (Fe-BTC and UiO-66) was synthesized, characterized, and studied its catalytic performances by earlier workers [67]. The activity of ZrSiW/UiO-66 was found to be much better than ZrSiW/Fe-BTC catalyst, which may be associated with the ZrSiW/UiO-66 nano-hybrids that possess highly acidic character. The experimental conditions were optimized. In the reaction mechanism, carbonyl group from fatty acids abstracts proton and gets adsorbed on the surface of catalyst. Here, carbonyl group gets protonated, followed by attack of alcohol molecule to the carbonyl group, which is present in the liquid phase, and finally the water molecules get removed to produce FAME.

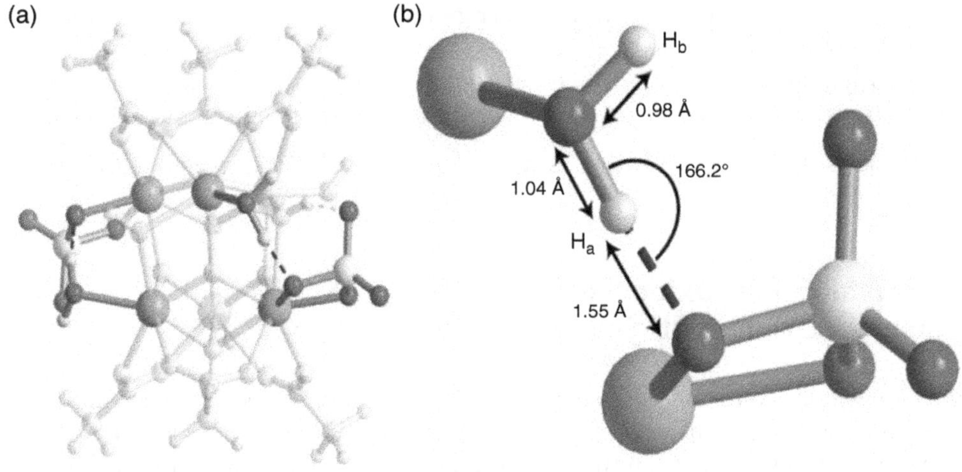


**Figure 7.6** Strongly acidic site in Zr-doped hydrated MOF (Ha is involved in hydrogen bond, Zr is indicated by blue; O indicated by red; S indicated by yellow; H indicated by white, whereas the atoms that are not directly part of the active site are represented by light gray). *Source:* Ref. [66].

Liu et al. [68] focused their research especially on the methods to access catalytic sites in MOF. The main problem arises when the substrates are larger than the pores of MOF that means the bulkier sterically hindered substrates. Research was carried out on introducing small substrates into the pores in order to make use of the "buried" catalytic sites. Those studies make use of porosity of MOF and catalytic sites only. Liu et al. focused on the crystal facets of the MOFs. They had transformed sterically hindered {111} facets to easily diffusible cube with only {100} facets and observed much enhanced catalytic activities in biodiesel production. The cubic crystals help in 90% conversion of fatty acids ($C_{12}$–$C_{22}$) in comparison with less than 22% using octahedral crystals.

The application of heterogeneous acid catalysts, chromium(III) terephthalate, and cobalt(II) terephthalate MOFs, to increase the yield of biodiesel production, was studied successfully by Marso et al. [69]. Surprisingly catalysts remained can be reused for 10 cycles. The increased recoverability may be due to the stable MOF structures and their applicability at the room temperature, which make the separation process easier. Rod-shaped Cu-benzene-1,3,5-tricarboxylic acid(CuBTc) MOF was also used as acid catalyst in the transesterification reaction. CuBTc provides unsaturated open sites for coordination with molecules of methanol and triglycerides. The mechanism can be stepped out as follows:

**Step 1:** Methanol molecules due to dipole–dipole interaction form coordinate bond with the central metal atom Cu from CuBTc, giving rise to M- CuBTc.

**Step 2:** Triglyceride now gets attached to central metal Cu atom of the previously mentioned M- CuBTc particle. The interaction is initiated by delocalization of electron between two oxygen atoms of the triglyceride, forming a resonance. This forms resonating intermediate structure MG- CuBTc.

**Step 3:** Methanol and triglyceride are in the close vicinity promoting the interaction. The nucleophilic oxygen atom from methanol hydroxyl group then attacks the electrophilic carbon at the triglyceride ester group, resulting in transesterification reaction and glycerol as a by-product.

**Step 4:** The produced biodiesel now gets detached from the catalyst through delocalization of electrons at the oxygen atoms, resulting in the formation of main product of biodiesel.

The process continues at its own, increasing the catalytic efficiency and reusability of the reaction.

## 7.9 Metal/Metal Oxide Supported on Magnetic Nanoparticles

Though heterogeneous catalysts are very much useful, still the powdered nature helps them either to coagulate or to wash out away in the separation process. This limits their catalytic activity and use in the further application at the industrial level [70]. To overcome such difficulties, magnetic nanoparticles were preferred loaded over supported catalysts [71, 72]. It makes the separation and regeneration process of catalyst more faster and easier by using an external magnetic field, thereby reducing the efforts of other conventional methods of separation [73]. The darker side of using nanoparticle is due to magnetic dipole–dipole attraction; it can form clusters, hence reducing their dispersion in the reaction mixture. Beside this, the widespread use of magnetic catalyst in the biodiesel production is well known [74–76], reporting the easier removal of used magnetic catalyst [77–79] and use of magnetic stabilized bed [80, 81], allowing catalyst to align in the reactor for further application,

Silveira et al. [82] synthesized magnetite particles based on catalyst ($K_2CO_3$, c-$Al_2O_3$, sepiolite) to catalyze the biodiesel formation, obtaining a yield of 88% at the end. The greatest advantage of such process is that such magnetic catalyst was separated after each reaction step. Such type of recovery was helpful in its reuse up to many cycles. Recently, Changmai et al. [83] reported the synthesis and application of *Citrus sinensis* (orange) peel ash-coated magnetic nanoparticles as a heterogeneous catalyst for biodiesel production from waste cooking oil. The magnetic catalyst, apart from $Fe_3O_4$ nanoparticles, contains highly basic elements such as K (8.64%) and Ca (4.46%) in high amount, which are responsible for the high activity of this catalyst. A high biodiesel yield of 98% was obtained with 6 wt% catalyst within 3 h. Physicochemical properties of synthesized biodiesel meet the ASTM standards.

Sarno et al. [84] studied biodiesel production from tomato waste using snowman-like $Au/Fe_3O_4$ catalyst (Figure 7.7). A phase transfer process is predicted during the reaction, in which $Au/Fe_3O_4$ from organic phase gets transferred to the interface between hexane and water layer containing acid moiety. At the interface ligand gets exchanged and $Fe_3O_4$ gets transferred to the aqueous phase. After some time, carboxylic group from acid binds to $Au/Fe_3O_4$ surface, which can be noticeable observed from the color change of acid to yellow due to transfer of group. At the end, such nanoparticles get capped with hydrophobic oleic acid (having more affinity for magnetic surface), increasing its catalytic efficiency due to polarity of gold-doped $Fe_3O_4$ surface. Such surface gets easily coupled with lipase, without requiring further tedious procedures and complex reagents, through physical adsorption, only showing enhanced properties.

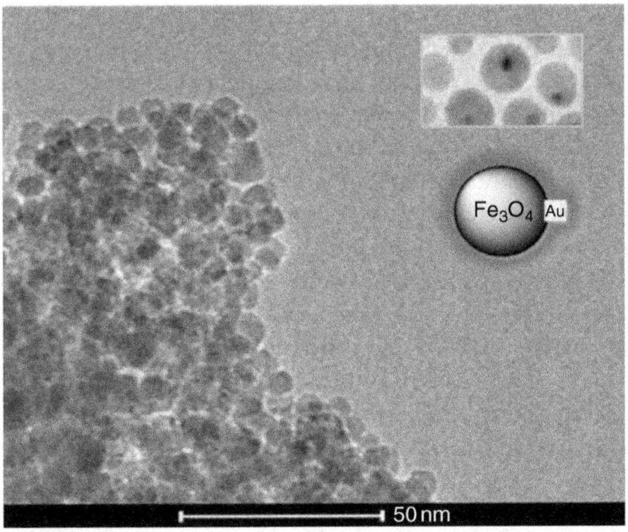


**Figure 7.7** TEM images of Au@$Fe_3O_4$ nanoparticles showing quasispherical $Fe_3O_4$ NPs (about 8 nm diameter) supporting faceted Au NPs. *Source:* Ref. [84].

Rhizopus oryzae lipase (ROL) was supported on the silica magnetic nanoparticles (MNPs) used for biodiesel production [85]. ROL was immobilized on these supports via physical adsorption and covalent attachment, increasing the enzyme loading efficiency from 67.8 to 82.4% on mesoporous silica. The evaluations of hydrolytic activity and kinetic parameters revealed that MNPs@MS-AP-GA is a suitable nanomaterial with the highest biodiesel production of 84.6% compared to other nanobiocatalysts. Ferrites contain several metal oxides with oxygen ions in a closed packed structure and cations in its interstitial spaces. However, different ferrites may be structured by replacing the divalent iron ion by another divalent metal ion, (Mn, Co, Ni, Zn, Ca, Mg, and Cd) or two different divalent ions at the same time. Any substitution will affect electrical, magnetic, and catalytic properties. The thirst for easily recoverable, more stable magnetic catalyst in biodiesel production leads to study the use of SrO doped over the surface of cobalt ferrite ($CoFe_2O_4$) [86]. $SrO/CoFe_2O_4$ (5 : 1), with 96% FAME and reusability up to four cycles, was observed. The formation of strontium hydroxide can be avoided using acetone medium for the reaction.

It can be concluded that the performance of any catalyst is dopant, and the nature of surface active sites as well as pore size is sensitive. Greener technology leads to newer area of research, satisfying the increased demand of energy from low sources.

## 7.10 Summary

Homogenous catalysts are replaced by solid (heterogeneous) catalysts to overcome its disadvantages. The highly porous surface area of catalyst favors the rate of biodiesel production. However, small particles of metal and metal oxide can cause pollution and environmental problems. Such particles must be supported on stable surface to provide them strength and help in increasing the yield. Thus production of efficient catalyst should be the primary focus before starting any reaction. Surfaces of alumina, ZnO, zeolite, silica, alumina, biochar, and iron oxide are known as a promising support for the production of biodiesel. It provides maximum strength and support for binding metal and metal oxides like Ag, Au, Cu, CaO, $K_2O$, NaOH, $K_2CO_3$, etc. The active sites generated may be acidic or basic depending on the feedstock and the oil content in it. The "basicity of the active sites" is directly proportional to the transesterification activity, whereas the "acidity of the active sites" decides the esterification activity of the catalyst. The esterification activity increases with increasing acidity of the catalyst. Controlled synthesis method is significant to improve the effectiveness of the required supported catalyst. In addition, it is very important to investigate the optimization of process parameters to get better yield.

The development of a novel supported catalyst having both acid and basic active centers residing on its magnetic surface will have a promising future in biodiesel production technologies. It is believed that such newly introduced catalysts will take a central position in the near future and help to produce biodiesel through eco-friendly and economically viable processes. This will meet the current social demand for clean energy sources.

## References


1 Othman, M.F., Adam, A., Najafi, G., and Mamat, R. (2017). Green fuel as alternative fuel for diesel engine: a review. *Renew. Sustain. Energy Rev.* 80: 694–709.

2 Brito, G.M., Chicon, M.B., Coelho, E.R.C. et al. (2020). Eco-green biodiesel production from domestic waste cooking oil by transesterification using LiOH into basic catalysts mixtures. *J. Renew. Sustain. Energy* 12 (4): 043101.

3 Martinez-Guerra, E. and Gude, V.G. (2017). Assessment of sustainability indicators for biodiesel production. *Appl. Sci.* 7 (9): 869.

4 Srivastava, N., Srivastava, M., Gupta, V.K. et al. (2018). Recent development on sustainable biodiesel production using sewage sludge. *3 Biotech* 8 (5): 1–1.

5 Estevez, R., Aguado-Deblas, L., Bautista, F.M. et al. (2019). Biodiesel at the crossroads: a critical review. *Catalysts* 9 (12): 1033.

6 Çeli̇k, M. and Bayindirli, C. (2020). Enhancement performance and exhaust emissions of rapeseed methyl ester by using n-hexadecane and n-hexane fuel additives. *Energy* 202: 117643.

7 Pathak, G., Das, D., Rajkumari, K., and Rokhum, L. (2018). Exploiting waste: towards a sustainable production of biodiesel using: *Musa acuminata* peel ash as a heterogeneous catalyst. *Green Chem.* 20 (10): 2365–2373.

8 Changmai, B., Sudarsanam, P., and Rokhum, L. (2019). Biodiesel production using a renewable mesoporous solid catalyst. *Ind. Crops Prod.* 145: 111911.

9 Kumar, L.R., Yellapu, S.K., Tyagi, R.D., and Zhang, X. (2019). A review on variation in crude glycerol composition, bio-valorization of crude and purified glycerol as carbon source for lipid production. *Bioresour. Technol.* 293: 122155.

10 Singh, D., Sharma, D., Soni, S.L. et al. (2021). A comprehensive review of physicochemical properties, production process, performance and emissions characteristics of 2nd generation biodiesel feedstock: *jatropha curcas*. *Fuel* 285: 119110.

11 Ruhul, A.M., Kalam, M.A., Masjuki, H.H. et al. (2015). State of the art of biodiesel production processes: a review of the heterogeneous catalyst. *RSC Adv.* 5 (122): 101023–101044.

**12** Deshpande, B.D., Agrawal, P.S., Yenkie, M.K.N., and Dhoble, S.J. (2020). Prospective of nanotechnology in degradation of waste water: a new challenges. *Nano-Struct. Nano-Objects* 22: 100442.

**13** Changmai, B., Vanlalveni, C., Ingle, A.P. et al. (2020). Widely used catalysts in biodiesel production: a review. *RSC Adv.* 10 (68): 41625–41679.

**14** Chakrabarti, P.P. and Prasad, R.B.N. (2012). Biodiesel production from *Jatropha curcas* oil. In: *Jatropha, Challenges for a New Energy Crop*, vol. 1 (ed. N. Carels, M. Sujatha and B. Bahadur), 3–583. New York, NY: Springer https://doi.org/10.1007/978-1-4614-4806-8_25.

**15** Lu, X., Ng, Y.H., and Zhao, C. (2014). Gold nanoparticles embedded within mesoporous cobalt oxide enhance electrochemical oxygen evolution. *ChemSusChem.* 7 (1): 82–86.

**16** Védrine, J.C. (2017). Heterogeneous catalysis on metal oxides. *Catalysts* 7 (11): 341.

**17** Terörde, R.J., Van den Brink, P.J., Visser, L.M. et al. (1993). Selective oxidation of hydrogen sulfide to elemental sulfur using iron oxide catalysts on various supports. *Catal. Today.* 17 (1–2): 217–224.

**18** Mejía, C.H., Vogt, C., Weckhuysen, B.M., and de Jong, K.P. (2020). Stable niobia-supported nickel catalysts for the hydrogenation of carbon monoxide to hydrocarbons. *Catal. Today* 343: 56–62.

**19** Abdel-Mageed, A.M., Rungtaweevoranit, B., Parlinska-Wojtan, M. et al. (2019). Highly active and stable single-atom Cu catalysts supported by a metal-organic framework. *J. Am. Chem. Soc.* 141 (13): 5201–5210.

**20** Dutta, K., Chaudhari, V., Li, C.J., and Kopyscinski, J. (2020). Methane conversion to ethylene over GaN catalysts. Effect of catalyst nitridation. *Appl. Catal. A Gen.* 595: 117430.

**21** Ahn, S., Thornburg, N.E., Li, Z. et al. (2016). Stable metal-organic framework-supported niobium catalysts. *Inorg. Chem.* 55 (22): 11954–11961.

**22** Arzamendi, G., Campo, I., Arguiñarena, E. et al. (2007). Synthesis of biodiesel with heterogeneous NaOH/alumina catalysts: comparison with homogeneous NaOH. *Chem. Eng. J.* 134 (1–3): 123–130.

**23** Ebiura, T., Echizen, T., Ishikawa, A. et al. (2005). Selective transesterification of triolein with methanol to methyl oleate and glycerol using alumina loaded with alkali metal salt as a solid-base catalyst. *Appl. Catal. A Gen.* 283 (1–2): 111–116.

**24** Bota RM, Houthoofd K, Grobet PJ, Jacobs PA. Alkali metal supported on mesoporous alumina as basic catalysts for fatty acid methyl esters preparation. *Studies in Surface Science and Catalysis.* 2010: 175: 775–778. DOI: https://doi.org/10.1016/S0167-2991(10)75158-2.

**25** Alonso, D.M., Mariscal, R., Moreno-Tost, R. et al. (2007). Potassium leaching during triglyceride transesterification using K/$\gamma$-$Al_2O_3$ catalysts. *Catal. Commun.* 8 (12): 2074–2080.

**26** Xie, W. and Li, H. (2006). Alumina-supported potassium iodide as a heterogeneous catalyst for biodiesel production from soybean oil. *J. Mol. Catal. A Chem.* 255 (1–2): 1–9.

**27** Noiroj, K., Intarapong, P., Luengnaruemitchai, A., and Jai-In, S. (2009). A comparative study of KOH/$Al_2O_3$ and KOH/NaY catalysts for biodiesel production via transesterification from palm oil. *Renew. Energy* 34 (4): 1145–1150.

**28** Lukić, I., Krstić, J., Jovanović, D., and Skala, D. (2009). Alumina/silica supported $K_2CO_3$ as a catalyst for biodiesel synthesis from sunflower oil. *Bioresour. Technol.* 100 (20): 4690–4696.

**29** Vyas, A.P., Subrahmanyam, N., and Patel, P.A. (2009). Production of biodiesel through transesterification of Jatropha oil using $KNO_3/Al_2O_3$ solid catalyst. *Fuel* 88 (4): 625–628.

**30** Zabeti, M., Mohd, W., Wan, A., and Aroua, M.K. (2010). Biodiesel production using alumina-supported calcium oxide: an optimization study. *Fuel Process. Technol.* 91 (2): 243–248.

**31** Pasupulety, N., Gunda, K., Liu, Y. et al. (2013). Production of biodiesel from soybean oil on $CaO/Al_2O_3$ solid base catalysts. *Appl. Catal. A Gen.* 452: 189–202.

**32** Marinković, D.M., Stanković, M.V., Veličković, A.V. et al. (2016). Calcium oxide as a promising heterogeneous catalyst for biodiesel production: current state and perspectives. *Renew. Sustain. Energy Rev.* 56: 1387–1408.

**33** Morris, S.M., Fulvio, P.F., and Jaroniec, M. (2008). Ordered mesoporous alumina-supported metal oxides. *J. Am. Chem. Soc.* 130 (45): 15210–15216.

**34** Da Costa Evangelista, J.P., Gondim, A.D., Di Souza, L., and Araujo, A.S. (2016). Alumina-supported potassium compounds as heterogeneous catalysts for biodiesel production: a review. *Renew. Sustain. Energy Rev.* 59: 887–894.

**35** Marinkovic, M., Stojkovic, N., Vasic, M. et al. (2016). Synthesis of biodiesel from sunflower oil over potassium loaded alumina as heterogeneous catalyst: the effect of process parameters. *Hem. Ind. Ind.* 70 (6): 639–648.

**36** Mierczynski, P., Mosinska, M., Szkudlarek, L. et al. (2021). Biodiesel production on monometallic pt, pd, ru, and ag catalysts supported on natural zeolite. *Materials (Basel)* 14 (1): 48.

**37** Lawan, I., Garba, Z.N., Zhou, W. et al. (2020). Synergies between the microwave reactor and CaO/zeolite catalyst in waste lard biodiesel production. *Renew. Energy* 145: 2550–2560.

**38** Lestari, W.A., Istiqomah, A., Krisdiyanto, D., and Afrianto, A.W. (2020). Preparation of Ni/Zeolite as esterification catalyst in biodiesel production based on off-grade crude palm oil. *IOP Conf. Ser.: Mater. Sci. Eng.* 858: 1(012015).

**39** Li, Z., Ding, S., Chen, C. et al. (2019). Recyclable Li/NaY zeolite as a heterogeneous alkaline catalyst for biodiesel production: process optimization and kinetics study. *Energy Convers. Manag.* 192: 335–345.

**40** Singh, D., Bhoi, R., Ganesh, A., and Mahajani, S. (2014). Synthesis of biodiesel from vegetable oil using supported metal oxide catalysts. *Energy Fuels* 28 (4): 2743–2753.

**41** Ariffin, M.I., Abdullah, N., Bahruji, H. et al. (2019). Optimization of biodiesel production from *Chlorella Vulgaris* using CuO, NiO and MgO on zeolite 4A catalysts. *J. Phys. Conf. Ser.* 1349: 1(012065).

**42** Rao, A.V.R.K., Dudhe, P., and Chelvam, V. (2021). Role of oxygen defects in basicity of Se doped ZnO nanocatalyst for enhanced triglyceride transesterification in biodiesel production. *Catal. Commun.* 149: 106258.

**43** Baskar, G., Aberna Ebenezer Selvakumari, I., and Aiswarya, R. (2018). Biodiesel production from castor oil using heterogeneous Ni doped ZnO nanocatalyst. *Bioresour. Technol.* 250: 793–798.

**44** Laskar, I.B., Rokhum, L., Gupta, R., and Chatterjee, S. (2019). Zinc oxide supported silver nanoparticles as a heterogeneous catalyst for production of biodiesel from palm oil. *Environ. Prog. Sustain. Energy* 39 (3): e13369.

**45** Saxena, V., Sharma, S., and Pandey, L.M. (2019). Fe(III) doped ZnO nano-assembly as a potential heterogeneous nano-catalyst for the production of biodiesel. *Mater. Lett.* 237: 232–235.

**46** Borah, M.J., Devi, A., Borah, R., and Deka, D. (2019). Synthesis and application of Co doped ZnO as heterogeneous nanocatalyst for biodiesel production from non-edible oil. *Renew. Energy* 133: 512–519.

**47** Istadi, I., Anggoro, D.D., Buchori, L. et al. (2015). Active acid catalyst of sulphated Zinc oxide for transesterification of soybean oil with methanol to biodiesel. *Procedia Environ. Sci.* 23: 385–393.

**48** Yan, S., Salley, S.O., and Simon Ng, K.Y. (2009). Simultaneous transesterification and esterification of unrefined or waste oils over ZnO-$La_2O_3$ catalysts. *Appl. Catal. A Gen.* 353 (2): 203–212.

**49** Toledo Arana, J., Torres, J.J., Acevedo, D.F. et al. (2019). One-step synthesis of CaO-ZnO efficient catalyst for biodiesel production. *Int. J. Chem. Eng.* 2019: 1–7.

**50** Albuquerque, M.C.G., Jiménez-Urbistondo, I., Santamaría-González, J. et al. (2008). CaO supported on mesoporous silicas as basic catalysts for transesterification reactions. *Appl. Catal. A Gen.* 334 (1–2): 35–43.

**51** Lani, N.S., Ngadi, N., and Inuwa, I.M. (2020). New route for the synthesis of silica-supported calcium oxide catalyst in biodiesel production. *Renew. Energy* 156: 1266–1277.

**52** Lani, N.S., Ngadi, N., Yahya, N.Y., and Rahman, R.A. (2017). Synthesis, characterization and performance of silica impregnated calcium oxide as heterogeneous catalyst in biodiesel production. *J. Clean. Prod.* 146: 116–124.

**53** Melero, J.A., Bautista, L.F., Iglesias, J. et al. (2012). Zr-SBA-15 acid catalyst: optimization of the synthesis and reaction conditions for biodiesel production from low-grade oils and fats. *Catal. Today* 195 (1): 44–53.

**54** Chen, X.R., Ju, Y.H., and Mou, C.Y. (2007). Direct synthesis of mesoporous sulfated silica-zirconia catalysts with high catalytic activity for biodiesel via esterification. *J. Phys. Chem. C.* 111 (50): 18731–18737.

**55** Lee, J., Kim, K.H., and Kwon, E.E. (2017). Biochar as a catalyst. *Renew. Sustain. Energy Rev.* 77: 70–79.

**56** Vakros, J. (2018). Biochars and their use as transesterification catalysts for biodiesel production: a short review. *Catalysts* 8 (11): 562.

**57** Arata, K. (1996). Preparation of superacids by metal oxides for reactions of butanes and pentanes. *Appl. Catal. A Gen.* 146 (1): 3–32.

**58** Tatsumi, T., Matsuhashi, H., and Arata, K. (1996). A study of the preparation procedures of sulfated zirconia prepared from zirconia gel. The effect of the pH of the mother solution on the isomerization activity of n-pentane. *Bull. Chem. Soc. Japan* 69 (5): 1191–1194.

**59** Matsuhashi, H., Miyazaki, H., Kawamura, Y. et al. (2001). Preparation of a solid superacid of sulfated tin oxide with acidity higher than that of sulfated zirconia and its applications to aldol condensation and benzoylation. *Chem. Mater.* 13 (9): 3038–3042.

**60** Furuta, S., Matsuhashi, H., and Arata, K. (2004). Catalytic action of sulfated tin oxide for etherification and esterification in comparison with sulfated zirconia. *Appl. Catal. A Gen.* 269 (1–2): 187–191.

61 Dehkhoda, A.M., West, A.H., and Ellis, N. (2010). Biochar based solid acid catalyst for biodiesel production. *Appl. Catal. A Gen.* 382 (2): 197–204.

62 di Bitonto, L., Reynel-Ávila, H.E., Mendoza-Castillo, D.I. et al. (2020). Synthesis and characterization of nanostructured calcium oxides supported onto biochar and their application as catalysts for biodiesel production. *Renew. Energy* 382 (2): 197–204.

63 Xuan, W., Zhu, C., Liu, Y., and Cui, Y. (2012). Mesoporous metal–organic framework materials. *Chem. Soc. Rev.* 41 (5): 1677–1695.

64 Xie, W. and Wan, F. (2019). Immobilization of polyoxometalate-based sulfonated ionic liquids on UiO-66-2COOH metal-organic frameworks for biodiesel production via one-pot transesterification-esterification of acidic vegetable oils. Chem. *Eng. J.* 365: 40–50.

65 Jamil, U., Husain Khoja, A., Liaquat, R. et al. (2020). Copper and calcium-based metal organic framework (MOF) catalyst for biodiesel production from waste cooking oil: a process optimization study. *Energy Convers. Manag.* 215: 112934.

66 Trickett, C.A., Osborn Popp, T.M., Su, J. et al. (2019). Identification of the strong Brønsted acid site in a metal–organic framework solid acid catalyst. *Nat. Chem.* 11 (2): 170–176.

67 Zhang, Q., Lei, D., Luo, Q. et al. (2020). Efficient biodiesel production from oleic acid using metal-organic framework encapsulated Zr-doped polyoxometalate nano-hybrids. *RSC Adv.* 10 (15): 8766–8772.

68 Liu, Y., Liu, S., He, D. et al. (2015). Crystal facets make a profound difference in polyoxometalate-containing metal-organic frameworks as catalysts for biodiesel production. *J. Am. Chem. Soc.* 137 (39): 12697–12703.

69 Marso, T.M.M., Kalpage, C.S., and Udugala-Ganehenege, M.Y. (2020). Application of chromium and cobalt terephthalate metal organic frameworks as catalysts for the production of biodiesel from calophyllum inophyllum oil in high yield under mild conditions. *J. Inorg. Organomet. Polym. Mater.* 30 (4): 1243–1265.

70 De Lima, A.L., Ronconi, C.M., and Mota, C.J.A. (2016). Heterogeneous basic catalysts for biodiesel production. *Catal. Sci. Technol.* 6 (9): 2877–2891.

71 Rokhum, L. and Pathak, G. (2019). Synthesis, characterization and catalytic activity of magnetic KI@$Fe_3O_4$ nanoparticles for Henry reaction under solvent free conditions. *Catal. Lett.* 149 (10): 2887–2898.

72 Rajkumari, K. and Rokhum, L. (2020). A sustainable protocol for production of biodiesel by transesterification of soybean oil using banana trunk ash as a heterogeneous catalyst. *Biomass Convers. Biorefin.* 10 (4): 839–848.

73 Netto, C.G.C.M., Toma, H.E., and Andrade, L.H. (2013). Superparamagnetic nanoparticles as versatile carriers and supporting materials for enzymes. *J. Mol. Catal. B: Enzym.* 85: 71–92.

74 Alves, M.B., Medeiros, F.C.M., Sousa, M.H. et al. (2014). Cadmium and tin magnetic nanocatalysts useful for biodiesel production. *J. Braz. Chem. Soc.* 25: 2304–2313.

75 Cubides-Roman, D.C., Pérez, V.H., de Castro, H.F. et al. (2017). Ethyl esters (biodiesel) production by Pseudomonas fluorescens lipase immobilized on chitosan with magnetic properties in a bioreactor assisted by electromagnetic field. *Fuel* 196: 481–487.

76 Zhang, F., Fang, Z., and Wang, Y.T. (2015). Biodiesel production directly from oils with high acid value by magnetic $Na_2SiO_3$@$Fe_3O_4$/C catalyst and ultrasound. *Fuel* 150: 370–377.

77 Xie, W. and Zang, X. (2016). Immobilized lipase on core-shell structured $Fe_3O_4$-MCM-41 nanocomposites as a magnetically recyclable biocatalyst for interesterification of soybean oil and lard. *Food Chem.* 194: 1283–1292.

**78** Xie, W. and Wan, F. (2018). Basic ionic liquid functionalized magnetically responsive $Fe_3O_4$@HKUST-1 composites used for biodiesel production. *Fuel* 220: 248–256.

**79** Quah, R.V., Tan, Y.H., Mubarak, N.M. et al. (2019). An overview of biodiesel production using recyclable biomass and non-biomass derived magnetic catalysts. *J. Environ. Chem. Eng.* 7 (4): 103219.

**80** Chen, G., Liu, J., Yao, J. et al. (2017). Biodiesel production from waste cooking oil in a magnetically fluidized bed reactor using whole-cell biocatalysts. *Energy Convers. Manag.* 138: 556–564.

**81** Hajar, M. and Vahabzadeh, F. (2016). Biolubricant production from castor oil in a magnetically stabilized fluidized bed reactor using lipase immobilized on $Fe_3O_4$ nanoparticles. *Ind. Crops Prod.* 94: 544–556.

**82** Silveira Junior, E.G., Justo, O.R., Perez, V.H. et al. (2018). Extruded catalysts with magnetic properties for biodiesel production. *Adv. Mater. Sci. Eng.* 2018: 3980967.

**83** Rajkumari, K., Changmai, B., Meher, K. et al. (2021). A reusable magnetic nanocatalyst for bio-fuel additives: the ultrasound-assisted synthesis of solketal. *Sustain. Energy Fuels* 5 (8): 2362–2372.

**84** Sarno, M. and Iuliano, M. (2019). Highly active and stable $Fe_3O_4$/Au nanoparticles supporting lipase catalyst for biodiesel production from waste tomato. *Appl. Surf. Sci.* 474: 135–146.

**85** Esmi, F., Nematian, T., Salehi, Z. et al. (2021). Amine and aldehyde functionalized mesoporous silica on magnetic nanoparticles for enhanced lipase immobilization, biodiesel production, and facile separation. *Fuel* 291: 120126.

**86** Falcão, M.S., Garcia, M.A.S., De Moura, C.V.R. et al. (2018). Synthesis, characterization and catalytic evaluation of magnetically recoverable $SrO/CoFe_2O_4$ nanocatalyst for biodiesel production from babassu oil transesterification. *J. Braz. Chem. Soc.* 29: 845–855.

# 8

## Mixed Metal Oxide Catalysts in Biodiesel Production

*Brandon Lowe[1], Jabbar Gardy[1], Kejun Wu[1,2], and Ali Hassanpour[1]*

[1] *School of Chemical and Process Engineering, University of Leeds, Leeds, UK*
[2] *School of Chemical and Biological Engineering, Zhejiang University, Hangzhou, P.R. China*

### 8.1 Introduction

In our collective attempts to reduce global greenhouse gas emissions, the development of renewable biodiesels offers a potential route to decarbonizing the transport industry. Biodiesel is predominantly produced by the transesterification and/or esterification-catalyzed reactions of biomass-derived fats and oils [1], as illustrated in Figure 8.1. Produced alkyl esters are purified, separated, and can then be mixed with conventional fossil fuels as a biodiesel blend suitable for current engine technology.

As previously discussed in Chapters 1–4, a variety of biodiesel feedstocks exist, for example, vegetable oils, animal fats, waste cooking oil (WCO), nonedible oils, and algal-based oils. The conversion of cheap, low quality used cooking oil (UCO)/WCO feedstocks is especially desirable. This could create a sustainable circular economy while preventing competition with food supplies. A wide range of catalyst types are under development and are summarized in Figure 8.2. An estimated 90% of current biodiesel production uses homogeneous base catalysts [2]. Base catalysts are generally economically superior, since higher reactivity minimizes the catalyst required. However, they are unsuitable for converting WCO to biodiesel due to high content of water, free fatty acids (FFA), and other impurities that promote catalyst deactivation and soap formation [3]. While homogeneous acid catalysts are more tolerant to low quality feedstocks, this is at the expense of slower reactions and acidified fuel. Since homogeneous catalysts cannot be easily separated, a high-performance cleanup of biofuel would be required prior to use. Solid heterogeneous catalysts instead offer simpler purification and easier separability for reuse [4].

Known challenges for heterogeneous catalysts are low surface area, small pore size distributions, stability issues, and lower reactivity, leading to suboptimal reaction times and temperatures, which nevertheless must be addressed [1]. Other catalytic approaches, such as the use of enzymes and ionic liquids, are currently still too expensive for biodiesel production [5]. Hence, there is a major drive to develop better heterogeneous catalysts that can

*Biodiesel Production: Feedstocks, Catalysts, and Technologies*, First Edition. Edited by Samuel Lalthazuala Rokhum, Gopinath Halder, Suttichai Assabumrungrat, and Kanokwan Ngaosuwan.

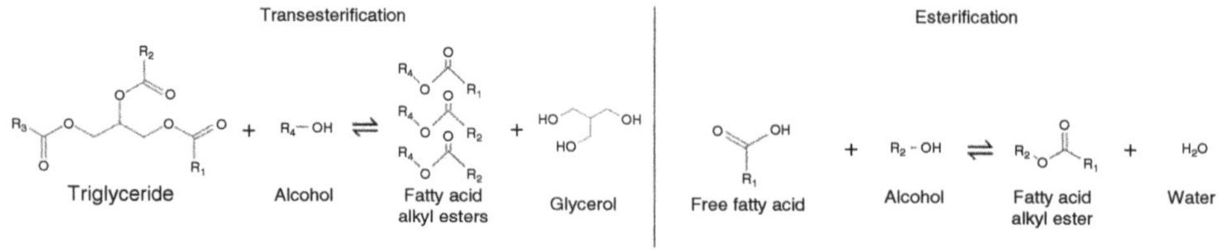


**Figure 8.1** Transesterification and esterification reactions.

maintain both high reactivity, stability, and recovery for continuous industrial application. This chapter will focus on mixed metal oxides (MMOs), a particularly interesting class of heterogeneous catalyst, with a diverse range of applications illustrated in Figure 8.3 [6]. MMO materials are defined here as the combination of multiple different metals and metal oxide species mixed together. This is different to the metal oxide catalysts discussed in Chapter 6, which contain only a singular metal oxide species. The synergistic effect of mixing several species together can lead to several advantages. Structural and textural properties such as surface area [7], pore size [8], and stability [9] could be improved. Enhanced catalytic activity via improved acidity [10], basicity [11], or reactivity [12] may also be achieved, or even resistance to lower quality feedstocks without pretreatment [3]. Optimization of such properties may lead to improved biodiesel yields at reduced reaction times or temperatures. Inclusion of metallic species such as oxides of iron could even improve recovery and separability by external magnetic fields [13]. MMO catalysts can be synthesized through a variety of routes [14], coprecipitation, impregnation, encapsulation, and combustion synthesis to name just a few. Calcination of layered double hydroxides (LDH) such as hydrotalcite (HT) [15] or exploiting the mixed metals and oxides naturally occurring in biomass is additional route to produce new catalysts. In 2012, Borges et al. [16] reviewed a wide variety of heterogeneous catalysts for biodiesel synthesis. More recent reviews by Gardy et al. [1] and Vasić et al. [17] continue to highlight developments in MMO materials. There is a wealth of research underway to develop superior MMO catalysts, with many new or improved catalysts synthesized since prior reviews. Therefore, an update into the state of the field is appropriate.

## 8.2 Previous Research

Before exploring the most recent literature, a brief overview of prior work should be discussed. A selection of previous research is summarized in Table 8.1, sorted alphabetically by type of feedstock used, symbols and abbreviations used throughout are also listed later in Table 8.5. The primary goal of past work has been either to develop new MMO catalysts, optimizing current MMO materials, or to investigate synthesis and reaction conditions. While there has been work undertaken to improve recyclability, stability issues such as leaching and poisoning still need to be addressed. Regeneration attempts explored so far mainly include washing, drying, and sometimes recalcination. Although a wide variety of feedstocks have been explored, the vast majority of experiments to date were performed in batch mode. Different challenges may arise for continuous mode, which need to be addressed, as the industry is shifting toward continuous processes and intensification.

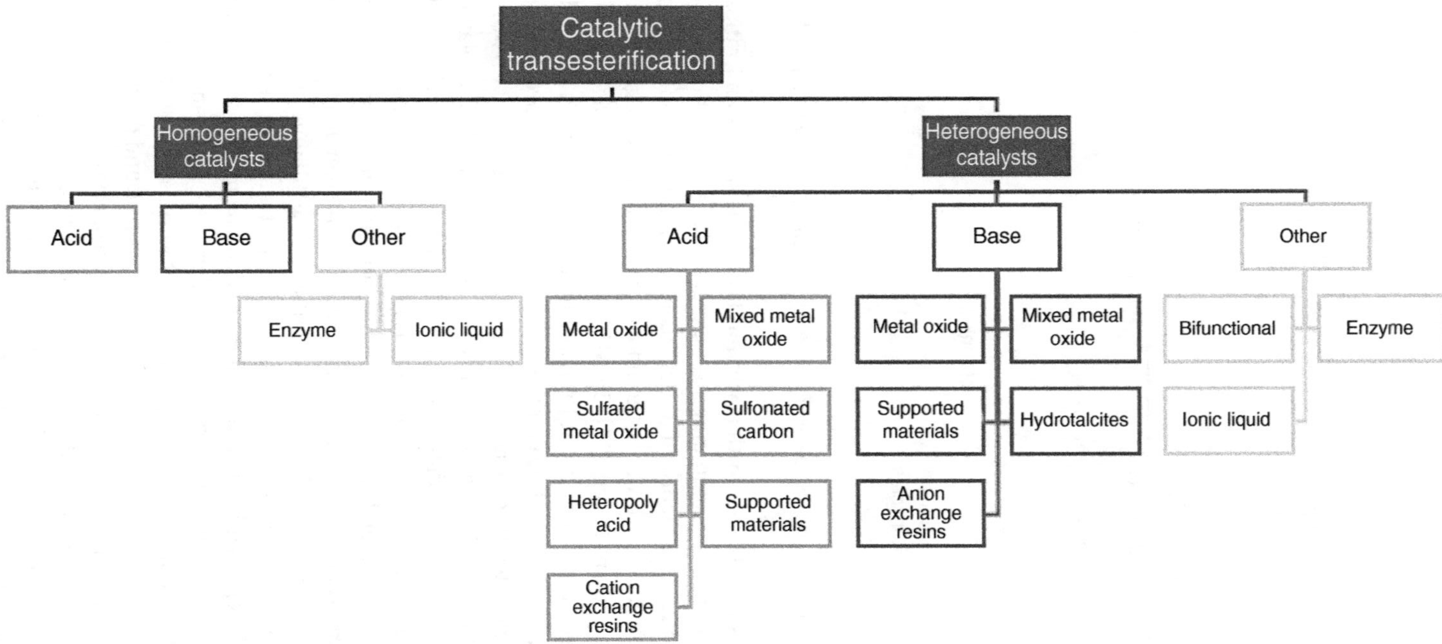


**Figure 8.2** Summary of catalyst types currently being researched. *Source:* Adapted from Ref. [1].

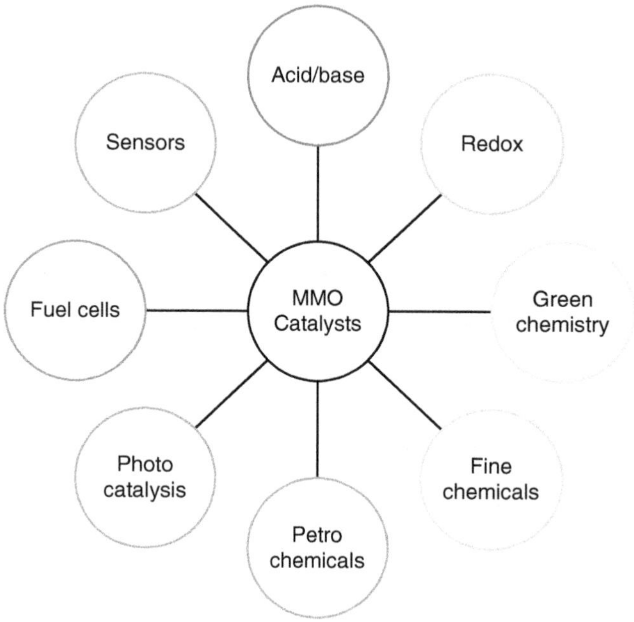


**Figure 8.3** A summary showing the diverse uses for MMO catalysts. *Source:* Based on Ref. [6].

Previous work has clearly shown that varying the metal species and ratios used during MMO synthesis can lead to predominantly acidic, basic, or bifunctional catalytic activity. The surface structures and theorized initial mechanisms are shown in Figure 8.4.

The transesterification pathway is thought to proceed by hydrogen removal from the alcohol to form an alkoxide ion [25]. In MMO base catalysis this can proceed via the electron-rich oxygen atom acting as a Brønsted base [26] or a Lewis base [5]. For sulfated acidic surfaces, Anuradha et al. [21] found that Lewis acid sites favor transesterification, while Brønsted acid sites favor esterification. This result agrees with work from Gardy et al. [3] that demonstrated a predominantly Brønsted acid catalyst displaying high resistance to FFA impurities via esterification. Jamil et al. [8] also reported how acidic and basic sites selectively promote esterification and transesterification, respectively. Combining highly acidic and basic metal oxides, or doping amphoteric oxides to tune surface site behavior, can therefore produce bifunctional catalysts. Prabu et al. [12] reported how varying Zn content allowed for tunable basicity, while Reyna-Villanueva et al. [15] found the Mg/Al cation ratio to be a key factor determining catalytic activity. Lee et al. [27] and Xie et al. [9] have both reported improved catalytic activity by mixing metals and their oxides together. Greater stability has been observed by Gardy et al. [3] and Alhassan et al. [7]. While Borah et al. [4] and Jamil et al. [8] noted that MMO catalysts could achieve higher surface area or pore size, respectively, this is not always true and depends strongly on what metal or ratios are utilized. Excessive amounts of any species are usually counterproductive. Both Boro et al. [11] and Ambat et al. [20] observed excess dopant blocking active sites, with Prabu et al. [12] noting a reduction in biodiesel yield and pore size. Both reaction and synthesis conditions can have a significant impact on catalyst performance. Unfortunately, these variables appear to be highly catalyst dependent and should therefore be investigated and optimized for all new MMO catalysts.

**Table 8.1** Summary of some previous developments in mixed metal oxide catalysts.

| Feedstock | A:O ratio[a] | Catalyst | Catalyst loading (wt%) | Reaction time (min) | Particle size (nm) | Acidity (A) basicity (B) (mmol $g^{-1}$) | T (°C) | Yield Y conversion C | Reusability (runs) | Amount leached | Regen. method[b] | References |
|---|---|---|---|---|---|---|---|---|---|---|---|---|
| Canola oil | 16.14 : 1 | $KOH/Fe_3O_4/\gamma\text{-}Al_2O_3$ | 6.45 | 326 | 11.9 | — | 65 | 97.4% Y | 1 | — | W, D, R | [18] |
| Cotton seed oil | 24 : 1 | 5-Na/ZnO/SBA-15 | 12 | 240 | ~16 | — | 65 | > 98–93% Y | 4 | — | — | [5] |
| Date palm kernel oil | 15 : 1 | $Mn/MgO\text{–}ZrO_2$ | 3 | 240 | 18–39 | — | 90 | 96.4–>90% Y | 5 | Zn: 7.7% Co: 2.33% | W, D | [8] |
| Jatropha oil | 9.88 : 1 | $SO_4\text{–}ZrO_2/Al_2O_3$ | 7.61 | 240 | — | 2.1232 A 3.4775 B | 150 | 90% Y | — | Mg: 0.96% Mn: 0.63% | W, D, R | [19] |
| *Mesua ferrea* oil | 9 : 1 | Co/ZnO | 2.5 | 180 | ~27.8 | — | 60 | ~98% C | 2 | K: ~0.24 Ce: ~0.15 (mg $l^{-1}$) | W, D, R | [4] |
| Rapeseed oil | 7 : 1 | $Fe_3O_4\text{-}CeO_2\text{-}25K$ | 3 | 120 | 20–33.9 | — | 60 | 96.13–80.94% Y | 5 | — | None | [20] |
| Rapeseed oil fatty acids | 1 ml $g^{-1}$ | $SO_4^{2-}/ZrO_2\text{–}TiO_2/La^{3+}$ | 5 | 300 | — | — | 60 | 95–>90% C | 5 | $SO_4^{2-}$: 3.4% | W, D | [10] |
| Soybean oil | 20 : 1 | Sulfated $Fe_2O_3\text{–}TiO_2$ | 15 | 120 | 89–103 | — | 100 | ~100–~80% C | 4 | Ca: 21 Mo: 13 (ppm) | W, R | [21] |
| Soybean oil | 50 : 1 | $CaO\text{–}MoO_3\text{–}SBA\text{-}15$ | 6 | 3000 | — | 0.18 B | ~65 | 83.2–77.2% C | 5 | — | — | [9] |

*(Continued)*

**Table 8.1** (Continued)

| Feedstock | A:O ratio[a] | Catalyst | Catalyst loading (wt%) | Reaction time (min) | Particle size (nm) | Acidity (A) basicity (B) (mmol $g^{-1}$) | T (°C) | Yield Y conversion C | Reusability (runs) | Amount leached | Regen. method[b] | References |
|---|---|---|---|---|---|---|---|---|---|---|---|---|
| Soybean oil | 20 : 1 | Mg/Al/Zn/ SBA-15 | — | 50 (residence time) | $5–8 \times 10^5$ | 0.088 A 0.036 B | 225 | 85–90% Y | 200hr (continuous) | 46.9% acid sites, 38.5% base sites | W, R / W, D | [12] |
| Soybean oil jatropha oil | 15 : 1 | $CaFe_2O_4–Ca_2Fe_2O_5$ | 4 | 30 | 200 and $1–2 \times 10^3$ | 9.087 B | 100 | 83.5%78.2% | 1 | Ca: 0.061% | W | [13] |
| Sunflower oil | 12 : 1 | $CaO/\gamma$-$Al_2O_3$ | 0.5 | 300 | $1–3 \times 10^6$ | — | 60 | 94.3–76.4% Y | 2 | — | — | [22] |
| Sunflower oil | 23 : 1 | LDH-derived Mg/Al | 3.5 | 60 | — | 0.27 B | 65 | 98.59% C | — | — | — | [15] |
| Vegetable oil | 12 : 1 | $Cs–Ca/SiO_2–TiO_2$ | 2 | 120 | 45 | 5.19 B | 60 | ~98% Y | 4 | Si: 4.50, Na: 0.68, Zn: 0.60 (ppm) | W, D, R | [23] |
| WCO | 6 : 1 | Ba-CaO(T striatula shell) | 1 | 180 | — | 0.2375 B | 65 | >98–~90% C | 3 | — | W, D | [11] |
| WCO | 20 : 1 | $Fe_2O_3–MnO–SO_4^{2-}/ZrO_2$ | 3 | 240 | 11.5–21 | 5.4724 A 1.519 B | 180 | ~96.5% Y | 6 | S: 50 (ppm) | W, D | [7] |
| WCO | 10 : 1 | $SO_4/Fe–Al–TiO_2$ | 3 | 150 | <50 | 1.18 A | 90 | 96–>90% Y | 10 | — | W, R | [3] |
| WCO | 21 : 1 | $Zn_{1-x}Mn_xFe_2O_4$ $(0 \leq x \leq 0.5)$ | 4 | 50 | 7.49–9.65 | — | 65 | 98.3–93.8% Y | 10 | — | W, D | [14] |
| WCO | 20 : 1 | Zn–CaO(eggshell) | 5 | 240 | ~34.2 | — | 65 | 96.74–~80% C | 4 | — | W, D | [24] |

[a] MeOH unless stated otherwise.
[b] Regeneration route included washing (W), drying (D), or recalcination (R).

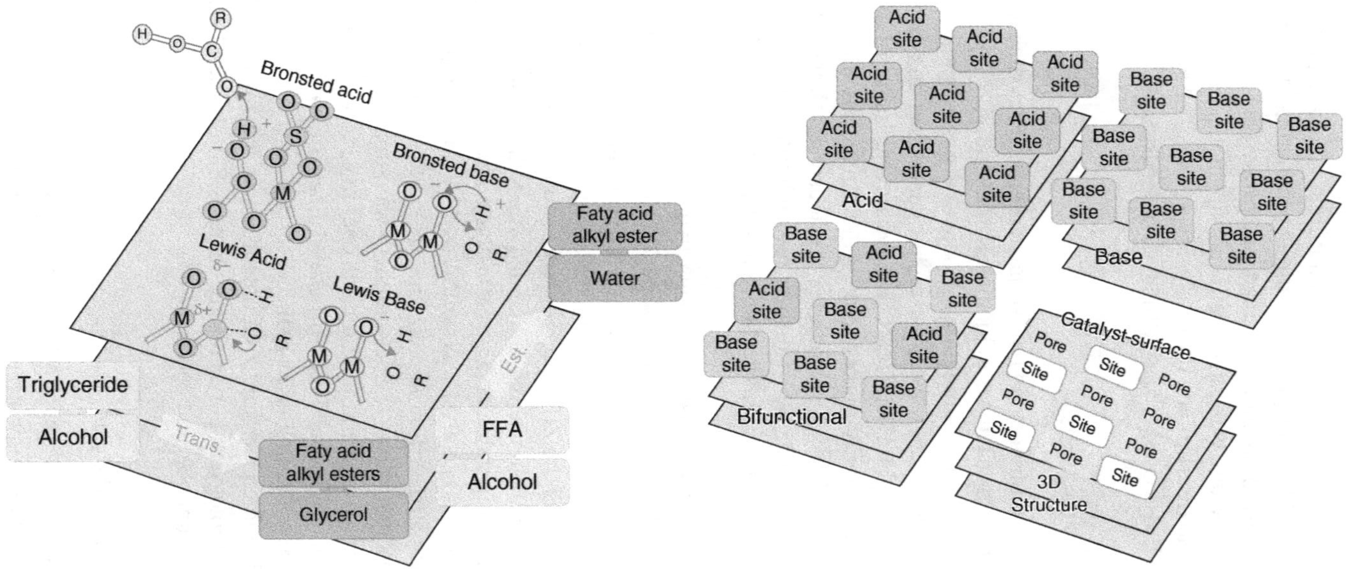


**Figure 8.4** Diagram of surface structures for acidic, basic, and bifunctional MMO catalysts.

## 8.3 State of the Art

To determine the current state of the field, a literature review was performed, focusing on recent developments in MMO catalysts for biodiesel production. Acidic, basic, and bifunctional catalysts were all explored, alongside magnetic catalysts and those derived from biomass and LDH sources. Where categorization was ambiguous, the most significant active site behavior was used to discriminate the catalyst type. Table 8.2 provides a key summary of all studies discussed, also categorized by the type of feedstock.

### 8.3.1 Solid Acid MMO Catalysts

The application of various sulfated and non-sulfated metal oxides such as Ti, Zn, Zr and Sn have been explored for use as catalysts in biodiesel production [2, 17], demonstrating both Lewis and Brønsted type acidity. Sulfonation introduces Brønsted sites which can preferentially remove FFA impurities often found at higher concentrations in impure feedstocks like UCO/WCO. Moreover, solid acid MMO catalysts avoid issues of soap formation via saponification [3]. This naturally makes acidic MMO materials highly desirable for the conversion of low grade feedstocks to biodiesel. Previously developed MMO acid catalysts are displayed earlier in Table 8.1. While further research into developing new catalysts in this area is strongly encouraged, in this section, a few examples of solid acid MMO catalysts are discussed in detail.

Fan et al. [38] performed sulfonation through two different approaches, a post gifting $ZrO_2$–$TiO_2$–$SO_3H$ catalyst and an impregnation $ZrO_2$–$TiO_2$–$SO_4^{2-}$ catalyst. The impregnated catalyst showed higher activity, achieving 95.3% yield compared with 93.1% for the post gifted. The post gifted catalyst, however, was more stable, likely due to stronger covalent bonding. After five uses the $ZrO_2$–$TiO_2$–$SO_3H$ nanorod remained at high activity with 85.1% yield; this catalyst was hence recommended due to the balance of high activity and reusability. Magnetically separable acidic MMOs have also been developed. A magnetic $Fe_3O_4$ core, Mg–Al–$SO_4$ shell catalyst was synthesized by Gardy et al. [62] via coprecipitation, encapsulation, and surface functionalization. The catalyst displayed moderate superacidity and high active site loading. HT development was suppressed by Al/Mg ratio of six. The shell improved surface area and stability, and over five recycles negligible deactivation or leaching occurred. Mapossa et al. [42] synthesized magnetic $NiFe_2O_4$ and $Ni_{0.3}Zn_{0.7}Fe_2O_4$ by combustion synthesis. Zn inclusion reduced surface area, however, increased Lewis acidity, zeta potential, and conversion. Zeta potential was the key contributor to catalytic activity. Through morphology and crystallinity changes, Zn doping also increased saturation magnetization, remanence, and coercivity.

### 8.3.2 Solid Base MMO Catalysts

While singular basic metal oxides such as CaO and MgO have been studied for applications in biodiesel production, the combination of these solid bases with other metal oxide species has more recently been explored to improve stability, structural properties or achieve greater separability via external magnetic field. Examples of MMO solid base catalysts previously investigated are discussed here.

**Table 8.2** Summary of some recent developments in mixed metal oxide catalysts.

| Feedstock | A:O ratio[a] | Catalyst | Catalyst loading (wt%) | Reaction time (min) | Particle size (nm) | Acidity (A) basicity (B) (mmol $g^{-1}$) | T (°C) | Yield (Y) conversion (C) | Reusability (runs) | Amount leached (%) | Regen. method[b] | References |
|---|---|---|---|---|---|---|---|---|---|---|---|---|
| *B. aegyptiaca* oil | 12 : 1 | BaO–$MoO_2$ | 4.5 | 120 | — | 0.621 B | 65 | 97.8–>80% Y | 5 | Ba: 11.97% Mo: 4.32% | W, D | [28] |
| Canola oil | 6 : 1 | $NdAlO_3$ | 10 | 300 | ~100 | 1.13 B | 200 | 75–50% Y | 6 | — | W, D, R | [29] |
| Canola oil | 12 : 1 | KOH/$Ca_{12}Al_{14}O_{33}$ | 4 | 240 | 24 | 0.392 B | 65 | 85.9–75.3% Y | 4 | — | W, R | [30] |
| Canola oil | 15 : 1 | ZnO/$BiFeO_3$ | 4 | 360 | 31.27 | 0.1576 B | 65 | 95.43–92.08% Y | 5 | — | W, D | [31] |
| Chicken fat | 15 : 1 | CaO/$CuFe_2O_4$ | 3 | 240 | — | — | 70 | 94.52% Y | — | — | — | [32] |
| Cooking oil | 6 : 1 | LDH-derived Co/Fe | 2 | 20 | — | — | 65 | 96% C | — | — | — | [33] |
| Eucalyptus oil | 6 : 1 | Zn-eggshell CaO | 5 | 150 | — | — | 65 | 93.8–>88% Y | 5 | — | W, R | [34] |
| *M. oleifera* oil | 0.3 : 1 | CuO-conch shell CaO | 4 | 150 | 37.54 | — | 65 | 95.24–>90% C | 5 | — | W, D | [35] |
| Palm oil | 6 : 1 | Cr/Ca and Cr/Zn | 1 | 60 | — | — | 60 | 27.8% and 7.4% Y | — | — | — | [36] |
| Palm oil | 15 : 1 | Cr–Ti | 2.5 | 180 | — | — | 55 | — | — | — | — | [37] |
| Palmitic acid | 20 : 1 | ZrO2–$TiO_2$–$SO_3H$ | 5 | 240 | 4000 × 100 | 1.9 A | 100 | 93.1–85.1% Y | 5 | — | W, D | [38] |
| Rapeseed oil | 24 : 1[c] | LDH-derived Mg–Fe | 1 | 240 | — | — | 120 | 97.5% Y | — | — | — | [39] |
| RBD palm olein | 30 : 1 | ZnO/CaO | 7.5 | 120 | 62–85 | 25.81 B | 56.9 | 86.99–83.87% Y | 2 | Ca: 50.9% | W, R | [40] |
| Soybean oil | 9 : 1 | LDH-derived CaAl and MgAl | 1 | 240 | — | 0.12 B and 0.11 B | 65 | 95.1% and 5.5% Y | — | — | — | [41] |
| Soybean oil | 12 : 1 | $NiFe_2O_4$ and $Ni_{0.3}Zn_{0.7}Fe_2O_4$ | 2 | 60 | 12–13 and 17–20 | 0.189 A and 0.191 A | 180 | 49% and 94% Y | — | — | — | [42] |

(*Continued*)

**Table 8.2** (Continued)

| Feedstock | A:O ratio[a] | Catalyst | Catalyst loading (wt%) | Reaction time (min) | Particle size (nm) | Acidity (A) basicity (B) (mmol $g^{-1}$) | T (°C) | Yield (Y) conversion (C) | Reusability (runs) | Amount leached (%) | Regen. method[b] | References |
|---|---|---|---|---|---|---|---|---|---|---|---|---|
| Soybean oil<br>Castor oil<br>Soybean oil<br>Castor oil | 6 : 1<br>MeOH<br>6 : 1<br>BuOH | $MgO/\gamma$-$Al_2O_3$<br>1.5-Zn/Mg<br>1.5-Zn/Mg<br>0.5-Zn/Mg | 5 | 120–360 | — | — | 60<br><br>80 | 57% Y<br>76% Y<br>66% Y<br>99–80% Y | —<br>—<br>—<br>3 | — | W, D, R | [43] |
| Soybean oil methyl acetate | 6 : 1 | Ba–MgAl | 4 | 48030 | 66 | 4.499 B | 7050 | 71.5% Y~80% Y | −3 | — | None | [44] |
| Spirulina oil | 18 : 1 | Ba/Ca/Zn | 2.5 | 120 | — | 2.65 B | 65 | 98.94–~80% C | 5 | — | W, D, R | [45] |
| Sunflower oil | 12 : 1 | $MgO/MgAl_{0.4}Fe_{1.6}O_4$ | 3 | 180 | 9.5 | — | 110 | 93.2–85.9% C | 5 | — | W, D | [46] |
| Sunflower oil | 12 : 1 | $MgO$–$MgFe_2O_4$ | 4 | 180 | 7.2–45.9 | — | 110 | 92.9–80.9% Y | 5 | — | W, D | [47] |
| Sunflower oil | 15 : 1 | LDH-derived Ca–Mg–Al | 2.5 | 360 | — | — | 60 | 95% Y | — | — | — | [48] |
| Vegetable oil | 8 : 1 | $Li_2ZrO_2$ | 6 | 120 | $2$–$5 \times 10^4$ | — | 65 | 99–90% C | 7 | — | — | [49] |
| Waste fish oil | 12 : 1 | $CaO$–$Ca_3Al_2O_6$ | 10 | 120 | — | — | 65 | >95% C | 7 | — | W, D | [50] |
| WCO | 14 : 1 | Sr/Ce | 2 | 120 | — | 1.35 B | 65 | 99.5–~90% C | 4 | Sr: 5.98%<br>Ce: 3.83% | W, D, R | [51] |
| WCO | 9 : 1 | LDH-derived Zn/MgAl(O) | 3 | 180 | ~30 | 0.62 B | 65 | 78.45–65.74% Y | 5 | Zn: 1.16%<br>Mg: 2.80%<br>Al: 3.05% | W | [52] |
| WCO | 20 : 1 | $Cu/Zn/\gamma$-$Al_2O_3$ | 10 | 120 | ~10 | — | 65 | 89.5–82.64% Y | 6 | — | None | [26, 53] |
| WCO | 12 : 1 | Anthill-eggshell-Ni-Co | 3 | 120 | <30 | — | 70 | 90.23–77.05% Y | 4 | — | W, D | [54] |
| WCO | 21 : 1 | $Zn_{0.5}Cu_{0.5}Fe_2O_4$ | 4 | 40 | ~5 | — | 65 | 98.9–96% Y | 4 | — | W, D | [55] |

| | | | | | | | | | | | | |
|---|---|---|---|---|---|---|---|---|---|---|---|---|
| WCO | 12 : 1 | Rice husk/$K_2O$/ Fe2O3 | 4 | 240 | — | 24.55 A<br>4.43 B | 75 | 98.6-~80% Y | 5 | — | W, D, R | [56] |
| WCO | 30 : 1 | LDH-derived 3 : 1 MgAl | 5 | 1440 | — | — | 65 | 87.23% Y | — | — | — | [57] |
| WCO | 11 : 1 | Sr-Ti 4 : 1 | 1 | 80 | 350 | 2.89 B | 65 | 98–83% C | 8 | Ti: 0.01%<br>Sr: 2.09% | W, D | [58] |
| WCO | 15 : 1 | α- $Fe_2O_3/Al_2O_3$ | 1 | 180 | — | — | 65 | >80% Y | — | — | — | [59] |
| WCO | 18 : 1 | $Zn_{0.5}Mg_{0.5}$–$ZnFe_2O_4$ | 3 | 30 | 8.16–10 | — | 65 | 99.9–94% Y | 10 | — | W, D | [25] |
| WCO | 6 : 1 | *Citrus sinensis* peel ash/$Fe_3O_4$ | 6 | 180 | 12–13 | 0.17 B | 65 | 98.7–91% Y | 9 | K: 0.53%<br>Ca: 0.26% | W, D | [60] |
| WCO Corn oil | 10 : 1 | $ZrO_2$–CuO and $ZrO_2$–$SrO_2$ | 5 | 180 | 44 and 52 | — | 120 | 87–83% and 84–79% Y<br>86–76% and 90–88% Y | 4 | — | W | [61] |
| WCO oleic acid | 9 : 1 | $SO_4$/Mg–Al–$Fe_3O_4$ | 4 | 300150 | Core: 20–150<br>Shell: 5–15 | 2.35 A | 95 | ~98.5% Y88% Y | 5— | Mg: 1.554<br>Al: 0.802<br>Fe: 0.459<br>($\mu g\ l^{-1}$) | W, R | [62] |
| WCO *S. obliquus* lipid | 9 : 1 | Carica papaya stem | 2 | 180 | — | — | 60 | 95.23–90.5%<br>C93.33% C | 5— | — | W, D | [63] |

[a] MeOH unless stated otherwise.
[b] Regeneration route included washing (W), drying (D), or recalcination (R).
[c] Equal parts MeOH:BuOH.

Banerjee et al. [51] produced a new gel combustion Sr/Ce basic catalyst. Optimal ratio was 3 : 1, beyond this excessive basicity promoted saponification. Conversion was greatest at high calcination temperatures, forming $Sr_2(CeO_4)$ active sites of higher porosity and surface area. Reduced activity after four recycles was predominantly poisoning as only slight leaching was observed. A better $NdAlO_3$ catalyst was developed by Dionicio-Navarrete et al. [29] by sol–gel route, improving surface area over previous flash combustion synthesis. Reducing catalyst particle size and greater methanol concentration gave higher yield; however, high temperature was still required to improve miscibility. Activity loss after six recycles was potentially from high pressure nanoparticle clustering, alongside possible moisture, $CO_2$, or glycerin poisoning. Sulaiman et al. [26, 53] developed a wet impregnation Cu/Zn/$\gamma$-$Al_2O_3$ catalyst. Cu/Zn loading increased active site dispersion, while alumina coating raised surface area, with optimal loading of up to 10/90 wt% and three layers, respectively. Excess of either reduced performance. Excess calcination temperature led to agglomeration and was most important variable affecting catalytic activity, followed by metal loading and number of alumina coatings.

Calcination of LDH produced various new MMO base catalysts. Hájek et al. [39] used an LDH-derived calcinated Mg/Fe catalyst to test the effect of solvent choice on reaction performance. The catalyst was produced by coprecipitation route with a Mg/Fe ratio of 2.9. Best performance was achieved with equal parts of methanol to butanol. This was likely due to more miscible two-phase catalyst liquid system presenting increased reaction rate, compared with the standard three-phase catalyst–methanol–oil system. Various Mg containing LDH-derived catalysts were also recently developed. Rosset et al. [41] produced LDH CaAl and MgAl coprecipitated catalysts and compared to pure oxides. Both CaO and CaAl achieved high fatty acid methyl ester (FAME) yield. While Ca catalysts performed better than Mg ones, for both a similar trend was observed. Pure oxides had highest activity followed by MMO then uncalcinated LDH. This correlates to active site basic strength. Sedaghat-Hoor and Anbia [52] synthesized coprecipitated MgAlO LDH-derived catalyst, followed by impregnation with ZnO and compared its activity against fully coprecipitated ZnMgAlO catalyst. Zn was introduced to increase surface area and catalytic activity through greater basicity. Impregnation gave higher surface area, better stability, and less Zn leaching, the main cause of activity loss over five recycles. Even after several uses, impregnated catalyst was superior to pure MgAlO. Dahdah et al. [48] observed similar results, best catalyst being 40 wt% calcinated Ca then impregnated to the $Mg_4Al_2$ HT support. Ca was introduced to increase basicity; however, coprecipitation trapped CaO mainly in bulk even at high dopant levels. Calcinating HT support decreased activity due to damaged crystal structure. Both thermal treatment and preparation route were important factors, alongside basicity. Lima-Corrêa et al. [44] developed LDH-derived calcinated MgAl doped with K, Ba, Sr, and La through sol–gel route at Mg/dopant ratio of 1.5. Sr catalyst had more reactive active sites; however, Ba-doped catalyst had greater active site density. Although both highly active, ultimately Ba-MgAl had greater stability and likely suffered less leaching over three recycles. Highly active catalysts had little surface area; hence surface chemistry dominated textural properties. Nur et al. [57] produced calcinated MgAl LDH-derived catalyst by alkali free coprecipitation route. Catalyst was free of alkali impurity, demonstrating successful synthesis route. Coprecipitation displayed optimal catalyst growth and activity for Mg/Al ratio of three.

Several CaO MMO base catalysts have been developed recently. Nayebzadeh et al. [30] produced $CaO–Al_2O_3$ catalysts through various routes, followed by impregnation of KOH. Best performance from microwave (MW)-assisted solution combustion route. The produced $Ca_{12}Al_{14}O_{33}$-KOH catalyst achieved four recycles, exhibited optimal crystallinity, mean pore size, basicity, and catalytic activity, alongside highest Cal/Al (1.34) and K/Al (1.05) ratio. Combustion synthesis $CaO–Ca_3Al_2O_6$ catalyst was performed by Papargyriou et al. [50]. Optimal Ca/Al ratio of six had highest basicity. Although possessing similar catalytic performance to CaO, the MMO catalyst had greater stability and withstood two extra reuses. Calcium diglyceride intermediate improved mass transfer and raised catalytic activity, however, led to Ca leaching and deactivation through hydration and carbonization. Sierra-Cantor et al. [40] tested coprecipitated Ca–Zn MMOs of ratio 3.12 and 4.64. Lower ratio had slightly higher activity but more leaching. Increasing Zn reduced basicity but raised porosity, surface area, and Ca dispersion. More stable 4.64 ratio only withstood two recycles before leaching. A:O (alcohol to oil) ratio was deemed the most important operating parameter. CaO/ZnO/BaO wet impregnation catalysts were investigated by Singh et al. [45] for ratios 1 : 1 : 1–1 : 1 : 4. High Ba loading increased basicity but reduced surface area, with 1 : 1 : 3 the optimum ratio. Catalytic activity strongly depended on basic site strength. Five recycles at high conversion before activity reduction dropped due to product poisoning and sintering during regeneration.

A variety of MMO base catalysts were derived from biomass. Rahman et al. [34] synthesized Zn- and Cu-doped calcinated eggshell by wet impregnation method. Dopants increased surface area, and Zn–CaO catalyst had greatest superbasicity and FAME yield. Catalyst loading and A:O ratio had little effect. Activity drop after five uses was likely from organic poisoning, structural change during regeneration, and partial filtering of active species. Yusuff et al. [54] synthesized novel anthill-eggshell catalyst coprecipitated with Ni and Co. Anthill support introduced small amounts of additional oxides. Surface area, pore volume, and pore diameter improved with calcination. Activity loss after four recycles was likely due to structural deformation from methyl/methylene poisoning. Niju et al. [35] coprecipitated CuO with conch shell CaO, with approximate Cu/Ca ratio of two. CaO provided strong basic catalytic activity, while CuO added stability, high porosity, and surface area. Five recycles were achieved before active site loss. Gohain et al. [63] produced a new calcinated papaya stem catalyst containing group one and two metal oxides. $K_2O$ was main species responsible for basicity. Five recycles at high conversion were achieved; activity decrease was due to K leaching, $Ca(OCH_3)_2$ and absorbed ester pore blocking, and mass losses. Seffati et al. [32] precipitated eggshell CaO with magnetic $CuFe_2O_4$ and achieved slightly higher yield at shorter reaction time to previous CaO catalysts. Increased density improved mechanical and thermal strength, and high porosity was achieved, although less than pure CaO. Changmai et al. [60] developed novel *Citrus sinensis* peel shell coprecipitated onto $Fe_3O_4$ magnetic core. Strongly basic K and Ca species are mainly responsible for high activity. The core enhanced dispersion and stability and facilitated magnetic separation. Only small amounts of K and Ca leaching was observed, with little loss of functional groups or structure after nine reuses.

Other magnetic base MMO catalysts include superparamagnetic coprecipitated ZnO/$BiFeO_3$ developed by Salimi et al. [31], which achieved greater saturation magnetization compared to previous work. Introduction of ZnO promoted basic sites, improving catalytic activity and FAME yield. Five reuses were achieved at high yield, with activity reduction likely due to loss of basic sites during reaction or separation. Ashok et al. [25, 55] synthesized $Zn_{1-x}Cu_xFe_2O_4$ and $Zn_{1-x}Mg_xFe_2O_4$ ferromagnetic catalysts by sonification-assisted MW and MW-assisted combustion methods, respectively. Increasing dopant percentage had little effect on yield, however, reduced reaction time. Zn/Dopant ratio of one gave optimal performance. Doping affected magnetic properties through surface defects, oxygen vacancies, and cationic redistribution. Both catalysts showed high saturation magnetization and magnetic moment. Mg-doped catalyst had greater stability in reusability studies. Naderi and Nayebzadeh [46] doped magnetic $MgFe_2O_4$ with Al by solution combustion method. Al doping increased thermal stability, surface area, and porosity. Optimal Al/Fe ratio of 0.4 was observed to balance improved structural and textural properties with excess Al hindering magnetic separability. Activity reduction over five recycles was likely due to MgO leaching, surface poisoning, and pore blockage. $MgFe_2O_4$ was also doped with MgO by Alaei et al. [47] by various routes. High surface area alone did not improve performance; combustion synthesis gave optimal yield due to the combined effects of increased porosity, active site distribution, pore diameter, and surface area. Reduction in activity after five recycles was caused by impurity pore blocking and small amounts of leaching.

### 8.3.3 Solid Bifunctional MMO Catalysts

As discussed earlier, deliberate combination of both acidic and basic metal oxides together can achieve bifunctional MMO catalysts. Table 8.1 features some bifunctional catalysts that have been previously explored, examples of which are further conferred in this section.

Afsharizadeh and Mohsennia [61] synthesized $ZrO_2$ mixed oxides of Sr and Cu through Pechini sol–gel and coprecipitation routes. Coprecipitated catalysts performed better due to higher metal content, more basic sites, and greater surface area from reduced agglomeration. Optimal Zr/dopant ratio was two. Another $ZrO_2$ bifunctional catalyst doped with Li, Na, or K was developed by Dai et al. [49]. $Li_2ZrO_2$ catalyst displayed highest conversion and greatest FFA tolerance; 20% FFA still achieved 80% conversion. Li doping greatly increased surface basicity and catalytic activity. Leaching mainly decreased performance over seven recycles. Ali Bashah et al. [36] synthesized Cr/Ca and Cr/Zn catalysts by coprecipitation. Ca doping led to larger pore volumes, diameters, and more basic sites, aiding transesterification performance. Zn doping had less favorable textural properties, although increased Lewis acidity sites would tolerate higher FFA feeds. Samin et al. [37] created Cr/Ti catalyst by sol–gel method. Two-hour calcination led to optimal surface area, pore diameter, and volume. Jamil et al. [28] developed bifunctional BaO–$MoO_2$ catalyst by coprecipitation. Increasing Ba/Mo ratio up to six improved surface area; excess Ba then started blocking active sites. Basic BaO leached out more than acidic $MoO_2$ species; however, catalyst was stable up to five uses. Navas et al. [43] produced bifunctional coprecipitated MMO of Zn/Mg on $Al_2O_3$ support. Highest yield was observed using castor oil, butanol, and Zn/Mg ratio of 1.5. Butanol improved miscibility of castor oil, while higher FFA content facilitated both basic MgO transesterification and amphoteric ZnO esterification. For lower FFA soybean oil and methanol, increasing Zn content disfavored FAME yield. Replacing methanol with butanol again demonstrated 1.5 Zn/Mg ratio to be optimal. This ratio produced greatest surface area and increased catalytic activity from higher basic site density. Sahani et al. [58] synthesized Sr/Ti bifunctional MMO by sol–gel method. Optimal Sr/Ti ratio of four was observed, Ti providing Lewis acidity, while Sr increased surface basicity and therefore catalytic activity. Increased Sr ratios reduced surface area by agglomeration. Catalytic deactivation through eight recycles was due to structural deformation and surface passivation; no significant leaching was observed. An LDH-derived bifunctional catalyst was developed by Ortega et al. [33] through coprecipitation of Co and Fe followed by calcination. The presence of O, Fe, Na, and Co in various oxidation states likely allowed for the bifunctional behavior. The Co/Fe ratio of four was the highest tested and was found to be most optimal. This may potentially be due to superior crystalline structure and porosity, as well as the presence of both $CoFe_2O_4$ and $CoNa_xO_2$ species alongside simpler $CoO_2$ and CoO oxides.

New magnetic bifunctional catalysts were also developed recently. Widayat et al. [59] impregnated hematite precipitated from iron sands onto $Al_2O_3$ and ZSM5 supports. ZSM5 support achieved superior yield, surface area, and pore volume. While $Al_2O_3$ increased surface area, negligible improvement to yield was potentially due to $Fe_2O_3$ overloading, reducing pore volume. A superparamagnetic bifunctional rice husk catalyst was also developed by Hazmi et al. [56] by wet impregnation of $K_2O$ and Fe. Interaction between dopants created porous structure exhibiting high surface area and activity with optimal 20% $K_2O$ and 5% Fe loadings. Excess $K_2O$ impeded mass transfer, while excess Fe potentially neutralized basic sites and reduced surface area by agglomeration. Activity loss after five recycles was mainly from reactant poisoning.

## 8.4 Discussion

For new MMO catalysts to be commercially viable, the produced biodiesel must adhere to international standards, such as ASTM D6751 or EN14214. However, less than half of recent literature test FAME properties to standards, with some only checking a few properties. Even a highly active and stable catalyst would be unsuitable for industrial application if such standards cannot be guaranteed. Reusability is both economically and environmentally favorable, minimizing regeneration downtime, cost of replacement, and quantity of catalytic waste produced. No matter how promising new catalyst performance may appear, stability studies are essential for assessing industrial applicability. As expected, the biodiesel yield regularly decreased during recycles through leaching, mass loss, poisoning, and other deactivation effects. Assessing property changes of the catalyst postreaction can determine the dominant causes of deactivation, informing future research on stability improvement. On noticing deactivation, unfortunately some researchers only speculated potential causes, and did not investigate further. Around half of studies performing reusability studies, however, did compare differences in pre/postreaction catalyst, particularly with regard to leaching. In particular Papargyriou et al. [50] provided detailed discussion into mechanisms of leaching and deactivation. While attempts at catalyst regeneration were performed, with some comparing regenerated properties to fresh catalyst, techniques were limited to washing, drying, and recalcination. Very few studies have reported attempts to chemically regenerate deactivated catalytic sites or leached species. A literature review by Melero et al. [64] explored biodiesel production from acidic heterogeneous catalysts, with regeneration steps mainly being solvent washing or recalcination. However, reference was made to work by Russbueldt et al. [65] using HCl to regenerate sulfonic acid ion exchange resins. Another review of solid acid catalysts for biodiesel production by Sharma et al. [66] noted the potential for regenerating sulfated zirconia by $H_2SO_4$ washing, with López et al. [67], highlighting previous research into this by Jitputti et al. [68] and Ni and Meunier [69]. Chen et al. [70] reported successful regeneration of a solid acid catalyst prepared from a glucose–starch mixture. Both methanol and cyclohexane washing failed to improve spent catalyst activity; however, significant regeneration was observed when washed with diluted (5%) and concentrated (98%) $H_2SO_4$ solutions. This regeneration method was explored further by Fu et al. [71] who used a 98% concentrated $H_2SO_4$ solution to successfully regenerate a β-cyclodextrin-derived solid acid carbon catalyst. Both studies suggested that further research into the mechanism of $H_2SO_4$ regeneration should be undertaken. More recent work by Gardy [72] demonstrated the successful chemical regeneration of metal oxide $Ti(SO_4)O$ catalyst through impregnation method with sulfuric acid. The regenerated catalyst achieved an FAME yield of 93.51% even after five cycles. Therefore, while some exploration of chemical regeneration beyond solvent washing and recalcination has occurred, future work on MMO catalysts should continue to explore this area further.

The importance of optimizing reaction and preparation variables was regularly noted, with most studies either testing for optimal conditions or using those previously reported, such as a temperature limit of 65 °C based on methanol evaporation. Extremes of any condition were detrimental. Some catalysts still required excessive temperatures [29] or reaction times [57] unfortunately. Reaction conditions like A:O ratio and catalyst loading sometimes dominated [29, 40], however, in other studies had little effect [34] and instead preparation conditions [53] were more influential. Since the dominating variable often depends on specific catalyst, in agreement with past work, optimized reaction and formation conditions should always be investigated and experimentally verified for new catalysts. Dahdah et al. [48] highlighted how thermal and preparation routes are important factors, and Gardy et al. [62] suggested further research into calcination temperature. Different synthesis routes or calcination temperatures imparted desirable or detrimental properties. Many studies adjusted dopant ratios or dopant used and observed notable catalyst performance variation. Naderi et al. [46] noted how metal doping ratios can significantly affect magnetic separability; analysis by vibrating sample magnetometry (VSM) would allow numerical quantification of such impact. Yet only half of magnetic catalysts reported in the literature were analyzed this way. Future use of analytical verification alongside experimental tests is encouraged to aid future research. The impact of magnetic catalyst agglomeration over repeated uses and continuous successful magnetic recovery should also be explored further in future studies. Table 8.3 summarizes what properties were influenced by inclusion of a given metal/metal oxide. Note that several highlighted species (in **bold**) were reported to both aid and hinder certain properties; there are several potential reasons for these disagreements. First, the amphoteric nature of certain species may help or hinder overall catalytic activity, depending on the interactions with other present active sites. Note that each study

**Table 8.3** Summary of how certain metals/metal oxides can influence catalyst properties.

| Property | Improved by | Hindered by | References |
|---|---|---|---|
| Acidity | Co, Fe, Mo, Ti, Zn, Zr | — | [28, 33, 36, 42, 49, 56, 58] |
| Active site density | Ba, Ca, Zn | — | [36, 43, 44] |
| Active site dispersion | **Cu**, Fe, Zn | **Cu**, Fe, K | [26, 40, 53, 60] |
| Basicity | **Al**, Ba, Ca, Co, **Fe**, K, Li, **Mg**, Nd, Sr, **Zn** | **Al**, **Fe**, **Mg**, **Zn** | [28–31, 33–35, 40, 41, 43–45, 48–52, 56, 58, 60, 63] |
| Coercivity | **Cu**, **Mg**, Zn | **Cu**, **Mg** | [25, 42, 55] |
| FFA tolerance | Li, Zn | — | [36, 43, 49] |
| Magnetic moment | Cu, Mg | — | [25, 55] |
| Magnetic separability | Fe | Al | [1, 25, 31, 32, 42, 46, 47, 55, 56, 59, 60] |
| Mass transfer | Ca | K | [50, 56] |
| Pore size | **Al**, Ca | **Al**, Zn | [36, 46] |
| Porosity | Al, Co, **Cu**, **Fe**, **K**, Zn | **Cu**, **Fe**, **K** | [32, 33, 35, 40, 46, 56, 59] |
| Reaction time | Cu, Fe, Mg | — | [25, 32, 55] |
| Remanence | **Cu**, **Mg**, Zn | **Cu**, **Mg** | [25, 42, 55] |
| Saturation magnetization | Bi, Cu, Mg, Zn | — | [25, 31, 42, 55] |
| Stability | Al, **Ba**, **Ca**, Cu, Fe, **Mg**, **Zn** | **Ba**, **Ca**, K, Li, **Mg**, Sr, **Zn** | [1, 25, 28, 32, 35, 40, 44, 46, 49, 50, 52, 60, 63] |
| Surface area | Al, **Ba**, Ca, Cu, **Fe**, K, Mg, **Zn** | **Ba**, **Fe**, Sr, **Zn** | [1, 26, 28, 34, 35, 40, 42–46, 52, 53, 56, 58, 59] |
| Yield/conversion | Ba, Ca, **Cu**, Fe, Li, **Mg**, Sr, **Zn** | Al, **Cu**, **Mg**, K, La, **Sr**, **Zn** | [31, 32, 34, 36, 42–44, 49, 51, 57] |

used different combinations of metals and metal oxides, hence leading to different chemical environments and active site interactivity. Additionally, an excess of any species can be detrimental as previously mentioned, and the ratios utilized often varied among studies. It is therefore too simplistic to assign general catalyst properties to any given species; the interactions between all metals and metal oxides present is complex and part of the reason why new catalyst development can be such a challenge.

Acidic and basic characteristics can determine activity, FFA impurity resistance, and preference to a specific reaction pathway. Fan et al. [38] reported that Bronsted sites mainly catalyzed the esterification reaction, in agreement with previous work [21]. A quarter of studies neglected to investigate strength, site density, or type of acid/base sites present; therefore, contributions to catalytic performance and potential underlying reaction

**Table 8.4** Kinetic data from recent studies.

| TOF ($s^{-1}$) | Activation energy ($KJ\ mol^{-1}$) | Rate constant[a] ($s^{-1}$) | Reaction order[b] | Biodiesel selectivity | References |
|---|---|---|---|---|---|
| $5.6 \times 10^{-3}$ | — | — | — | — | [28] |
| $2.1 \times 10^{-4}$ | — | — | — | — | [40] |
| — | — | — | — | 94–~100% | [43] |
| $1.0 \times 10^{-2}$ | — | — | — | — | [44] |
| — | 48.029 | $2 \times 10^{-4}$ | 1 | — | [45] |
| — | 17.04 | $4.2 \times 10^{-5}$ | 1 | — | [51] |
| — | 38.78 | $4.483 \times 10^{-4}$–$1.722 \times 10^{-3}$ | 1 | — | [55] |
| $28 \times 10^{-2}$ | 29.67 | $9.2 \times 10^{-4}$ | 1 | — | [58] |
| — | 52 | $6.25 \times 10^{-4}$–$3.97 \times 10^{-3}$ | 1 | — | [25] |
| — | 34.4 | $2.34 \times 10^{-3}$ | 1 | — | [60] |

[a] At reported optimum reaction temperature.
[b] Pseudo first order.

mechanisms are unknown. Recently there has been a lack of research into acidic MMO catalysts, with base and bifunctional materials favored potentially due to improved reactivity, tunable chemical environment, and moderate FFA resistance, respectively. However, in order to avoid expensive pretreatment and facilitate cheap feedstock usage, acidic MMO materials must still be considered for future research.

A variety of feedstocks have been explored [43, 45, 61], including various cheap, low quality WCO sources, advantageous for reasons discussed previously. Increased FFA concentration in WCO can both improve [43, 61] and reduce [61] FAME yield depending on acidity, while other impurities like water can affect basic sites [51] and hence reduce activity. While some researchers pretreated to remove FFA content [57, 59], resistant catalysts were also synthesized that achieved one-step transesterification and esterification. Gardy et al. [62] recommend mechanistic study of these resistant catalysts, alongside testing a variety of fatty acid chain lengths. This is wise, since WCO is a mixture of many different triglycerides, and individual impacts are not yet fully understood. A few studies measured other useful catalyst characteristics such as turnover frequency (TOF) [28, 44], environment factor (E-factor) [58], and zeta potential or performed mechanistic [26], lifecycle [40], kinetic and thermodynamic [25] studies. Table 8.4 displays kinetic data from recent studies. Wider variety in investigation techniques is welcomed as this provides further information to guide catalyst development and optimization. Several studies reported optimizing both catalyst and reaction conditions through modeling such as Box–Behnken design (BBD) [53], response surface methodology (RSM), and central composite design (CCD) [35, 58]. These models agreed strongly with experimental results and highlighted a potential way to generate new research questions.

Ashok et al. [14, 25, 55] have explored the $Zn_{0.5}X_{0.5}ZnFe_2O_4$ range of magnetic catalysts, with the most recent Mg dopant demonstrating exceptional stability and improvements on previous Mn and Cu dopants. New dopants for this series of catalysts could be explored further, such as group two metals. While alkali earth metals such as Ca and Mg have been widely explored to date, extended research involving basic Ba or Sr should be investigated. For example, the Sr/Ti MMO developed by Sahani et al. [58] demonstrated a highly reusable catalyst, capable of high performance with only 1% loading under mild reaction conditions. Both magnetic and biomass-derived MMOs demand further research; the potential benefits of improved separability and sustainability, respectively, are highly desirable. The impressive *C. sinensis* peel ash catalyst synthesized by Changmai et al. [60], for example, achieved high performance and stability at low A:O ratios.

Several researchers recommend that more work is required for successful industrial application [30], reduced leaching [40], and catalyst optimization [37]. Further research is also required for continuous reaction setups and new regeneration approaches, as previously highlighted. Future studies should therefore explore these areas to demonstrate potential for scale-up. Adapting from batch to continuous presents an opportunity for process intensification, reducing costs and the number of process units required. More sustainable biodiesel production may also be achieved through increased efficiency and reductions in waste generation and energy consumption. Nevertheless, continuous operation brings new issues that must be addressed. While continuous production may allow for greater operational control and a biodiesel product that consistently meets standards, affordable feedstocks such as WCO can have large variations in quality between batches; hence batch operation may be more suitable. Sustained biodiesel production also requires sustainable feedstocks that are high in supply, economically viable, and possess a lifecycle with low environmental impact. Therefore, while there has been a lot of progress developing novel MMO catalysts, unfortunately there remain key challenges on the path to full scale industrial use.

## 8.5 Conclusion

New catalysts are continuously being developed to facilitate production of greener and more sustainable fuels. Mixing metals and their oxides together can address problems heterogeneous catalysts currently face, with each species working in synergy for superior catalytic performance. Despite significant development of a diverse range of MMO catalysts, several key issues remain. Formation and reaction conditions need optimizing for each new catalyst, through various modeling, analytical, and experimental methods. Produced biodiesel must adhere to international standards. Catalyst recyclability still requires further research, particularly with regard to mitigating catalytic poisoning and active species leaching. Novel regeneration methods beyond washing, drying, and recalcination should be investigated. Resistant catalysts that can handle low quality, economical, and environmentally sustainable feedstocks should continue to be explored. Finally, the journey from proof of concept to full scale industrial application will require evidence of successful continuous reaction, separation, catalyst recovery, and regeneration.

## 8.6 Symbols and Nomenclature

**Table 8.5** Symbols and abbreviations used throughout this chapter.

| | |
|---|---|
| α, γ | Alpha/gamma phase |
| β | β Anomer |
| A:O | Alcohol-to-oil ratio |
| BBD | Box–Behnken design |
| CCD | Central composite design |
| FAME | Fatty acid methyl ester |
| FFA | Free fatty acid |
| HT | Hydrotalcite |
| LDH | Layered double hydroxide |
| MMO | Mixed metal oxide |
| RBD | Refined, bleached, and deodorized (palm olein) |
| RSM | Response surface methodology |
| TOF | Turnover frequency |
| UCO/WCO | Used/waste cooking oil |
| VSM | Vibrating sample magnetometry |

## References


1 Gardy, J., Rehan, M., Hassanpour, A. et al. (2019). Advances in nano-catalysts based biodiesel production from non-food feedstocks. *Journal of Environmental Management* 249: 109316.

2 Gardy, J., Hassanpour, A., Lai, X., and Ahmed, M.H. (2016). Synthesis of $Ti(SO_4)O$ solid acid nano-catalyst and its application for biodiesel production from used cooking oil. *Applied Catalysis A: General* 527: 81–95.

3 Gardy, J., Osatiashtiani, A., Céspedes, O. et al. (2018). A magnetically separable $SO_4$/ Fe-Al-$TiO_2$ solid acid catalyst for biodiesel production from waste cooking oil. *Applied Catalysis B: Environmental* 234: 268–278.

4 Borah, M.J., Devi, A., Borah, R., and Deka, D. (2019). Synthesis and application of Co doped ZnO as heterogeneous nanocatalyst for biodiesel production from non-edible oil. *Renewable Energy* 133: 512–519.

5 Malhotra, R. and Ali, A. (2019). 5-Na/ZnO doped mesoporous silica as reusable solid catalyst for biodiesel production via transesterification of virgin cottonseed oil. *Renewable Energy* 133: 606–619.

6 Gawande, M.B., Pandey, R.K., and Jayaram, R.V. (2012). Role of mixed metal oxides in catalysis science – versatile applications in organic synthesis. *Catalysis Science & Technology* 2 (6): 1113–1125.

7 Alhassan, F.H., Rashid, U., and Taufiq-Yap, Y.H. (2015). Synthesis of waste cooking oil-based biodiesel via effectual recyclable bi-functional $Fe_2O_3$–MnO–$SO_4^{2-}$/$ZrO_2$ nanoparticle solid catalyst. *Fuel* 142: 38–45.

**8** Jamil, F., Al-Muhtaseb, A.'a., Myint, M.T.Z. et al. (2018). Biodiesel production by valorizing waste *Phoenix dactylifera* L. Kernel oil in the presence of synthesized heterogeneous metallic oxide catalyst (Mn@MgO-$ZrO_2$). *Energy Conversion and Management* 155: 128–137.

**9** Xie, W. and Zhao, L. (2014). Heterogeneous CaO–$MoO_3$–SBA-15 catalysts for biodiesel production from soybean oil. *Energy Conversion and Management* 79: 34–42.

**10** Li, Y., Zhang, X.-D., Sun, L. et al. (2010). Fatty acid methyl ester synthesis catalyzed by solid superacid catalyst $SO_4^{2-}/ZrO_2–TiO_2/La^{3+}$. *Applied Energy* 87 (1): 156–159.

**11** Boro, J., Konwar, L.J., Thakur, A.J., and Deka, D. (2014). Ba doped CaO derived from waste shells of *T striatula* (TS-CaO) as heterogeneous catalyst for biodiesel production. *Fuel* 129: 182–187.

**12** Prabu, M., Manikandan, M., Kandasamy, P. et al. (2019). Synthesis of biodiesel using the Mg/Al/Zn hydrotalcite/SBA-15 nanocomposite catalyst. *ACS Omega* 4 (2): 3500–3507.

**13** Xue, B.-j., Luo, J., Zhang, F., and Fang, Z. (2014). Biodiesel production from soybean and Jatropha oils by magnetic $CaFe_2O_4$–$Ca_2Fe_2O_5$-based catalyst. *Energy* 68: 584–591.

**14** Ashok, A., Kennedy, L.J., and Vijaya, J.J. (2019). Structural, optical and magnetic properties of $Zn_{1-x}Mn_xFe_2O_4$ ($0 \leq x \leq 0.5$) spinel nano particles for transesterification of used cooking oil. *Journal of Alloys and Compounds* 780: 816–828.

**15** Reyna-Villanueva, L.R., Dias, J.M., Medellín-Castillo, N.A. et al. (2019). Biodiesel production using layered double hydroxides and derived mixed oxides: the role of the synthesis conditions and the catalysts properties on biodiesel conversion. *Fuel* 251: 285–292.

**16** Borges, M.E. and Díaz, L. (2012). Recent developments on heterogeneous catalysts for biodiesel production by oil esterification and transesterification reactions: a review. *Renewable and Sustainable Energy Reviews* 16 (5): 2839–2849.

**17** Vasić, K., Podrepšek, G.H., Knez, Ž., and Leitgeb, M. (2020). Biodiesel production using solid acid catalysts based on metal oxides. *Catalysts* 10 (2): 237. (Referencing pages 5–13).

**18** Ghalandari, A., Taghizadeh, M., and Rahmani, M. (2019). Statistical optimization of the biodiesel production process using a magnetic core-mesoporous shell KOH/$Fe_3O_4$@γ-$Al_2O_3$ nanocatalyst. *Chemical Engineering and Technology* 42 (1): 89–99.

**19** Yee, K.F., Lee, K.T., Ceccato, R., and Abdullah, A.Z. (2011). Production of biodiesel from *Jatropha curcas* L. oil catalyzed by $SO_4^{2-}/ZrO_2$ catalyst: effect of interaction between process variables. *Bioresource Technology* 102 (5): 4285–4289.

**20** Ambat, I., Srivastava, V., Haapaniemi, E., and Sillanpää, M. (2019). Nano-magnetic potassium impregnated ceria as catalyst for the biodiesel production. *Renewable Energy* 139: 1428–1436.

**21** Anuradha, S., Raj, K.J.A., Vijayaraghavan, V.R., and Viswanathan, B. (2014). Sulphated $Fe_2O_3$-$TiO_2$ catalysed transesterification of soybean oil to biodiesel. *Indian Journal of Chemistry - Section A Inorganic, Physical, Theoretical and Analytical Chemistry* 53 (12): 1493–1499.

**22** Marinković, D.M., Avramović, J.M., Stanković, M.V. et al. (2017). Synthesis and characterization of spherically-shaped CaO/γ-$Al_2O_3$ catalyst and its application in biodiesel production. *Energy Conversion and Management* 144: 399–413.

**23** Feyzi, M. and Shahbazi, E. (2015). Catalytic performance and characterization of Cs–Ca/$SiO_2$–$TiO_2$ nanocatalysts for biodiesel production. *Journal of Molecular Catalysis A: Chemical* 404–405: 131–138.

**24** Borah, M.J., Das, A., Das, V. et al. (2019). Transesterification of waste cooking oil for biodiesel production catalyzed by Zn substituted waste egg shell derived CaO nanocatalyst. *Fuel* 242: 345–354.

**25** Ashok, A., Ratnaji, T., John Kennedy, L. et al. (2021). Magnetically recoverable Mg substituted zinc ferrite nanocatalyst for biodiesel production: process optimization, kinetic and thermodynamic analysis. *Renewable Energy* 163: 480–494.

**26** Sulaiman, N.F., Lee, S.L., Toemen, S., and Bakar, W.A.W.A. (2020). Physicochemical characteristics of Cu/Zn/$\gamma$-$Al_2O_3$ catalyst and its mechanistic study in transesterification for biodiesel production. *Renewable Energy* 156: 142–157.

**27** Lee, A.F. and Wilson, K. (2015). Recent developments in heterogeneous catalysis for the sustainable production of biodiesel. *Catalysis Today* 242: 3–18.

**28** Jamil, F., Al-Riyami, M., Al-Haj, L. et al. (2021). Waste balanites aegyptiaca seed oil as a potential source for biodiesel production in the presence of a novel mixed metallic oxide catalyst. *International Journal of Energy Research* 45 (12): 17189–17202.

**29** Dionicio-Navarrete, M., Dinorah Arrieta-Gonzalez, C., Quinto-Hernandez, A. et al. (2019). Synthesis of $NdAlO_3$ nanoparticles and evaluation of the catalytic capacity for biodiesel synthesis. *Nanomaterials* 9 (11): 1545. (Referencing pages 2–18).

**30** Nayebzadeh, H., Saghatoleslami, N., Haghighi, M., and Tabasizadeh, M. (2019). Catalytic activity of KOH–CaO–$Al_2O_3$ nanocomposites in biodiesel production: impact of preparation method. *International Journal of Self-Propagating High-Temperature Synthesis* 28 (1): 18–27.

**31** Salimi, Z. and Hosseini, S.A. (2019). Study and optimization of conditions of biodiesel production from edible oils using ZnO/$BiFeO_3$ nano magnetic catalyst. *Fuel* 239: 1204–1212.

**32** Seffati, K., Honarvar, B., Esmaeili, H., and Esfandiari, N. (2019). Enhanced biodiesel production from chicken fat using CaO/$CuFe_2O_4$ nanocatalyst and its combination with diesel to improve fuel properties. *Fuel* 235: 1238–1244.

**33** Gutiérrez-Ortega, N., Ramos-Ramírez, E., Serafín-Muñoz, A. et al. (2019). Use of Co/Fe-mixed oxides as heterogeneous catalysts in obtaining biodiesel. *Catalysts* 9 (5): 403. (Referencing pages 3–14).

**34** Rahman, W.U., Fatima, A., Anwer, A.H. et al. (2019). Biodiesel synthesis from eucalyptus oil by utilizing waste egg shell derived calcium based metal oxide catalyst. *Process Safety and Environmental Protection* 122: 313–319.

**35** Niju, S., Raj, F.R., Anushya, C., and Balajii, M. (2019). Optimization of acid catalyzed esterification and mixed metal oxide catalyzed transesterification for biodiesel production from *Moringa oleifera* oil. *Green Processing and Synthesis* 8 (1): 756–775.

**36** Ali Bashah, N.A., Luin, A., Jalaluddin, I.A. et al. (2019). Characteristics of chromium based mixed oxide catalyst in biodiesel production. *Journal of Physics: Conference Series* 1349 (1): 012143. (Referencing pages 2–6).

**37** Samin, M.S., Wan, Z., Yazid, N. et al. (2020). Effect of sythesis conditions of Cr-Ti mixed oxides on FAME and catalyst characteristics. *IOP Conference Series: Materials Science and Engineering* 864 (1): 012027. (Referencing pages 2–4).

**38** Fan, M., Si, Z., Sun, W., and Zhang, P. (2019). Sulfonated $ZrO_2$-$TiO_2$ nanorods as efficient solid acid catalysts for heterogeneous esterification of palmitic acid. *Fuel* 252: 254–261.

**39** Hájek, M., Vávra, A., and Mück, J. (2020). Butanol as a co-solvent for transesterification of rapeseed oil by methanol under homogeneous and heterogeneous catalyst. *Fuel* 277: 118239. (Referencing pages 2–7).

**40** Sierra-Cantor, J.F., Parra-Santiago, J.J., and Guerrero-Fajardo, C.A. (2019). Leaching and reusing analysis of calcium–zinc mixed oxides as heterogeneous catalysts in the biodiesel production from refined palm oil. *International Journal of Environmental Science and Technology* 16 (2): 643–654.

41 Rosset, M. and Perez-Lopez, O.W. (2019). FTIR spectroscopy analysis for monitoring biodiesel production by heterogeneous catalyst. *Vibrational Spectroscopy* 105: 102990. (Referencing pages 2–5).

42 Mapossa, A.B., Dantas, J., Silva, M.R. et al. (2020). Catalytic performance of $NiFe_2O_4$ and $Ni_{0.3}Zn_{0.7}Fe_2O_4$ magnetic nanoparticles during biodiesel production. *Arabian Journal of Chemistry* 13 (2): 4462–4476.

43 Navas, M.B., Ruggera, J.F., Lick, I.D., and Casella, M.L. (2020). A sustainable process for biodiesel production using Zn/Mg oxidic species as active, selective and reusable heterogeneous catalysts. *Bioresources and Bioprocessing* 7 (1): 4. (Referencing pages 3–11).

44 Lima-Corrêa, R.A.B., Castro, C.S., Damasceno, A.S., and Assaf, J.M. (2020). The enhanced activity of base metal modified MgAl mixed oxides from sol-gel hydrotalcite for ethylic transesterification. *Renewable Energy* 146: 1984–1990.

45 Singh, R., Kumar, A., and Chandra Sharma, Y. (2019). Biodiesel production from microalgal oil using barium-calcium-zinc mixed oxide base catalyst: optimization and kinetic studies. *Energy and Fuels* 33 (2): 1175–1184.

46 Naderi, F. and Nayebzadeh, H. (2019). Performance and stability assessment of Mg-Al-Fe nanocatalyst in the transesterification of sunflower oil: effect of Al/Fe molar ratio. *Industrial Crops and Products* 141: 111814. (Referencing pages 2–8).

47 Alaei, S., Haghighi, M., Rahmanivahid, B. et al. (2020). Conventional vs. hybrid methods for dispersion of MgO over magnetic Mg–Fe mixed oxides nanocatalyst in biofuel production from vegetable oil. *Renewable Energy* 154: 1188–1203.

48 Dahdah, E., Estephane, J., Haydar, R. et al. (2020). Biodiesel production from refined sunflower oil over Ca–Mg–Al catalysts: effect of the composition and the thermal treatment. *Renewable Energy* 146: 1242–1248.

49 Dai, Y.M., Li, Y.Y., Lin, J.H. et al. (2021). Applications of $M_2ZrO_2$ (M = Li, Na, K) composite as a catalyst for biodiesel production. *Fuel* 286: 119392. (Referencing pages 2–6).

50 Papargyriou, D., Broumidis, E., de Vere-Tucker, M. et al. (2019). Investigation of solid base catalysts for biodiesel production from fish oil. *Renewable Energy* 139: 661–669.

51 Banerjee, S., Sahani, S., and Chandra Sharma, Y. (2019). Process dynamic investigations and emission analyses of biodiesel produced using Sr–Ce mixed metal oxide heterogeneous catalyst. *Journal of Environmental Management* 248: 109218. (Referencing pages 2–15).

52 Sedaghat-Hoor, S. and Anbia, M. (2019). ZnO impregnated MgAl(O) catalyst with improved properties for biodiesel production: the influence of synthesis method on stability and reusability. *Particulate Science and Technology* 37 (7): 893–899.

53 Sulaiman, N.F., Wan Abu Bakar, W.A., Toemen, S. et al. (2019). In depth investigation of bi-functional, $Cu/Zn/\gamma\text{-}Al_2O_3$ catalyst in biodiesel production from low-grade cooking oil: optimization using response surface methodology. *Renewable Energy* 135: 408–416.

54 Yusuff, A.S., Adeniyi, O.D., Azeez, S.O. et al. (2019). Synthesis and characterization of anthill-eggshell-Ni-Co mixed oxides composite catalyst for biodiesel production from waste frying oil. *Biofuels, Bioproducts and Biorefining* 13 (1): 37–47.

55 Ashok, A., Ratnaji, T., John Kennedy, L., and Judith Vijaya, J. (2020). Magnetically separable $Zn_{1-x}Cu_xFe_2O_4$ ($0 \leq x \leq 0.5$) nanocatalysts for the transesterification of waste cooking oil. *Advanced Powder Technology* 31 (6): 2573–2585.

**56** Hazmi, B., Rashid, U., Taufiq-Yap, Y.H. et al. (2020). Supermagnetic nano-bifunctional catalyst from rice husk: synthesis, characterization and application for conversion of used cooking oil to biodiesel. *Catalysts* 10 (2): 225.

**57** Nur Fazira Edlina, I., Nazrizawati, A.T., Erma Hafiza, I.A.Z., and Noraini, H. (2020). MgAl mixed oxide derived alkali-free hydrotalcite for transesterification of waste cooking oil to biodiesel. *ASM Science Journal* 13.

**58** Sahani, S., Roy, T., and Sharma, Y.C. (2020). Smart waste management of waste cooking oil for large scale high quality biodiesel production using Sr-Ti mixed metal oxide as solid catalyst: optimization and E-metrics studies. *Waste Management* 108: 189–201.

**59** Widayat, Hadiyanto, Putra, D.A. et al. (2020). Waste cooking oil processing for fatty acid methyl ester and mono glycerides production with magnetite catalyst. *Food Research* 4 (S1): 220–226.

**60** Changmai, B., Rano, R., Vanlalveni, C., and Rokhum, L. (2021). A novel *Citrus sinensis* peel ash coated magnetic nanoparticles as an easily recoverable solid catalyst for biodiesel production. *Fuel* 286: 119447.

**61** Afsharizadeh, M. and Mohsennia, M. (2019). Catalytic synthesis of biodiesel from waste cooking oil and corn oil over zirconia-based metal oxide nanocatalysts. *Reaction Kinetics, Mechanisms and Catalysis* 128 (1): 443–459.

**62** Gardy, J., Nourafkan, E., Osatiashtiani, A. et al. (2019). A core-shell $SO_4$/Mg-Al-$Fe_3O_4$ catalyst for biodiesel production. *Applied Catalysis B: Environmental* 259: 118093.

**63** Gohain, M., Laskar, K., Paul, A.K. et al. (2020). Carica papaya stem: a source of versatile heterogeneous catalyst for biodiesel production and C–C bond formation. *Renewable Energy* 147: 541–555.

**64** Melero, J.A., Iglesias, J., and Morales, G. (2009). Heterogeneous acid catalysts for biodiesel production: current status and future challenges. *Green Chemistry* 11 (9): 1285–1308.

**65** Russbueldt, B.M.E. and Hoelderich, W.F. (2009). New sulfonic acid ion-exchange resins for the preesterification of different oils and fats with high content of free fatty acids. *Applied Catalysis A: General* 362 (1): 47–57.

**66** Sharma, Y.C., Singh, B., and Korstad, J. (2011). Advancements in solid acid catalysts for ecofriendly and economically viable synthesis of biodiesel. *Biofuels, Bioproducts and Biorefining* 5 (1): 69–92.

**67** López, D.E., Goodwin, J.G., Bruce, D.A., and Furuta, S. (2008). Esterification and transesterification using modified-zirconia catalysts. *Applied Catalysis A: General* 339 (1): 76–83.

**68** Jitputti, J., Kitiyanan, B., Rangsunvigit, P. et al. (2006). Transesterification of crude palm kernel oil and crude coconut oil by different solid catalysts. *Chemical Engineering Journal* 116 (1): 61–66.

**69** Ni, J. and Meunier, F.C. (2007). Esterification of free fatty acids in sunflower oil over solid acid catalysts using batch and fixed bed-reactors. *Applied Catalysis A: General* 333 (1): 122–130.

**70** Chen, G. and Fang, B. (2011). Preparation of solid acid catalyst from glucose–starch mixture for biodiesel production. *Bioresource Technology* 102 (3): 2635–2640.

**71** Fu, X.-b., Chen, J., Song, X.-l. et al. (2015). Biodiesel production using a carbon solid acid catalyst derived from β-cyclodextrin. *Journal of the American Oil Chemists' Society* 92 (4): 495–502.

**72** Gardy, J. (2017). Biodiesel production from used cooking oil using novel solid acid catalysts. PhD thesis. University of Leeds.

# 9

# Nanocatalysts in Biodiesel Production

*Avinash P. Ingle*[1], *Rahul Bhagat*[2], *Mangesh P. Moharil*[1], *Samuel Lalthazuala Rokhum*[3,4], *Shreshtha Saxena*[1], *and S. R. Kalbande*[5]

[1] *Biotechnology Centre, Department of Agricultural Botany, Dr. Panjabrao Deshmukh Agricultural University, Akola, Maharashtra, India*
[2] *Department of Biotechnology, Government Institute of Science, Aurangabad, Maharashtra, India*
[3] *Hamid Yusuf Department of Chemistry, University of Cambridge, Cambridge, UK*
[4] *Department of Chemistry, National Institute of Technology, Silchar, Assam, India*
[5] *Department of Unconventional Energy Sources and Electrical Engineering, Dr. Panjabrao Deshmukh Krishi Vidyapeeth, Akola, Maharashtra, India*

## 9.1 Introduction

The massive usage of fossil fuels due to globalization and the continuously increasing population has raised many environmental concerns like climate change, global warming, emissions of greenhouse gases, environmental pollution, diminishing fossil fuel reserves, high fuel prices, etc. [1]. Therefore, the demand for renewable and alternative energy sources such as bioenergies has taken a new dimension. Bioenergy like biofuels are considered as one of the important sources of renewable forms of energy due to their cost-effective and eco-friendly nature, and, hence, this form of energy has taken center stage in replacing the conventional fossil-based energy sources [2]. Biofuels are fuels that are produced from materials of biological origin (biomass). Such biomass can be obtained from agriculture, forest, industrial and domestic waste, etc. Usually, biofuels are in liquid (e.g. bioethanol, biodiesel) and gaseous form (e.g. biohydrogen, biomethane) [3, 4]. These fuels have attracted a great deal of attention worldwide, because it does not lead to a net increase in carbon dioxide levels and produces low amounts of sulfur. Among the various biofuels previously mentioned, special focus has been given to production of biodiesel in this chapter.

Biodiesel is environment friendly and nontoxic, which can be directly used in standard diesel-based vehicles without any engine modification and without blending with conventional diesel [5, 6]. As far as biodiesel production is concerned, it can be produced from a variety of feedstocks such as edible oils (soybean, sunflower, rapeseed, linseed, safflower, palm oil, olive oil, etc.), nonedible oils (petroleum-based oils and animal fat), waste cooking oils (WCO), algal oils, and the oils extracted from seeds of energy crops like *Jatropha curcas*, *Pongamia pinnata*, etc. The approaches of producing biodiesel from the previously

*Biodiesel Production: Feedstocks, Catalysts, and Technologies*, First Edition. Edited by Samuel Lalthazuala Rokhum, Gopinath Halder, Suttichai Assabumrungrat, and Kanokwan Ngaosuwan.

mentioned nonfood sources were found to be economically viable and robust [7]. However, various factors like the production cost, plenty availability of feedstocks, process complications, etc. can strongly influence biodiesel production at the commercial level [8, 9]. The literature available indicated that among all these feedstocks, different oils (e.g. rapeseed oil, soybean, oil, palm oil, used cooking oil, etc.) are considered to be the most promising feedstocks [10]. According to the report of "Renewable Carbon News," the use of feedstocks has been increasing all over the world year by year due to an increase in the production of biodiesel [both fatty acid methyl ester (FAME) and hydrotreated vegetable oil (HVO)]. The percentages of feedstocks used remained virtually unchanged from 2017, i.e. palm oil (35%), soybean oil (26%), rapeseed oil (16%), used cooking fats (11%), animal fats (7%), and other fats (6%) (Figure 9.1) (https://renewable-carbon.eu/news/global-biodiesel-production-is-increasing/). Similarly, various reports available including the report of IFP Energies Nouvelles, which claimed that the volume of biodiesel consumed in the transport sector, have been increasing constantly since 2011 due to its characteristics features. Therefore, a continuous increase in the production of biodiesel has been recorded all over the world. In Europe, the demand for biodiesel has increased with the growth in diesel consumption. Consumption rose from almost 13 Mtoe in 2017 to 15 Mtoe in 2018. Figure 9.2 shows the evolution of biodiesel production (FAME and HVO) by zone.

To date, different approaches have been investigated and proposed for the production of biodiesel, which mainly includes transesterification, pyrolysis, micro-emulsification, dilution, the supercritical fluid method, etc. However, among these transesterification of oils is the most accepted and widely used all over the world [11, 12]. As far as biodiesel production via transesterification is concerned, the reaction can be performed using different catalytic processes such as acid-catalyzed processes, base-catalyzed processes, enzyme-catalyzed processes, and use of supercritical conditions. The main conventional catalysts used in the transesterification of triglycerides are homogeneous catalysts, heterogeneous catalysts, biocatalysts (enzymes), etc. [1]. All these conventional catalysts have their own advantages and disadvantages, which have been pointed out in many articles.

Considering the different limitations of conventional catalysts, researchers working in this area try to search the alternative strategies for the production of biodiesel. In this

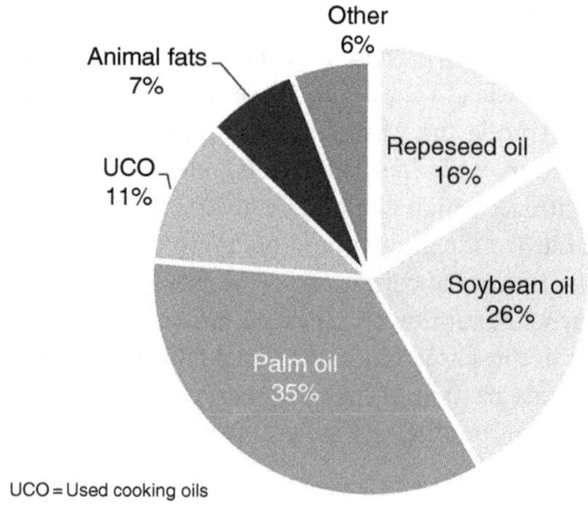


**Figure 9.1** Different feedstock used in FAME HVO biodiesel production. *Source:* https://www.ufop.de/english/news/chart-week/archive-chart-week/chart-week-2019/

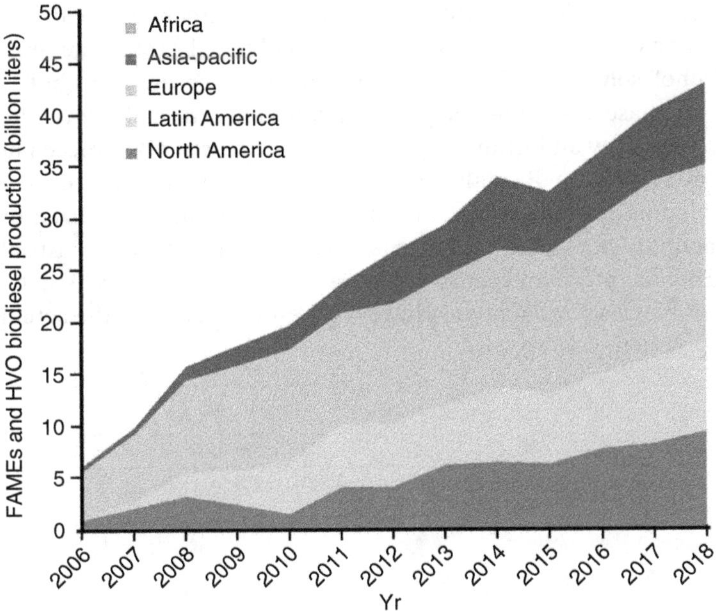


**Figure 9.2** Year-wise global production of biodiesel (FAME and HVO). *Source:* www.ifpenergiesnouvelles.com / last accessed October 23, 2021.

context, nanotechnology-based approaches, i.e. application of nanocatalysts, are found to be promising [13, 14]. The application of nanocatalysts (like magnetic nanocatalysts) not only reduces the cost involved in biodiesel production but also enhances the yield of biodiesel using mild reaction conditions. Nanocatalysts have several unique properties like high reactivity, good selectivity, enhanced yield, and the capability to reuse the same nanocatalysts for more than one transesterification reaction. Moreover, they also have several advantages over conventional catalysts like an enhanced rate of mixing with reactants, required less time for reaction, and easy and fast separation of product from the reaction mixture [14]

Considering these facts, the present chapter is aimed to discuss the various general transesterification reactions, conventional catalysts used in biodiesel production, and their limitations. Moreover, the possible application of nanotechnology in general and different nanocatalysts in particular in transesterifications has been briefly discussed.

## 9.2 Transesterification of Vegetable Oils

As mentioned previously, different methods can be used for biodiesel production, but the transesterification process was found to be most promising, and, hence, this process has been discussed here. The transesterification process is a reversible reaction. It is carried out by reacting fatty acids (oils) with alcohol in the presence of a suitable catalyst to obtain FAMEs (biodiesel) and glycerol. A strong base like sodium hydroxide, sodium methoxide, and potassium hydroxide or a strong acid such as hydrochloric acid and sulfuric acid can be used as a catalyst [15]. The catalyst is usually used to start the reaction and acts as an alcohol solubilizer; it is essentially required because alcohol is sparingly soluble in the oil phase, also non-catalyzed reactions are extremely slow. The catalyst helps in alcohol solubility and ultimately facilitates the reaction to proceed at a reasonable rate [16]. The separation of catalysts from the reaction mixture is the most difficult. After the completion of the reaction, the catalyst needs to be neutralized or removed with a large amount of hot water, which generates a large amount of industrial wastewater and also makes the process expensive. Figure 9.3 represents the schematic overview of the process of biodiesel production. However, Figure 9.4 showed the actual reaction involved in the production of FAMEs.

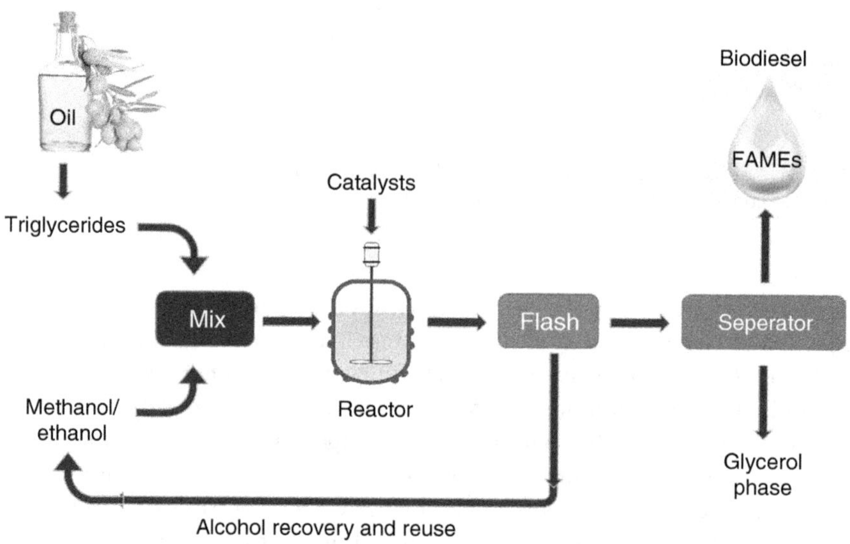


**Figure 9.3** Schematic representation of overall biodiesel production process.

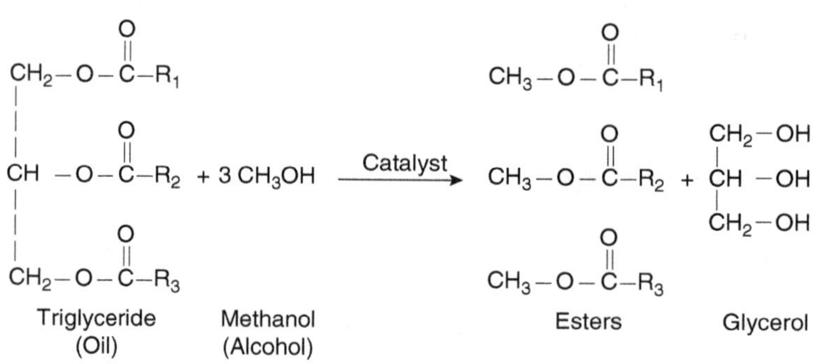


**Figure 9.4** The general chemical reaction showing transesterification of triglycerides.

## 9.3 Conventional Catalysts Used in Biodiesel Production: Advantages and Limitations

Various catalysts being used in the production of biodiesel are broadly categorized into homogeneous catalysts, heterogeneous catalysts, and biocatalysts. All these catalysts have their own merits and demerits. Figure 9.5 shows the classification of different conventional catalysts used in biodiesel production.

### 9.3.1 Homogeneous Catalysts

Homogeneous catalysis involves sequential reactions, which are catalyzed by using different kinds of homogeneous catalysts having the same phase as the reaction system have. These catalysts are widely used in the production of biodiesel because they are simple, possess high selectivity, turnover frequency, reaction rate, etc. [17]. These catalysts are usually subclassified as acid and base (alkali) catalysts. The important acid catalysts like sulfuric, hydrochloric, phosphoric, and sulfonic acids can be effectively used in the production of biodiesel through esterification and transesterification processes. Among these sulfonic acid is most preferably used [18]. However, the base or alkali catalysts such as sodium methoxide, sodium hydroxide, potassium methoxide, and potassium hydroxide are widely used. Among these sodium hydroxide is preferred over potassium hydroxide because of its high and quick solubility in methanol [19]. Moreover, its other features like high purity, low cost, and requirement of considerably low amount as compared to potassium hydroxide make them suitable catalysts for transesterification. In addition, alkali metal alkoxides are found to be more promising than hydroxides [20]. If the comparison is made among the homogeneous acid and base catalysts, the base catalysts facilitate high yields of biodiesel (FAMEs and FAEEs [Fatty acid ethyl esters]), at mild reaction conditions and only one

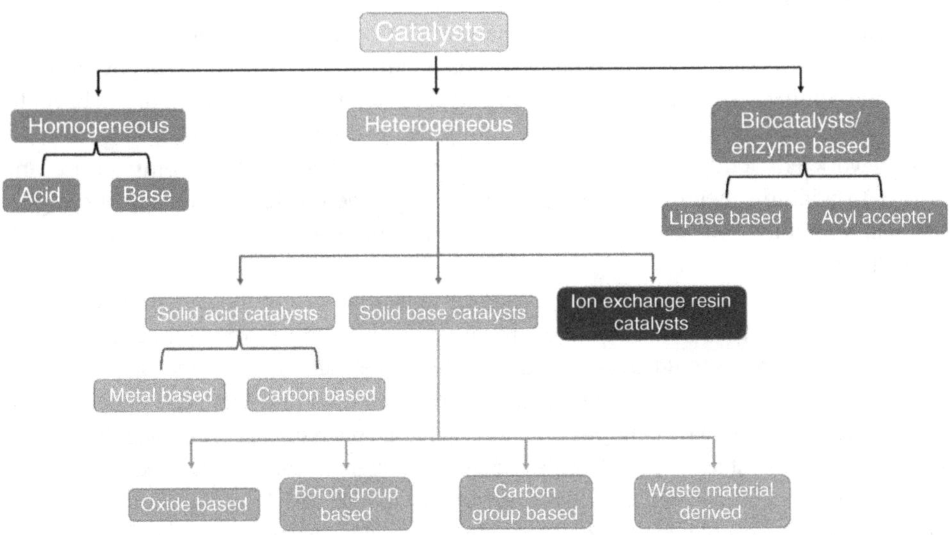


**Figure 9.5** The different conventional catalysts used in biodiesel production.

hour of reaction time [21]. However, homogeneous acid catalysts require high temperature and pressure and also a long time for the conversion of biodiesel [22].

Despite having the previously mentioned advantages, homogeneous catalysts have certain disadvantages, which are mainly associated with the formation of some amount of water due to the interaction of catalysts with methanol even after using water-free vegetable oils and alcohol. The presence of water and free fatty acids (FFAs) leads to soap formation by hydrolysis of the triglycerides, which reduces the biodiesel yield and affects the quality of the product [11]. Moreover, it was also demonstrated that the recovery of biodiesel and its separation from glycerin becomes difficult due to the formation of soap [23]. The separation and purification process often produces large quantities of wastewater, leading to pollution and environmental contamination [13]. These important limitations of homogeneous catalysts have limited the scope for the development and commercialization of biodiesel biorefineries.

### 9.3.2 Heterogeneous Catalysts

The different limitations, difficulties, and challenges associated with homogeneous catalysts created a pressing need to search for the most effective, selective, and eco-friendly alternative catalysts, which can potentially change the current scenario of biodiesel refinery. In this context, various heterogeneous catalysts become the center of attraction for global researchers due to their number of suitable characteristics features, and properties [24]. The most important is the phase of the heterogeneous catalysts; it is different from the phase of reactants and products. This unique property of heterogeneous catalysts allows easy and quick separation of catalysts from the reaction mixture using simple separation techniques like filtration, centrifugation, and sedimentation [25]. The easy and rapid separation of the catalyst from the reaction mixture significantly minimizes the time required for postreaction treatment and also reduces the process cost involved in biodiesel production [26]. Heterogeneous catalysts mediated transesterification does not generate chemical wastewater (which is required for separation of catalysts and other reaction products in case of homogeneous catalysis), which ultimately avoids pollution and also stops the environmental contamination, making the process green and eco-friendly [27]. These catalysts have the ability to tolerate a higher amount of FFA and moisture content. Moreover, these are also found to work efficiently even at harsh reaction conditions like high temperature and pressure. In addition, heterogeneous catalysts allow usage of alcohols having the higher molecular weight and are found to be efficient in reactions where homogeneous catalysts are usually inactive.

Heterogeneous catalysts are generally designed to coat or entrap the active molecules on the surface or inside the pores of a solid support such as silica, alumina, or ceria. An ideal heterogeneous catalyst should possess some features like high stability, mesoporous, inexpensive, and benign [17]. These catalysts are broadly classified into three categories such as solid acid catalysts, solid base catalysts, and ion exchange resin catalysts. Among these, solid base heterogeneous catalysts have been extensively studied by the scientific community all over the world due to their higher catalytic activity compared with the solid acid heterogeneous catalysts and ion exchange resin catalysts [16, 28]. To date, a variety of solid base catalysts have been successfully exploited in transesterification, which mainly includes CaO, MgO, SrO, $KNO_3/Al_2O_3$, $K_2CO_3/Al_2O_3$, $KF/Al_2O_3$, Li/CaO, KF/ZnO, basic hydrotalcite, etc. [23].

### 9.3.3 Biocatalysts

Biocatalysts or enzyme-based catalysts are another category of catalysts commonly used in biodiesel production. Biocatalysts have the ability to overcome the limitations of both homogeneous and heterogeneous catalysts because they have their biocompatibility, biodegradability, and eco-friendly nature. These usually consist of free lipase, traditionally immobilized lipase (lipase immobilized on nonmagnetic material), and lipase immobilized on magnetic nanoparticles (MNPs) (nanobiocatalysts) [29]. The enzyme lipases are reported to have promising catalytic activity and stability in nonaqueous media. Lipases can be easily obtained from various microbial systems like bacteria (*Bacillus thermoleovorans, Chromobacterium viscosum, Pseudomonas aeruginosa, Pseudomonas cepacia, Pseudomonas fluorescens, Pseudomonas putida, Staphylococcus hyicus*, etc.), fungi, and yeasts (*Aspergillus niger, Fusarium oxysporum, Geotrichum candidum, Fusarium heterosporum, Humicola lanuginose, Candida cylindracea, Candida rugose, Penicillium cyclopium, Penicillium roqueforti*, number of *Rhizopus* species, etc.) [30]. The biocatalysts reported to have higher catalytic activity and, hence, enhance biodiesel conversion yield up to 97–98% that is possible to achieve. But the high cost of the enzyme restricted its use in biocatalysts mediated transesterification at large-scale production. Moreover, due to the application of free enzyme (lipase), recovery of the enzyme is not possible. Therefore, it is not possible to reuse the biocatalysts in more than one cycle of transesterification, and, similarly, a gradual decrease in enzyme activity is a certain limitation in using a free enzyme as biocatalysts [31].

## 9.4 Role of Nanotechnology in Biodiesel Production

Nanotechnology is the most emerging branch of science getting dimensions in all fields like agriculture, medicine, food industry, biofuel industry, etc. due to its noteworthy applications. The nanomaterials that are considered as basic pillars of nanotechnology possess extraordinary and unique properties, which enable their promising uses in the biofuel industry also [32]. Nanomaterials can play a pivotal role in providing cost-effective and process-efficient technologies to improve the production of biofuels like biodiesel [33, 34]. According to Ramsurn and Gupta [35], the basic aim behind using nanotechnology in the biofuel industry is to apply both scientific (green and catalytic chemistry) and engineering solutions together in the quest for eco-friendly energy sources. To date, extensive efforts have been taken to develop and fabricate nanomaterials including advanced functional catalysts for biodiesel production where they can act as promising modern tools [36]. Currently, nanotechnology has attracted a deal of attention for the optimization of biodiesel production through nanotechnology-based catalysts, i.e. nanocatalysts. This can lead to the development of more efficient, economically viable, durable, and stable nanocatalysts holding the potential to achieve higher product quality and yields.

## 9.5 Different Nanocatalysts in Biodiesel Production

As mentioned earlier the unique feature of nanomaterials such as exceptionally higher catalytic activity, greater stability, high degree of crystallinity, better adsorption capacity, durability, and efficient storage makes them more suitable candidate as far as biodiesel production is concerned. In addition, nanocatalysts like magnetic nanocatalysts can be easily recovered and reused in multiple cycles of transesterification [37]. Various nanocatalysts such as metal and metal oxide-based nanocatalysts [e.g. titanium dioxide, calcium oxide, magnesium oxide, and strontium oxide [38–40], magnetic nanocatalysts, carbon-based nanocatalysts nanozeolites, nanoferrite, nanoclay, etc. [17] have been developed with high catalytic performance for biodiesel production. All these nanocatalysts have been discussed in the chapter and represented in Figure 9.6. Figure 9.5 shows the schematic illustration of different nanocatalysts used in biodiesel production.

### 9.5.1 Metal-Based Nanocatalysts

Among the various heterogeneous nanocatalysts, metal oxide nanocatalysts like alkali earth oxides catalysts were found to have significant catalytic activity in the transesterification of different feedstocks including vegetable oils [9, 41]. These nanocatalysts exhibit numerous extraordinary characteristics like high basic strength, easy availability, environmentally friendly, economic viability, and long catalyst lifetime, which generally decides the fate of transesterification reaction and resulting yield of biodiesel [42]. The important solid metal oxide nanocatalysts include calcium oxide (CaO)-based nanocatalysts, magnesium oxide (MgO)-based nanocatalysts, zirconium oxide ($ZrO_2$)-based nanocatalysts, zinc oxide (ZnO)-based nanocatalysts, and titanium dioxide ($TiO_2$)-based nanocatalysts [1, 9, 43]. Among these metal-based nanocatalysts, CaO nanocatalysts have been extensively studied.

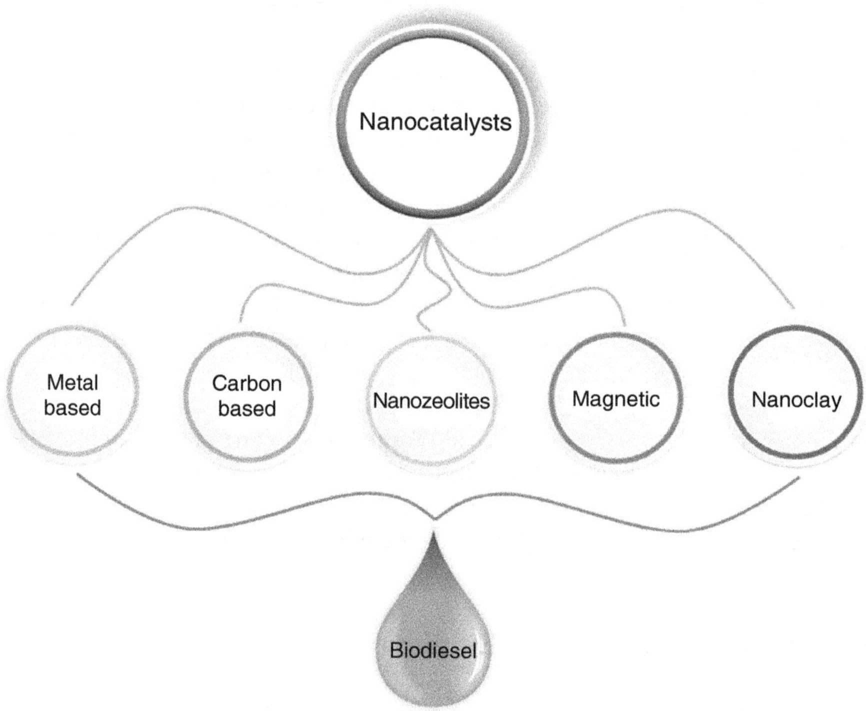


**Figure 9.6** Schematic illustration of different nanocatalysts used in biodiesel production.

The catalytic activity of CaO nanocatalysts is usually dependent upon their basicity, and there are different ways by which such basicity can be adapted such as cationic defect site in the lattice and surface terminations [44]. Bharti et al. [45] demonstrated the synthesis of CaO nanocatalysts using the sol–gel method and further evaluated their potential in catalytic transesterification of soybean oil. The maximum biodiesel yield i.e. was recorded at 97.61% on varied reaction conditions like catalyst loading of 3.675 wt%, oil, the molar ratio (alcohol to oil) of 11:1, and reaction temperature of 60 °C for 2 h. Similarly, Degfe et al. [46] reported promising applications of CaO nanocatalysts in the production of biodiesel from WCO.

Like Cao nanocatalysts, MgO-based catalysts are also reported to have an efficient catalytic in converting different biomasses into biodiesel. Amirthavalli and Warrier [47] studied the catalytic efficiency of MgO nanocatalyst synthesized via the sol–gel method in the conversion of WCO into biodiesel through transesterification. The maximum biodiesel yield of 80% was recorded at selected reaction conditions. In another study, the conversion of *Moringa oleifera* seeds oil in biodiesel through transesterification reaction using MgO nanocatalysts was proposed by Esmaeili et al. [48]. The authors claimed that it is possible to achieve a yield of 93.69% using these catalysts. Apart from these, $ZrO_2$-based nanocatalysts also attracted great interest due to their promising application in biodiesel refineries. Afsharizadeh and Mohsennia [49] demonstrated that mixed $ZrO_2$ nanocatalysts, i.e. $ZrO_2$-$SrO_2$ and $ZrO_2$-CuO, can be promisingly used production of biodiesel from WCO. Similarly, ZnO and $TiO_2$ nanocatalysts can also play important role in biodiesel production. Table 9.1 showed the list of different metal-based nanocatalysts having potential applications in biodiesel production.

### 9.5.2 Carbon-Based Nanocatalysts

Recently, much research focus is given on the use of carbon-based nanocatalysts in biodiesel production due to various advantages including higher thermal and chemical stability, ample availability and cost effectiveness, recovery, and reusability of catalyst [50]. The carbon-based nanocatalysts include carbon nanotubes, carbon dot, graphite, fullerene, carbon nanofibers, buckyball, carbon black, and carbon nanohorns (CHNs) [51, 52]. Moreover, nano-sized carbon materials including sulfonated carbon nanotubes, basic and acidic activated carbon, and supported carbon materials can also be used as heterogeneous catalysts for efficient biodiesel production [53, 54]. Zhang et al. [55] demonstrated the use of sulfonated acid-functionalized solid acid catalysts in biofuel production. Besides, several researchers claimed that carbon-based nanocatalysts are more promising catalytic agents in the biodiesel production process [1, 56]. Carbon-based nanomaterials are increasingly used as solid acid catalysts for biodiesel production using diverse feedstocks. Abdulkareem-Alsultan et al. [57] studied the catalytic potential of carbon-based nanocatalysts in biodiesel production from low-cost feedstock and further concluded that carbon-based catalysts are highly effective in the production of diesel range hydrocarbons.

Moreover, researchers around the globe reported the preparation and use of carbon-based nanocatalysts derived from various carbon forms including activated carbon [58], amorphous carbon [59], graphene oxide and graphene [60], CHNs [61], carbon dots [54], carbon nanotubes [44], and biochar [62] in biodiesel production technologies. However, Sano et al. [63] investigated the catalytic potential of CHNs dispersed with a calcium ferrite

**Table 9.1** Different metal-based nanocatalysts used in the production of biodiesel from different feedstocks. *Source:* Ref. [1]/With permission of American Chemical Society.

| Type of catalysts | Feedstocks | Type of alcohol | Alcohol:oil molar ratio, mol $mol^{-1}$ | Catalyst loading/ amount (wt% to the oil) | Temp °C | Time (h) | Yield (%) |
|---|---|---|---|---|---|---|---|
| **CaO-based nanocatalysts** | | | | | | | |
| CaO | Soybean oil | Methanol | 27 : 1 | CaO/1.13 | Room | 24 | 99 |
| CaO | Soybean oil | Methanol | 15 : 5 | CaO/0.056g | Room | 6 | 99 |
| CaO | Poultry fat | Methanol | 10 : 3 | CaO/1 mmol | Room | 6 | 100 |
| CaO | Soybean oil | Methanol | 6 : 1–10 : 1 | CaO/1–3 | 50–65 | 1 | 96.6 |
| CaO | Palm olein | Methanol | 15 : 1 | CaO/7 | 60 | 0.75 | 95.7 |
| CaO | Soybean oil | Methanol | 9 : 1 | CaO/3 | 65 | 3 | 98.72 |
| CaO | Canola oil | Methanol | 6 : 1–12 : 1 | CaO/2–4 | 60–65 | 2 | 99.85 |
| CaO | | Methanol | 6 : 1–12 : 1 | CaO/0.1–0.5 | 60–65 | 1 | 100 |
| CaO | Jatropha oil | Methanol | 5.15 : 1 | CaO/0.02 | 60 | 133.1 min | 95.8 |
| CaO | Soybean oil | Methanol | 11 : 1 | CaO/ 3.675 | 60 | 2 | 97.61 |
| CaO | Waste cooking oil | Methanol | 1 : 8 | CaO/1 | 50 | 1.5 | 96 |
| Lithium ion-impregnated CaO | Used cottonseed oil | Methanol | 3 : 1–18 : 1 | LiC/CaO/1–8 | 35–65 | 0.75 | 100 |
| | Karanja oil | Methanol | 3 : 1–12 : 1 | $Li^+$/CaO/1–10 | 35–65 | 1 | 99 |
| | Jatropha oil | Methanol | 3 : 1–12 : 1 | $Li^+$/CaO/1–10 | 35–65 | 2 | 99 |
| Potassium ion-impregnated CaO | Waste cottonseed oil | Methanol | 6 : 1–21 : 1 | K/CaO/2.5–12.5 | 35–65 | 1.25 | >99 |
| | Karanja oil | Methanol | 6 : 1–21 : 1 | K/CaO/2.5–12.5 | 35–65 | 2 | >99 |
| | Jatropha oil | Methanol | 6 : 1–21 : 1 | K/CaO/2.5–12.5 | 35–65 | 2.5 | >99 |
| KF/CaO | Chinese tallow seed oil | Methanol | 6 : 1–12 : 1 | KF/CaO/1–5 | 50–70 | 2 | 96 |
| Microporous solid base KF/CaO | Rapeseed oil | Methanol vapor | 6 : 1–12 : 1 | KF/CaO | 70–90 | — | 93.7 |

| | | | | | | | |
|---|---|---|---|---|---|---|---|
| Zinc-doped CaO | Waste cottonseed oil | Methanol | 3 : 1–18 : 1 | Zn/CaO/1–10 | 35–65 | 0.75 | 99 |
| Zirconium-impregnated CaO | Jatropha oil | Methanol | 3 : 1–18 : 1 | Zr/CaO/1–6 | 35–65 | 1.83 | 99 |
| | Jatropha oil | Ethanol | 3 : 1–24 : 1 | Zr/CaO/1–6 | 45–75 | 7.1 | 99 |
| Potassium carbonate ($K_2CO_3$)-impregnated CaO | Canola oil | Methanol | 6 : 1–18 : 1 | $K_2CO_3$/CaO/1–7 | 25–65 | 8 | 97.67 |
| Potassium fluoride-impregnated CaO/NiO | Waste cottonseed oil | Methanol | 3 : 1–18 : 1 | KF/(CaO/NiO)/1–6 | 35–75 | 4 | 99 |
| Mesoporous KF/CaO–MgO | Rapeseed oil | Methanol | 12 : 1 | KF/(CaO/MgO)/3 | 70 | 3 | 95 |
| Ferromagnetic KF/CaO–$Fe_3O_4$ | Stillingia oil | Methanol | 12 : 1 | KF/(CaO/$Fe_3O_4$)/1 | 65 | 3 | 95 |
| Graphite oxide-loaded CaO | Soybean oil | Methanol | 0.83 : 1 | CaO/graphite oxide/8 | 60 | 2 | 98.3 |
| | Soybean oil | Methanol | 0.83 : 1 | CaO/graphite oxide/8 | 60 | 2 | 97.8 |
| Magnetic $Fe_3O_4$ oxide-loaded CaO | Jatropha oil | Methanol | 15 : 1 | CaO/$Fe_3O_4$/2 | 70 | 4 | 99 |
| Calcium aluminate onto $Fe_3O_4$ | Rapeseed oil | Methanol | 15 : 1 | (Ca/Al)/$Fe_3O_4$/6 | 65 | 3 | 98.71 |
| Low-cost material containing CaO | Jatropha oil | Methanol | 9 : 1 | *Musa balbisiana* Colla underground stem ash (10% CaO)/5 | 60–275 | 1 | 98 |
| **MgO-based nanocatalysts** | | | | | | | |
| MgO | Waste cooking oil | Methanol | 10 : 1 | 2 | 60 | 2 | 80 |
| MgO | *Moringa oleifera* seeds oil | Methanol | 12 : 1 | 1 | 45 | 4 | 93.69 |
| MgO | Waste cooking oil | Methanol | 24 : 1 | 2 | 65 | 1 | 93.3 |
| Magnetic MgO/$MgFe_2O_4$ spinel nanocatalyst | Sunflower oil | Methanol | 12 : 1 | 3 | 110 | 3 | 92.5 |

(*Continued*)

**Table 9.1** (Continued)

| Type of catalysts | Feedstocks | Type of alcohol | Alcohol:oil molar ratio, mol $mol^{-1}$ | Catalyst loading/amount (wt% to the oil) | Temp °C | Time (h) | Yield (%) |
|---|---|---|---|---|---|---|---|
| $MgO/MgAl_2O_4$ | Sunflower oil | Methanol | 12 : 1 | 3 | 110 | 3 | 95 |
| $MgO$-$La_2O_3$ | Sunflower oil | Methanol | 18 : 1 | 3 | 338 K | 5 | 97.7 |
| **$ZrO_2$-based nanocatalysts** | | | | | | | |
| S-$ZrO_2$ | Soybean oil | Methanol | 1 : 20 (methanolysis) | 5 | 120 | 1 | 98.6 |
| Sulfated zirconia (SZ) | Soybean oil | Methanol | 1 : 20 (ethanolysis) | 5 | 120 | 1 | 92. |
| $ZrO_2$ loaded with $C_4H_4O_6HK$ | Soybean oil | Methanol | 16 : 1 | 6 | 60 | 2 | 98.03 |
| Zirconia/titania | Cooking oil | Methanol | — | — | — | — | 100 |
| Zirconia modified with KOH | *Silybum marianum* oil | Methanol | 15 : 1 | 6 | 60 | 2 | 90.8 |
| $ZrO_2/Al_2O_3$ nanocatalyst | Oleic acid | Methanol | 8 : 1 | 0.2 | 340 K | 2 | 90.11 |
| $CdO/ZrO_2$ | Soybean oil | Methanol | 40 : 1 | 7 | 135 | 3 | 97 |
| $ZrO_2$-$SrO_2$ (Pechini sol–gel method) | Waste cooking oil | Methanol | 10 : 1 | 5 | 120 | 3 | 68 |
| | Corn oil | Methanol | 10 : 1 | 5 | | 3 | 59 |
| $ZrO_2$-$SrO_2$ (coprecipitation methods) | Waste cooking oil | Methanol | 10 : 1 | 5 | 120 | 3 | 92 |
| | Corn oil | Methanol | 10 : 1 | 5 | | 3 | 84 |
| $ZrO_2$-$CuO$ (coprecipitation methods) | Waste cooking oil | Methanol | 10 : 1 | 5 | 120 | 3 | 87 |
| | Corn oil | Methanol | 10 : 1 | 5 | | 3 | 79 |
| $ZrO_2$-$CuO$ (coprecipitation methods) | Waste cooking oil | Methanol | 10 : 1 | 5 | 120 | 3 | 90 |
| | Corn oil | Methanol | 10 : 1 | 5 | 120 | 3 | 88 |

| | | | | | | | |
|---|---|---|---|---|---|---|---|
| $Bi_2O_3$-supported $ZrO_2$ | *Nannochloropsis* sp. lipid | Methanol | 1 : 90 (g $ml^{-1}$) | 20 | 80 | 6 | 73.21 |
| **ZnO-based nanocatalysts** | | | | | | | |
| Cu-doped ZnO | Waste cooking oil | Methanol | 5 : 1 | 4 | 60 | 40 min | 98 |
| | Waste cooking oil | Methanol | 8 : 1 | 12 | 55 | 50 | 97.71 |
| | Neem oil | Methanol | 10 : 1 | 10 | 55 | 1 | 73.95 |
| Ferromagnetic (Iron (II)-doped) ZnO | Castor oil | Methanol | 12 : 1 | 14 | 50 | 55 min | 91 |
| Iron (II)-doped ZnO | Pongamia oil | Methanol | 10 : 1 | 12 | 50 | 55 min | 93 |
| Heteropoly acid coated ZnO | Madhuca indica oil | Methanol | — | — | 55 ± 5 | 5 | 95 |
| Mn-doped ZnO | Mahua oil | Methanol | 7 : 1 | 8 | 50 | 50 min | 97 |
| Ni-doped ZnO | Castor oil | Methanol | 8 : 1 | 11 | 55 | 1 | 95.20 |
| Ag–ZnO nanoparticles | Simarouba oil | Methanol | | - | | | 84.5 |
| Co-doped ZnO | *Mesua ferrea* oil | Methanol | 9 : 1 | 2.5 | 60 | 3 | 98.03 |
| **$TiO_2$-based nanocatalysts** | | | | | | | |
| $Ti(SO_4)O$ | Used cooking oil | Methanol | 9 : 1 | 1.5 | 75 | 3 | 97.1 |
| $TiO_2/PrSO_3H$ | Used cooking oil | Methanol | 15 : 1 | 4.5 | 60 | 9 | 98.3 |
| $TiO_2$ nanoparticles | Olive oil | Methanol | 10 : 1 | — | 120 | 4 | 91.2 |

$Ca_2Fe_2O_5$ ($Ca_2Fe_2O_5$-CNH) in the conversion of triglyceride to FAMEs. The improved yield of 100% was recorded by authors within 3 h of reaction time and also found that catalyst can be reused up to three cycles, whereas Poonjarernsilp et al. [61] reported 90% of the yield using oil mixture of palmitic acid and tributyrin feedstock at reaction conditions such as methanol-to-oil molar ratio 20 : 1, at 60 °C temperature, and reaction time of 3 h. Moreover, Ballotin et al. [64] tested the applicability of catalysts composed of sulfonated carbon nanostructures embedded in an amorphous carbon matrix in the esterification of oleic acid with methanol. The conversion of 95% of oleic acid was reported with catalyst loading of 2.5 wt%, 1 : 10 oleic acid-to-methanol molar ratio, at 100 °C. Moreover, sulfonated nanographene has received much attention for biodiesel due to its high catalytic efficiency. The 98% of biodiesel yield was achieved using sulfonated graphene catalyst at optimal process conditions of methanol-to-oil ratio 20 : 1, over 10 wt% of catalyst loading at 100 °C in 840 min [60].

Recently, Jume et al. [65] studied biodiesel production from WCO using graphene oxide-doped metal oxides nanocomposite (GO@ZrO2-SrO) catalyst. The highest FAME yield of 91% was recorded with reaction conditions of the material ratio of 1 : 0.5 (w/w) of GO: $ZrO_2$-SrO, oil-to-methanol ratio (1 : 4), at 120 °C in 90 min. This heterogeneous catalyst can be used effectively for the production of biodiesel via transesterification of WCO. Furthermore, Ibrahim et al. [56] examined the synthesis and application of $Na_2O$ (20 wt%) / CNTs catalysts in biodiesel production. Maximum 97% of FAME yield was recorded with methanol-to-oil molar ratio 20 : 1, 3 wt% of catalyst concentration at 65 °C for 3 h of reaction time. Moreover, the catalysts can be recycled up to three subsequent reaction cycles. Furthermore, Shuit and Tan [66] reported a yield of 93.4% at 170 °C in 180 min using 2% of catalyst amount and 20 : 1 ratio of methanol to oil assisted by multiwalled carbon nanotubes (MWCT) upon alcoholysis of palm fatty acid distillate reported the *trans*-esterification of triglycerides using sulfonated multiwalled CNT as nanocatalysts. Similarly, Guan et al. [67] reported a biodiesel yield of 97.8% with 3.7 wt% catalysts amount at 150 °C and 1 h reaction time.

### 9.5.3 Zeolites/Nanozeolites

Zeolites are the class of microporous crystalline aluminosilicates made up of silicon (Si), aluminum (Al), oxygen (O), and microporous crystalline materials [68]. Zeolites exhibit key features including profuse acid sites, distinctive ion-exchange property, and greater hydrothermal stability; therefore, these materials have been widely used in adsorption, separation, and catalysis [69]. In the recent past, zeolites have been tested for their potential role in biodiesel production due to their acidic nature, high surface area, shape selectivity, and distinctive molecular sieving properties [70]. Several research articles highlighted the applicability of the different zeolites such as Li/NaY [71], MgO/ZSM-5 [72], CaO/ZSM-5 [73], and $La_2O_3$/NaY [74] for efficient transesterification reactions. However, 3D basic zeolites have certain limitations including lesser base density and mass transfer constraint for bulky triglycerides (TGs) and FAME; thus basic zeolites displayed lower catalytic activity in transesterification reactions [75, 76]. In this context, the synthesis of nanoscale zeolites is an effective approach for overcoming the limitations of basic zeolites. Nanozeolites are hydrophobic supports exhibiting key properties like higher external surface area, strong acid sites, and superior dispersion capability in aqueous and organic solvents, hence showing excellent catalytic efficiency in transesterification reactions [9, 36]. Amalia et al. [77] found that the use of 70% KOH/zeolite catalyst in transesterification of castor oil at 55 °C and 7 h of reaction time offered 92.11% biodiesel yield, whereas Saeedi et al. [78] demonstrated the catalytic function of imidazolate zeolite doped with potassium (KNa/ZIF 8) catalyst in transesterification of soybean oil. It was recorded that there are more than 98% yield with KNa/ZIF 8 catalyst under optimum reaction conditions of methanol-to-oil ratio of 3 : 1 and 3 h of the processing time. However, Li et al. [71] demonstrated the efficacy of Li/NaY zeolite catalysts with different $Li_2CO_3$ to NaY zeolite content in the transesterification of castor oil with ethanol. The results obtained recorded the biodiesel yield of 98.6% while using 18 : 1 methanol/oil ratio, catalyst loading 3 wt% and 75 °C temperature for 2 h. Likewise, NaY zeolite-supported $La_2O_3$ catalysts yielded 84.6% fatty acid ethyl ester with reaction conditions like the molar ratio of ethanol to oil 15 : 1, 10 wt% catalyst, at 70 °C temperature for 50 min [74]. Additionally, Mostafa et al. [79] synthesized K-La nanocatalysts supported on zeolite ZSM-5 and tested for biodiesel production from soybean oil. They reported 90% FAMEs yield with methanol-to-oil molar ratio of 12 : 1 at 60 °C for 3 h. The studied K-La/ZSM nanocatalysts showed higher catalytic activity, which might be due to the high number of basic sites. Pang et al. [68] reported the striking performance of 2D Na/ITQ-2 zeolite catalyst in the transesterification of bio-derived oil to produce biodiesel. However, MgO/ZSM-5 catalysts provide 92.1% biodiesel yield at reactions conditions like the molar ratio of ethanol to oil 15/1, catalyst amount 3 wt%, at 75 °C and reaction time of 1 h. Further, it was found that these nanocatalysts are stable and catalytically active even after five cycles [69]. Recently, Mohebbi et al. [80] obtained 98% biodiesel yield from WCO under optimum conditions like methanol/oil 1 : 7 M assisted with high silica $MoO_3$/B-ZSM-5 nanocatalyst concentration 3 wt% at 160 °C for 6 h, whereas HZSM-5 (nanosheets) yielded 95.12% biodiesel from linoleic acid feedstock with methanol-to-oil molar ratio of 6 : 1, catalyst amount 10 wt% at 180 °C for 4 h [81]. The nanomagnetic-based zeolite catalyst CaO/NaY-$Fe_3O_4$ offers a greater conversion efficiency of 95.37% using canola oil as feedstock and also shows greater stability at high temperatures and easier separation process [82].

Brito et al. [83] reported promising catalytic efficacy of the zeolite Y with different $Al_2O_3$ content for transesterification of waste oil. Hassani et al. [84] reported the highest conversion of triglyceride (46%) under optimum process conditions like methanol/WCO molar ratio 2.6–6.0, 2–10 h of reaction time at 50–85 °C temperature. Similarly, 95.1% of biodiesel yield from sunflower oil feedstock was found under conditions of molar ratio of methanol/oil 6 : 1, at 60 °C temperature, the reaction time of 7 h, and reusability up to three cycles [85]. Besides, Doyle et al. [86] investigated the catalytic property of zeolite Y with Si/Al ratio of 3.1, in the esterification of oleic acid. The authors reported 85% the maximum oleic acid conversion of 85% with ethanol/oleic acid ratio of 6 : 1 M, 5 wt% catalyst loading, and 70 °C for 1 h reaction time, whereas nanozeolite-enzyme complexes have been developed and being used as excellent catalysts for the improved production of biodiesel. It is well documented that lipase enzyme catalyst immobilized on a solid support such as nanozeolites resulted in higher catalytic efficiency over to free enzyme [87]. Another study by the authors reported that biocatalyst immobilized on amino-functionalized nanozeolitic iron oxide composite support (EMNZ-X/F) gives 83.1% of ethyl esters under reaction conditions of ethanol to microalgae oil molar ratio 5 : 1 M, at 45 °C for 48 h and 3 wt% catalyst loading [88]. Furthermore, *Thermomyces lanuginosus* lipase (TLL) enzyme immobilized on the nanozeolites supports functionalized with (3-Aminopropyl)trimethoxy-silane (APTMS) and GA offers improved biocatalysis, with FAEEs yields of 93% and stability during five successive cycles of use [89]. Hence it can be concluded that zeolites and nanozeolites can act as efficient heterogeneous catalysts for biodiesel production.

### 9.5.4 Magnetic Nanocatalysts

Nowadays, the nanocatalysts derived from magnetic nanoparticles have received substantial attention and are greatly employed for efficient biodiesel production from cheap feedstocks. The magnetic nanocatalysts have certain benefits such as easier separation and recyclability and their acidity [44]. The widespread applicability of the magnetic nanocatalysts is due to economically affordable, raw material saving, extraordinary efficiency, and higher reaction speed, thereby plummeting the processing time [90]. The use of magnetic solid base nanocatalysts are highly advantageous as it possesses combined properties of both homogeneous (high specific area) and heterogeneous (easy separation from reagents) catalysts, hence averting the catalyst loss and dropping the overall costs of production [91]. For instance, magnetic $MgO/MgFe_2O_4$ spinel nanocatalysts, synthesized using microwave-assisted combustion methods, serve as promising alternate heterogeneous catalysts for biodiesel synthesis [92]. Similarly, Alaei et al. [93] synthesized magnetic $MgO/MgFe_2O_4$ nanocatalyst using combustion methods having structures with a larger pore (above 10 nm) and higher surface area (97.8 $m^2\,g^{-1}$). This catalyst showed greater performance in biodiesel production (91.2% conversion) upon conditions such as methanol-to-oil molar ratio of 12 : 1 catalyst concentration 4 wt% reaction temperature of 110 °C for 4 h.

Nevertheless, Feyzi and Norouzi [94] synthesized magnetic $Ca/Fe_3O_4@SiO_2$ nanocatalysts using a combination of sol–gel and incipient wetness impregnation methods and evaluated its potential for biodiesel production. The biodiesel yield of 97% was obtained with operational conditions like methanol-to-oil molar ratio of 15/1 at 65 °C with mechanical stirring for 5 h. The study done by Rahimi et al. [95] reported the economically viable production of biodiesel with $MgO/Fe_2O_3$-$SiO_2$ core–shell magnetic nanocatalysts using *Camelina sativa* seed oil feedstock. The enhanced biodiesel efficiency (99%) was reported in transesterification of *C. sativa* seed oil using methanol/oil 12 : 1 M, the catalyst amount of 4.9 wt%, reaction temperature 70 °C, and reaction time 4.1 h.

Recently, Ashok et al. [96] prepared magnetically separable $Mg^{2+}$-doped zinc ferrite nanocatalysts using microwave-assisted combustion process and used them for biodiesel synthesis from WCO. The said nanocatalysts showed better magnetic moment and exhibited higher saturation magnetization property, which is very essential for magnetic separation of the catalyst from the reaction medium. The authors reported biodiesel conversion of 99.9% under conditions like 3 wt% ZnMgF5 nanocatalyst, methanol–oil molar ratio of 18 : 1 at 65 °C for 30 min. Further, it was noted that magnesium-doped zinc ferrite catalyst retained 94% of biodiesel yield even after 10 cycles of recovery. Moreover, Dantas et al. [97] evaluated the effect of $Cu^{2+}$ ions on magnetism, morphology, and structure of nanoferrite catalyst ($Ni_{0.5}Zn_{0.5}Fe_2O_4$) and also studied its consequences, leading to alteration of catalytic efficacy in the transesterification. The $Cu^{2+}$-doped $Ni_{0.5}Zn_{0.5}Fe_2O_4$ catalyst showed promising catalytic activity as it was evident from improved biodiesel yield within the range of 5.5–85%. However, Liu et al. [98] evaluated the potential of nanomagnetic catalyst, $K/ZrO_2/\gamma\text{-}Fe_2O_3$ for biodiesel production with methanol-to-oil molar ratio 10 : 1, catalyst concentration of 5 wt%, and 65 °C for 3 h. The maximum biodiesel yield of about 93.6% was achieved, and the said catalyst was reused for 6 successive cycles without any loss of activity. Moreover, the nanomagnetic $CaO/Fe_3O_4$ catalyst yielded 69.7% of biodiesel under the optimum reaction process using methanol/oil molar ratio 20 : 1, 10 wt% of nanocatalyst at 65 °C for 5 h [99]. Additionally, magnetic mesoporous nanocatalyst ($KOH/Fe_3O_4@\gamma\text{-}Al_2O_3$) resulted in 97.4% biodiesel yield from transesterification of canola oil at 65 °C for 2 h of reaction time, catalyst concentration of 6.5 wt%, and methanol/ oil 16.2 : 1 M [100].

In another study, Shaker and Elhamifar [101] developed a highly active and recyclable nanocatalyst viz. sulfonic acid-containing magnetic methylene-based organosilica with core–shell structure ($Fe_3O_4@OS–SO_3H$). Further, this nanocomposite was used for efficient biodiesel production via esterification of carboxylic acids with alcohols. Besides, the optimum biodiesel yield of 96.13% was recorded by using oil-to-methanol ratio 1 : 7, catalyst amount 4.5 wt%, at 65 °C for 2 h [102]. Likewise, Xie et al. [103] employed magnetically recyclable solid catalysts ($Fe_3O_4$/MCM-41 composites) for biodiesel synthesis. These magnetic nanocomposites displayed a strong magnetic response and also showed better catalytic efficiency in the transesterification of soybean oil for biodiesel production with the oil conversion reaching 99.2% with methanol/oil 25 : 1 M, catalyst amount of 3 wt% at 65 °C, and reflux of methanol after 8 h. The catalytic potential of magnetic nanoparticles [$Fe_3O_4$/ZnMg(Al)O] was investigated for efficient biodiesel production [104]. The study done was done Feyzi et al. [105] reported a biodiesel yield of 94.8% using magnetic $Cs/Al/Fe_3O_4$ nanocatalyst in transesterification reaction of sunflower oil. Besides, Hu et al. [106] reported 95% FAME yield from stillingia oil using nanomagnetic catalyst $KF/CaO\text{-}Fe_3O_4$ upon optimal conditions.

From the available literature, it was evident that the in situ transesterification process has economic issues unlike biodiesel production from microalgae that is cost-effective as lipid extraction and transesterification reaction carried out in one step [107, 108]. In this context, Safakisha et al. [109] cultivated the microalgae using wastewater as an inexpensive medium. The microalgal biomass was further exploited for the synthesis of biodiesel using $SO_4^{2-}/Fe_3O_4\text{-}Al_2O_3$ magnetic nanocatalyst. The studied catalyst resulted in 87.6% biodiesel yield under study conditions of catalyst amount of 8 wt%, 9 ml methanol $g^{-1}$ microalgae, at 120 °C, and 4 h of reaction time. The said nanocatalyst retains its activity up to the fifth cycle of use. A similar study was performed by Farrokheh et al. [90] utilizing the two different magnetic nanocatalysts viz. $CaO/KOH\text{-}Fe_3O_4$ and $KF/KOH\text{-}Fe_3O_4$ for preparation of biodiesel using microalgae strains, namely, *Chlorella vulgaris* and *Spirulina platensis*. The results obtained in the study indicated that *C. vulgaris* microalgae are superior over *S. platensis*, and $KF/KOH\text{-}Fe_3O_4$ is the best preferable catalyst for biodiesel production by the electrolysis method. The nanomagnetic catalyst yielded 96.8% biodiesel under study conditions like catalyst loading of 1.5 wt% methanol/oil 1 : 6 M, at 25 °C, for 2 h of reaction time, and stirring speed of 400 rpm. Furthermore, this nanocatalyst is recyclable up to 10 times without a significant loss of activity. The magnetic $K/Fe_2O_3\text{-}Al_2O_3$ core–shell nanocatalyst converted 95.6% of microalgae lipids to esters under reactions such as methanol-to-dry biomass 12 ml $g^{-1}$, 4 wt% of magnetic catalyst loading, and at 65 °C for 6 h of reaction time and also showed better reusability [110]. Recently, several research groups reviewed applications and the performance of different functionalized magnetic nanocatalysts and feedstock used for biodiesel production [17, 111, 112].

### 9.5.5 Nanoclays

The use of nanoclays as a catalyst in biodiesel production is increasingly demanding in recent days. It is due to the extraordinary properties of nanoclays; additionally these materials are available at lower cost and required lesser time for transesterification reactions. Numerous nanoclays have been used in diverse chemical reactions including transesterification. Montmorillonite is one of the commonly used nanoclays; structurally it is composed of compact plate aluminosilicate. Clays with montmorillonite group exhibit higher chemical activity due to nanometer dimension layered structure [113]. Various groups studied the applicability of clays including naturally occurring or acid-modified clays (K10 or KSF) [114], acidic silica [115], modified montmorillonite [116], and bentonite [117] for biodiesel synthesis. But many studies reported a lower percentage (below 50%) of methyl esters in obtained biodiesel [114, 118]. However, Halek [113] reported the synthesis of different nanocatalyst viz. CaO/Cloisite15A, CaO/NaMMT, and CaO/K10MMT using calcium oxide/nanoclay catalysts. These nanocatalysts were further utilized for the production of biodiesel from waste oil. It was found that biodiesel produced using these nanocatalyst shows higher content of methyl esters (over 95%,) thereby indicating the greater efficacy of synthesized nanocatalysts. Moreover, Sultana et al. [119] successfully exploited bentonite nanoclay catalyst for transesterification of sal oil to biodiesel at high temperature 443–493 K. They reported 96.5% conversion of sal oil under optimum study conditions.

### 9.5.6 Other Nanocatalysts

Numerous other nanomaterial-based catalysts have been developed and exploited for the production of biodiesel. These nanocatalysts exhibited distinguished catalytic efficacy along with diverse applications. Some of the widely used other notable nanocatalysts include lanthanum phosphate foam [120] and metal-organic framework (MOF) nanocatalyst [121]. Rezania et al. [120] demonstrated the function of LaPO foam as a heterogeneous catalyst in the transesterification of highly acidic WCO. These nanocatalysts displayed excellent stability, superior catalytic activity, and easy separation. The FAME yield of 91% under process conditions like methanol/ oil 5 : 1 and 2.5 wt% of LaPO at 90 °C for 120 min was obtained. MOF is a class of polyporous material that control its topological structure and pore size so as to meet the requirements of the catalytic reactions. The MOF-based nanocatalyst offers excellent performance in biodiesel production and other related biorefineries [121]. Several studies reported the use of MOF-based nanocomposite for biodiesel production. For instance, Zhang et al. [55] synthesized Sn1.5PW/Cu-BTC, which is MOF Cu-BTC-supported Sn (II)-substituted Keggin heteropoly nanocomposite, and further studied its performance in the esterification reaction. The studied nanocatalyst provided oleic acid conversion of 87.7% under optimum reaction conditions. Further, it was found that nanocatalyst can be reused up to seven cycles with greater efficacy of 80% after three cycles. Moreover, NiHSiW/UiO-66 nanocatalyst showed greater catalytic activity in the conversion of FFA to biodiesel [122]. Jeon et al. [123] also stated the utility of core–shell nanostructured heteropoly acid-functionalized framework MOF as a bifunctional heterogeneous catalyst in the improved biodiesel production process. The HPA-functionalized ZIF-8 catalyst displayed a higher FAME conversion of 98.02% and good reusability. However, Negm et al. [124] reported the conversion of 99.9% for methyl oleate with Me/OLA of 6 : 1 M, catalyst concentration 4 wt%, and at time of 2 h using heterogeneous polymer-heteropoly acid nanocatalyst.

## 9.6 Conclusion

Biofuels like biodiesel attract a great deal of attention from the global scientific community due to their several benefits over fossil fuels. This form of renewable bioenergy can serve as a promising substitute for conventional diesel and has the potential to bring noteworthy revolutions in the transport sector. Nanotechnology was found to play a pivotal role in biofuel industries through the development of nanocatalysts using different nanomaterials, which has the ability to overcome most of the limitations associated with conventional homogeneous and heterogeneous catalysts. Nanocatalysts due to their unique and extraordinary features like high reactivity, enhanced catalytic performance, selectivity, economic viability, and eco-friendly nature is gaining importance in biodiesel production. Therefore, different nanocatalysts like metal-based nanocatalysts, carbon-based nanocatalysts, magnetic nanocatalysts, nanoferrites, nanoclays, etc. can be successfully used for higher production of biodiesel from a variety of feedstocks.

## Acknowledgment

API is highly thankful to Science and Engineering Research Board (SERB), Department of Science and Technology, Government of India, New Delhi, for providing financial assistance in the form of Ramanujan Fellowship (RJF/2019/000044).

## References

1 Ingle, A.P., Philippini, R., Martiniano, S.E. et al. (2020a). Application of metal oxide nanostructures as heterogeneous catalysts for biodiesel production. In: *Advanced Heterogeneous Catalysts Volume 1: Applications at the Nano-Scale ACS Symposium Series* (ed. P. Sudarsanam and L. Singh), 261–289. Washington, DC: American Chemical Society.

2 Bano, S., Ganie, A.S., Sultana, S. et al. (2020). Fabrication and optimization of nanocatalyst for biodiesel production: an overview. *Front. Energy Res.* 8: 579014. https://doi.org/10.3389/fenrg.2020.579014.

3 Ganesan, R., Manigandan, S., Samuel, M.S. et al. (2020). A review on prospective production of biofuel from microalgae. *Biotechnol. Rep.* 27: e00509. https://doi.org/10.1016/j.btre.2020.e00509.

4 Adegboye, M.F., Ojuederie, O.B., Talia, P.M., and Babalola, O.O. (2021). Bioprospecting of microbial strains for biofuel production: metabolic engineering, applications, and challenges. *Biotechnol. Biofuels* 14: 5. https://doi.org/10.1186/s13068-020-01853-2.

5 Abraham, B.C. and Kuriakose, S. (2020). Biodiesel mixtures with additives as alternative fuels for direct injection diesel engine. *AIP Conf. Proc.* 2263: 020008. https://doi.org/10.1063/5.0017126.

6 Kaya, C. and Kökkülünk, G. (2020). Biodiesel as alternative additive fuel for diesel engines: an experimental and theoretical investigation on emissions and performance characteristics. *Energy Sources, Part A* https://doi.org/10.1080/15567036.2020.1774685.

7 Adekunle, A.S., Oyekunle, J.A.O., Oduwale, A.I. et al. (2020). Biodiesel potential of used vegetable oils transesterified with biological catalysts. *Energy Rep.* 6: 2861–2871.

8 Perez, V.H., Junior, E.G.S., Cubides, C. et al. (2014). Trends in biodiesel production: present status and future directions. In: *Biofuels in Brazil*, 281–302. Cham, Switzerland: Springer.

9 Ingle, A.P., Chandel, A.K., Philippini, R. et al. (2020b). Advances in nanocatalysts mediated biodiesel production: a critical appraisal. *Symmetry* 12: 256. https://doi.org/10.3390/sym12020256.

10 Zulqarnain, Ayoub, M., Yusoff, M.H.M. et al. (2021). A comprehensive review on oil extraction and biodiesel production technologies. *Sustainability* 13: 788. https://doi.org/10.3390/su13020788.

11 Ogunkunle, O. and Ahmed, N.A. (2019). A review of global current scenario of biodiesel adoption and combustion in vehicular diesel engines. *Energy Rep.* 5: 1560–1579.

12 Rizwanul-Fattah, I.M., Ong, H.C., Mahlia, T.M.I. et al. (2020). State of the art of catalysts for biodiesel production. *Front. Energy Res.* 8: 101. https://doi.org/10.3389/fenrg.2020.00101.

13 Bharti, M.K., Chalia, S., Thakur, P. et al. (2021). Nanoferrites heterogeneous catalysts for biodiesel production from soybean and canola oil: a review. *Environ. Chem. Lett.* https://doi.org/10.1007/s10311-021-01247-2.

14 Najeeb, J., Akram, S., Mumtaz, M.W. et al. (2021). Nanobiocatalysts for biodiesel synthesis through transesterification: a review. *Catalysts* 11: 171. https://doi.org/10.3390/catal11020171.

15 Gebremariam, S.N. and Marchetti, J.R. (2017). Biodiesel production technologies: review. *AIMS Energy* 5 (3): 425–457.

16 Vasić, K., Podrepšek, G.H., Knez, Z., and Leitgeb, M. (2020). Biodiesel production using solid acid catalysts based on metal oxides. *Catalysts* 10: 237. https://doi.org/10.3390/catal10020237.

17 Changmai, B., Vanlalveni, C., Ingle, A.P. et al. (2020). Widely used catalysts in biodiesel production: a review. *RSC Adv.* 10: 41625.

18 Fukuda, H., Kondo, A., and Noda, H. (2001). Biodiesel fuel production by transesterification of oils. *J. Biosci. Bioeng.* 92: 405–416.

19 Jitputti, J., Kitiyanan, B., Rangsunvigit, P. et al. (2006). Transesterification of crude palm kernel oil and crude coconut oil by different solid catalysts. *Chem. Eng. J.* 116: 61–66.

20 Saydut, A., Kafadar, A.B., Aydin, F. et al. (2016). Effect of homogeneous alkaline catalyst type on biodiesel production from soybean [*Glycine max* (L.) Merrill] oil. *Indian J. Biotechnol.* 15: 596–600.

21 Vyas, A.P., Verma, J.L., and Subrahmanyan, N. (2010). A review on FAME production processes. *Fuel* 89: 1–9.

22 Marchetti, J.M., Pedernera, M.N., and Schbib, N.S. (2011). Production of biodiesel from acid oil using sulfuric acid as catalyst: kinetics study. *Int. J. Low-Carbon Technol.* 6: 38–43.

23 Thangaraj, B., Solomon, P.R., Muniyandi, B. et al. (2019). Catalysis in biodiesel production: a review. *Clean Energy* 3 (1): 2–23.

24 Farooq, M., Ramli, A., and Subbarao, D. (2013). Biodiesel production from waste cooking oil using bifunctional heterogeneous solid catalysts. *J. Clean. Prod.* 59: 131–140.

25 Diamantopoulos, N. (2015). Comprehensive review on the biodiesel production using solid acid heterogeneous catalysts. *J. Thermodyn. Catal.* 6: 1–8.

26 Wilson, K. and Lee, A.F. (2012). Rational design of heterogeneous catalysts for biodiesel synthesis. *Catal. Sci. Technol.* 2: 884–897.

27 Faruque, M.O., Razzak, S.A., and Hossain, M.M. (2020). Application of heterogeneous catalysts for biodiesel production from microalgal oil: a review. *Catalysts* 10: 1–25.

28 Hossain, M.N., Bhuyan, M.S.U.S., Alam, A.H.M.A., and Seo, Y.C. (2019). Optimization of biodiesel production from waste cooking oil using S-$TiO_2$/SBA-15 heterogeneous acid catalyst. *Catalysts* 9: 67. https://doi.org/10.3390/catal9010067.

29 Rai, M., Ingle, A.P., Pandit, R. et al. (2019). Emerging role of nanobiocatalysts in hydrolysis of lignocellulosic biomass leading to sustainable bioethanol production. *Catal. Rev.* 61: 1–26.

30 Akoh, C.C., Chang, S.W., Lee, G., and Shaw, J.F. (2007). Enzymatic approach to biodiesel production. *J. Agri. Food Chem.* 55: 8995–9005.

31 Xu, Y., Du, W., Liu, D., and Zeng, J. (2003). A novel enzymatic route for biodiesel production from renewable oils in a solvent-free medium. *Biotechnol. Lett.* 25: 1239–1241.

32 Rai, M., Santos, J.C.D., Soler, M.F. et al. (2016). Strategic role of nanotechnology for production of bioethanol and biodiesel. *Nanotechnol. Rev.* 5 (2): 231–250.

33 Serrano, E., Rus, G., and García-Martínez, J. (2009). Nanotechnology for sustainable energy. *Renew. Sustain. Energy Rev.* 13 (9): 2373–2384.

34 Sekhon, B.S. (2014). Nanotechnology in agri-food production: an overview. *Nanotechnol. Sci. Appl.* 7: 31–53.

35 Ramsurn, H. and Gupta, R.B. (2013). Nanotechnology in solar and biofuels. *ACS Sustain. Chem. Eng.* 1 (7): 779–797.

36 Nizami, A. and Rehan, M. (2018). Towards nanotechnology-based biofuel industry. *Biofuel Res. J.* 18: 798–799.

37 García-Martínez, N., Andreo-Martínez, P., Quesada-Medina, J. et al. (2017). Optimization of non-catalytic transesterification of tobacco (*Nicotiana tabacum*) seed oil using supercritical methanol to biodiesel production. *Energy Convers. Manage.* 131: 99–108.

38 Liu, X., He, H., Wang, Y., and Zhu, S. (2007). Transesterification of soybean oil to biodiesel using SrO as a solid base catalyst. *Catal. Commun.* 8: 1107–1111.

39 Verziu, M., Cojocaru, B., Hu, J. et al. (2008). Sunflower and rapeseed oil transesterification to biodiesel over different nanocrystalline MgO catalysts. *Green Chem.* 10: 373–381.

40 Gardy, J., Hassanpour, A., Lai, X. et al. (2017). Biodiesel production from used cooking oil using a novel surface functionalized $TiO_2$ nano-catalyst. *Appl. Catal. B Environ.* 207: 297–310.

41 Ingle, A.P. (2021a). Zinc-based nanocatalysts in biodiesel production. In: *Nano- and Biocatalysts for Biodiesel Production* (ed. A.P. Ingle), 143–155. UK: Wiley.

42 Wagner, G.W., Koper, O.B., Lucas, E. et al. (2000). Reactions of VX, GD and HD with nanosize CaO: autocatalytic dehydrogenation of HD. *J. Phys. Chem. B* 104: 5118–5123.

43 Ingle, A.P. (2021b). *Nano and Biocatalysts for Biodiesel Production*, 368. UK: Wiley.

44 Gardy, J., Rehan, M., Hassanpour, A. et al. (2019). Advances in nano-catalysts based biodiesel production from non-food feedstocks. *J. Environ. Manag.* 249: 109316. https://doi.org/10.1016/j.jenvman.2019.109316.

45 Bharti, P., Singh, B., and Dey, R.K. (2019). Process optimization of biodiesel production catalyzed by CaO nanocatalyst using response surface methodology. *J. Nanostruct. Chem.* 9: 269–280.

46 Degfe, T.A., Mamo, T.T., and Mekonnen, Y.S. (2019). Optimized biodiesel production from waste cooking oil (WCO) using calcium oxide (CaO) nanocatalyst. *Sci. Rep.* 9: 18982. https://doi.org/10.1038/s41598-019-55403-4.

47 Amirthavalli, V. and Warrier, A.R. (2019). Production of biodiesel from waste cooking oil using MgO nanocatalyst. *AIP Conf. Proc.* 2115: 030609. https://doi.org/10.1063/1.5113448.

48 Esmaeili, H., Yeganeh, G., and Esmaeilzadeh, F. (2019). Optimization of biodiesel production from *Moringa oleifera* seeds oil in the presence of nano-MgO using Taguchi method. *Int. Nano Lett.* 9: 257–263.

49 Afsharizadeh, M. and Mohsennia, M. (2019). Catalytic synthesis of biodiesel from waste cooking oil and corn oil over zirconia-based metal oxide nanocatalysts. *Reaction Kin. Mecha. Catal.* 128: 443–459.

50 Bahuguna, A., Kumar, A., and Krishnan, V. (2019). Carbon support based heterogeneous nanocatalysts: synthesis and application toward organic transformation reactions. *Asian J. Org. Chem.* 8 (8): 1263–1305.

51 Mofijur, M., Arafat Siddiki, S.Y., Ahmed, M.B. et al. (2020). Effect of nanocatalysts on the transesterification reaction of first, second and third generation biodiesel sources- a mini-review. *Chemosphere* 270: 128642.

52 Bhagat, R., Panakkal, H., Gupta, I., and Ingle, A. (2021). Carbon based nanocatalyst in biodiesel production. In: *Nano and Biocatalyst* (ed. A.P. Ingle), 183–198. UK: Wiley.

53 Liu, R., Wang, X., Zhao, X., and Feng, P. (2008). Sulfonated ordered mesoporous carbon for catalytic preparation of biodiesel. *Carbon* 46 (13): 1664–1669.

54 Macina, A., De Medeiros, T.V., and Naccache, R. (2019). A carbon dot-catalyzed transesterification reaction for the production of biodiesel. *J. Mater. Chem.* 7: 23794–23802.

55 Zhang, Q., Zhang, Y., Deng, T. et al. (2020a). Sustainable production of biodiesel over heterogeneous acid catalysts. In: *Biomass, Biofuels, Biochemicals: Recent Advances in Development of Platform Chemicals* (ed. S. Saravanamurugan, A. Pandey, H. Li and A. Riisager), 407–432. UK: Elsevier.

56 Ibrahim, M.L., Nik Abdul Khalil, N.N.A., Islam, A. et al. (2020). Preparation of $Na_2O$ supported CNTs nanocatalyst for efficient biodiesel production from waste-oil. *Energy Convers. Manage.* 205: 112445.

57 Abdulkareem-Alsultan, G., Asikin-Mijan, L., Islam, R., and Yap, T. (2019). A review on thermal conversion of plant oil (edible and inedible) into green fuel using carbon-based nanocatalyst. *Catalysts* 9 (4): 350.

58 Gnanaserkhar, S., Asikin-Mijan, N., AbdulKareem-Alsultan, G. et al. (2020). Biodiesel production via simultaneous esterification and transesterification of chicken fat oil by mesoporous sulfated Ce supported activated carbon. *Biomass Bioenergy* 141: 105714.

59 Hara, M. (2009). Environmentally benign production of biodiesel using heterogeneous catalysts. *ChemSusChem* 2 (2): 129–135.

**60** Borah, M.J., Devi, A., Saikia, R.A., and Deka, D. (2018). Biodiesel production from waste cooking oil catalysed by in-situ decorated $TiO_2$ on reduced graphene oxide nanocomposite. *Energy* 158: 881–889.

**61** Poonjarernsilp, C., Sano, N., and Tamon, H. (2015). Simultaneous esterification and transesterification for biodiesel synthesis by a catalyst consisting of sulfonated single-walled carbon nanohorn dispersed with $Fe/Fe_2O_3$ nanoparticles. *Appl. Catal. A-Gen.* 497: 145–152.

**62** Gholizadeh, M., Hu, X., and Liu, Q. (2019). Progress of using biochar as a catalyst in thermal conversion of biomass. *Rev. Chem. Eng.* 35: 1–35.

**63** Sano, N., Yamada, K., Tsunauchi, S., and Tamon, H. (2017). A novel solid base catalyst for transesterification of triglycerides toward biodiesel production: carbon nanohorn dispersed with calcium ferrite. *Chem. Eng. J.* 307: 135–142.

**64** Ballotin, F.C., Perdigão, L.T., Rezende, M.V.B. et al. (2019). Bio-oil: a versatile precursor to produce carbon nanostructures in liquid phase under mild conditions. *New J. Chem.* 43: 2430–2433.

**65** Jume, B.H., Gabris, M.A., Nodeh, H.R. et al. (2020). Biodiesel production from waste cooking oil using a novel heterogeneous catalyst based on graphene oxide doped metal oxide nanoparticles. *Renew. Energy* 162: 2182–2189.

**66** Shuit, S.H. and Tan, S.H. (2014). Biodiesel production via esterification of palm fatty acid distillate using sulphonated multi-walled carbon nanotubes as a solid acid catalyst: process study, catalyst reusability and kinetic study. *Bioenergy Res.* 8 (2): 605–617.

**67** Guan, Q., Li, Y., Chen, Y. et al. (2017). Sulfonated multi-walled carbon nanotubes for biodiesel production through triglycerides transesterification. *RSC Adv.* 7 (12): 7250–7258.

**68** Pang, H., Yang, G., Li, L., and Yu, J. (2020). Efficient transesterification over two-dimensional zeolites for sustainable biodiesel production. *Green Energy Environ.* 5 (4): 405–413.

**69** Qu, H., Ma, Y., Li, B., and Wang, L. (2020a). Hierarchical zeolites: synthesis, structural control, and catalytic applications. *Emergent Mater.* 3: 225–245.

**70** Al-Ani, A., Darton, R.J., Sneddon, S., and Zholobenko, V. (2018). Nanostructured zeolites: the introduction of intracrystalline mesoporosity in basic Faujasite-type catalysts. *ACS Appl. Nano Mater.* 1: 310–318.

**71** Li, Z., Ding, S., Chen, C. et al. (2019). Recyclable Li/NaY zeolite as a heterogeneous alkaline catalyst for biodiesel production: process optimization and kinetics study. *Energy Convers. Manag.* 192: 335–345.

**72** Qu, S., Chen, C., Guo, M. et al. (2020b). Synthesis of MgO/ZSM-5 catalyst and optimization of process parameters for clean production of biodiesel from *Spirulina platensis*. *J. Clean. Prod.* 276: 123382.

**73** Lawan, I., Garba, Z.N., Zhou, W. et al. (2020). Synergies between the microwave reactor and CaO/zeolite catalyst in waste lard biodiesel production. *Renew. Energy* 145: 2550–2560.

**74** Du, L., Ding, S., Li, Z. et al. (2018). Transesterification of castor oil to biodiesel using NaY zeolite-supported $La_2O_3$ catalysts. *Energy Convers. Manag.* 173: 728–734.

**75** Prech, J., Pizarro, P., Serrano, D.P., and Cejka, J. (2018). From 3D to 2D zeolite catalytic materials. *Chem. Soc. Rev.* 47: 8263–8306.

**76** Narasimhan, M., Chandrasekaran, M., Govindasamy, S., and Aravamudhan, A. (2021). Heterogeneous nanocatalysts for sustainable biodiesel production: a review. *J. Environ. Chem. Eng.* 9 (1): 104876.

77 Amalia, S.S., Khalifah, N., Baroroh, H. et al. (2021). Biodiesel production from castor oil using heterogeneous catalyst KOH/zeolite of natural zeolite Bandung Indonesia. *AIP Conf. Proc.* 1: 080016.

78 Saeedi, M., Fazaeli, R., and Aliyan, H. (2016). Nanostructured sodium–zeolite imidazolate framework (ZIF-8) doped with potassium by sol-gel processing for biodiesel production from soybean oil. *J. Sol-Gel Sci. Technol.* 77: 404–415.

79 Mostafa, F., Zinatizadeh, L., Akbar, A. et al. (2018). Catalytic performance and characterization of promoted K-La/ZSM-5 nanocatalyst for biodiesel production. *Iran J. Chem. Eng.* 37 (2): 33–44.

80 Mohebbi, S., Rostamizadeh, M., and Kahforoushan, D. (2020). Effect of molybdenum promoter on performance of high silica $MoO_3$/B-ZSM-5 nanocatalyst in biodiesel production. *Fuel* 266: 1–8.

81 Fawaz, E.G., Salam, D.A., Pinard, L., and Daou, T.J. (2019). Catalysis science and technology crystal morphologies of HZSM-5 zeolites for the catalyst effectiveness. *Catal. Sci. Technol.* 9: 5456–5471.

82 Firouzjaee, M.H. and Taghizadeh, M. (2017). Optimization of process variables for biodiesel production using the nanomagnetic catalyst CaO/NaY-$Fe_3O_4$. *Chem. Eng. Technol.* 40 (6): 1140–1148.

83 Brito, A., Borges, M.E., and Otero, N. (2007). Zeolite Y as a heterogeneous catalyst in biodiesel fuel production from used vegetable oil. *Energy Fuels* 21: 3280–3283.

84 Hassani, M., Najafpour, G.D., Mohammadi, M., and Rabiee, M. (2014). Preparation, characterization and application of zeolite-based catalyst for production of biodiesel from waste cooking oil. *J. Sci. Ind. Res.* 73: 129–133.

85 Ramos, M.J., Casas, A., Rodrıguez, L. et al. (2008). Transesterification of sunflower oil over zeolites using different metal loading: a case of leaching and agglomeration studies. *Appl. Catal. A-Gen.* 346: 79–85.

86 Doyle, A.M., Albayati, T.M., Abbas, A.S., and Alismaeel, Z.T. (2016). Biodiesel production by esterification of oleic acid over zeolite Y prepared from kaolin. *Renew. Energy.* 97: 19–23.

87 de Vasconcellos, A., Miller, A.H., Aranda, D.A.G., and Nery, J.G. (2018). Biocatalysts based on nanozeolite-enzyme complexes: effects of alkoxysilane surface functionalization and biofuel production using microalgae lipids feedstock. *Colloids Surf., B* 165: 150–157.

88 de Vasconcellos, A., Miller, A.H., Silva, T.C., and Nery, J.G. (2014). Magnetically separable nanozeolite as a novel enzyme support and application for transesterification of microalgae oil to produce biodiesel. *13 Brazilian SBPM Meeting*, Brazil.

89 Nery, J.G., Vasconcellos, A. de, Miller, A.H., and Aranda, D.A. (2019). Surface chemical modulation of Biocatalysts based on nanozeolite-enzyme complexes aiming the production of biofuel production using non edible lipids feedstocks. *Proceedings of the 18. Brazil MRS Meeting*, Camboriu, SC, Brazil (22–26 September 2019). Brazil: Sociedade Brasileira de Pesquisa de Materiais.

90 Farrokheh, A., Tahvildari, K., and Nozari, M. (2020). Biodiesel production from the *Chlorella vulgaris* and *Spirulina platensis* microalgae by electrolysis using CaO/KOH-$Fe_3O_4$ and KF/KOH-$Fe_3O_4$ as magnetic nanocatalysts. *Biomass Convers. Biorefin.* https://doi.org/10.1007/s13399-020-00688-z.

91 Liu, C., Lv, P., Yuan, Z. et al. (2010). The nanometer magnetic solid base catalyst for production of biodiesel. *Renew. Energy* 35: 1531–1536.

**92** Amani, T., Haghighi, M., and Rahmani Vahid, B. (2019). Microwave-assisted combustion design of magnetic Mg-Fe spinel for MgO-based nanocatalyst used in biodiesel production: influence of heating-approach and fuel ratio. *J. Ind. Eng. Chem.* 80: 43–52.

**93** Alaei, S., Haghighi, M., Toghiani, J., and Rahmani Vahid, B. (2018). Magnetic and reusable $MgO/MgFe_2O_4$ nanocatalyst for biodiesel production from sunflower oil: influence of fuel ratio in combustion synthesis on catalytic properties and performance. *Ind. Crops Prod.* 117: 322–332.

**94** Feyzi, M. and Norouzi, L. (2016). Preparation and kinetic study of magnetic $Ca/Fe_3O_4@SiO_2$ nanocatalysts for biodiesel production. *Renew. Energy* 94: 579–586.

**95** Rahimi, T., Kahrizi, D., Feyzi, M. et al. (2021). Catalytic performance of $MgO/Fe_2O_3$-$SiO_2$ core-shell magnetic nanocatalyst for biodiesel production of *Camelina sativa* seed oil: optimization by RSM-CCD method. *Ind. Crops Prod.* 159: 113065.

**96** Ashok, A., Ratnaji, T., John Kennedy, L. et al. (2021). Magnetically recoverable Mg substituted zinc ferrite nanocatalyst for biodiesel production: process optimization, kinetic and thermodynamic analysis. *Renew. Energy* 163: 480–494.

**97** Dantas, J., Leal, E., Mapossa, A.B. et al. (2017). Magnetic nanocatalysts of $Ni_{0.5}Zn_{0.5}Fe_2O_4$ doped with Cu and performance evaluation in transesterification reaction for biodiesel production. *Fuel* 191: 463–471.

**98** Liu, K., Wang, R., and Yu, M. (2017). Biodiesel production from soybean oils by a novel nano-magnetic solid base catalyst ($K/ZrO_2/\gamma$-$Fe_2O_3$). *RSC Adv.* 7: 51814–51821.

**99** Ali, M.A., Al-Hydary, A.I., and Al-Hattab, A.T. (2017). Nano-magnetic catalyst CaO-$Fe_3O_4$ for biodiesel production from date palm seed oil. *Chem. React. Eng. Catal.* 12 (3): 460–468.

**100** Ghalandari, A., Taghizadeh, M., and Rahmani, M. (2018). Statistical optimization of biodiesel production process using magnetic core-mesoporous shell $KOH/Fe_3O_4@\gamma$-$Al_2O_3$ nanocatalyst. *Chem. Eng. Technol.* 42 (1): 89–99.

**101** Shaker, M. and Elhamifar, D. (2020). Sulfonic acid supported on magnetic methylene-based organosilica as an efficient and recyclable nanocatalyst for biodiesel production via esterification. *Front. Energy Res.* 8: https://doi.org/10.3389/fenrg.2020.00078.

**102** Ambat, I., Srivastava, V., Haapaniemi, E., and Sillanpää, M. (2019). Nano-magnetic potassium impregnated ceria as catalyst for the biodiesel production. *Renew. Energy* 139: 1428–1436.

**103** Xie, W., Han, Y., and Wang, H. (2018). Magnetic $Fe_3O_4$/MCM-41 composite-supported sodium silicate as heterogeneous catalysts for biodiesel production. *Renew. Energy* 125: 675–681.

**104** Chen, Y., Liu, T., He, H., and Liang, H. (2018). $Fe_3O_4$/ZnMg(Al)O magnetic nanoparticles for efficient biodiesel production. *Appl. Organomet. Chem.* 32: e4330.

**105** Feyzi, M., Hassankhani, A., and Rafiee, H. (2013). Preparation and characterization of $CsAlFe_3O_4$ nano-catalysts for biodiesel production. *Energy Convers. Manage.* 71: 62–68.

**106** Hu, S., Guan, Y., Wang, Y., and Han, H. (2011). Nano-magnetic catalyst KF/CaO–$Fe_3O_4$ for biodiesel production. *Appl. Energy* 88: 2685–2690.

**107** Akubude, V.C., Nwaigwe, K.N., and Dintwa, E. (2019). Production of biodiesel from microalgae via nanocatalyzed transesterification process: a review. *Mater. Sci. Technol.* 2: 216–225.

**108** Kim, B., Heo, H.Y., Son, J. et al. (2019). Simplifying biodiesel production from microalgae via wet in situ transesterification: a review in current research and future prospects. *Algal Res.* 41: 101557.

**109** Safakish, E., Nayebzadeh, H., Saghatoleslami, N., and Kazemifard, S. (2020). Comprehensive assessment of the preparation conditions of a separable magnetic nanocatalyst for biodiesel production from algae. *Algal Res.* 49: 101949.

**110** Kazemifard, S., Nayebzadeh, H., Saghatoleslami, N., and Safakish, E. (2019). Application of magnetic alumina-ferric oxide nanocatalyst supported by KOH for in-situ transesterification of microalgae cultivated in wastewater medium. *Biomass Bioenergy* 129: 105338.

**111** Rajkumari, K. and Rokhum, L. (2021). Functionalized magnetic nanocatalyst in biodiesel production. In: *Nano and Biocatalyst* (ed. A.P. Ingle), 183–198. UK: Wiley.

**112** Tamjidi, S., Esmaeili, H., and Moghadas, K.B. (2021). Performance of functionalized magnetic nanocatalysts and feedstocks on biodiesel production: a review study. *J. Clean. Prod.* 305: 127200.

**113** Halek, F., Aghamohammadi, N., and Mohamadi, F. (2019). Biodiesel production from waste edible oil with heterogeneous catalysts (nanoclay-based nanocatalysts). *Arab. J. Sci. Eng.* 44: 9919–9924.

**114** Zanette, A.F., Barella, R.A., SBC, P. et al. (2011). Screening, optimization and kinetics of *Jatropha curcas* oil transesterification with heterogeneous catalysts. *Renew. Energy* 36 (2): 726–731.

**115** Shah, K., Parikh, J., Dholakiya, B., and Maheria, K. (2014). Fatty acid methyl ester production from acid oil using silica sulfuric acid: process optimization and reaction kinetics. *Chem. Pap.* 68 (4): 472–483.

**116** Zatta, L., Ramos, L.P., and Wypych, F. (2013). Acid-activated montmorillonites as heterogeneous catalysts for the esterification of lauric acid with methanol. *Appl. Clay Sci.* 80–81: 236–244.

**117** Jeenpadiphat, S. and Tungasmita, D.N. (2014). Esterification of oleic acid and high acid content palm oil over an acid-activated bentonite catalyst. *Appl. Clay Sci.* 87: 272–277.

**118** Kansedo, J., Lee, K.T., and Bhatia, S. (2009). *Cerbera odollam* (sea mango) oil as a promising non-edible feedstock for biodiesel production. *Fuel* 88 (6): 1148–1150.

**119** Sultana, N., Hajra, B., Saxena, V.K., and Guria, C. (2015). Optimal synthesis of Sal (*Shorea robusta*) oil biodiesel using recycled bentonite nanoclay at high temperature. *Energy Fuels* 30 (1): 386–397.

**120** Rezania, S., Korrani, Z.S., Gabris, M.A. et al. (2021). Lanthanum phosphate foam as novel heterogeneous nanocatalyst for biodiesel production from waste cooking oil. *Renew. Energy* 176: 228–236.

**121** Cong, W.J., Nanda, S., Li, H. et al. (2021). Metal–organic framework-based functional catalytic materials for biodiesel production: a review. *Green Chem.* 23: 2595–2618.

**122** Zhang, Q., Ling, D., Lei, D. et al. (2020b). Green and facile synthesis of metal-organic framework Cu-BTC-supported Sn (II)-substituted Keggin heteropoly composites as an esterification nanocatalyst for biodiesel production. *Front. Chem.* 8: 129.

**123** Jeon, Y., Chi, W.S., Hwang, J. et al. (2019). Core-shell nanostructured heteropoly acid-functionalized metal-organic frameworks: bifunctional heterogeneous catalyst for efficient biodiesel production. *Appl. Catal. B: Environ.* 242: 51–59.

**124** Negm, N.A., Betiha, M.A., Alhumaimess, M.S. et al. (2019). Clean transesterification process for biodiesel production using heterogeneous polymer-heteropoly acid nanocatalyst. *J. Clean. Prod.* 238: 117854.

# 10

# Sustainable Production of Biodiesel Using Ion-Exchange Resin Catalysts

*Naomi Shibasaki-Kitakawa and Kousuke Hiromori*

*Department of Chemical Engineering, Tohoku University, Sendai, Japan*

## 10.1 Introduction

In Part 1, the characteristics of various oils used as raw materials for fatty acid methyl esters, a biodiesel fuel, were introduced. Vegetable oils are biosynthesized in the form of triglycerides, but once the oils are pressed, their hydrolysis to free fatty acids (FFAs) by lipase enzymes begins. Similarly, heat treatment causes the degradation of triglycerides to FFAs. Therefore, the FFA content of crude oils varies depending on the storage temperature and time elapsed after pressing. As FFAs are biotoxins, their content in food is regulated, requiring their removal to a content below the standard value. Therefore, a large amount of by-product oils containing FFAs is generated in edible oil refining. Furthermore, refined edible oils are broken down again by heating and cooking, resulting in FFAs. The FFA content of waste cooking oil varies depending on the conditions under which it was used. Therefore, the FFA content of the oils cannot be controlled.

As industrial biodiesel production employs homogeneous alkali-catalyzed transesterification of triglycerides, only oils with a low FFA content (typically, <1 wt%) can be used to reduce soap by-products [1]. However, owing to the high cost of such refined oils, the use of inexpensive oils containing FFAs has been recommended. Therefore, the development of a versatile biodiesel production technology that can use oils with various FFA contents as raw materials is desirable. For this purpose, technology that both converts FFAs into biodiesel by esterification and converts triglycerides into biodiesel by transesterification must be developed. Various catalysts have been developed by many researchers to accelerate the conversion of FFAs, or triglycerides, into fuels. In Part 2, these proposed catalysts are categorized by material, with each chapter describing their features and issues in detail. This chapter focuses on one of the solid catalysts, namely, ion-exchange resins.

*Biodiesel Production: Feedstocks, Catalysts, and Technologies*, First Edition. Edited by Samuel Lalthazuala Rokhum, Gopinath Halder, Suttichai Assabumrungrat, and Kanokwan Ngaosuwan.

## 10.2 Features of Ion-Exchange Resin Catalysts

Ion-exchange resins have traditionally been used as separation materials in water treatment. Resins can be acidic or basic depending on their functional groups, with commercially available examples exhibiting various characteristics, including different polymer materials in the backbone, acid or base strengths, degrees of cross-linking, porosity, and particle sizes. These resins swell in the presence of solvents and reactants, exhibiting a large volume change, with the degree of swelling (adsorption properties) varying depending on the affinity for each component [2–4]. Independent control of these properties is an important and attractive prospect (for example, many properties in zeolites change in a dependent manner). Furthermore, resins are inexpensive and already mass produced. Therefore, a large-scale biodiesel production plant can be built without also needing to build a catalyst preparation plant. This is a great advantage for practical use.

The use of ion-exchange resins as catalysts for the synthesis of fatty acid esters was first investigated around 2000. A cation-exchange resin was first used as a strong acid catalyst to convert FFAs in oil into fatty acid esters. Bondioli [5] noted that ion-exchange resins as catalysts for fatty acid ester synthesis offer the following advantages:

1) Ion-exchange resins are solid acids or bases and, if a batch method is used, can be easily separated from the system after reaction completion.
2) By packing the resin into a reaction tower and passing through the solution to be treated, highly productive continuous reactions can be conducted. This results in a simplified and more economical process.
3) Ion-exchange resins allow continuous usage more often than homogeneous acid or base methods, resulting in lower running costs.
4) Ion-exchange resins are insoluble acids or bases that can be used to perform operations such as neutralization or reconcentration after reaction completion.
5) Owing to their molecular sieve action, resins result in few by-products and can give higher selectivity.
6) Ion-exchange resins are less corrosive than the acids and bases generally used. Therefore, the choice of equipment materials presents little problem, facilitating lower investment costs.

To date, many studies have been conducted using ion-exchange resins as catalysts for fatty acid ester synthesis. About 80% of these are related to cation-exchange resins, while 16% are related to anion-exchange resins. In the following sections, we will summarize the findings of various studies on each resin type and discuss their advantages and problems as catalysts.

## 10.3 Cation-Exchange Resin Catalyst

### 10.3.1 Notes of Caution When Comparing the Activity of Resins with Different Properties

Initially, resin was used for the pretreatment of FFAs in cheaper oils, such as used cooking oil, to convert them into fatty acid esters by esterification [6]. The FFA content of waste cooking oil is usually less than 5%. Subsequently, the application of resin to the esterification of acid oils with FFA contents of more than 90%, a by-product of the edible oil refinery, was investigated [7]. Generally, FFAs become liquid at temperatures above 40 °C and are completely soluble in methanol, while triglycerides dissolve only 10 wt% or less of methanol at 50 °C [8]. Therefore, when the esterification of FFAs in oils is conducted, the addition of a larger amount of methanol (>15 wt%, or a methanol molar ratio of >4 at 50 °C) will cause phase separation. Care should be taken to ensure that the reaction solution is homogeneous or heterogeneous, as this significantly affects the reaction rate.

The resin backbone mainly comprises a copolymer of styrene and divinylbenzene, with the degree of cross-linking varying depending on the amount of divinylbenzene. A higher degree of cross-linking results in a denser mesh structure (micropores) and greater mass transfer resistance. The basic physical structure of resins is gel type, in addition to a porous-type structure that contains physical holes (macropores). Many studies have been conducted to compare the esterification activity of various ion-exchange resins [9–12]. However, caution must be exercised when comparing the activities of resins with different structures. Slow esterification rates in experiments can be due to various factors, such as slow incorporation of reactants into the resin, slow mass transfer rates within the resin, or a low number of acid sites.

Accordingly, it is useful to conduct experiments in which conditions other than the target parameter are kept constant. Some researchers have conducted esterification experiments on FFAs using gel-type and porous-type resins, reporting that the porous-type resin was more active [9, 11, 13]. Although the ion-exchange capacities of porous-type resins were lower than those of gel-type resins, these higher activities were attributed to the greater effect of mass transfer on the reaction. Several studies have also compared the same resin in particle and powder forms [6, 9, 11]. The results showed that powdered resin tended to have a higher reaction rate, indicating a larger mass transfer effect within the resin. We also compared resins with similar properties and different degrees of cross-linking, as shown in Table 10.1. In this case, a resin with a small degree of cross-linking is positively affected by its small mass transfer resistance and negatively affected by its small ion exchange capacity. Figure 10.1 shows an increased consumption rate of FFAs and formation rate of fatty acid

**Table 10.1** Physical properties of cation-exchange resins.

| | Cation-exchange resin | | |
|---|---|---|---|
| **Properties** | **Diaion PK208LH** | **Diaion PK228** | **Diaion HPK25** |
| Functional group | Sulfonic acid | Sulfonic acid | Sulfonic acid |
| Ionic form | $H^+$ | $H^+$ | $H^+$ |
| Type | Porous | Porous | Highly porous |
| Cross-linking density [%] | 4 | 14 | 25 |
| Particle size [mm] | 0.40–0.60 | 0.40–0.60 | 0.40–0.60 |
| Apparent density [g dm$^{-3}$-resin] | 780 | 810 | 740 |
| Min ion-exchange capacity [meq cm$^{-3}$-resin] | 1.2 | 2.05 | 0.8 |
| Max operating temperature | 120 | 120 | 120 |

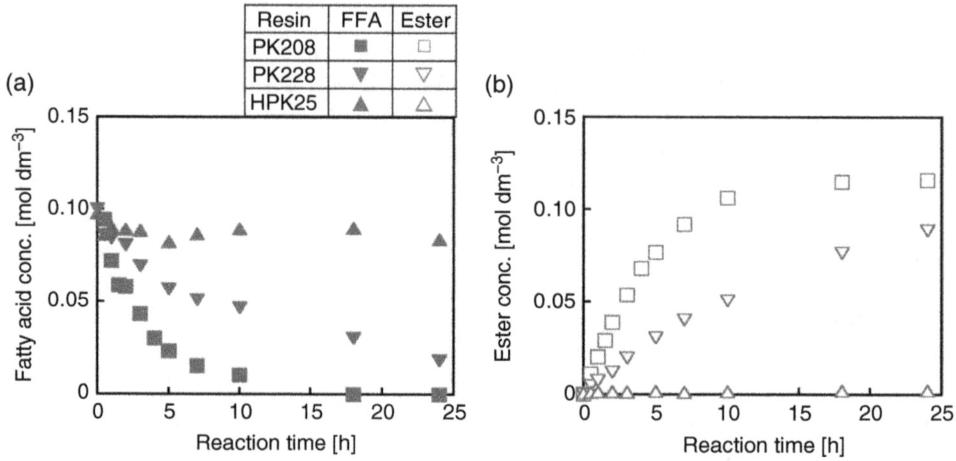


**Figure 10.1** Comparison of esterification behavior of free fatty acids (FFA) using various cation-exchange resins (50 °C; FFA/ethanol molar ratio, 1 : 10; catalyst concentration, 40 wt%; 150 spm).

esters for resins with a lower degree of cross-linking. This clearly indicates that mass transfer resistance has a significant effect on the esterification rate. Therefore, resins preferred for the esterification of FFAs can be concluded to have properties that reduce the effect of mass transfer resistance, namely, a lower degree of cross-linking, porous-type structure, and small particle size.

### 10.3.2 Reversible Reduction of Resin Catalytic Activity by Water

Another major issue concerning cation-exchange resins is the effect of water on catalytic activity. Some studies have reported water-induced inhibition of activity and explained the mechanisms involved [10, 14]. We have also studied the effect of water content in the resin and reaction solution on the esterification of FFAs in batch and continuous systems [15, 16]. Commercially available ion-exchange resins are sold in a water-swollen state, requiring pretreatment to remove water from the resin for use in oil solution. Many studies have achieved this by washing the resin with alcohol, followed by drying. The dried resin is then added to the reaction solution at the start of the experiment. However, the reaction does not start smoothly because the resin must first absorb the reaction solution and swell. Therefore, we proposed replacing water in the swollen state with alcohol by packing the column with water-swollen resin and supplying it with alcohol [15]. Furthermore, the effect of the amount of alcohol supplied during pretreatment was investigated by monitoring the water content in the effluent from the column. As shown in Figure 10.2, the water content in the effluent from the column, as an indicator of resin water content, was rapidly reduced by the alcohol supply. Figure 10.3 shows the concentration profiles of reactants and products for esterification conducted using resins with different water contents. No differences in the reactant consumption and product generation behavior were observed using resins with water contents below 20 wt%. When resin with a 35% water content was used, the final concentration reached was similar to that obtained using resin with a lower water content but with more gradual changes in the reactant and product concentrations. This might be due to either fewer acid sites being involved in the esterification or a lower

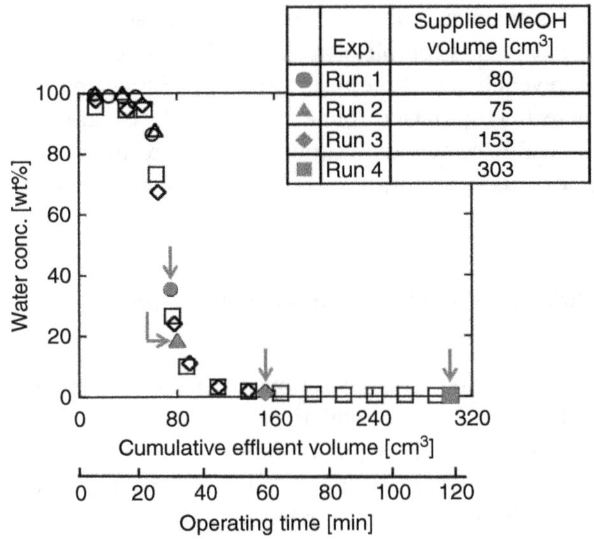


**Figure 10.2** Variation of water concentration in effluent from column packed with cation-exchange resin (Diaion PK208LH) during supply of methanol in pretreatment (arrows show the end point of each pretreatment). *Source:* Shibasaki-Kitakawa et al. [15].

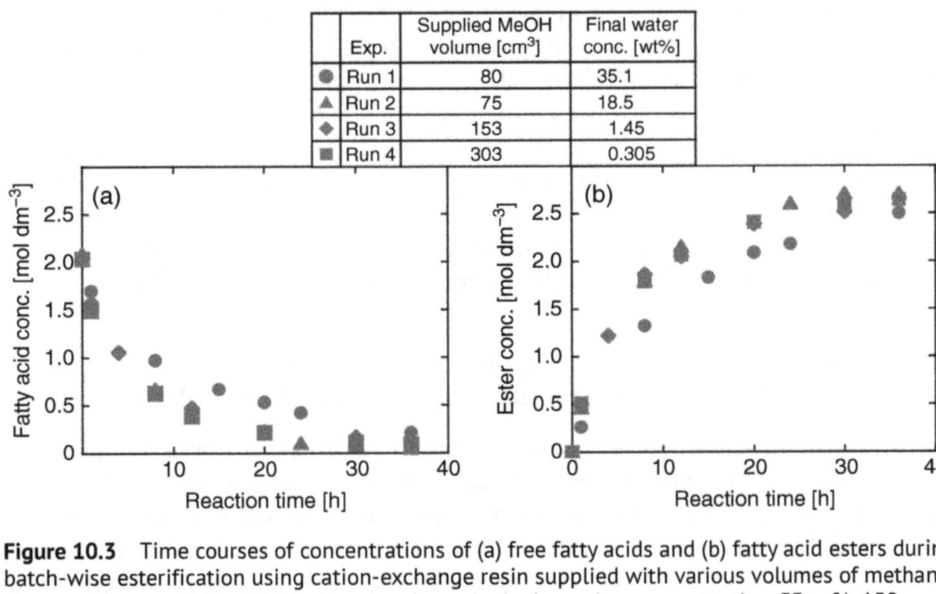


**Figure 10.3** Time courses of concentrations of (a) free fatty acids and (b) fatty acid esters during batch-wise esterification using cation-exchange resin supplied with various volumes of methanol as pretreatment (50 °C; FFA/methanol molar ratio, 1 : 1; catalyst concentration, 33 wt%; 150 spm). *Source:* Shibasaki-Kitakawa et al. [15].

concentration of reactants in the resin. This suggested that water coordination made it difficult for FFAs to access the active site or that the presence of water in the resin inhibited the incorporation of fat-soluble FFAs into the resin.

The catalytic activity of the continuous esterification process has also been shown to gradually decrease with prolonged operating time, due to the accumulation of by-product water in the resin. When the resin bed is again supplied with alcohol, water accumulated in the resin flows out, and the catalytic activity is completely restored. Therefore, the inhibition of catalytic activity by water is reversible, and the resin can be used repeatedly by removing accumulated water from the resin [16]. In fact, Tohoku University's venture company has installed a continuous esterification system with cation-exchange resin, and this regeneration method allows the resin to be used continuously for more than one year without any loss of activity [17].

### 10.3.3 Search for Operating Conditions for Maximum Productivity Rather than Maximum Catalytic Activity

In almost all studies reported, the effects of operating conditions, such as temperature, molar ratio, reaction time, and catalyst amount, on the catalytic activity have been examined. The goal should be to gain knowledge for the design of an industrial continuous production process that takes the greatest advantage of resin catalysts. The reaction temperature should be set accounting for the heat resistance of the resin and boiling point of the alcohol used. High-pressure conditions should be avoided to reduce environmental and cost loads. The heat-resistance temperature of cation-exchange resins is usually around 120 °C, which is above the boiling points of methanol and ethanol. When examining the reactant molar ratios, it should be noted that, in a solvent-free system, increasing the alcohol concentration will decrease the FFA concentration. Many researchers have used the conversion of FFAs to evaluate catalytic activity. However, when considering industrial biodiesel production, lower FFA concentrations result in lower fatty acid ester production amounts obtained per batch. This means that, even if the alcohol concentration increases to improve conversion, the actual production amount of biodiesel is often lower. Again, the objective of optimizing the operating conditions is to maximize biodiesel productivity, which is the amount produced per hour per unit catalyst amount. When the maximum productivity has been determined by optimizing the operating conditions, this value can be used to determine the amount of catalyst required to achieve the desired production amount [18, 19].

### 10.3.4 Challenges Regarding One-Step Reaction with Simultaneous Esterification and Transesterification Catalyzed by Cation-Exchange Resin

Acid catalysts promote not only the esterification of FFAs but also the transesterification of triglycerides. Therefore, some studies have been conducted to synthesize fatty acid esters from cheaper oils with high FFA contents through a one-step reaction using both functions [20–23]. Paterson et al. [22] first screened commercial ion-exchange resins as catalysts in the transesterification of triolein and methanol. They selected the most active resin (Amberlyst 15) and achieved 97 mol% conversion of triolein by optimizing the operating conditions. Subsequently, various amounts of FFAs (0–15 wt%) and water (0–2 wt%) were added to the reactants, but this did not significantly affect the conversion, and the FFA and water contents at the end of the reaction were similar under all conditions. This showed that FFAs and water were consumed and produced and that the resin was considered to catalyze the hydrolysis of triglycerides and esterification of FFAs, in addition to the transesterification of triglycerides. In contrast, Cabrai et al. [23] focused on a porous, strongly acidic resin (Amberlyst 45) with very high heat resistance (170 °C), discussing its catalytic activity for the transesterification of triglycerides and methanol or ethanol. The ethanolysis of waste oil using this resin gave fatty acid ester yields similar to those from commercial refined vegetable oil (for both, oil/ethanol ratio = 1 : 18). Furthermore, water had little effect on the reaction rate, and no significant effect on the product yield. This resin requires a high temperature and high ethanol molar ratio to obtain a high biodiesel yield but has the advantage of being water tolerant and showing stable high activity under repeated use. Despite these many advantages, the rate of triglyceride transesterification using cation-exchange resins is much slower than that using conventional homogeneous basic catalysts, and its application in industrial processes remains difficult.

## 10.4 Anion-Exchange Resin Catalysts

### 10.4.1 Requirements for High Catalytic Activity in the Transesterification of Triglycerides

Vicente et al. [24] reported that anion-exchange resins showed no catalytic activity for the transesterification of triglycerides. In contrast, we discovered that anion-exchange resins have high activity for transesterification [25]. This section considers why we were able to find this activity. The operation conditions reported by Vicente et al. [24] were a methanol/triglyceride molar ratio of 6 : 1 at 60 °C for 8 h. As shown in Figure 10.4a, in the system using a conventional homogeneous alkali catalyst, such as NaOH, triglycerides and methanol did not dissolve at a typical methanol/triglyceride molar ratio of 6 : 1 at 60 °C. In this case, the catalyst is dissolved in alcohol, and the reaction proceeds at the liquid–liquid interface. When a solid catalyst, such as a resin, is used in this way, with the reaction liquid phase separated, the reaction is a three-phase system (liquid–liquid–solid; Figure 10.4b). The reaction proceeds only when both reactants meet at the active site within the resin, making this reaction very difficult. Under the experimental conditions used by Vicente et al. [24], the reaction system would have been in this state. In contrast, we used ethanol to avoid phase separation of the reaction solution. Triglycerides and ethanol dissolve each other, resulting in a liquid–solid two-phase system (Figure 10.4c). Therefore, the reaction proceeded rapidly at the liquid–solid interface, similar to the liquid–liquid two-phase system of conventional homogeneous catalytic systems.

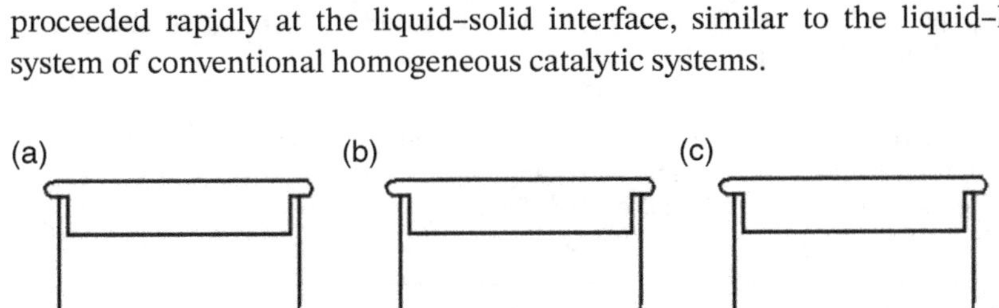


**Figure 10.4** Phase conditions of each reaction system: (a) two-phase system with conventional homogenous alkali catalyst, (b) three-phase system with heterogeneous resin catalyst, and (c) two-phase system with heterogeneous resin catalyst under specific conditions.

However, as internationally standardized biodiesel is a fatty acid methyl ester, methanol has to be used as the alcohol. Therefore, Tsuji et al. [8] performed dissolution experiments of triglycerides and methanol at 50 °C, which is below the heat resistance temperature of the anion exchange resin. As shown in Table 10.2 and Figure 10.5, when the methanol/triglyceride molar ratio was less than 3.9 : 1, the reaction solution was found to be homogeneous after stirring at 50 °C for more than 30 min. Batch and continuous transesterification experiments were then conducted under molar ratios that resulted in a homogeneous phase, showing that complete conversion could be achieved. In particular, as shown in Table 10.3, when the methanol/triglyceride molar ratio of the feed solution

**Table 10.2** Phase conditions of methanol/triglyceride mixtures. *Source:* Tsuji et al. [8].

| MeOH [g] | Triglyceride [g] | Molar ratio of MeOH/triglyceride | Phase condition |
|---|---|---|---|
| 0.00 | 40.1 | 0.0 : 1 | Single phase |
| 4.37 | 40.1 | 3.0 : 1 | Single phase |
| 5.65 | 40.1 | 3.9 : 1 | Single phase |
| 6.15 | 40.1 | 4.2 : 1 | Single phase |

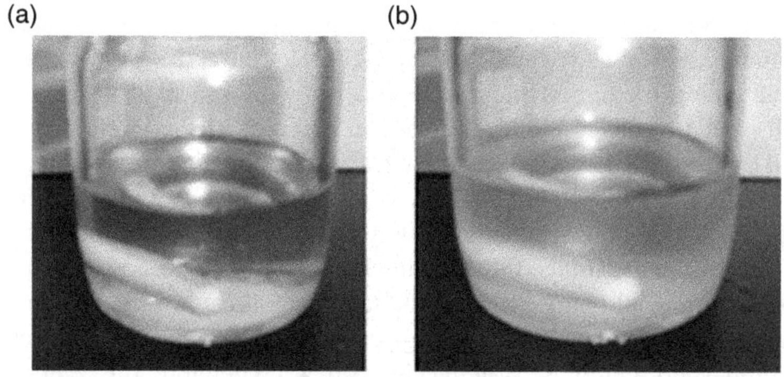

**Figure 10.5** Photographs of phase conditions of methanol/triglyceride mixtures at 50 °C with molar ratios of (a) 3.9 : 1 and (b) 4.2 : 1.

**Table 10.3** Steady-state triglyceride conversions during continuous biodiesel production in a column reactor packed with anion-exchange resin using various MeOH/triglyceride molar ratios (50 °C; DiaionPA306s, 50 g; feed flow rate, 0.16 $cm^3\ min^{-1}$).

| Molar ratio of MeOH/triglyceride | Triglyceride conversion [%] |
|---|---|
| 6.0 : 1 | No progress |
| 3.9 : 1 | 96.3 |
| 3.0 : 1 | 99.7 |
| 2.5 : 1 | 99.7 |

was varied in continuous experiments, transesterification proceeded almost completely, even under conditions where the amount of methanol was less than the stoichiometric ratio. This was probably due to the resin easily adsorbing the more polar methanol in the oil solution and the methanol swelling state from pretreatment being retained, which was also used in the reaction. As a result, transesterification proceeded irreversibly, and complete conversion was easily achieved using the anion-exchange resin catalyst.

Accordingly, the requirements for anion-exchange resin to exhibit high catalytic activity in the transesterification of triglycerides are as follows: pretreatment with sufficient substitution of the active OH group and swelling with alcohol, homogenization of the reaction solution using a stoichiometric methanol/triglyceride molar ratio, and using a reaction temperature below the heat resistance temperature of the resin.

### 10.4.2 Analysis of Previous Studies

Since our report of the high catalytic activity of anion-exchange resin for the transesterification of triglycerides, several studies have been conducted using this resin. These studies were analyzed based on the aforementioned requirements for obtaining high activity. In all cases, the reaction temperature was below 60 °C, accounting for the heat resistance of the resin.

Studies in which pretreatment with sufficient substitution of the active OH groups and swelling with alcohol were performed, and the reaction solution was a homogeneous phase with ethanol, showing a high activity of over 90% [26, 27]. Similarly, when methanol was used and hexane was added to make the reaction solution homogeneous, high conversions of over 95% were obtained [28, 29]. In contrast, when experiments were conducted with methanol at molar ratios where the reaction solution was heterogeneous, the conversion was only 60–70% [30, 31]. Furthermore, in a study in which commercially available water-swollen resins were used without pretreatment, the conversion was as low as 30%, despite the reaction solution being homogeneous with ethanol [32]. This was presumably due to incorporation of the triglyceride into the water-swollen resin being difficult.

Only a few studies have discussed the properties of resins. For example, Figure 10.6 shows our results using various resins listed in Table 10.4. The rates of triglyceride consumption and ester formation were slower for PA306s, PA306, PA308, and HPA25, in that order. This indicated that the reaction rate was greater using resins with a lower degree of cross-linking and smaller particle size. A resin manufacturer was asked to investigate the extent to which triglyceride molecules with a molecular weight of approximately 1000 were incorporated into the resin. The results showed that triglyceride molecules were incorporated into the center of the resin. Furthermore, Figure 10.6 shows a direct comparison of the transesterification activity of cation- and anion-exchange resins, with the cation-exchange resins showing a much slower rate. Kim et al. [30] also investigated the effect of the properties of anion-exchange resins, finding that porous-type resins with lower ion-exchange capacity were more active than gel-type resins and that resins with less cross-linking and smaller particle sizes were more active. In conclusion, the preferred resin properties for the transesterification of triglycerides are those that reduce the effect of mass transfer resistance, namely, low cross-linking, a porous-type structure, and a small particle size, as also observed for the esterification of FFAs.

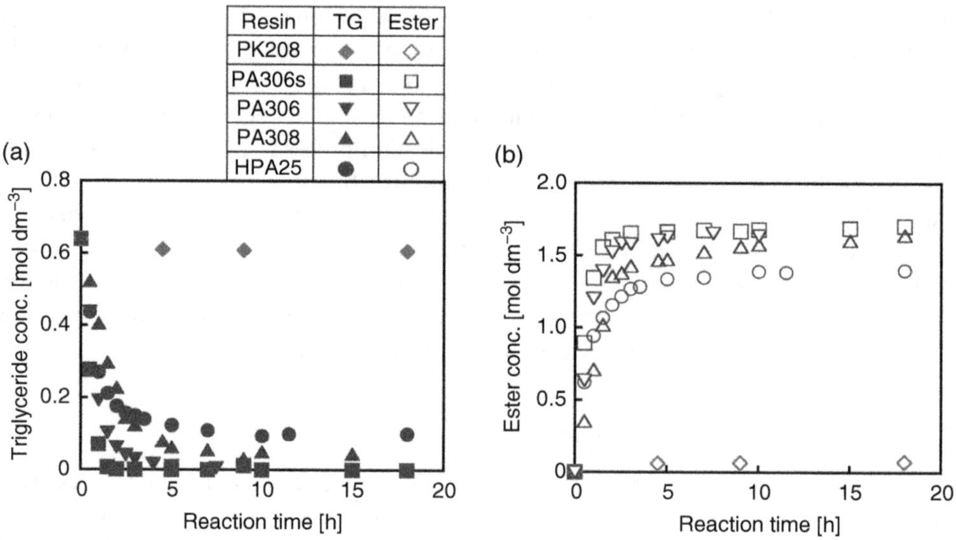


**Figure 10.6** Comparison of transesterification behavior of triglyceride (TG) using a cation-exchange resin and various anion-exchange resins (50 °C; TG/ethanol molar ratio, 1 : 10; catalyst concentration, 28 wt%; 150 spm).

**Table 10.4** Physical properties of anion-exchange resins.

| | Anion-exchange resin | | | |
|---|---|---|---|---|
| **Properties** | **Diaion PA306s** | **Diaion PA306** | **DiaionPA308** | **Diaion HPA25** |
| Functional group | Trimethyl ammonium | Trimethyl ammonium | Trimethyl ammonium | Sulfonic acid |
| Ionic form | $OH^-$ | $OH^-$ | $OH^-$ | $OH^-$ |
| Type | Porous | Porous | Porous | Highly porous |
| Cross-linking density [%] | 3 | 3 | 4 | 25 |
| Particle size [mm] | 0.15–0.25 | 0.40–0.60 | 0.40–0.60 | 0.40–0.60 |
| Apparent density [g dm$^{-3}$-resin] | 720 | 720 | 710 | 680 |
| Min ion-exchange capacity [meq cm$^{-3}$-resin] | 0.8 | 0.8 | 1.0 | 0.5 |
| Max operating temperature | 60 | 60 | 60 | 60 |

Regarding the operating factors, as in the case of cation-exchange resins, it is important to search for optimum conditions that maximize productivity. Accordingly, the reaction temperature should be as high as possible below the heat resistance temperature of the resin, and the alcohol/triglyceride molar ratio and catalyst amount should be set to maximize productivity per time and catalyst amount. We determined the maximum productivity as a scale-up parameter in bench-scale continuous biodiesel production by optimizing the operating conditions [18]. Based on this value, we designed and constructed a pilot-scale continuous production system and showed that the target production amount (50 l day$^{-1}$) was actually achieved [19].

### 10.4.3 Decreased Catalytic Activity and Regeneration Method

OH groups, which are responsible for the catalytic activity of the anion-exchange resin, undergo ion exchange with fatty acid residues in the reaction system, resulting in a gradual decrease in activity [25, 28, 29]. The fatty acid residues in the raw oils were FFAs and triglycerides. The ease of ion exchange was determined from the size of the pKa value for each component. Therefore, ion exchange with resin OH groups proceeds preferentially with FFAs with low pKa (about 4.8 [33]). The selectivity of ion exchange with fatty acid residues of triglycerides was low, reported to be about 3% [18].

In these studies, methods to regenerate the resins with reduced activity were also explored. For example, the results of an early our study are shown in Figure 10.7. To conduct the regeneration method for repeated use, the resin was recovered by filtration after the batch experiment and then treated using the following three steps: (i) first-stage regeneration, in which the fatty acid residues were replaced with weak acid groups in a weak acid ethanol solution; (ii) second-stage regeneration, in which the weak acid groups were replaced with hydroxyl ions in NaOH solution and washed with pure water; and (iii) washing with ethanol to restore the initial swollen state. Figure 10.7 shows that if operation (ii) is conducted directly without performing operation (i), the activity recovered is only half that of the original active group. This was attributed to the formation of a salt (soap) between the desorbed fatty acid groups and sodium in the solution. To avoid this, a

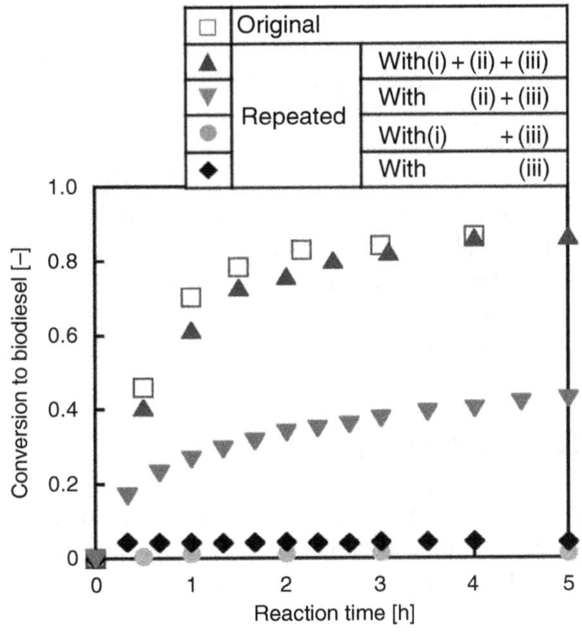


**Figure 10.7** Comparison of catalytic activity for triglyceride transesterification of anion-exchange resin (Diaion PA306s) regenerated by various operations (50 °C; TG/ethanol molar ratio, 1 : 10; catalyst concentration, 28 wt%; 150 spm).

two-step ion-exchange procedure comprising steps (i) and (ii) was performed, resulting in almost 100% replacement. NaOH was used as an aqueous solution to achieve efficient ion exchange. However, in the esterification of triglycerides, water in the resin needs to be replaced by alcohol, meaning that the load will be reduced if no water is used during the regeneration.

Regeneration has been reported using NaOH directly dissolved in methanol, but has not resulted in high catalytic activity. This is probably due to the low solubility of NaOH in methanol and the low amount of OH groups in the solution. We measured the OH substitution ratio in NaOH aqueous methanol solutions by varying the water content. Without water, the OH substitution ratio was approximately halved, but when the water content increased above 20%, the OH substitution ratio was equivalent to that with 100% water content. Ren et al. [28] used a KOH–alcohol solution that, owing to the high solubility of KOH in alcohol, might be able to achieve a high replacement ratio without water. In addition to the esterification system with cation-exchange resin, the aforementioned venture company has also installed a continuous transesterification system with anion-exchange resins, which has likewise been regenerated and used for over a year without any loss of activity [17].

### 10.4.4 Additional Functions Unique to Anion-Exchange Resins

The ease with which OH groups of anion-exchange resins are exchanged with fatty acid groups has led to many investigations into their use in the removal of FFAs from oils. As mentioned previously, there are strict standards for the presence of FFAs in foods, and technology that can selectively remove FFAs is greatly needed. In the edible oil refinery, the classical method for saponification and separation of FFAs by addition of NaOH is still used. Therefore, the removal of FFAs by anion-exchange resins is expected to be a new technique. Debom et al. [34] further proposed a process for the simultaneous deacidification and transesterification of triglycerides using an anion-exchange resin.

Anion-exchange resins have also been found to selectively adsorb by-product glycerine [18, 35]. This is due to the resin tending to adsorb more polar components in the oil solution. In other words, FFAs are chemically adsorbed by ion exchange, while glycerine is physically adsorbed. Owing to this property, as previously mentioned, the resin maintains a methanol-swollen state, and the reverse reaction of transesterification is inhibited. The adsorption behavior of resins toward glycerine has also been observed in some studies [28, 36–38]. Therefore, the continuous biodiesel production process using resins eliminates the need for time-consuming phase separation to remove glycerine, which reduces the separation load and improves fuel quality.

## 10.5 Summary

This chapter focuses on the catalytic activity of cation-exchange and anion-exchange resins in the synthesis of biodiesel (fatty acid esters). In addition to their high catalytic activity for FFA esterification and triglyceride transesterification, ion-exchange resins have the ability to physically adsorb highly polar methanol and glycerol. This property makes it possible to construct simple production processes, which do not require the pretreatment and purification steps essential in conventional processes using homogeneous catalysts. For solid catalysts such as resins, mass transfer and reaction occur together, meaning that great care must be taken when discussing phenomena based on experimental results. It is hoped that this review will contribute to the further development of research using resins in future.

## References

**1** Canakci, M. and Van Gerpen, J. (2001). Biodiesel production from oils and fats with high free fatty acids. *Transactions of the American Society of Agricultural Engineers* 44: 1429–1436.

**2** Lotero, E., Liu, Y., Lopez, D.E. et al. (2005). Synthesis of biodiesel via acid catalysis. *Industrial and Engineering Chemistry Research* 44: 5353–5363. https://doi.org/10.1021/ie049157g.

**3** Vitiello, R., Li, C., Russo, V. et al. (2017). Catalysis for esterification reactions: a key step in the biodiesel production from waste oils. *Rendiconti Lincei-Matematica e Applicazionii* 28: 117–123. https://doi.org/10.1007/s12210-016-0570-2.

**4** Trombettoni, V., Lanari, D., Prinsen, P. et al. (2018). Recent advances in sulfonated resin catalysts for efficient biodiesel and bio-derived additives production. *Progress in Energy and Combustion Science* 65: 136–162. https://doi.org/10.1016/j.pecs.2017.11.001.

**5** Bondioli, P. (2004). The preparation of fatty acid esters by means of catalytic reactions. *Topics in Catalysis* 27: 77–82. https://doi.org/10.1023/B:TOCA.0000013542.58801.49.

**6** Tesser, R., Di Serio, M., Guida, M. et al. (2005). Kinetics of oleic acid esterification with methanol in the presence of triglycerides. *Industrial & Engineering Chemistry Research* 44: 7978–7982. https://doi.org/10.1021/ie050588o.

**7** Haas, M.J. (2005). Improving the economics of biodiesel production through the use of low value lipids as feedstocks: vegetable oil soapstock. *Fuel Process Technology* 86: 1087–1096. https://doi.org/10.1016/j.fuproc.2004.11.004.

**8** Tsuji, T., Kubo, M., Shibasaki-Kitakawa, N., and Yonemoto, T. (2009). Is excess methanol addition required to drive transesterification of triglyceride toward complete conversion? *Energy Fuels* 23: 6163–6167. https://doi.org/10.1021/ef900622d.

**9** Ozbay, N., Oktar, N., and Tapan, N. (2008). Esterification of free fatty acids in waste cooking oils (WCO): role of ion-exchange resins. *Fuel* 87: 1789–1798. https://doi.org/10.1016/j.fuel.2007.12.010.

**10** Russbueldt, B.M.E. and Hoelderich, W.F. (2009). New sulfonic acid ion-exchange resins for the preesterification of different oils and fats with high content of free fatty acids. *Applied Catalysis A: General* 362: 47–57. https://doi.org/10.1016/j.apcata.2009.04.019.

**11** Feng, Y., He, B., Cao, Y. et al. (2010). Biodiesel production using cation-exchange resin as heterogeneous catalyst. *Bioresource Technology* 101: 1518–1521. https://doi.org/10.1016/j.biortech.2009.07.084.

**12** Hu, X., Wei, T., Liao, A., and Tong, Z. (2016). Use of acid cation-exchange resin for catalytic conversion of soybean acid oil to biodiesel. *Journal of Material Cycles and Waste Management* 18: 123–131. https://doi.org/10.1007/s10163-014-0317-z.

**13** Fu, J., Chen, L., Lv, P. et al. (2015). Free fatty acids esterification for biodiesel production using self-synthesized macroporous cation exchange resin as solid acid catalyst. *Fuel* 154: 1–8. https://doi.org/10.1016/j.fuel.2015.03.048.

**14** Kouzu, M., Nakagaito, A., and Hidaka, J.-S. (2011). Pre-esterification of FFA in plant oil transesterified into biodiesel with the help of solid acid catalysis of sulfonated cation-exchange resin. *Applied Catalysis A: General* 405: 36–44. https://doi.org/10.1016/j.apcata.2011.07.026.

**15** Shibasaki-Kitakawa, N., Hiromori, K., Ihara, T. et al. (2015). Production of high quality biodiesel from waste acid oil obtained during edible oil refining using ion-exchange resin catalysts. *Fuel* 139: 11–17. https://doi.org/10.1016/j.fuel.2014.08.024.

**16** Siddique, N., Suzue, M., Kato, M. et al. (2021). Process optimization for continuous production of sustainable biodiesel from completely non-edible biomass, lignocellulosic hydrous ethanol and waste fatty acids. *Fuel* 2891: 119884. https://doi.org/10.1016/j.fuel.2020.119884.

**17** Phytochem Product Inc (2021). Technology: the ion exchange resin method. https://phytochem-products.co.jp/en/tech/ (accessed 10 June 2021).

**18** Shibasaki-Kitakawa, N., Kanagawa, K., Nakashima, K., and Yonemoto, T. (2013). Simultaneous production of high quality biodiesel and glycerin from Jatropha oil using ion-exchange resins as catalysts and adsorbent. *Bioresource Technology* 142: 732–736. https://doi.org/10.1016/j.biortech.2013.05.111.

**19** Shibasaki-Kitakawa, N., Oyanagi, T., Borjigin, S. et al. (2016). Pilot-scale production of high quality biodiesel from vegetable oils with high free fatty acid content using ion-exchange resin catalysts. *Kagaku Kogaku Ronbunshu* 42: 30–36. https://doi.org/10.1252/kakoronbunshu.42.30.

**20** Melero, J.A., Bautista, L.F., and Morales, G. (2010a). Biodiesel production over arenesulfonic acid-modified mesostructured catalysts: optimization of reaction parameters using response surface methodology. *Topics in Catalysis* 53: 795–804. https://doi.org/10.1007/s11244-010-9465-0.

**21** Melero, J.A., Bautista, L.F., Morales, G. et al. (2010b). Biodiesel production from crude palm oil using sulfonic acid-modified mesostructured catalyst. *Chemical Engineer Journal* 161 (3): 23–331. https://doi.org/10.1016/j.cej.2009.12.037.

**22** Paterson, G., Issariyakul, T., Baroi, C. et al. (2013). Ion-exchange resins as catalyst in transesterification of triolein. *Catalysis Today* 212: 157–163. https://dx.doi.org/10.1016/j.cattod.2012.10.013.

**23** Cabral, N.M., Lorenti, J.P., Plass, W., and Gallo, J.M.R. (2020). Solid acid resin Amberyst 45 as a catalyst for the transesterification of vegetable oil. *Frontiers in Chemistry* 8: 1–10. https://doi.org/10.3389/fchem.2020.003050.

**24** Vicente, G., Coteron, A., Martinez, M., and Aracil, J. (1998). Application of the factorial design of experiments and response surface methodology to optimize biodiesel production. *Industrial Crops and Products* 8: 29–35. https://doi.org/10.1016/S0926-6690(97)10003-6.

**25** Shibasaki-Kitakawa, N., Honda, H., Kuribayashi, H. et al. (2007). Biodiesel production using anionic ion-exchange resin as heterogeneous catalyst. *Bioresource Technology* 98: 416–421. https://doi.org/10.1016/j.biortech.2005.12.010.

**26** Marchetti, J.M. and Errazu, A.F. (2010). Biodiesel production from acid oils and ethanol using a solid basic resin as catalyst. *Biomass and Bioenergy* 34: 272–277. https://doi.org/10.1016/j.biombioe.2009.10.016.

**27** Bejarano-Alva, I.J., Hirata, G.A.M., Sampaio, K.A. et al. (2020). High conversion of palm olein to ethyl esters using a strong anion exchange resin: study of the operational parameters. *Journal of the Brazilian Chemical Society* 31: 1401–1412. https://doi.org/10.21577/0103-5053.20200025.

**28** Ren, Y., He, B., Yan, F. et al. (2012). Continuous biodiesel production in a fixed bed reactor packed with anion-exchange resin as heterogeneous catalyst. *Bioresource Technology* 113 (19): 22. https://doi.org/10.1016/10.1016/j.biotech.2011.10.103.

**29** He, B., Ren, Y., Cheng, Y., and Li, J. (2012). Deactivation and in situ regeneration of anion exchange resin in the continuous transesterification for biodiesel production. *Energy & Fuels* 26: 3897–3902. https://doi.org/10.1021/ef300524w.

**30** Kim, M., Salley, S.O., and Ng, K.Y.S. (2008). Transesterification of glycerides using a heterogeneous resin catalyst combined with a homogeneous catalyst. *Energy & Fuels* 22: 3594–3599. https://doi.org/10.1021/ef800443x.

**31** Co, C.E.T., Tan, M.C., Diamante, J.A.R. et al. (2011). Internal mass-transfer limitations on the transesterification of coconut oil using an anionic ion exchange resin in a packed bed reactor. *Catalysis Today* 174: 54–58. https://doi.org/10.1016/j.cattof.2011.02.065.

**32** Keneni, Y.G., Hvoslef-Eide, A.K.(.T.)., and Marchetti, J.M. (2020). Optimization of the production or biofuel from Jatropha oil using a recyclable anion-exchange resin. *Fuel* 278: 118253. https://doi.org/10.1016/j.fuel.2020.118253.

**33** Spector, A.A. (1975). Fatty acid binding to plasma albumin. *Journal of Lipid Research* 16: 165–179.

**34** Deboni, T.M., Hirata, G.A.M., Shimamoto, G.G. et al. (2018). Deacidification and ethyl biodiesel production from acid soybean oil using a strong anion exchange resin. *Chemical Engineering Journal* 222: 686–696. https://doi.org/10.1016/j.cej.2017.09.107.

**35** Shibasaki-Kitakawa, N., Tsuji, T., Kubo, M., and Yonemoto, T. (2011). Biodiesel production from waste cooking oil using anion-exchange resin as both catalyst and adsorbent. *BioEnergy Reserach* 4: 287–293. https://doi.org/10.1007/s12155-011-9148-0.

**36** Wall, J.A., Van Gerpen, J.H., and Thompson, J.C. (2011). Soap and glycerin removal from biodiesel using waterless processes. *Transactions of the ASABE* 54: 535–541.

**37** Gallego Morales, L.J., Cardeno Lopez, F., and Alberto Rios, L. (2013). Refining the glycerin phase from palm-oil biodiesel by ion-exchange with cationic resins. *Journal of Chemical Technology and Biotechnology* 88: 1978–1983. https://doi.org/10.1002/jctb.4054.

**38** Li, R., Liang, N., Ma, X. et al. (2019). Study on the adsorption behavior of glycerin from fatty acid methyl esters by a tertiary amine-type anion exchange resin. *Journal of Chromatography A* 1586: 62–71. https://doi.org/10.1016/j.chroma.2018.11.079.

# 11

# Advances in Bifunctional Solid Catalysts for Biodiesel Production

*Bishwajit Changmai[1], Michael Van Lal Chhandama[2], Chhangte Vanlalveni[3], Andrew E.H. Wheatley[4], and Samuel Lalthazuala Rokhum[1,4]*

[1] *Department of Chemistry, National Institute of Technology Silchar, Silchar, Assam, India*
[2] *Department of Biotechnology, School of Sciences (Block-I), JAIN (Deemed-to-be University), Bengaluru, Karnataka, India*
[3] *Department of Botany, Mizoram University, Aizawl, Mizoram, India*
[4] *Hamid Yusuf Department of Chemistry, University of Cambridge, Cambridge, UK*

## 11.1 Introduction

Increasing global energy demand coupled with the depletion of fossil fuels reservoirs and rapid rise in $CO_2$ emission has intensified the interest in developing biofuels' production, an alternative to fossil fuels from renewable and sustainable resources [1, 2]. In this regard, abundant and renewable biomass's catalytic valorization into a sustainable energy source, fine chemicals, and materials is considered a promising approach by both carbon neutral and zero-waste economy concepts [3, 4]. It is well known that catalysts served a significant role in removing the scientific and engineering constraints in biodiesel production, like in the petrochemical industry [5]. Despite the high reactivity of homogeneous catalysts, it possesses some severe problems such as wastewater generation, equipment corrosion, and catalyst reusability issue [6, 7]. In this context, solid catalysts are considered "green" sustainable alternative to the homogeneous catalyst. It does not produce toxic waste, reusable, noncorrosive to the equipment, and easy to separate from the reaction mixture. Nowadays, solid catalyst-mediated transformation of biomass feedstocks to biofuels is a growing field in modern chemistry globally [8, 9]. Hence, developing novel and efficient solid catalysts for biofuels' industrial-scale production from biomass feedstocks is a significant challenge.

Biodiesel, also called fatty acid methyl ester (FAME), is promising liquid biofuel and can be produced from edible and nonedible oils, waste cooking oil (WCO), animal fats, etc. [10, 11]. Interestingly, biodiesel possesses several advantages over fossil fuels, such as being renewable, reducing $CO_2$ emission, and nontoxic [12, 13]. Besides, biodiesel properties are similar to conventional diesel, which can be directly utilized in diesel engines without modification [14, 15]. Furthermore, it also prevents the emission of unburned hydrocarbons and $SO_2$ during combustion due to the presence of oxygen in high amounts [16, 17].

*Biodiesel Production: Feedstocks, Catalysts, and Technologies*, First Edition. Edited by Samuel Lalthazuala Rokhum, Gopinath Halder, Suttichai Assabumrungrat, and Kanokwan Ngaosuwan.

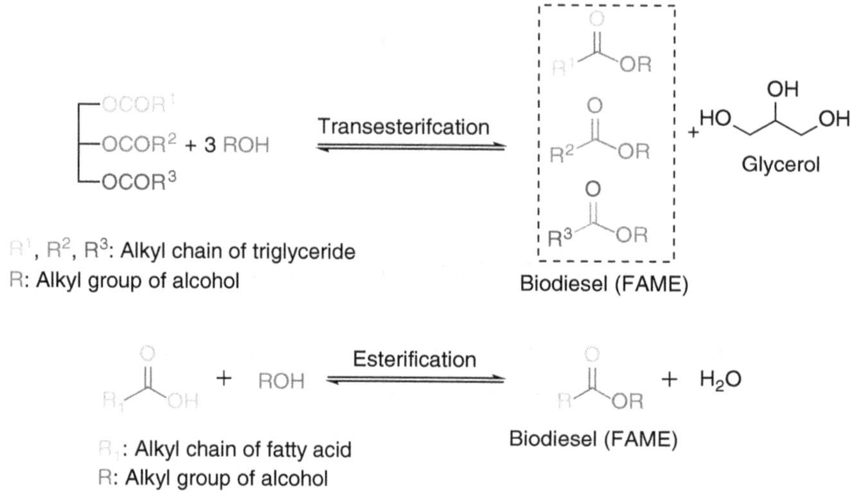


**Scheme 11.1** Biodiesel production via transesterification and esterification.

Biodiesel is the transesterification and esterification product of long-chain triglycerides and free fatty acids, respectively (Scheme 11.1) [18]. Both solid acid and base catalysts can efficiently catalyze the transesterification reaction [9, 19]. Despite the high activity of the base catalyst, it is inefficient for the esterification reaction as it forms soap with the FFA, whereas acidic catalyst can facilitate both esterification and transesterification reaction [13, 20]. Currently, above 95% of commercial biodiesel is produced from the edible oils, which led to the generation of food *vs* fuel dilemma in the developing countries [6, 21]. On the contrary, utilization of nonedible oils as a raw material for biodiesel production can solve the problem due to the wide availability and low cost [22, 23]. However, the presence of a high amount of FFA in nonedible oils impedes solid base catalyst utilization [24, 25]. In this context, instead of using a two-step process involving esterification of FFA with acid catalyst followed by transesterification of triglyceride with base catalyst, the bifunctional catalyst having both the characteristics of acid and base could be an excellent choice for the trans(esterification) of nonedible oils as it eliminates the two-step process and complex steps involved in the product isolation and separation [26, 27]. It is reported that the Brønsted acid sites facilitate the esterification reaction, whereas Lewis acid sites play a cooperative role in the esterification of carboxylic acids [28, 29]. To date, a wide variety of review articles are available in the field of biodiesel production using heterogeneous catalysts [30–34]. However, the application of solid bifunctional catalyst for the trans(esterification) reaction is hardly reviewed [26, 27]. In this book chapter, we emphasize the discussion of different types of bifunctional catalysts utilized for the trans(esterification) of nonedible oils.

## 11.2 Application of Solid Bifunctional Catalyst in Biodiesel Production

### 11.2.1 Acid–Base Bifunctional Catalysts

Currently, biodiesel production from nonedible oils has high FFA content attracting immense attention as it can potentially solve the competition of edible food *vs.* fuel dilemma and economic feasibility [22, 35]. Solid base catalysts are highly efficient in the transesterification of triglycerides to form biodiesel. Still, it creates soap in the presence of FFA, thus requiring a two-step process such as esterification with an acid catalyst followed by transesterification, making the process lengthy and costly [36, 37]. On the other hand, the solid acid catalyst can facilitate the trans(esterification) reaction in a single reaction vessel, but they display comparatively less efficiency toward transesterification reaction, unlike solid base catalyst [38, 39]. Considering that, solid acid–base bifunctional catalysts are highly preferable as they show high activity toward trans(esterification) reaction to form biodiesel. Different solid acid-base bifunctional catalysts employed for biodiesel production are given in Table 11.1.

#### 11.2.1.1 Oxides of Acid–Base

Due to the presence of both acidic and basic sites, $La_2O_3$ is widely utilized for biodiesel production from nonedible oils in a single-step process [40]. However, the requirement of high methanol/oil molar ratio (30 : 1) and temperature (200 °C) to achieve a moderate yield of 56% and poor reusability due to the active site leaching limits its practical application in industrial-scale production of biodiesel. In this regard, $La_2O_3$ is employed as catalyst support for various metal oxides such as MgO, ZnO, $Al_2O_3$, CaO, and $ZrO_2$ to enhance the catalytic activity toward trans(esterification) reaction. In this context, $La_2O_3$ was loaded with Zn in a different molar ratio (La/Zn: 1 : 1, 1 : 3, 1 : 9) and utilized as a catalyst for the trans(esterification) of WCO. The La/Zn molar ratio of 1 : 3 showed the highest biodiesel yield of 96% under the optimized reaction conditions. The increased activity of the catalyst could be attributed to the strong interaction between Zn and La. Besides, La promotes Zn's homogeneous distribution on the surface, thus enhancing the surface acid and basic sites [41]. Similarly, a series of MgO-$La_2O_3$ catalysts were prepared via the coprecipitation method followed by calcined at different temperatures (723–923 K) and employed to convert sunflower oil to biodiesel [42]. It is reported that 60 wt% Mg/La (w.r.t. La) calcined at 873 K is the optimum catalyst, showing 96.9% yield under the optimized reaction conditions. The catalyst was reused for four consecutive reaction cycles, with no significant loss of the active sites. In another study, $La_2O_3$ was loaded with different amounts of $Bi_2O_3$ (1–7 wt% w.r.t. $La_2O_3$) and employed for the trans(esterification) of *Jatropha curcas* oil (JO) and achieved a high oil conversion of 93% [43]. $Bi_2O_3$ loading increased the surface acidity of the catalyst, while the basic strength decreased in the process. The catalyst's excellent activity is accounted for the catalyst's high surface area, resulting due to the good dispersion of $Bi_2O_3$ on the $La_2O_3$ surface. Nonetheless, the catalyst showed 83% oil conversion even after three reaction cycles, indicating the catalyst's high stability.

CaO is widely investigated as a solid base catalyst for biodiesel production due to its high basic strength, availability, and low cost [44]. Despite its excellent activity toward biodiesel production, it cannot be considered a potential catalyst for industrial-scale biodiesel production because of leaching and low stability [45]. Hence, it is essential to modify CaO-based materials with other materials to improve their strength and reusability. To this end, CaO-based binary metal oxides having both acid–base characteristics could be an excellent choice. Lee et al. [46] reported a series of CaO-$La_2O_3$ catalyst with various Ca/La atomic ratios and employed for the trans(esterification) of *Jatropha curcas* oil (JCO), showing 98.7% biodiesel yield under the optimized reaction conditions. The high activity of the catalyst compared to the bulk CaO and $La_2O_3$ is due to the homogeneous dispersion of CaO on the composite surface, resulted in high acidic and basic strength and stability. Similarly, CaO loaded on $Al_2O_3$ showed excellent stability and activity toward trans(esterification) of

**Table 11.1** Solid acid–base bifunctional catalyst used for biodiesel/FAME production.

| Entry | Catalyst | Feedstock | Reaction conditions[a] | Yield (%) | References |
|---|---|---|---|---|---|
| 1 | $ZnO-La_2O_3$ | WCO | 40 : 1, 3, 3, 200 | 96 | Yan et al. [41] |
| 2 | $MgO-La_2O_3$ | Sunflower oil | 18 : 1, 3, 4, 65 | 96.9 | Feyzi et al. [42] |
| 3 | $Bi_2O_3-La_2O_3$ | JCO | 15 : 1, 2, 4, 150 | 93[b] | Nizah et al. [43] |
| 4 | $CaO-La_2O_3$ | JCO | 25 : 1, 3, 3, 160 | 98.7 | Lee et al. [46] |
| 5 | $CaO-Al_2O_3$ | Waste palm oil | 9 : 1, 4, 4, 65 | 89 | Elias et al. [47] |
| 6 | $CaO-Al_2O_3$ | Waste sunflower oil | 9 : 1, 4, 4, 65 | 98 | Elias et al. [47] |
| 7 | $CaO-Fe_2(SO_4)_3$ | JCO | 6 : 1, 5, 3, 60 | 100[b] | Endalew et al. [36] |
| 8 | $CaO-MoO_3$-SBA15 | Soybean oil | 50 : 1, 6, 50, 65 | 83.2[b] | Xie et al. [48] |
| 9 | $Mn-MgO-ZrO_2$ | *Phoenix dactylifera* L. kernel oil | 15 : 1, 3, 4, 90 | 96.4 | Jamil et al. [49] |
| 10 | CuO-SrO | Hempseed oil | 12 : 1, 10, 0.5, 60 | 96 | Su et al. [50] |
| 11 | $SrO-ZnO-Al_2O_3$ | WCO | 10 : 1[c], 15, 5, 75 | 95.7 | Al-Saadi et al. [51] |
| 12 | CaO-(Zr)MCM-41 | WCO | 9 : 1, 5, 6, 70 | 88.5[b] | Dehghani et al. [52] |
| 13 | Mg-(MCM-41) | WCO | 8 : 1, 10, 3, 80 | ≥92 | Pirouzmand et al. [53] |
| 14 | Co-(MCM-41) | WCO | 8 : 1, 10, 3, 80 | ≥92 | Pirouzmand et al. [53] |
| 15 | Zn-(MCM-41) | WCO | 8 : 1, 10, 3, 80 | ≥92 | Pirouzmand et al. [53] |
| 16 | [CTA]-MCM-41 | WCO | 8 : 1, 10, 3, 80 | 93 | Pirouzmand et al. [53] |
| 17 | K-(ITQ-6) | WFSO | 20 : 1, 5, 24, 180 | 80 | Macario et al. [54] |
| 18 | SiC-NaOH-GO | Rapeseed oil | 12 : 1, 5, 6[d], 65 | 96 | Loy et al. [55] |
| 19 | SiC-NaOH-GO | OA | 12 : 1, 5, 6[d], 65 | 92 | Loy et al. [55] |
| 20 | SiC-NaOH-GO | *Chlorella vulgaris* lipid | 12 : 1, 5, 6[d], 65 | 92 | Loy et al. [55] |
| 21 | Lysine/HPA | ESGSO | 9 : 1, 9, 12, 65 | 93 | Wang et al. [56] |
| 22 | $Zn_8@Fe-C_{400}$ | JCO | 40 : 1, 7, 4, 160 | 100[b] | Wang et al. [57] |

[a] Methanol/oil molar ratio, catalyst loading (wt%), reaction time (h), reaction temperature (°C).
[b] Conversion (%).
[c] Ethanol/oil molar ratio.
[d] Reaction time (min).

waste palm oil and waste sunflower oil with high FAME yield of 89 and 98%, respectively [47]. Furthermore, the activity and stability of CaO toward trans(esterification) was also increased upon the interaction with the solid acid catalyst $Fe_2(SO_4)_3$ due to both basic–acidic properties, where CaO facilitates the transesterification reaction and $Fe_2(SO_4)_3$

facilitates the esterification reaction. Besides, the composites' catalytic activity was compared with a different catalyst such as $La_2O_3/ZnO$, $La_2O_3/Al_2O_3$, and $La_{0.1}Ca_{0.9}MnO_3$; it was found that the activity lies in the order of $La_2O_3/Al_2O_3 < La_{0.1}Ca_{0.9}MnO_3 < La_2O_3/ZnO < CaO–Fe_2(SO_4)_3$. The catalysts' moderate activity was $La_2O_3/ZnO$, $La_2O_3/Al_2O_3$, and $La_{0.1}Ca_{0.9}MnO_3$ toward trans(esterification) of JCO be attributed to the lack of basic strength. However, the activity can be improved by applying high temperature, pressure, time, and reactant molar ratio [36]. Another CaO-based acid–base bifunctional catalyst, $CaO–MoO_3$–SBA15, was synthesized in an environmentally benign process and utilized to transesterification of soybean oil to produce FAME. The influence of $CaO–MoO_3$ loading and calcination temperature was investigated as they play a vital role in the oil conversion to biodiesel, and it was found that 40 wt% $CaO–MoO_3$ loading and 823 K calcination temperature showed the highest activity with oil conversion of 83.2% under the optimized reaction conditions. The catalyst displayed excellent stability with recyclability up to five consecutive cycles with 77% oil conversion. The authors claimed that the catalyst is insensitive to the FFA and water content present in the oil [48].

Jamil et al. [49] reported $Mn–MgO@ZrO_2$ for the trans(esterification) of waste *Phoenix dactylifera* L. kernel oil, achieving 96.4% biodiesel yield. The modification of amphoteric $ZrO_2$ with strongly basic MgO increased the moderate basic sites indicated by $CO_2$-TPD results. The results also showed that increasing MgO amount increased the average basic sites, which was further improved upon impregnation of Mn on $MgO@ZrO_2$ as demonstrated in Figure 11.1. Furthermore, the surface area, pore volume, and pore diameter were increased compared to bulk $ZrO_2$ upon impregnation of Mn and MgO, which could be the primary reason for the catalyst's high activity. In another study, alkaline earth metal oxides (AMO) loaded with Cu were utilized to transesterification of hempseed oil to produce biodiesel. Among all Cu/AMO, Cu/SrO displayed the highest activity (96% FAME yield) as CuO is homogeneously distributed on the SrO surface along with the high basicity of SrO. Besides, the latter's surface area is higher than the other Cu/AMO due to the smaller particle size (50–100 nm), thus enabling higher reactant molecules to adsorb [50]. Furthermore, to enhance the Lewis acidity and stability of SrO-based catalyst, Saadi et al. [51] reported $SrO–ZnO/Al_2O_3$ for the trans(esterification) of corn oil having high FFA content, achieving 95.1% oil conversion under the moderate reaction conditions. The activation energy for the trans(esterification) reaction was reported as 15.84 kJ $mol^{-1}$, indicating the catalyst's great activity. The same catalyst was further employed for the trans(esterification) of WCO having FFA content of 18%, showing 95.7% oil conversion under the optimized reaction conditions.

#### 11.2.1.2 Acid–Base Hydrides

Apart from the acid–base metal oxides, acid–base hydrides are also investigated widely for the synthesis of biodiesel. Dehghani et al. [52] incorporated Zr into the structure of MCM-41, followed by loaded with CaO via sonochemical dispersion and impregnation method and employed for the trans(esterification) of WCO to produce biodiesel. The sonochemical setup for loading of CaO on MCM-41 is displayed in Figure 11.2. The sonochemical dispersion resulted in the homogeneous scattering of CaO over Zr/MCM-41 surface compared with the impregnation method, suggested by energy-dispersive X-ray spectroscopy (EDX) elemental mapping analysis (Figure 11.3). Thus, the catalyst prepared via the

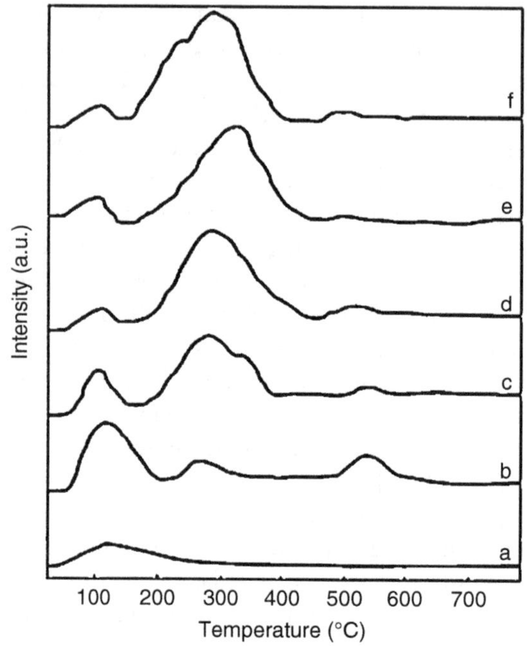


**Figure 11.1** $CO_2$-TPD profile for pure $ZrO_2$ (a), modified catalyst with Mg/Zr ratio of 0.2–0.5 (b–e) and Mn–Mg–$ZrO_2$ (0.4 Mg:Zr) (f). *Source:* Ref. [49].

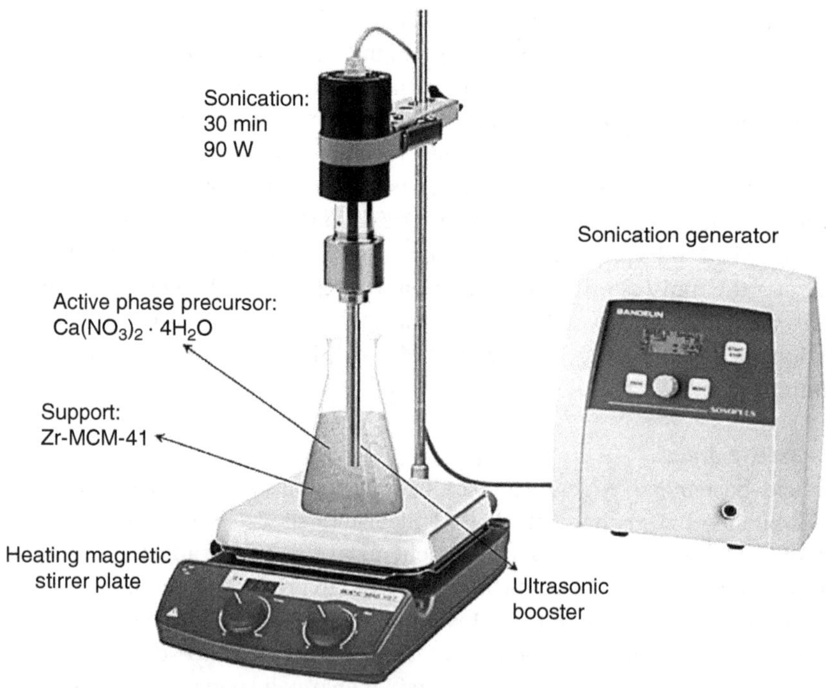


**Figure 11.2** Sonochemical setup for CaO dispersion on MCM-41. *Source:* Ref. [52].

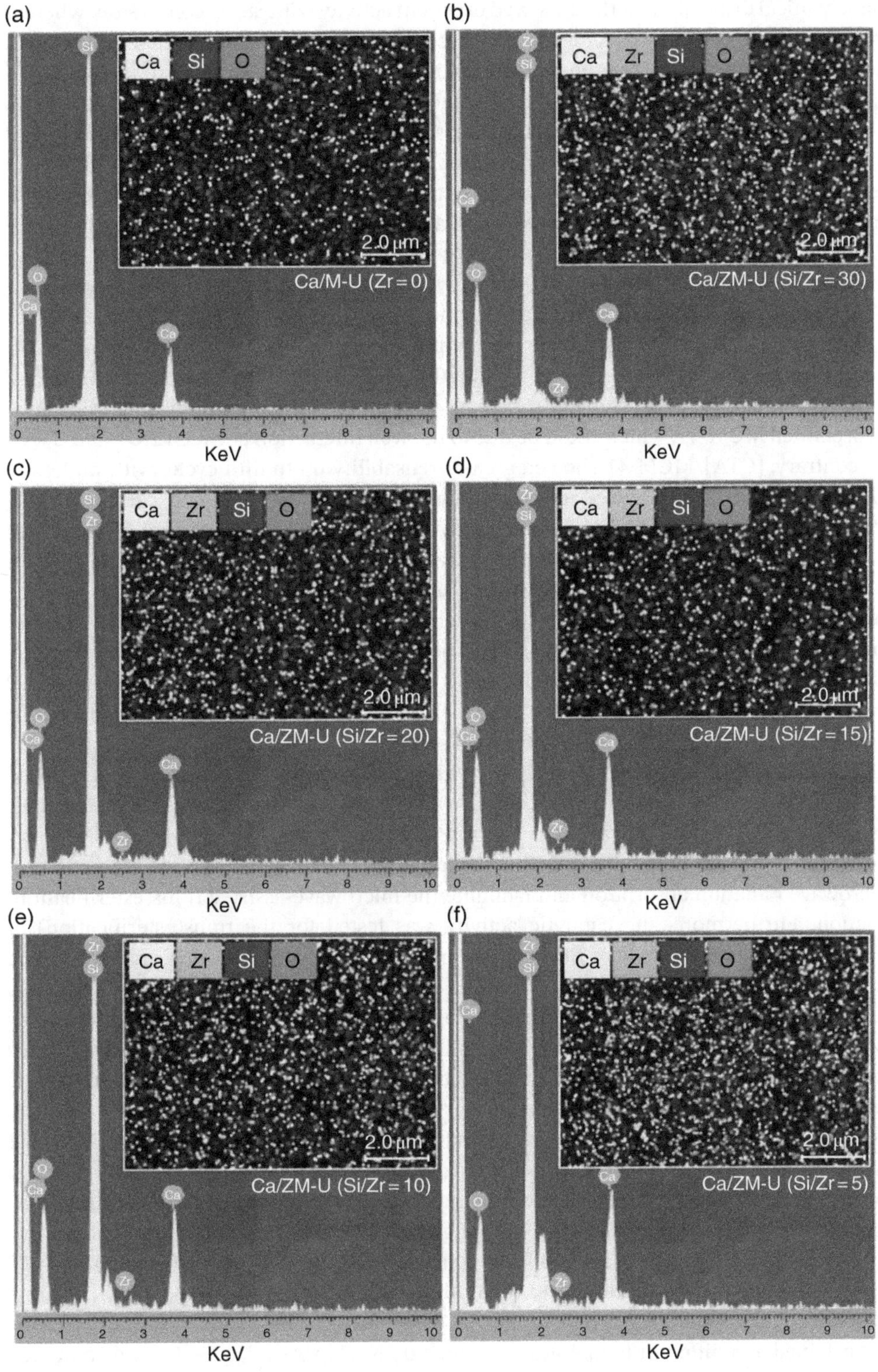


**Figure 11.3** EDX analysis of sono-enhanced CaO-dispersed over Zr-doped MCM-41 nanocatalysts with various Si/Zr ratios: (a) Ca/M-U (Zr = 0), (b) Ca/ZM-U (Si/Zr = 30), (c) Ca/ ZM-U (Si/Zr = 20), (d) Ca/ZM-U (Si/Zr = 15), (e) Ca/ZM-U (Si/Zr = 10), and (f) Ca/ZM-U (Si/Zr = 5). *Source:* Ref. [52].

sonochemical dispersion method showed excellent activity with 88.5% conversion, whereas only 83.7% conversion was achieved in the catalyst prepared through the impregnation method. The effect of Si/Zr atomic ratio on the transformation of WCO to biodiesel was also examined, and it was found that Si/Zr molar ratio of 10 : 1 showed the highest activity, which can be attributed to the increased surface area of the catalyst. Besides, the catalyst prepared via the sonochemical method showed more excellent reusability compared with the un-sonochemical one. Similarly, different metals such as Mg (weak Lewis acid), Co (moderate Lewis acid), and Zn (strong Lewis acid) were incorporated into the structure of MCM-41 to examine the catalytic activity toward the trans(esterification) of WCO to produce FAME, and it was found that every catalyst showed great activity with ≥92% FAME yield under the optimized reaction conditions, which could be due to the existence of both acidic and basic sites. In the same study, they compared the efficiency of all the metal-incorporated catalysts with cetyltrimethylammonium [CTA]-loaded MCM-41 and found that the latter one showed the same activity (93% biodiesel yield) as the metal incorporated MCM-41, which could be due to the well interaction between $SiO^-$ and $CTA^+$. On contrary, [CTA]/MCM-41 showed greater reusability up to fifth cycle, with no loss of $CTA^+$ site as they are stabilized in the micelles [53]. In another study, Macario et al. [54] prepared potassium-loaded ITQ-6 and employed for the trans(esterification) of waste fruit seed oil (WFSO). Comparison of the catalytic activity of K-ITQ-6 with strong acid catalysts such as USY, BEA, FAU-X, and weak acid MCM-41 and ITQ-6 revealed that the first one showed the highest activity with 80% biodiesel yield. The occurrence of Brønsted acid sites and basic sites facilitates the trans(esterification) reaction without soap formation. Unfortunately, the catalyst's reusability is not significant as it showed only 35% biodiesel yield after the second run, attributed to the leaching of basic sites, but the activity can be regenerated by treating with fresh potassium source.

Recently, graphene oxide (GO) modified with silicon carbide (SiC) and NaOH, employed for the microwave-assisted trans(esterification) of a binary mixture of rapeseed oil and oleic acid, achieved 96 and 92% transesterification and esterification yield, respectively, under the optimized reaction conditions. The introduction of SiC to the catalyst is to enhance microwave radiation absorption and facilitate the microwave-assisted trans(esterification) reaction. Furthermore, the catalytic activity was tested for the trans(esterification) of *Chlorella vulgaris* lipid with an FFA content of 20% and showed 92% FFA conversion and achieved an overall biodiesel yield 81% [55]. A new type of hybrid acid–base bifunctional catalyst was synthesized by fabrication of heteropolyacid (HPA) with an amino acid having two amine groups (Lysine) (Figure 11.4) and employed for trans(esterification) of *Eruca Sativa Gars* seed oils (ESGSO) to biodiesel. The high acidity (3.6 mmol $g^{-1}$) and moderate basicity (0.2 mmol $g^{-1}$) facilitates the transesterification reaction and achieved 93% biodiesel yield under the optimized reaction conditions. Tailored acid–base characteristics and easy handling are the key advantages of this catalyst [56]. To improve the stability and reusability, magnetic nanoparticles are gaining immense attention for their easy separation by using an external magnet. To this end, Wang et al. [57] reported magnetically separable acid–base amphoteric $Zn_8$@Fe-C400 nanoparticles for the trans(esterification) of JCO, achieving 100% biodiesel yield under the optimized reaction conditions. Various alkaline oxides such as $Na_2SiO_3$, MgO, $Al_2O_3$, and ZnO were mixed with the acidic group –COOFe and calcined at a different temperature to generate Na/Mg/Al/$Zn_n$@Fe-$C_T$ and compared

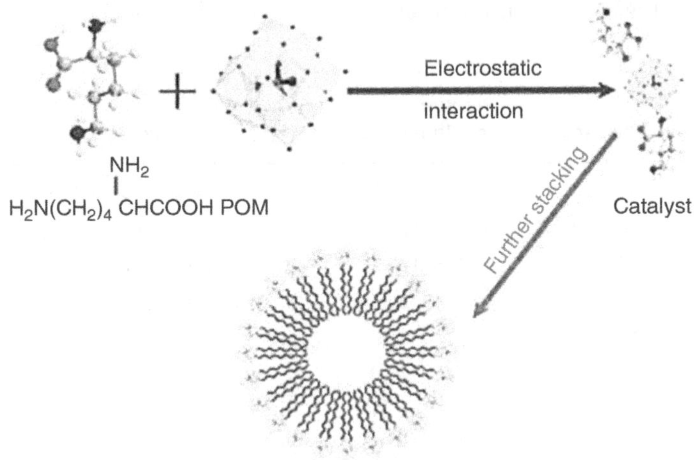


**Figure 11.4** Schematic representation of the synthesis of lysine/HPA catalyst. *Source:* Ref. [56].

their catalytic activity toward the conversion of JCO to biodiesel. $Na_n$@Fe-$C_{600}$ showed excellent catalytic activity with a biodiesel yield of 99.4%; however, leaching of Na was occurred in this process, which led to the saponification of FFA. Similarly, Mg8@Fe-$C_{600}$ showed high activity with a biodiesel yield of ≥99%, but it is unfeasible for industrial application due to the catalyst's poor stability. On the other hand, Al8@Fe-$C_{600}$ showed the lowest activity due to weak basic and acid density. The best biodiesel yield of 100% was achieved in case of $Zn_8$@Fe-$C_{400}$ due to its high acid–base density. Apart from that, the influence of calcination temperature on the incorporation of ZnO into –COOFe structure suggested that at 400 °C, the concentration of ZnO is highest in the –COOFe framework, thus improving the biodiesel yield.

### 11.2.2 Bifunctional Acid Catalyst

Apart from the acid–base metal oxide/hydride catalyst, Brønsted–Lewis acid bifunctional catalysts are immensely employed for biodiesel production via trans(esterification) of low-grade oils having high FFA content as Bronsted acid site of the catalyst facilitates the esterification reaction, whereas Lewis acid promotes the transesterification reaction [27]. In this section, we are discussing various bifunctionalized acid catalysts for biodiesel production (Table 11.2).

#### 11.2.2.1 Bifunctional Brønsted–Lewis Acid Oxides

A bifunctionalized acid oxide catalyst $SO_4/TiO_2$ (sulfated super acid) was reported for the transesterification and trans(esterification) of soybean oil and canola oil to produce biodiesel, achieving 40 and 25% biodiesel yield, respectively [58]. The authors examined the effect of sulfuric acid concentration on the catalytic activity by varying it from 0.9 to 3.7 wt% (w.r.t. $TiO_2$) and found that 0.9 wt% $H_2SO_4$ loaded on $TiO_2$ showed the highest catalytic activity as it consisted of high surface area (266 $m^2$ $g^{-1}$) and pore diameter (5.2 nm). To improve the stability and reusability of sulfated catalyst, Kaur et al. [59] utilized

**Table 11.2** Different Brønsted–Lewis acid bifunctional solid catalyst used for biodiesel/FAME production.

| Entry | Catalyst | Feedstock | Reaction conditions[a] | Yield (%) | References |
|---|---|---|---|---|---|
| 1 | $SO_4/TiO_2$ | Soybean oil | 6 : 1, 5, 1, 120 | 40 | Almeida et al. [58] |
| 2 | $SO_4/TiO_2$ | Castor oil | 6 : 1, 5, 1, 120 | 25 | Almeida et al. [58] |
| 3 | $Ce/ZrO_2$-$TiO_2$ | OA | 6 : 1, 5, 1, 65 | ˃98 | Kaur et al. [59] |
| 4 | $SO_4/Fe_2O_3$-$Al_2O_3$-$TiO_2$ | WCO | 10 : 1, 3, 2.5, 90 | 96 | Gardy et al. [60] |
| 5 | $Nb_2O_5$/SBA-15 | Propanoic acid | 30 : 1, 7, 6, 60 | 80[b] | Silva et al. [61] |
| 6 | 20-W/$TiO_2$-$SiO_2$-700 | WCSO | 30 : 1, 5, 4, 65 | 98 | Kaur et al. [62] |
| 7 | $SO_4$/kaolin | OA | 60 : 1, NR, 4, 160 | 98.9[b] | Kaur et al. [62] |
| 8 | PTA-MIL-53(Fe) | OA | 16 : 1[c], 30, 15[d], RT | 96 | Nikseresht et al. [63] |
| 9 | Fe(III)/$H_3PMo_{12}O_{40}$ | OA | 10 : 1, 5, 3, 70 | 89.2[b] | Nikseresht et al. [63] |
| 10 | $Ti_{0.6}H_{0.6}PW_{12}O_{40}$ | WCO | 21 : 1, 4.6, 12, 65 | 59.2[e] 94.7[f] | Nikseresht et al. [63] |
| 11 | CTA-$H_4PTiW_{11}O_{40}$ | Palmitic acid | 20 : 1, 0.25[g], 65, 6 | 91.8 | Zhao et al. [64] |
| 12 | H3PW11TiO40 | Triacetin | 6 : 1, 4, 4, 65 | 92[b] | Tao et al. [65] |
| 13 | Cs-STA/Zr-KIT6 | Casto oil | 27 : 1, 7, 5, 120 | 98 | Gopinath et al. [66] |

[a] Methanol/oil molar ratio, catalyst loading (wt%), reaction time (h), reaction temperature (°C).
[b] Conversion (%).
[c] Ethanol/oil molar ratio.
[d] Reaction time (min).
[e] Transesterification yield.
[f] Esterification yield.
[g] Catalyst loading (g).

$Ce/ZrO_2$-$TiO_2$ to support sulfate group impregnation and employed for the esterification of oleic acid. The reported catalyst showed 100% acid conversion to biodiesel under the optimized reaction conditions. Acidity measurement revealed that it lies between $3.1 < pKa < 4.8$, a key reason for the catalyst's high activity. The catalyst showed excellent stability and reusability up to the fifth cycle, with no significant loss of active sulfate sites and thus the catalyst activity. The catalyst activity was also tested for vegetable oils with high FFA content. It found that it can convert ˃98% FFA content to its corresponding methyl ester, suggesting the catalyst activity is not affected by the triglycerides. Recently, a core–shell magnetic sulfated catalyst $SO_4$/Fe–Al–$TiO_2$ was reported for the trans(esterification) of WCO to achieve biodiesel. At first, a mixture of rutile and anatase phase of $TiO_2$ was modified with $Al_2O_3$ to introduce a buffer phase, followed by functionalized with hematite to induce magnetic character to the catalyst, and, finally, it was functionalized with sulfate group to enhance the Brønsted acidity. TEM displayed that the particles of the catalyst, having the size of <50 nm with lattice fringes at 0.356 nm, corresponded to rhombohedral structure of $Al_2(SO_4)_3$, [110], [110], [012], planes of anatase, $TiO_2$, rutile $TiO_2$, and rhombohedral structure of hematite. Besides, Figure 11.5c and d displayed the catalyst's core–shell

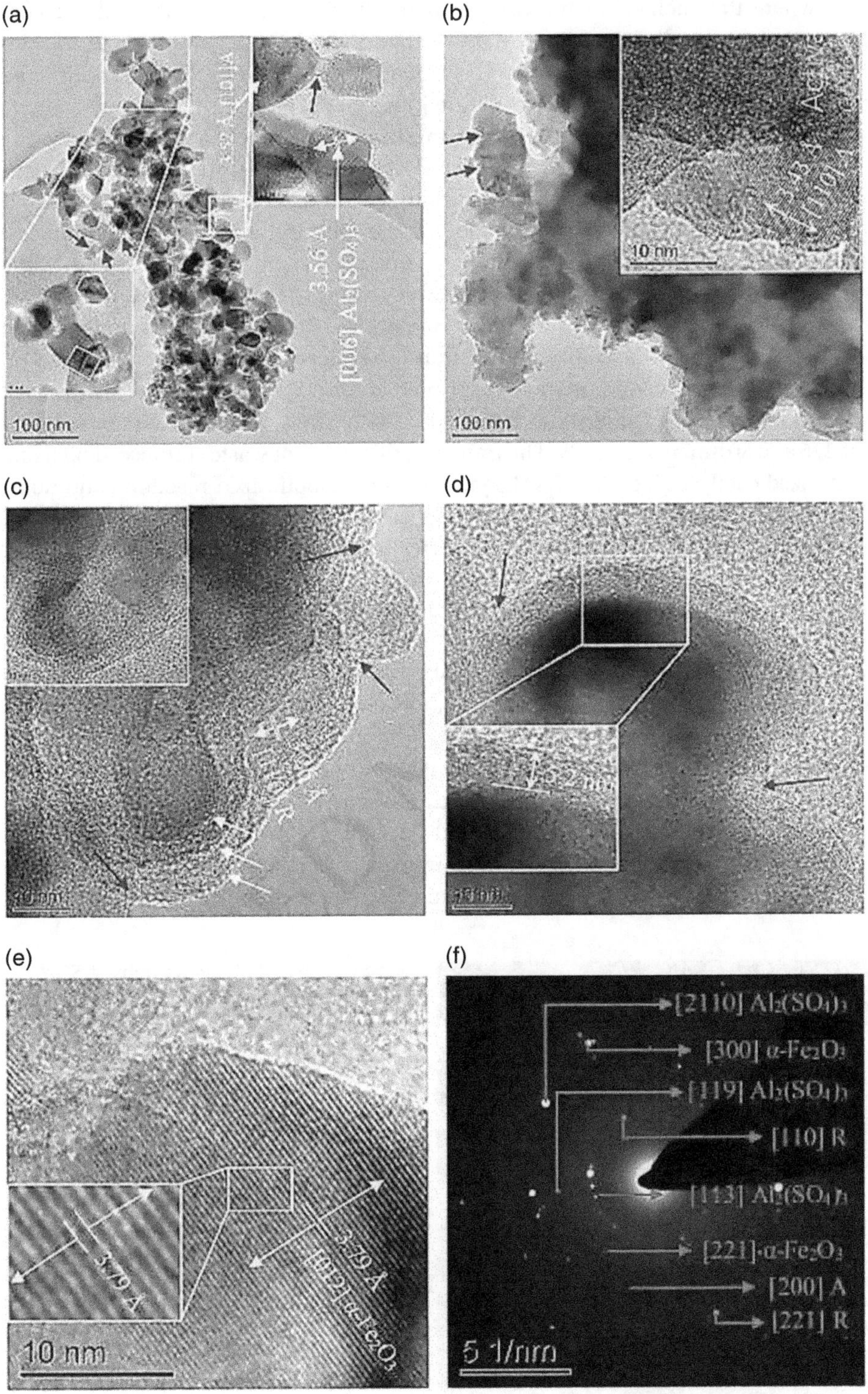


**Figure 11.5** TEM micrographs (a–e) and SAED pattern (f) of $SO_4$/Fe–Al–$TiO_2$. *Source:* Ref. [60].

structure, where the shell was comprised of both alumina and hematite and the core contains the anatase/rutile $TiO_2$. The catalyst showed excellent activity with 95.6% biodiesel yield under the optimized reaction conditions and showed excellent stability and reusability due to the catalyst's magnetic nature [60].

A series of $Nb_2O_5$ loaded on SBA-15 was synthesized for propanoic acid esterification to produce biodiesel. It is reported that at low calcination temperature (500 °C), the loading of $Nb_2O_5$ is low; however, the Brønsted/Lewis acidity ratio was increased at that particular temperature, thus showing the highest activity. On the contrary, at higher calcination temperature (800 °C), the Bronsted/Lewis acidity decreased, which can be attributed to the recrystallization of $Nb_2O_5$, thus displaying lower activity [61]. In another study, the mesoporous $Ti/SiO_2$ matrix was utilized to support the impregnation of tungstate to improve the stability and reusability toward trans(esterification) of waste cotton seed oil (WCSO). Scanning electron microscopy (SEM) micrograph (Figure 11.6) of the synthesized catalyst displayed that the catalyst has a flowerlike structure with petals with length and width of 226 and 80 nm, respectively. The trans(esterification) of waste cottonseed oil using the synthesized catalyst gives 98% FAME yield under the optimized reaction conditions. Besides, it displayed high reusability up to four consecutive cycles, with no significant depreciation in the catalyst activity [62]. Waste kaolin was calcined at two different temperatures of 850 and 950 °C, followed by treated with sulfuric acid to form $SO_4$/kaolin catalyst, which was employed for the esterification of oleic acid to create biodiesel. The catalyst calcined at 950 °C and treated with 4 M sulfuric acid possesses a very high surface area of 406 $m^2\ g^{-1}$ and an acidity of 0.24 mmol $g^{-1}$, thus showing a maximum biodiesel yield 98.9% under the optimized reaction conditions [62].

#### 11.2.2.2 Heteropolyacid-Based Bifunctional Catalyst

HPAs, a class of polyoxometalates, are mainly utilized as an acid catalyst for various organic transformations as they are easy to handle and reusability nature [67–70]. However, the primary drawbacks of HPA consist of solubility in polar solvents and minimal surface area, and having only Bronsted acidic sites and thus restricts their potential application in

**Figure 11.6** SEM micrographs of $SO_4$/kaolin. *Source:* Ref. [62].

industrial processes [71, 72]. To this end, incorporating Lewis acidic sites to the Bronsted acidic sites enhances the overall acidity and enlarges their catalytic application.

Nikseresht et al. [63] reported polytungstic acid (PTA) encapsulated in Fe-based metal–organic framework (MOF) MIL-53 catalyst (PTA/MIL-53(Fe)) for the ultrasonic-assisted esterification of oleic acid to form biodiesel, achieving 96% under the optimized reaction conditions. The presence of MOF in the catalyst enhanced the surface area ($239\,m^2\,g^{-1}$) and stability, due to which the catalyst displayed excellent activity. Besides, the catalyst showed high reusability up to the fifth cycle, with no significant depreciation in the catalytic activity. Similarly, Fe(III) was impregnated on phosphomolybdic acid ($H_3PMo$) and utilized as a solid Lewis–Bronsted acid catalyst for oleic acid esterification. The incorporation of Fe(III) improved the surface area, porosity, and stability and, most importantly, restricts the solubility in organic polar solvents, thus increasing the catalyst's heterogeneous nature. The catalyst showed excellent activity with 89.2% oil conversion under the optimized reaction conditions. However, the leaching of catalyst active sites was observed during the reusability test and obtained only 70.2% conversion after the third run [63]. In another study, HPA was loaded with various cations of Ti, Zn, Sn, Al, Ir, In, Zr, Ce, La, and Zn and utilized to convert WCO with FFA content of 26.89 wt% to biodiesel. Acidity measurement via titration revealed that the acidity of the different cation-loaded HPA follows the order: $Ti_{0.6}H_{0.6}PW > Sn_{0.6}H_{0.6}PW > Al_{0.8}H_{0.6}PW > Fe_{0.8}H_{0.6}PW > In_{0.8}H_{0.6}PW > Zr_{0.6}H_{0.6}PW > Ce_{0.8}H_{0.6}PW > La_{0.8}H_{0.6}PW > Zn_{1.2}H_{0.6}PW$. From the acidity order, it is observed that the catalyst having moderate Lewis acidity improved the overall catalyst acidity compared to bare HPA. The catalyst $Ti_{0.6}H_{0.6}PW$ showed the highest efficiency with 59.2 and 94.7% oil conversion for the transesterification and esterification reaction, respectively [63]. Zhao et al. [64] demonstrated a Lewis–Bronsted HPA modified with a surfactant cetyltrimethyl ammonium (C16TA) catalyst for the esterification of palmitic acid to biodiesel. At first, titanium ion was introduced as Lewis acid site to the HPA to enhance the catalyst's total acidity, followed by a C16TA surfactant to form Lewis–Bronsted-surfactant HPA ((C16TA)$H_4TiPW_{11}O_{40}$). The combination of surfactant involves various advantages such as (i) interacting efficiently with the substrate molecules because of the presence of lipophilic tail and (ii) restricting the water molecules to interact with the catalyst because of the hydrophobic groups present in the catalyst, making the catalyst water tolerant and stable. A maximum biodiesel yield of 91.8% was obtained under the optimized reaction conditions due to the catalyst's mix acidity, and it can be reused up to six consecutive cycles with a slight leaching of the active sites. The incorporation effect of various metals having Lewis acid characteristics such as $Ti^{4+}$, $Cu^{2+}$, $Al^{3+}$, $Sn^{4+}$, $Fe^{3+}$, $Cr^{3+}$, $Zr^{4+}$, and $Zn^{2+}$ on to HPA catalyst for the trans(esterification) reaction was examined and found that the metals having mild Lewis acidity are mainly influence the total acidity of the catalyst. $H_5PW_{11}TiO_{40}$ catalyst showed the highest catalytic activity with maximum triacetin conversion of 92% and reusability up to five consecutive cycles, with no significant loss of catalytic activity [65]. Recently, cesium-incorporated silicotungstic acid (STA) loaded on Zr-KIT6 and Sn-KIT6 was employed for the biodiesel production from oleic acid and various nonedible oils such as castor oil, pongamia oil, and neem oil. It is reported that HPA supported on various metal oxides such as $ZrO_2$, $TiO_2$, $WO_3$, and $Nb_2O_5$ showed both esterification and transesterification simultaneously. Cs-STA/Zr-KIT6 showed highest catalytic activity toward both oleic acid and nonedible oils compared with Cs-STA/Sn-KIT6, which is due to the high acidity

(0.91 mmol $g^{-1}$) of the catalyst as incorporation of zirconium introduces additional Lewis acid sites to the catalyst. Besides, $Zr^{4+}$ is able to interact strongly with the Keggin unit of HPA, making the catalyst highly stable [66].

### 11.2.3 Biowaste-Derived Bifunctional Catalyst

Trans(esterification) of nonedible oils using heterogeneous inorganic catalyst possesses some significant drawbacks: high methanol/oil molar ratio, high reaction temperature, and time [41, 73]. Besides, the synthesis of catalysts from inorganic sources requires a complex reaction, limiting their trans(esterification) reaction [74]. The exploitation of biowaste material for the synthesis of bifunctional solid catalyst attracts immense attention due to cheap raw materials, easy availability, nontoxic, and biodegradable [75, 76]. Hence, the utilization of biowaste derived bifunctional catalysts for the trans(esterification) of nonedible oils to biodiesel makes the process cost-effective and, most importantly, solves the problem associated with waste disposal of huge quantities. Different biowaste-derived bifunctional catalysts employed for biodiesel production are discussed in this section (Table 11.3).

Akhabue et al. [77] reported corncob-based bifunctional solid catalyst for the trans(esterification) of neem seed oil (NSO) to produce biodiesel. The preparation of the catalyst follows three-step process: (i) the corncobs are dried under the sun, grounded to powder, carbonized incompletely at 280 °C followed by treated with sulfuric acid to form sulfonated corncob (SCC); (ii) dried poultry dropping was carbonized at 280 °C, treated with 0.1 M KOH followed by calcined at 900 °C for 6 h (KPD); (iii) finally, the SCC and KOH impregnated poultry dropping was mixed to form the catalyst (SCC-KPD). The catalyst showed excellent catalytic activity toward NSO trans(esterification) with biodiesel yield of 92.89% under the optimized reaction conditions. The reusability investigation of catalyst showed 89% biodiesel yield after two successive runs. Similarly, KOH impregnated SCC (KOH/corncob-activated carbon (AC)) was employed for the trans(esterification) of WCO having high FFA content for biodiesel production, achieving 97.8% yield under the optimized reaction conditions. The catalyst prepared with corncob/$H_2SO_4$ (weight/volume) ratio of 5 : 1, 600 °C temperature, and 1 h time is considered as the best one among all AC prepared as it possesses high surface area (627 $m^2$ $g^{-1}$). Moreover, the catalyst can be reused

**Table 11.3** Different biowaste-derived solid bifunctional catalysts for biodiesel production.

| Entry | Catalyst | Feedstock | Reaction conditions[a] | Yield (%) | References |
|---|---|---|---|---|---|
| 1 | SCC-KPD | NSO | 15 : 1. 2.6, 1.2, 62 | 92.9 | Akhabue et al. [77] |
| 2 | KOH/corncob-AC | WCO | 18 : 1, 1, 1, 45 | 97.8 | Naeem et al. [78] |
| 3 | RHC/KOH/Fe | WCO | 12 : 1, 4, 4, 75 | 98.6 | Hazmi et al. [79] |
| 4 | RHC-$K_2O$-Ni | WCO | 12 : 1, 4, 2, 65 | 98.2 | Hazmi et al. [80] |
| 5 | PKSAC-$K_2CO_3$-CuO | WCO | 12 : 1, 5, 4, 80 | 95 | Abdullah et al. [81] |
| 6 | PKSHAC-$K_2CO_3$-CuO | WCO | 12 : 1, 4, 2, 70 | 95.4 | Abdullah et al. [82] |
| 7 | CAWS-$SO_4$ | PFAD | 15 : 1, 5, 3, 80 | 97.4 | Nur et al. [83] |

[a] Methanol/oil molar ratio, catalyst loading (wt%), reaction time (h), reaction temperature (°C).

up to fifth consecutive cycles for the esterification reaction; however, it is inefficient for repeated transesterification cycles due to the easy leaching of KOH sites [78]. Rice husk (RH) was also utilized for the synthesis of solid catalyst for biodiesel production, where the RH was carbonized to form rice husk biochar (RHC) and then modified the latter with different amount of KOH and Fe, generating magnetic nano-bifunctional catalyst (RHC/KOH-Fe) and employed for the trans(esterification) of WCO. Among all the synthesized catalyst, RHC loaded with 20% KOH and 5% Fe (RHC/KOH20%-Fe5%) possesses the highest basicity (4.43 mmol $g^{-1}$), moderate surface area (57.89 $m^2$ $g^{-1}$), and acidity (24.59 mmol $g^{-1}$), thus showing great activity (98.6%) than other catalysts as a significant portion of biodiesel yield is dependent on the basic sites of the catalyst. The catalyst showed high stability and can be reused up to five consecutive reaction cycles without considerable catalyst activity loss. The catalyst's high reusability is due to the magnetic nature of the catalyst, which allows easy separation of the catalyst from the reaction mixture [79]. Similarly, another RH char-based magnetic bifunctional catalyst was prepared by loading with different amounts of $K_2O$ and NiO and employed for the trans(esterification) of WCO. FESEM micrographs (Figure 11.7) of the synthesized catalysts displayed an irregular shape, attributed to the aggregation of potassium and nickel on the carbon framework. It is observed that the pore size of the carbon framework increases in nickel amount due to the crystal growth. RHC/$K_2O$-20%/Ni-5% showed the highest biodiesel yield among synthesized catalyst series, which could be due to the moderate acidity (3.014 mmol $g^{-1}$), basicity (4.485 mmol $g^{-1}$), and surface area (32.50 $m^2$ $g^{-1}$) [80].

Another biowaste-derived solid bifunctional catalyst was prepared from waste palm kernel shell (PKS) and employed for the biodiesel production from WCO. The catalyst preparation process involves the calcination of the waste palm kernel shell to AC, followed by the loading of different amounts of $K_2CO_3$ and CuO to provide surface basicity and acidity, respectively. Among all the synthesized PKS-based catalyst, PKSAC-$K_2CO_3$(30%)CuO(5%) showed the highest activity with a biodiesel yield of 95%, which could be due to the highest basicity (8.86 mmol $g^{-1}$) and moderate acidity (27.02 mmol $g^{-1}$) and surface area (438.08 $m^2$ $g^{-1}$). Reusability test of the catalyst revealed that the catalyst's basicity was decreased to 3.11 mmol $g^{-1}$ after the fifth cycle; as a consequence the FAME yield also decreased to $<80\%$, suggesting that the catalyst activity is primarily dependent on the basicity of the catalyst [81]. In another study, PKS was carbonized hydrothermally to generate corresponding AC (PKSHAC), which was then treated with NaOH to increase the surface area from 3.57 to 3368.60 $m^2$ $g^{-1}$ followed by incorporation of different amount of $K_2CO_3$

**Figure 11.7** FESEM micrographs of (a) RHC/K2O-20%/Ni-1%, (b) RHC/K2O-20%/Ni-5%, and (c) RHC/K2O-20%/Ni-10%. *Source:* Ref. [80].

and $Cu(NO_3)_2$ to introduce surface acidity and basicity. The maximum FAME yield of 95.4% was achieved by using PKSHAC-$K_2CO_3$(20%)CuO(5%) catalyst, which could be due to the high surface area of the catalyst [82]. Recently, angel wing shell (AWS)-based solid bifunctional catalyst was synthesized by a two-step process: (i) calcination of dried AWS at 900 °C for 2 h to convert into CaO (CAWS) and (ii) sulfonation with different amount of sulfuric acid to introduce an acid character to the CAWS, termed as CAWS-$SO_4$ and employed for the trans(esterification) of palm fatty acid distillate (PFAD) to produce biodiesel. The sulfuric acid concentration varied from 3 to 11 M to study its effect on the catalyst's acidity and basicity. An increase in sulfuric acid concentration results increase in acidity of the catalyst as the acidity of CAWS-$SO_4$-5M and CAWS-$SO_4$-7M was found to be 2.14 and 4.73 mmol $g^{-1}$, respectively, as high amount of sulfuric acid concentration introduces high amount of sulfonic group. However, further increase in sulfuric acid concentration leads to the decrease in acidity of the catalyst as it damaged the catalyst surface. On the other hand, the basicity of the catalyst is highest (1.33 mmol $g^{-1}$) for CAWS-$SO_4$-3M due to the low concentration of sulfuric acid, and it is decreased subsequently with increase in sulfuric acid concentration. The maximum FAME yield of 97.4% was achieved by using CAWS-$SO_4$-7M catalyst, could be due to the high catalyst acidity, indicating that the catalyst activity for esterification reaction primarily depends on the acid sites present [83].

## 11.3 Summary and Concluding Remarks

Solid catalysts have validated brilliant prospects in biodiesel synthesis from edible and nonedible oils. Compared with a monofunctionalized solid catalyst, bifunctionalized solid catalyst improved the selectivity of the desired products generation and facilitates the catalytic transformation and product/catalyst extraction in one-pot process. This book chapter overviews the advances within the catalytic tactics to synthesize biodiesel from renewable biomass by employing solid bifunctional catalysts with acid–base and Lewis–Brønsted functionalities. The critical findings from this overview are summarized as follows:

1) It is highly desirable to make cheap, stable, environmentally benign, and efficient solid catalyst having both acid–base and dual acid characteristics for biodiesel synthesis.
2) Bifunctional catalyst is considered an efficient catalyst for biodiesel production from nonedible oils having high FFA content compared with the single functionality solid catalyst, where a tedious and two-step process is involved.
3) Introduction of surface hydrophobicity into the solid catalyst facilitates FAME anti-hydrolysis activity and restricts the water molecule to interact with the catalyst active sites, thus improving the catalyst stability and reusability.
4) Incorporating magnetic behavior to the solid bifunctional catalyst enhances the catalyst durability and reusability due to the accessibility of easy separation of the catalyst from the reaction mixture.
5) The biowaste-derived bifunctional solid catalyst for biodiesel production is a growing area as it is readily available, renewable, nontoxic, and biodegradable, making the process cost effective. Besides, it also solves the waste disposal problems.

## Acknowledgment

Science and Engineering Research Board (SERB), India, is thankfully acknowledged for the research fund (Grant No. SB/FT/CS-103/2013 and SB/EMEQ-076/2014).

## References


1. Mahmudul, H.M., Hagos, F.Y., Mamat, R. et al. (2017). *Renew. Sustain. Energy Rev. 72*: 497–509.
2. Changmai, B., Vanlalveni, C., Ingle, A.P. et al. (2020). *RSC Adv. 10*: 41625–41679.
3. Li, H., Fang, Z., Smith, R.L., and Yang, S. (2016). *Prog. Energy Combust. Sci. 55*: 98–194.
4. Bajwa, D.S., Peterson, T., Sharma, N. et al. (2018). *Renew. Sustain. Energy Rev. 96*: 296–305.
5. Lee, A.F., Bennett, J.A., Manayil, J.C., and Wilson, K. (2014). *Chem. Soc. Rev. 43*: 7887–7916.
6. Changmai, B., Rano, R., Vanlalveni, C., and Rokhum, L. (2021). *Fuel 286*: 119447.
7. Shahbaz, M., yusup, S., Inayat, A. et al. (2017). *Renew. Sustain. Energy Rev. 241*: 284–295.
8. Su, F. and Guo, Y. (2014). *Green Chem. 16*: 2934–2957.
9. Atadashi, I.M., Aroua, M.K., Abdul Aziz, A.R., and Sulaiman, N.M.N. (2013). *J. Ind. Eng. Chem. 19*: 14–26.
10. Pathak, G., Das, D., Rajkumari, K., and Rokhum, L. (2018). *Green Chem. 20*: 2365–2373.
11. Kouzu, M. and Hidaka, J.S. (2012). *Fuel 93*: 1–12.
12. Changmai, B., Sudarsanam, P., and Rokhum, L. (2019). *Ind. Crops Prod. 145*: 111911.
13. Sani, Y.M., Daud, W.M.A.W., and Abdul Aziz, A.R. (2014). *Appl. Catal. A Gen. 470*: 140–161.
14. Teo, S.H., Goto, M., and Taufiq-Yap, Y.H. (2015). *J. Supercrit. Fluids. 104*: 243–250.
15. Rajkumari, K. and Rokhum, L. (2020). *Biomass Conv. Biorefine. 10*: 839–848.
16. An, H., Yang, W.M., Chou, S.K., and Chua, K.J. (2012). *Appl. Energy 99*: 363–371.
17. Tatur, M., Nanjundaswamy, H., Tomazic, D. et al. (2009). *J. Fuels Lubr. 1*: 119–131.
18. Laskar, I.B., Rajkumari, K., Gupta, R. et al. (2018). *RSC Adv. 8*: 20131–20142.
19. Semwal, S., Arora, A.K., Badoni, R.P., and Tuli, D.K. (2011). *Bioresour. Technol. 102*: 2151–2161.
20. Mardhiah, H.H., Ong, H.C., Masjuki, H.H. et al. (2017). *Renew. Sustain. Energy Rev. 67*: 1225–1236.
21. Nabi, M.N. and Rasul, M.G. (2018). *Energy Convers. Manag. 169*: 326–333.
22. Atabani, A.E., Silitonga, A.S., Ong, H.C. et al. (2013). *Renew. Sustain. Energy Rev. 61*: 480–483.
23. Banković-Ilić, I.B., Stamenković, O.S., and Veljković, V.B. (2012). *Renew. Sustain. Energy Rev. 16*: 3621–3647.
24. Avhad, M.R. and Marchetti, J.M. (2015). *Renew. Sustain. Energy Rev. 50*: 696–718.
25. Nurfitri, I., Pragas, G., Hindryawati, N., and Yusoff, M.M. (2013). *Energy Convers. Manag. 74*: 395–402.
26. Mansir, N., Teo, S.H., Rashid, U. et al. (2018). *Renew. Sustain. Energy Rev. 82*: 3645–3655.
27. Zhang, H., Li, H., Hu, Y. et al. (2019). *Renew. Sustain. Energy Rev. 114*: 109296.

28 López, D.E., Suwannakarn, K., Bruce, D.A., and Goodwin, J.G. (2007). *J. Catal. 247*: 43–50.
29 Di Serio, M., Tesser, R., Dimiccoli, M. et al. (2005). *J. Mol. Catal. A Chem. 22*: 207–217.
30 Galadima, A. and Muraza, O. (2014). *Energy 78*: 72–83.
31 Boey, P.L., Maniam, G.P., and Hamid, S.A. (2011). *Chem. Eng. J. 88*: 283–288.
32 Chua, S.Y., Periasamy, L.A., Goh, C.M.H. et al. (2020). *J. Ind. Eng. Chem. 118*: 626–636.
33 Said, N.H., Ani, F.N., and Said, M.F.M. (2015). *J. Mech. Eng. Sci.* 1302–1311.
34 Dhawane, S.H., Kumar, T., and Halder, G. (2018). *Energy Convers. Manag. 167*: 176–202.
35 Demirbas, A., Bafail, A., Ahmad, W., and Sheikh, M. (2016). *Energy Explor. Exploit. 34*: 290–318.
36 Endalew, A.K., Kiros, Y., and Zanzi, R. (2011). *Energy 36*: 2693–2700.
37 Baskar, G. and Aiswarya, R. (2016). *Renew. Sustain. Energy Rev. 57*: 496–504.
38 Lokman, I.M., Rashid, U., and Taufiq-Yap, Y.H. (2015). *Chinese J. Chem. Eng. 23*: 1857–1864.
39 Lokman, I.M., Rashid, U., Taufiq-Yap, Y.H., and Yunus, R. (2015). *Renew. Energy 81*: 347–354.
40 Russbueldt, B.M.E. and Hoelderich, W.F. (2010). *J. Catal. 271*: 290–304.
41 Yan, S., Salley, S.O., and Simon Ng, K.Y. (2009). *Appl. Catal. A Gen. 353*: 203–212.
42 Feyzi, M., Hosseini, N., Yaghobi, N., and Ezzati, R. (2017). *Chem. Phys. Lett.* 677: 19–29.
43 Nizah, M.F.R., Taufiq-yap, Y.H., Rashid, U. et al. (2014). *Energy Convers. Manag. 88*: 1257–1262.
44 Liu, X., He, H., Wang, Y. et al. (2008). *Fuel 87*: 216–221.
45 Ngamcharussrivichai, C., Totarat, P., and Bunyakiat, K. (2008). *Appl. Catal. A Gen. 341*: 77–85.
46 Lee, H.V., Juan, J.C., and Taufiq-Yap, Y.H. (2015). *Renew. Energy 74*: 124–132.
47 Elias, S., Rabiu, A.M., Okeleye, B.I. et al. (2020). *Appl. Sci. 10*: 3153.
48 Xie, W. and Zhao, L. (2014). *Energy Convers. Manag. 79*: 34–42.
49 Jamil, F., Al-muhatseb, A.H., Tay, M. et al. (2018). *Energy Convers. Manag. 155*: 128–137.
50 Su, M., Yang, R., and Li, M. (2013). *Fuel 103*: 398–407.
51 Al-Saadi, A., Mathan, B., and He, Y. (2020). *Chem. Eng. Res. Des. 162*: 238–248.
52 Dehghani, S. and Haghighi, M. (2020). *Renew. Energy 153*: 801–812.
53 Pirouzmand, M., Mahdavi, M., and Ghasemi, Z. (2018). *Fuel 216*: 296–300.
54 Macario, A., Giordano, G., Onida, B. et al. (2010). *Appl. Catal. A Gen. 378*: 160–168.
55 Loy, A.C.M., Quitain, A.T., Lam, M.K. et al. (2019). *Energy Convers. Manag. 180*: 1013–1025.
56 Wang, X., Zhao, Q., Wang, H. et al. (2013). *Catal. Sci. Technol. 3*: 2204–2209.
57 Wang, Y.T., Fang, Z., Yang, X.X. et al. (2018). *Chem. Eng. J. 348*: 929–939.
58 de Almeida, R.M., Noda, L.K., Gonçalves, N.S. et al. (2008). *Appl. Catal. A Gen. 347*: 100–105.
59 Kaur, N. and Ali, A. (2015). *Renew. Energy 81*: 421–431.
60 Gardy, J., Osatiashtiani, A., Céspedes, O. et al. (2018). *Appl. Catal. B Environ. 234*: 268–278.
61 Silva, Â., Wilson, K., Lee, A.F. et al. (2017). *Appl. Catal. B Environ. 205*: 498–504.
62 Kaur, M., Malhotra, R., and Ali, A. (2018). *Renew. Energy 116*: 109–119.
63 Nikseresht, A., Daniyali, A., Ali-Mohammadi, M. et al. (2017). *Ultrason. Sonochem. 37*: 203–207.
64 Zhao, J., Guan, H., Shi, W. et al. (2012). *Catal. Commun. 20*: 103–106.
65 Tao, M., Xue, L., Sun, Z. et al. (2015). *Sci. Rep. 5*: 1–10.
66 Gopinath, S., Vinoth Kumar, P., Sahaya Murphin Kumar, P. et al. (2018). *Fuel 234*: 824–835.

**67** Romanelli, G. and Autino, J. (2009). *Mini. Rev. Org. Chem. 6*: 359–366.

**68** Li, J., Yang, Z., Li, S. et al. (2020). *J. Ind. Eng. Chem. 82*: 1–16.

**69** Da Silva, M.J. and Liberto, N.A. (2015). *Curr. Org. Chem.* https://doi.org/10.2174/1385272819666150907193100.

**70** Misono, M. (2009). *Catal. Today 144*: 285–291.

**71** Kozhevnikov, I.V. (2007). *J. Mol. Catal. A Chem. 263*: 112–114.

**72** Mai, C.T.Q., Ng, F.T.T., and Rempel, G.L. (2018). *Catalysis 30*: 176–222.

**73** Nisar, J., Razaq, R., Farooq, M. et al. (2017). *Renew. Energy 101*: 111–119.

**74** Chaveanghong, S., Smith, S.M., Smith, C.B. et al. (2018). *Renew. Energy 126*: 156–162.

**75** He, D., Zhao, W., Li, P. et al. (2019). *Appl. Surf. Sci. 465*: 303–312.

**76** Han, F., Yun, S., Shi, J. et al. (2020). *Appl. Catal. B Environ.* 273: 119004–119019.

**77** Akhabue, C.E., Osa-Benedict, E.O., Oyedoh, E.A., and Otoikhian, S.K. (2020). *Renew. Energy 152*: 724–735.

**78** Naeem, M.M., Al-Sakkari, E.G., Boffito, D.C. et al. (2021). *Fuel 283*: 118914.

**79** Hazmi, B., Rashid, U., Taufiq-Yap, Y.H. et al. (2020). *Catalysts 10*: 225–246.

**80** Hazmi, B., Rashid, U., Ibrahim, M.L. et al. (2021). *Environ. Technol. Innov. 21*: 101296.

**81** Abdullah, R.F., Rashid, U., Taufiq-Yap, Y.H. et al. (2020). *RSC Adv. 10*: 27183–27193.

**82** Abdullah, R.F., Rashid, U., Ibrahim, M.L. et al. (2021). *Renew. Sustain. Energy Rev. 137*: 110638.

**83** Nur, O., Rashid, U., and Sufri, M. (2019). *Renew. Energy 131*: 187–196.

# 12

# Application of Catalysts Derived from Renewable Resources in Production of Biodiesel

*Kanokwan Ngaosuwan[1], Apiluck Eiad-ua[2], Atthapon Srifa[3], Worapon Kiatkittipong[4], Weerinda Appamana[5], Doonyapong Wongsawaeng[6], Armando T. Quitain[7,8], and Suttichai Assabumrungrat[9,10]*

[1] *Division of Chemical Engineering, Faculty of Engineering, Rajamangala University of Technology Krungthep, Bangkok, Thailand*
[2] *College of Nanotechnology, King Mongkut's Institute of Technology Ladkrabang, Bangkok, Thailand*
[3] *Department of Chemical Engineering, Faculty of Engineering, Mahidol University, Nakhon Pathom, Thailand*
[4] *Department of Chemical Engineering, Faculty of Engineering and Industrial Technology, Silpakorn University, Nakhon Pathom, Thailand*
[5] *Department of Chemical and Materials Engineering, Faculty of Engineering, Rajamangala University of Technology Thanyaburi, Pathum Thani, Thailand*
[6] *Department of Nuclear Engineering, Faculty of Engineering, Chulalongkorn University, Bangkok, Thailand*
[7] *Faculty of Advanced Science and Technology, Kumamoto University, Kumamoto, Japan*
[8] *Center for International Education, Kumamoto University, Kumamoto, Japan*
[9] *Center of Excellence in Catalysis and Catalytic Reaction Engineering, Department of Chemical Engineering, Faculty of Engineering, Chulalongkorn University, Bangkok, Thailand*
[10] *Bio-Circular-Green-economy Technology & Engineering Center, BCGeTEC, Department of Chemical Engineering, Faculty of Engineering, Chulalongkorn University, Bangkok, Thailand*

## 12.1 Introduction

Recently, several heterogeneous catalysts have been successfully employed to replace homogeneous catalysts for biodiesel production as they offer several benefits such as easy separation, potential catalyst reuse, and significant reduction of water consumption especially in the purification step, thus enabling the process to be more environmentally friendly [1]. The most common heterogeneous catalysts for biodiesel production can be divided into two types: i.e. solid base and solid acid catalysts. Solid base catalysts or alkali catalysts such as calcium oxide (CaO), magnesium oxide (MgO), and strontium oxide (SrO) have been widely used to produce biodiesel via transesterification because they offer higher transesterification rate, mild reaction conditions, easy catalyst separation from the reaction mixture, and possible reusability and regeneration [2]. However, solid base catalysts face challenges such as the sensitivity to free fatty acid (FFA) content in oil feedstocks, generation of soap during transesterification of high FFA feedstocks, and decrease in their catalytic activity by leaching of active sites and adsorption of moisture from the atmosphere. For solid acid catalysts such as zirconium oxide ($ZrO_2$), titanium dioxide ($TiO_2$), sulfonic

*Biodiesel Production: Feedstocks, Catalysts, and Technologies*, First Edition. Edited by Samuel Lalthazuala Rokhum, Gopinath Halder, Suttichai Assabumrungrat, and Kanokwan Ngaosuwan.

ion-exchange resins (Amberlyst-15 and Nafion-NR-50), and heteropolyacids, they could be used in the pretreatment step to reduce FFA in the oil feedstocks via esterification, offering the great performance in terms of insensitivity to FFA and simultaneous esterification and transesterification processes [3, 4]. However, these catalysts require expensive raw materials and tedious synthetic procedures [5, 6]. Catalysts derived from renewable resources such as animals, plants, natural materials, and industrial wastes have become increasingly prevalent alternative sources. These could promote more economically feasible and sustainable biodiesel production process by reducing the catalyst cost and the amount of solid waste, as well as offering completion of $CO_2$ cycle.

The following provides a literature survey of catalysts derived from various renewable resources being used for biodiesel production, taking into consideration their compositions and characteristics (surface area, pore size, and basicity), preparation methods, and catalytic performance. Finally, advantages, disadvantages, and challenges of these types of catalysts are also discussed.

## 12.2 Potential Renewable Resources for Production of Biodiesel Catalysts

Many researchers have utilized various renewable resources such as shells, biomass, natural materials, and other wastes for biodiesel reaction catalysis. The following provides critical discussion on heterogeneous catalysts derived from different renewable resources under four categories: (i) animal-based, (ii) plant-based, (iii) natural-based, and (iv) industrial waste resources.

### 12.2.1 Animal Resources

Recently, calcium oxides (CaO) derived from bioresources have been extensively utilized as catalysts for biodiesel production due to their basicity [7]. CaO can be widely produced from the low-cost animal resources such as shells and bones via the calcination process under an air atmosphere in a temperature range of 500–1100 °C [7–9]. In the preparation process, the calcium carbonate ($CaCO_3$) in animal sources could be transformed into CaO with the release of $CO_2$ by calcination process at atmospheric pressure [10]. The utilization of CaO derived from animal resources does not only have economic and environmental impacts but also reduces the abundant agricultural and food wastes. **Figure 12.1** illustrates the process for the transformation of animal resources as the natural catalysts for biodiesel production including the calcination at high temperature and ranges of investigated

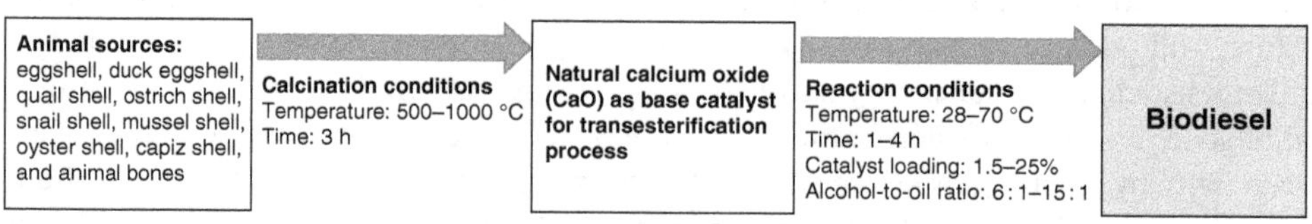


**Figure 12.1** Process for transformation of animal sources as natural catalysts for biodiesel production.

conditions to catalyze transesterification of oil feedstocks. In addition, the catalyst preparation method, reaction conditions, and biodiesel yield reported in the literatures are summarized in **Table 12.1**.

#### 12.2.1.1 Eggshells (Chicken and Ostrich)

Chicken and ostrich eggshells have been extensively investigated and employed for the transesterification process due to their availability in the agricultural and food industries. Several advantages obtained from using natural CaO catalysts are high yield of biodiesel, low catalyst leaching, simplicity of separation process, suitable basicity for biodiesel production, and nontoxicity of the catalysts [11–13, 17]. An activity of natural CaO catalysts derived from animal resources showed the complete triglyceride and/or fatty acids conversion with >90% yield at 60 °C under MeOH-to-oil molar ratio in a range of 6 : 1–12 : 1 [7–9, 11]. The CaO synthesized from waste chicken eggshell offered the highest biodiesel yield (>95%) via palm oil transesterification at 60 °C and 1 : 12 of MeOH-to-oil ratio within 90 min [11]. This was due to a high Ca content, large catalyst surface area, and small particle size after calcination at 800 °C for 2–4 h. Similarly, the catalyst synthesized from a low-cost waste chicken eggshell calcined in a range of 500–1000 °C for 3 h was employed for biodiesel production [8]. The optimized yield of more than 92% was obtained at reaction temperature and time of 57.5 °C and 3 h, respectively, and the properties of biodiesel were in the ranges of ASTM standard. Alternatively, the catalyst derived from daily food waste ostrich and chicken eggshells was calcined at 1000 °C for 4 h [7, 9]. It was found that the highest yields of biodiesel via combined transesterification and esterification of waste cooking oil were achieved at 98 and 96% for ostrich and chicken eggshells, respectively, under the same optimized conditions. This would suggest that the ostrich and chicken eggshells are suitable raw materials for producing calcium oxide (CaO) catalyst, an environmentally friendly and low-cost base oxide catalyst, for biodiesel production.

#### 12.2.1.2 Seashells (Snail, Mussel, Oyster, and Capiz)

To explore alternative bioresources, seashells (snail, mussel, oyster, and capiz) have been recently utilized for producing CaO catalyst [13–16, 18]. Laskar et al. [13] implemented the natural CaO catalyst derived from waste snail shell using calcination temperature of 800 °C for 3 h, and the catalyst was employed for a new unconventional biodiesel production using acetone as a cosolvent. The addition of acetone during the soybean oil transesterification resulted in the homogenous phases between the oil feed and cosolvent. Under the optimized conditions (28 °C, 2 h, a catalyst loading of 20 wt%, and a MeOH-to-oil molar ratio of 6 : 1), the highest biodiesel yield was achieved at 98% without catalyst degradation up to 11 cycles. Likewise, a mussel shell was selected for a CaO synthesis [14, 15]. A honeycomb-structured catalyst with high surface area was obtained after calcination of waste freshwater mussel shell at 900 °C for 4 h [14]. The transesterification of tallow oil at the reaction temperature of 70 °C, time of 1.5 h, a MeOH-to-oil molar ratio of 12 : 1, and of 5 wt% catalyst exhibited the yield of above 90% without catalyst deactivation after seven cycles. Recently, *Turel Kongreng* (river mussels) was calcined at 800 °C for 1 h; the properties of as-synthesized CaO were similar to those of the CaO samples prepared from the other sources, and the yield of 91% was obtained at 75 °C and 1 h [18]. Likewise, Yuliana et al. [16] utilized the waste capiz shell calcined at 900 °C for 2 h for transesterification of leather

**Table 12.1** Summary of the production of biodiesel using catalysts from different animal resources.

| | | Reaction conditions | | | | | | | |
|---|---|---|---|---|---|---|---|---|---|
| Catalyst resource | Calcination condition | Oil feedstock | Temperature (°C) | Time (h) | Catalyst loading (wt%) | Alcohol to oil | Biodiesel yield (%) | Reusability | References |
| Chicken eggshell | 800 °C, 2–4 h | Palm olein oil | 60 | 1.5 | 10 | 12 : 1 | >95 | n/a | [11] |
| Chicken eggshell | 500–1000 °C, 3 h | Soybean oil | 57.5 | 3 | 7 | 10 : 1 | >92 | 17.1% yield loss after fifth cycle | [8] |
| Ostrich and chicken eggshells | 1000 °C, 4 h | Waste cooking oil | 65 | 2 | 1.5 | 10 : 1 | 96–98 | 14% yield loss after fifth cycle | [7, 9] |
| Oyster shell | 700 °C, 3 h | Soybean oil | 65 | 5 | 25 | 6 : 1 | 95 | n/a | [12] |
| Waste snail shell | 800 °C, 3 h | Soybean oil | 28 | 2 | 20 | 6 : 1 | 98 | 29.6% yield loss after $12^{th}$ cycle | [13] |
| Freshwater mussel shell | 900 °C, 4 h | Tallow oil | 70 | 1.5 | 5 | 12 : 1 | >90 | 36.8% yield loss after $16^{th}$ cycle | [14] |
| Waste oyster shells | 1000 °C, 2 h | Waste cooking oil | 65 | 3 | 3–7 | 9 : 1 | 87.3 | n/a | [15] |
| Waste capiz shell | 900 °C, 2 h | Leather tanning waste | 60 | 4 | 3 | 6 : 1 | 93.4 | 33.7% yield loss after fourth cycle | [16] |
| Chicken bones | 900 °C, 4 h | Waste cooking oil | 65 | 4 | 5 | 15 : 1 | 89.3 | 33.3% yield loss after sixth cycle | [17] |
| River mussels | 800 °C, 1 h | Waste cooking oil | 75 | 1 | 10 | 10 : 1 | 91 | n/a | [18] |
| Chicken and fish bones | 1000 °C, 4 h | Waste cooking oil | 65 | 1.54 | 1.98 | 10 : 1 | 89.5 | 39.0% yield loss after fifth cycle | [19] |

tanning waste. The maximum yield was 93.4% at 60 °C, 5 h, an EtOH-to-oil feed molar ratio of 6 : 1 and a CaO loading of 3 wt%. Furthermore, Nakatani et al. [12] and Lin et al. [15] prepared the CaO by calcination of waste oyster shells at the temperature >700 °C. The biodiesel yield significantly increased during 2–3 h of reaction time and that the maximum yield was 87.3% at 65 °C and 3 h [15]. By using the full factorial design for soybean oil transesterification over CaO, the catalyst loading and reaction time were found to be 25% and 5 h, respectively, with a biodiesel yield of 98.4% [12]. It can be emphasized that the biowaste was appropriate for valuable CaO production due to its economic and environmentally friendly advantages.

#### 12.2.1.3 Bones

A hydroxyapatite ($Ca_5(PO_4)_3(OH)$) compound containing in waste animal hard tissues such as bones and quail beaks can be employed as materials for producing the biodiesel catalysts [17, 19, 20]. Farooq et al. [17] developed the catalyst by calcination of waste chicken bones at 900 °C and the phases of $Ca(OH)_2$ and CaO were detected. The maximum yield from the transesterification of waste cooking oil was obtained at 89.3% at 65 °C, 4 h, a MeOH-to-oil molar ratio of 15 : 1, and a 5 wt% catalyst. Likewise, an equal mass fraction of chicken bones and fish bones was calcined at 1000 °C for 4 h to generate calcium oxide [19]. The biodiesel yield at 89.5% was obtained at 65 °C, 1.54 h, 1 : 10 of an alcohol-to-oil molar ratio, and a 1.98% w/v of catalyst/oil. Alternatively, waste quail beaks were recently considered as a renewable source for synthesizing the natural hydroxyapatite catalyst [20]. A high crystallinity of hydroxyapatite was detected in addition to the functional groups of phosphate ($PO_4^{-3}$) and hydroxyl ($OH^-$) after calcination at 900 °C. The calcined hydroxyapatite catalyst gave the highest biodiesel yield of 96.7% at 65 °C and 4 h via transesterification of rapeseed oil.

### 12.2.2 Plant Resources

Plant residues are found in large quantities after various agricultural activities, such as harvesting and process agricultural product as postharvestings. Applications of catalytic materials from waste plant sources have recently gained a great deal of interest. The residues were explored for biodiesel production due to a significant amount of alkali or alkaline earth elements found in the burned ash from the plant resources. Normally, plant residues are mostly composed of carbon structure and silica, which can be transformed into high-performance carbon and silica materials, respectively. In recent years, many researchers have used low-cost biomass by transforming it through various post-handling processes into high-quality and value-added products. In this section, only the carbon-supported, silica-supported catalysts, and other compounds are summarized. **Figure 12.2** shows the production pathway, while **Table 12.2** shows biodiesel production using plant-derived catalysts.

#### 12.2.2.1 Carbon-Supported Catalysts

Activated carbon (AC) can generally be derived from the thermal decomposition of carbon-rich feedstock, such as coconut shells, wood, and palm shells. The commonly used methods of carbonization to prepare a carbon-based catalyst are pyrolysis and

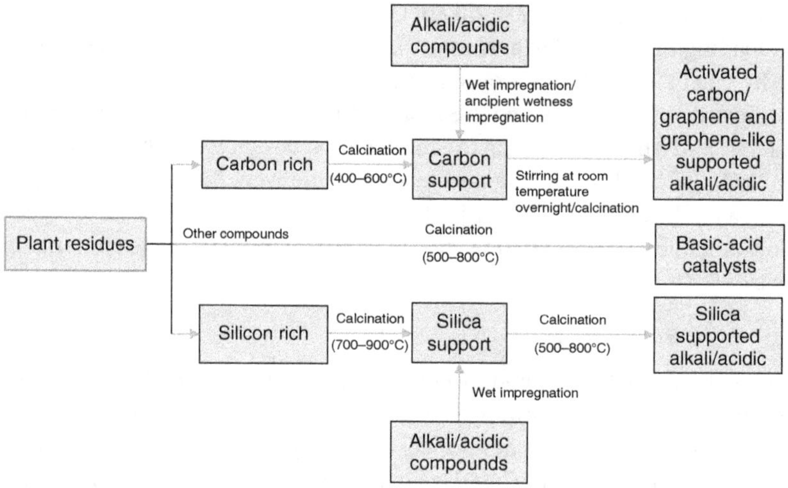


**Figure 12.2** Production pathways for various plants-derived alkali catalysts. *Source:* Modified from Ref. [1].

hydrothermal carbonization. Pyrolysis is mostly carried out in an inert atmosphere at a temperature range of 200–700 °C. As a result, high surface area carbon with large pore sizes, also having cracks and crevices, is obtained [21]. The AC-supported alkali (base) catalyst can be used for biodiesel production. For example, Baroutian et al. [22] impregnated AC derived from palm shell with potassium hydroxide (KOH). The transesterification was carried out under a catalyst loading of 30.3 wt%, a MeOH-to-oil molar ratio of 24 : 1, and reaction time of 1 h, achieving overall fatty acid methyl ester (FAME) yield of 98.03%. Furthermore, the catalyst prepared from sulfonation of a carbonized carbon material has captured the attention. Since it is stable, inexpensive, safer, and provides easier synthesis routes, the sulfonated carbon-based catalyst (SCBC) showed promising potential. In general, the temperature and time of carbonization including the temperature of sulfonation greatly affected on the textural properties of the prepared catalyst [23]. Using high temperatures that hinder the anchoring of sulfonic (-$SO_3H$) groups, the rigid carbon structure can be formed, thus reducing the density of active acid sites on the prepared catalyst [23, 24]. However, carbon catalyst synthesized at a lower pyrolysis temperature (400–500 °C) results in soft aggregated, cross-linked polymer generation that increases the active acid sites on the prepared catalyst [24]. In the synthesis of acid biochar used as a catalyst in the manufacture of biodiesel from jupati oil, Bastos et al. [25] utilized murumuru kernel shell waste as the precursor biomass. After the carbonization of the murumuru kernel shell, the catalyst was sulfonated in concentrated sulfuric acid. After the fourth reaction cycle, biodiesel yield dropped for about 13%. The progressive loss in the catalytic activity was little due to the leaching of acid sites.

Graphene and graphene-like carbon from biomass are interesting choices as nanoparticle support materials because of their high surface area, porous structure, high graphitization degree, and chemically stable surface by controlling the preparation method [26]. Graphene and graphene-like carbon show great performance in supercapacitors, catalysis, and other related applications. Graphene oxide is a graphitic one-layer product attached with many hydrophilic functional groups such as epoxy (C–O–C), carboxyl (C = O), and

**Table 12.2** Summary of the production of biodiesel using catalysts from different plant resources.

| | | | Reaction conditions | | | | | | | |
|---|---|---|---|---|---|---|---|---|---|---|
| **Support** | **Catalyst resource** | **Catalyst preparation** | **Oil feedstock** | **Temperature (°C)** | **Alcohol-to-oil molar ratio** | **Time (h)** | **Catalyst loading (wt%)** | **Biodiesel yield (%)** | **Reusability** | **References** |
| Carbon-supported catalysts | Palm kernel shell | Calcined at 600 °C, 1 h under $N_2$ and loaded with sulfuric acid | Jupati oil | 135 | 30 | 4 | 6% | 91.8 | 13% yield loss after fourth cycle | [25] |
| | Oil palm trunk | Pyrolysed at 400 °C, 4 h, then subjected to direct sulfonation with $H_2SO_4$ at 150 °C for 4 h | Palmitic acid | | 18 | 5 | 9 | 88.2 | 10.5–12.5% yield loss after sixth cycle | [37] |
| | Cork biochar | Calcined at 600 °C, 2 h, and activated with conc. $H_2SO_4$ | Waste cooking oil | 65 | 25 | 6 | 1.5 | 98.0 | ~24% yield loss after fifth cycle | [38] |
| | Waste sesamum indicum plants | Calcined at 550 °C, 2 h, under air flow | Sunflower oil | 65 | 12 | 0.67 | 7 | 98.9 | ~4% conversion loss after third cycle | [36] |
| | Waste brassica nigraplant | Calcined at 550 °C, 2 h, under air flow | Soybean oil | 65 | 12 | 0.42 | 7 | 98.7 | ~2% conversion loss after third cycle | [39] |
| Silica-supported catalysts | Rice husk biochar | Calcined at 700 °C, 3 h, under $N_2$. Then, RB-supported CaO was synthesized using a wet impregnation method | Vegetable oil | 65 | 9 | 3 | 8 | 93.4 | ~10% yield loss after sixth cycle | [35] |
| | Rice husk ash | Loaded with alkali metals silicate, then calcined at 500 °C for 3 h | Waste cooking oil | 65 | 9 | 3 | 3 | 98.2 | — | [34] |

hydroxyl (O–H) that can be used effectively as an acid catalyst in the biodiesel production process [27–29]. In addition, the graphene composite provides Brønsted acid sites for biodiesel production [27]. Ali et al. [30] reported that the novel graphene oxide/bentonite (GO-to-NaOH-bentonite) bifunctional heterogeneous catalyst showed excellent catalytic performance for simultaneous esterification and transesterification.

#### 12.2.2.2 Silica-Supported Catalysts

Normally, as a support material, silica is a regular preference as it has a high surface area and is commonly considered inert. $SiO_2$ also has the advantage of being abundant and inexpensive with silica produced using a conventional method [31]. Plant residues such as rice husk (silica content ~ 87–99%) are the main materials for silica support [32, 33]. Hindryawati et al. [34] reported a large potential of $SiO_2$ prepared from rice husk as a catalyst for the production of biodiesel. Three alkali metals (Li, Na, and K) were loaded and then calcined at 200–700 °C for 3 h. The resulting catalyst yielded biodiesel of 98.2% under optimum conditions as lithium silicate calcination temperature of 500 °C.

In addition, Zhao et al. [35] reported high efficiency and stability of Ca/Si composite catalysts from rice husk biochar for biodiesel production. The results showed that the behavior of the catalyst activity depended primarily on their pretreatment temperature, structure, and basicity. The biodiesel yield could reach 93.4% with the calcination temperature of 700 °C and a calcium to silica weight ratio (Ca:Si) of 3 : 1. In addition, up to 10 cycles were performed in a reusability study of the catalyst without a significant decrease in catalytic activity. The primary reason for the high stability was the appearance of the Ca–O–Si bond on the catalyst surface.

#### 12.2.2.3 Other Potential Elements from Plant Residues

Recently, a heterogeneous catalyst derived from the waste *Sesamum indicum* plant was reported for sunflower oil transesterification [36]. The dry waste *S. indicum* plant was completely burnt into ash in open air, and the obtained ash was calcinated further, at 550 °C for 2 h. With the presence of Na, Mg, Fe, Mn, Zn, Si, Sr, and Cl, the characterization revealed a high percentage of K (29.64 wt%) and Ca (33.80 wt%) as oxides and carbonates. Under optimized conditions of a methanol-to-oil molar ratio of 12 : 1 and a catalyst loading of 7% at reaction temperature of 65 °C, the catalyst exhibited excellent catalytic activity, producing a yield of 98.9% biodiesel in a short reaction time of only 40 min. The catalyst could be reused with 4% of biodiesel yield loss at the third cycle of reaction.

### 12.2.3 Natural Resources

#### 12.2.3.1 Dolomitic Rock (Calcined Dolomite and Modified Dolomite)

Dolomitic rock abundant in nature is mostly composed of layers of calcium and magnesium carbonates [40]. These structures can be decomposed into CaO and MgO when calcined above 750 °C. Both CaO and MgO are recognized as great catalyst precursors in the transesterification of lipids to biodiesel [41]. Regarding the use of calcined dolomite, Wilson et al. [42] showed the utilization of calcined dolomitic rock as a high potential catalyst in the biodiesel production from olive oil. The XRD results revealed that the natural dolomitic rock consisted of 77% dolomite and 23% magnesium calcite. In the catalyst preparation, the catalyst was prepared by the calcination of raw dolomitic rock at 900 °C. The structures of magnesium calcite in the dolomitic rock were successfully decomposed into MgO and CaO particles with increased surface area compared with the natural dolomitic rock. Hence, an excellent catalytic performance in transesterification of triglycerides to biodiesel was achieved by catalyst prepared from calcination of dolomite.

Similarly, Oguzhan Ilgen et al. [43] used a calcined dolomite as heterogeneous catalysts for the biodiesel production from canola oil. They also showed that the maximum catalytic activity was obtained at 850 °C. The highest biodiesel yield was above 91%, which obtained from the condition at 60 °C for 3 h with the catalyst amount of 3 wt% using a 6 : 1 molar ratio of MeOH-to-canola oil. Jindapon et al. [44] also reported the performance of calcined dolomite for biodiesel synthesis from palm oil. They found that the dolomite treated at 800 °C gave the highest basicity and catalytic activity, obtaining FAME yield of 99 wt%. In another study, even though the dolomite was successfully applied as a heterogeneous catalyst for biodiesel production, its activity was not better than the conventional liquid catalysts. Thus, various modification methods have been explored to improve the activity of dolomite-derived catalyst. Nur et al. [45] employed the catalyst including $SnO_2$ and ZnO-supported on calcined Malaysian dolomite. In this research, the calcined dolomite catalyst revealed a superior catalytic performance obtaining a maximum oil conversion of 99.98% at 65 °C with a 15 : 1 MeOH-to-oil molar ratio within 4 h. The relatively greater basicity of catalysts and increased surface area are ascribed to its higher activity compared with the unmodified catalyst. Also, Nui et al. [46] investigated the enhancement of transesterification activity of calcined dolomite with cerium doping ($CeO_2$ species) by different impregnation techniques. The highest biodiesel yield of 97.21% was obtained at the experimental condition of catalyst/oil mass ratio of 0.05 using a MeOH-to-oil molar ratio of 15 : 1 at 65 °C for 2 h over the modified calcined dolomite catalysts.

#### 12.2.3.2 Lime

Lime or hydrated lime (an inorganic compound) is commonly comprised of calcium hydroxide ($Ca(OH)_2$), which is produced by calcination of limestone, an organic compound that forms from the accumulation of shell, coral, algal, and fecal debris. After calcination, the lime is all decomposed into CaO structure [47]. Sánchez-Cantú et al. [48] investigated a raw hydrated lime ($Ca(OH)_2$) as a catalyst in transesterification of vegetable oil to biodiesel without any treatment process. They found that 100% conversion could be reached by using soybean oil as an oil precursor, while the conversion of 98% was obtained from the transesterification of castor oil at room temperature within 14 h. Similarly, Roschat et al. [49] reported that palm oil biodiesel production yield of above 97% was obtained over the CaO catalyst derived from hydrated lime calcination. Likewise, Dias et al. [50] evaluated CaO catalyst from a natural cheap lime in the biodiesel production from semirefined rapeseed oil. They suggested that the natural limes-derived CaO catalyst exhibited a high stability even with highly acidic oil feedstocks.

#### 12.2.3.3 Natural Clays

Clays such as bentonite and kaolin are abundant in nature with some prominence as inexpensive and relatively huge surface area that could provide as a catalyst or catalyst support material for transesterification in biodiesel production. Bentonite is a natural clay formed by volcanic ash, which is a two-dimensional nanomaterial with large specific surface area and high cation-exchange capacity [51]. Hence, bentonite-supported K salt catalysts were investigated for biodiesel, recently. For instance, Boz et al. [51] used KF, KOH, and $K_2CO_3$ supported on calcium/bentonite for biodiesel production from canola oil. The authors demonstrated that the KF-loaded catalyst exhibited a better activity, with biodiesel yield of 98.2%, than the catalysts prepared by KOH and $K_2CO_3$. This can be presumed that KF-loaded catalyst has a greater basicity. Recently, Soetaredjo et al. [52] studied the performance of KOH-doped bentonite catalyst for biodiesel synthesis. Bentonite from Pacitan was loaded with potassium hydroxide by impregnation to synthesize KOH/bentonite catalysts to produce biodiesel from palm oil. The maximum biodiesel yield using bentonite-supported KOH catalyst was approximately 90.7%.

Kaolin ($Al_2Si_2O_5(OH)_4$), the most common clay in the world, is not suitably employed as a catalyst or catalyst support material in biodiesel synthesis due to low surface area and lack of active sites. Before the application in catalytic process, kaolin must be modified by some treatment process, such as acid activation and calcination. The surface area, porosity, and quantity of acid sites of kaolin was significantly enhanced after the treatment by acid activation, which is the common technique for modification of kaolin before utilization as a catalyst or catalyst support [53]. Nascimento et al. [53] explored the catalyst synthesis from different Amazon kaolins (century and flint) for biodiesel synthesis from oleic acid by calcination at 950 °C with leaching in 4 M sulfuric acid solution to obtain the activated metakaolin. They supposed that the leached metakaolin in the century clay exhibited a low aluminum content, resulting in a higher acidity (250.5 $\mu$mol $g^{-1}$) and offering larger conversion than the catalyst sample prepared from flint kaolin.

#### 12.2.3.4 Zeolites

Zeolites have been widely used in the area of heterogeneous catalysis for several reactions as a catalyst or catalyst support because of their advantages such as large surface area, moderate acidity, etc. [54]. Recently, zeolites could be synthesized from natural mineral reagents including clays, ores, rocks, and ash residues from the solid fuels burning normally that contains large amounts of oxygen, aluminum, and silicon that have similar chemical compositions to aluminosilicate zeolites. Moreover, the renewable sources for the preparation of zeolite reveal much attention for cost effectiveness for zeolite-based catalysts [55, 56]. However, the zeolites were also successfully utilized in biodiesel production since these materials exhibit high performance in catalyzing the transesterification reaction. For example, Doyle et al. [57] have explored the production of faujasite (FAU-type zeolite) using an Irish shale rock by using the hydrothermal treatment process. Furthermore, they found that the prepared FAU zeolite achieved the catalyzing model test of oleic acid esterification for biodiesel production. In another similar work, clinoptilolite, an inexpensive material, is a natural zeolite successfully modified by KOH impregnation and tested in biodiesel production from waste cooking oil using a pilot microreactor. They claimed that the biodiesel production yield of 97.45% was achieved at 65 °C with catalyst loading of 8.1 wt% using a MeOH-to-oil volume ratio of 2.25 : 1 and reaction time of 13.4 min. Results indicated that the development of microchannels significantly decreased the reaction time by the increase in the contacting time of the mixture between oil feedstock and catalysts [58]. **Table 12.3** summarizes the optimal conditions of using natural sources to produce biodiesel.

**Table 12.3** Summary of the production of biodiesel using catalysts from different natural resources.

| | | Reaction conditions | | | | | | |
|---|---|---|---|---|---|---|---|---|
| Catalyst resource | Calcination condition | Oil feedstock | Temperature (°C) | Time (h) | Catalyst loading (wt%) | Alcohol to oil | Biodiesel yield (%) | References |
| Dolomitic rock | 900 °C, 3 h | Olive oil | 60 | 3 | n/a | n/a | 100 | [42] |
| Dolomitic rock | 850 °C, 2 h | Canola oil | 60 | 3 | 3 | 6 : 1 | >90 | [43] |
| Dolomitic rock | 800 °C, 3 h | Refined palm oil | 60 | 0.3 | 5 | 1 (vol) | >99 | [44] |
| Dolomitic rock | $SnCl_2$, $ZnCl_2$ 500 °C, 3 h | Palm oil | 60 | 2 | 1.6 | 0.42 (vol) | 100 | [45] |
| Dolomitic rock | $Ce(NO_3)_3$ $9H_2O$, 800 °C, 3 h | Palm oil | 65 | 2 | 0.05 | 15 : 1 | 88 | [46] |
| Lime | 500 °C | Wasted soybean oil | 60 | 2 | 3.8 | 0.5 (vol) | >98 | [48] |
| Lime | 700–800 °C, 3 h | Palm oil | 65 | 2 | 6 | 15 : 1 | >97 | [49] |
| Bentonite | 40 wt% KF/KOH/$K_2CO_3$ dried at 120 °C, 16 h; 500 °C, 3 h | Canola | 65 | 7 | 3 | 6 : 1 | 98.2/95.2/ 95.0 | [51] |
| Bentonite | 25 wt% KOH at 60 °C, 24 h; dried at 11 °C, 24 h; 400 °C, 5 h | Palm oil | 60 | 3 | 3 | 6 : 1 | >90 | [52] |
| Kaolin | 800 °C, 10 h; 90 °C for 1 h in NaOH; 500 °C, 6 h | Soybean oil/ palm oil | 60 | 7 | 1.7 | 15 : 1 | >95 | [59] |
| Brazilian kaolin | 950 °C, 4 h; 90 °C for 1 h in 4 M $H_2SO_4$; dried at 120 °C, 12 h; 400 °C, 2 h | Oleic acid | 160 | 4 | 5 | 60 : 1 | 98.9 | [53] |
| Clinoptilolite | 1 : 4 KOH/clinoptilolite stirred 60 °C for 24 h; 110 °C for 24 h; 400 °C, 5 h | Waste cooking oil | 65 | 13.4 (min) | 8.1 | 2.25 : 1 (vol) | >97 | [58] |

### 12.2.4 Industrial Waste Resources

Industrial wastes containing a suitable composition to be converted into metal oxides such as $CaCeO_3$, $CaZrO_3$, CaO, CuO, MgO, and ZnO could be interesting resources for biodiesel catalyst production [60]. It offers several advantages including cost effectiveness, eco-friendliness, and waste valorization [61]. There are myriad industrial waste resources used for the synthesis of biodiesel catalysts that can be categorized according to industry as follows:

#### 12.2.4.1 Food Industry Wastes

A waste material from refined edible oils processing is bleaching clay, which is used to decolorize oils and absorb impurities. Spent bleaching clay (SBC) consists of organic and inorganic substances: $SiO_2$, $Al_2O_3$, and other metallic oxides [62]. Moreover, brewer's spent yeast from the beer industry is often used because it contains a carbonaceous matter that can be a carbon precursor to obtain a suitable porous catalyst [61]. Ngaosuwan et al. [63] also used spent coffee ground as a support of sulfonated AC to produce biodiesel via esterification of caprylic acid.

#### 12.2.4.2 Mining Industry Wastes

There are several kinds of mining resources useful for production of biodiesel catalysts. The following are examples of wastes presented in the study of Galadima and Muraza [64]:

1) Red mud from aluminum ore processing, which contains iron (III) oxide and other oxides, metals, carbonates, and hydroxides. The sites of Ca, Fe, and Al enhance the catalyst performance.
2) Slags and tailings from iron and steel processing. For slags, they are composed of metal oxides ($Fe_2O_3$, $Fe_3O_4$, $TiO_2$, and $FeTiO_3$), silica, free metal species, and Fe, V, and Ti. Tailings comprise rock materials of small sizes and wastewater from the plant.
3) Geothermal wastes and carbide slag (CS). There are three main types of geothermal wastes: light gray waste materials, red waste materials, and clayey waste fraction. Light gray waste materials consist of $SiO_2$ and kaolinite ($Al_2Si_2O_5(OH)_4$). For red waste materials, they contain ankerite and dolomite as the major compositions. Oxides and silicate compounds are also found in these waste types. Lastly, clay waste fraction contains mainly feldspar and dolomite. As for the CS or carbide residue, it is generated during the processes of synthesis and upgrading in the calcium carbide residue industry. It is mainly composed of $Ca(OH)_2$ and can be modified into CaO for biodiesel production. Furthermore, it presents a source of several metal oxides, e.g. $Al_2O_3$, CaO, $Fe_2O_3$, and $SiO_2$.
4) Fly ash from coal combustion in coal-fired power stations and mining processes has been utilized as the raw material for zeolite production because it consists of aluminous and siliceous ($Al^{4+}$ and $Si^{4+}$) materials. Zeolite materials are widely used in various fields, comprising Al, O, and Si with some common metals such as K, Mg, and Na. Most importantly, the use of zeolites derived from fly ash promotes desirable diffusion during the reaction. Consideration on the porosity and design of nanoscale zeolite should also be made during the synthesis process.

The uses of industrial waste materials derived from various sources to produce biodiesel are summarized in **Table 12.4**. Catalysts can function with new oils or waste oils with high

**Table 12.4** Summary of the production of biodiesel using catalysts from different industrial wastes.

| Catalyst resource | Catalyst preparation | Catalyst property | Oil feedstock | Reaction conditions | Biodiesel yield (%) | Reusability | References |
|---|---|---|---|---|---|---|---|
| Raw sugar beet agro-industrial waste (RBIW) | Calcination at 800 °C | Surface area: 27.7 $m^2\ g^{-1}$ | Sunflower | Transesterification/ 4.5: 1 MeOH: oil molar ratio, 1 wt% catalyst, 75 °C, 60 min | 93 | Two cycles with yield above 80%. At third cycle, yield was less than 10% due to leaching of CaO catalyst in glycerol and losing the catalytic activity | [60] |
| Brewer's spent yeast (BSY) | Carbonization at 600 °C and sulfonation at 120 °C | Surface area: 889 $m^2\ g^{-1}$ | Palm fatty acid distillate (PFAD) | Esterification/ 21: 1 MeOH: oil molar ratio, 8 wt% catalyst, 65 °C, 180 min | 87.80 | Four cycles with yield above 80%. After that, it reduced to 74.2% due to leaching of active sites | [61] |
| Spent bleaching clay (SBC) ash | Calcination at 600 °C | Surface area: 0.02 $m^2\ g^{-1}$<br>basicity: 0.21 $mmol\ g^{-1}$ | Waste oil with high acid value:<br>• Soybean oil (SO)<br>• Spent bleaching clay oil (SBCO) | Transesterification/ 9: 1 MeOH: oil molar ratio, 4 wt% catalyst, 65 °C, 180 min | 99.1 for SO<br>95.8 for SBCO | Eight cycles with yield reduction to 95.93% due to leaching of $Na_2SiO_3$ in methanol | [62] |
| K-RAC on *Rhodotorula mucilaginosa* deoiled cake | Wet impregnation and calcination at 600 °C | Surface area: 234.66 $m^2\ g^{-1}$<br>basic strength: $15 < H < 18.4$ | • Yeast single cell oil (SCO)<br>• Waste cooking oil (WCO) | Transesterification/ 9: 1 MeOH: oil molar ratio, 3 wt% K-RAC, 60 °C, 180 min | 96.6 for SCO<br>98.66 for WCO | Five cycles with slight reduction of catalyst activation due to leaching of potassium active species from surface | [65] |
| Date pit powder | Carbonization at 500 °C + calcination at 450 °C | Surface area: 232.2 $m^2\ g^{-1}$ | Pits oil (oil extracted from date pits) | Transesterification/ 12: 1 MeOH: oil molar ratio, 4.5 wt% catalyst, 70 °C, 120 min | 98.2 | — | [66] |
| Lobster shell and eggshells | Calcination at 900 °C | — | *Camelina sativa* oil | Transesterification/ 12: 1 MeOH: oil molar ratio, 1 wt% catalyst, 65 °C, 180 min | 90 for eggshells<br>97 for lobster shell | — | [67] |
| (Pilot scale) kettle limescale | Calcination at 900 °C | Surface area: 29.3 $m^2\ g^{-1}$ | Waste cooking oil | Transesterification/ 1.7: 3 MeOH: oil volume ratio, 8.87 wt% catalyst, 61.7 °C, 15 min | 93.41 | Five cycles with yield reduction. Decrease in reaction rate was less than 2% due to calcium oxide extraction by methanol | [68] |

acid values. The catalyst can be used with waste oils through one-step transesterification without relying on esterification. Most of the results show high FAME yield, although in some cases the yield does not reach the biodiesel standard (96.5%). The catalysts can also be reused several times, reducing the production cost.

## 12.3 Advantages, Disadvantages, and Challenges of These Types of Catalyst for Biodiesel Production

This chapter comprehensively explores the applications of renewable resources including animal-based (mostly shells and bone), plant-based (biomass), natural-based (rocks), and industrial wastes to synthesize heterogeneous catalysts for biodiesel production. Various catalytic activities have been depicted, which are related to their renewable resources and the preparation methods. For instance, the alkali catalyst is mostly synthesized from the calcination of animal and industrial waste resources to generate the metal oxide as an active phase. On the contrary for plant-based and natural resources, they are usually utilized as a catalyst support by pyrolysis and are modified with the strong acid or strong base as a sulfonated functional group or potassium hydroxide, respectively. The alkali catalyst offers higher catalytic activity via transesterification, while the acid catalyst can simultaneously catalyze esterification and transesterification. For this reason, the applications of biodiesel catalysts derived from renewable resources have demonstrated excellent performances with high biodiesel production rates under mild reaction conditions at atmospheric pressure using various feedstocks such as refined vegetable oil, waste cooking oil, nonedible plant extracted oil, animal fat, and micro–macro algae. In addition, using these catalysts is indisputably important for environmental and economic sustainability. However, the continuous development of these catalysts has been performed on the laboratory scale, especially in a batch system. Considering the recent interest for implementation of the heterogeneous catalysts derived from renewable resources on a commercialized scale, the catalyst fabrication to achieve the ideal catalyst in terms of high active sites, stability, and cost effectiveness has been explored. The large-scale catalyst production should also be a concern as the composition of the material derived from renewable resources might vary depending on its source. Therefore, the challenge to develop the large-scale catalyst production process to achieve high catalytic performance in terms of activity and stability for a continuous process is a crucial issue. Further investigation should be carried out to bridge the gap between the laboratory scale and commercialization, for a greener and more sustainable future, which can be summarized as follows:

- Enhancement of catalytic activity and minimization of catalyst deactivation.

The use of a heterogeneous catalyst has shown inferior catalytic performance compared with a homogeneous catalyst especially for transesterification with certain issues like the formation of the three-phase system. Not only the immiscibility of the reactants between oil and alcohol but also the presence of a solid catalyst can generate a three-phase system, which can reduce the transesterification rate involving the mass transfer step. Therefore, the enhancement of the catalytic activity should be pointed out using the innovative catalyst design. The following characteristic parameters have a direct impact on the yield of biodiesel including (i) number of active sites (basicity of base catalysts and acidity of acid catalysts), (ii) surface morphology (hydrophobic surface is preferred to adsorb the oil reactant and to avoid catalyst poisoning by water and FFA), (iii) textural properties (high surface area with mesoporous structure larger than the triglyceride molecule), and (iv) interaction between active species and its support to avoid the leaching process. Regardless of using the heterogeneous catalyst, catalyst deactivation should be an important factor to consider because this factor not only reduces the catalytic activity but also influences the duration of continuous operation. As described previously, the deactivation cause of the biodiesel catalyst derived from renewable resources is mainly due to leaching of the active species such as $Ca^{+}$, $K^{+}$, and other alkali metals, as well as the deposition of organic material on the active site of the catalyst. Therefore, the sophisticated catalyst design such as a hierarchical shape, functionalization of cometal, or other chemicals and nanoscale catalyst is required to conform to the ideal green catalyst for biodiesel production.

- Continuous operation of biodiesel production process using solid catalyst.

Numerous fundamental research has proposed two steps: (i) synthesis of catalysts from renewable resources and (ii) investigation on their optimum operating condition for biodiesel production in the powder form. Taking the criteria of industrial feasibility into account, the pelletized catalyst is a promising catalyst shape to avoid slurry reactor operation-related problems for continuous operations. Further research should involve the effect of a binder that can interact with active species to prepare the pelletized catalyst. However, it can have a substantial effect on catalyst porosity inducing meso- and macroporous and dispersion of active species of the pelletized catalyst. Mechanical strength of the pelletized catalyst is also concerned for the continuous operation, which is related to the binder type, ratio of the active species and binder, particle size, and calcination step. The binder type also has a counterbalancing effect on the pelletized catalyst performance, which can decrease its catalytic activity, but the deactivation of this pelletized catalyst is also decreased.

In conclusion, the use of catalysts derived from renewable resources is a good approach for implementing the concept of circular economy involving biodiesel production process. Further elaborate research as mentioned earlier is required to shift toward using the heterogeneous catalysts derived from renewable resources for commercialized biodiesel production.

## Acknowledgment

The authors would like to acknowledge the supports from the "Research Chair Grant" National Science and Technology Development Agency (NSTDA).

## References

**1** Chua, S.Y., Periasamy, L.A.P., Goh, C.M.H. et al. (2020). Biodiesel synthesis using natural solid catalyst derived from biomass waste – a review. *Journal of Industrial and Engineering Chemistry* 81: 41–60.

2 Wang, S., Yan, W., and Zhao, F. (2020). Recovery of solid waste as functional heterogeneous catalysts for organic pollutant removal and biodiesel production. *Chemical Engineering Journal* 401: 126104.

3 Ravi, A., Gurunathan, B., Rajendiran, N. et al. (2020). Contemporary approaches towards augmentation of distinctive heterogeneous catalyst for sustainable biodiesel production. *Environmental Technology & Innovation* 19: 100906.

4 Gupta, J., Agarwal, M., and Dalai, A.K. (2020). An overview on the recent advancements of sustainable heterogeneous catalysts and prominent continuous reactor for biodiesel production. *Journal of Industrial and Engineering Chemistry* 88: 58–77.

5 Shimada, G.B. and Cestari, A. (2020). Synthesis of heterogeneous catalysts by the hydrolytic Sol-Gel method for the biodiesel production. *Renewable Energy* 156: 389–394.

6 Bharathiraja, B., Chakravarthy, M., Kumar, R.R. et al. (2014). Biodiesel production using chemical and biological methods – a review of process, catalyst, acyl acceptor, source and process variables. *Renewable and Sustainable Energy Reviews* 38: 368–382.

7 Tan, Y.H., Abdullah, M.O., Nolasco-Hipolito, C., and Ahmad Zauzi, N.S. (2017). Application of RSM and Taguchi methods for optimizing the transesterification of waste cooking oil catalyzed by solid ostrich and chicken-eggshell derived CaO. *Renewable Energy* 114: 437–447.

8 Goli, J. and Sahu, O. (2018). Development of heterogeneous alkali catalyst from waste chicken eggshell for biodiesel production. *Renewable Energy* 128: 142–154.

9 Tan, Y.H., Abdullah, M.O., Nolasco-Hipolito, C., and Taufiq-Yap, Y.H. (2015). Waste ostrich- and chicken-eggshells as heterogeneous base catalyst for biodiesel production from used cooking oil: catalyst characterization and biodiesel yield performance. *Applied Energy* 160: 58–70.

10 de Sousa, F.P., dos Reis, G.P., Cardoso, C.C. et al. (2016). Performance of CaO from different sources as a catalyst precursor in soybean oil transesterification: kinetics and leaching evaluation. *Journal of Environmental Chemical Engineering* 4 (2): 1970–1977.

11 Viriya-empikul, N., Krasae, P., Puttasawat, B. et al. (2010). Waste shells of mollusk and egg as biodiesel production catalysts. *Bioresource Technology* 101 (10): 3765–3767.

12 Nakatani, N., Takamori, H., Takeda, K., and Sakugawa, H. (2009). Transesterification of soybean oil using combusted oyster shell waste as a catalyst. *Bioresource Technology* 100 (3): 1510–1513.

13 Laskar, I.B., Deshmukhya, T., Bhanja, P. et al. (2020). Transesterification of soybean oil at room temperature using biowaste as catalyst; an experimental investigation on the effect of co-solvent on biodiesel yield. *Renewable Energy* 162: 98–111.

14 Hu, S., Wang, Y., and Han, H. (2011). Utilization of waste freshwater mussel shell as an economic catalyst for biodiesel production. *Biomass and Bioenergy* 35 (8): 3627–3635.

15 Lin, Y.-C., Amesho, K.T.T., Chen, C.-E. et al. (2020). A cleaner process for green biodiesel synthesis from waste cooking oil using recycled waste oyster shells as a sustainable base heterogeneous catalyst under the microwave heating system. *Sustainable Chemistry and Pharmacy* 17: 100310.

16 Yuliana, M., Santoso, S.P., Soetaredjo, F.E. et al. (2020). Utilization of waste capiz shell – based catalyst for the conversion of leather tanning waste into biodiesel. *Journal of Environmental Chemical Engineering* 8 (4): 104012.

17 Farooq, M., Ramli, A., and Naeem, A. (2015). Biodiesel production from low FFA waste cooking oil using heterogeneous catalyst derived from chicken bones. *Renewable Energy* 76: 362–368.

18 Singh, T.S. and Verma, T.N. (2019). An assessment study of using *Turel Kongreng* (river mussels) as a source of heterogeneous catalyst for biofuel production. *Biocatalysis and Agricultural Biotechnology* 20: 101185.

19 Tan, Y.H., Abdullah, M.O., Kansedo, J. et al. (2019). Biodiesel production from used cooking oil using green solid catalyst derived from calcined fusion waste chicken and fish bones. *Renewable Energy* 139: 696–706.

20 Khan, H.M., Iqbal, T., Ali, C.H. et al. (2020). Waste quail beaks as renewable source for synthesizing novel catalysts for biodiesel production. *Renewable Energy* 154: 1035–1043.

21 Hamza, M., Ayoub, M., Shamsuddin, R.B. et al. (2020). A review on the waste biomass derived catalysts for biodiesel production. *Environmental Technology & Innovation* 21: 101200.

22 Baroutian, S., Aroua, M.K., Raman, A.A.A., and Sulaiman, N.M.N. (2010). Potassium hydroxide catalyst supported on palm shell activated carbon for transesterification of palm oil. *Fuel Processing Technology* 91 (11): 1378–1385.

23 Abdullah, S.H.Y.S., Hanapi, N.H.M., Azid, A. et al. (2017). A review of biomass-derived heterogeneous catalyst for a sustainable biodiesel production. *Renewable and Sustainable Energy Reviews* 70: 1040–1051.

24 Kastner, J.R., Miller, J., Geller, D.P. et al. (2012). Catalytic esterification of fatty acids using solid acid catalysts generated from biochar and activated carbon. *Catalysis Today* 190 (1): 122–132.

25 Bastos, R.R.C., da Luz Corrêa, A.P., da Luz, P.T.S. et al. (2020). Optimization of biodiesel production using sulfonated carbon-based catalyst from an amazon agro-industrial waste. *Energy Conversion and Management* 205: 112457.

26 Kong, X., Zhu, Y., Lei, H. et al. (2020). Synthesis of graphene-like carbon from biomass pyrolysis and its applications. *Chemical Engineering Journal* 399: 125808.

27 Jume, B.H., Gabris, M.A., Nodeh, H.R. et al. (2020). Biodiesel production from waste cooking oil using a novel heterogeneous catalyst based on graphene oxide doped metal oxide nanoparticles. *Renewable Energy* 162: 2182–2189.

28 Loy, A.C.M., Quitain, A.T., Lam, M.K. et al. (2019). Development of high microwave-absorptive bifunctional graphene oxide-based catalyst for biodiesel production. *Energy Conversion and Management* 180: 1013–1025.

29 Vu, T.H.T., Nguyen, M.H., and Nguyen, M.D. (2019). Synthesis of acidic heterogeneous catalysts with high stability based on graphene oxide/activated carbon composites for the esterification of lactic acid. *Journal of Chemistry* 2019: 7815697.

30 Salinas, D., Sepúlveda, C., Escalona, N. et al. (2018). Sol–gel $La_2O_3$–$ZrO_2$ mixed oxide catalysts for biodiesel production. *Journal of Energy Chemistry* 27 (2): 565–572.

31 Xu, Y., Du, X.-h., Li, J. et al. (2019). A comparison of $Al_2O_3$ and $SiO_2$ supported Ni-based catalysts in their performance for the dry reforming of methane. *Journal of Fuel Chemistry and Technology* 47 (2): 199–208.

32 Krishnarao, R.V., Subrahmanyam, J., and Jagadish Kumar, T. (2001). Studies on the formation of black particles in rice husk silica ash. *Journal of the European Ceramic Society* 21 (1): 99–104.

33 Prasetyoko, D., Ramli, Z., Endud, S. et al. (2006). Conversion of rice husk ash to zeolite beta. *Waste Management* 26 (10): 1173–1179.

34 Hindryawati, N., Maniam, G.P., Karim, M.R., and Chong, K.F. (2014). Transesterification of used cooking oil over alkali metal (Li, Na, K) supported rice husk silica as potential solid base catalyst. *Engineering Science and Technology, an International Journal* 17 (2): 95–103.

35 Zhao, C., Yang, L., Xing, S. et al. (2018). Biodiesel production by a highly effective renewable catalyst from pyrolytic rice husk. *Journal of Cleaner Production* 199: 772–780.

36 Nath, B., Kalita, P., Das, B., and Basumatary, S. (2020). Highly efficient renewable heterogeneous base catalyst derived from waste *Sesamum indicum* plant for synthesis of biodiesel. *Renewable Energy* 151: 295–310.

37 Liu, X.-Y., Huang, M., Ma, H.-L. et al. (2010). Preparation of a carbon-based solid acid catalyst by sulfonating activated carbon in a chemical reduction process. *Molecules.* 15 (10): 7188–7196.

38 Bhatia, S.K., Gurav, R., Choi, T.-R. et al. (2020). Conversion of waste cooking oil into biodiesel using heterogenous catalyst derived from cork biochar. *Bioresource Technology* 302: 122872.

39 Nath, B., Das, B., Kalita, P., and Basumatary, S. (2019). Waste to value addition: utilization of waste Brassica nigra plant derived novel green heterogeneous base catalyst for effective synthesis of biodiesel. *Journal of Cleaner Production* 239: 118112.

40 Algoufi, Y.T., Kabir, G., and Hameed, B.H. (2017). Synthesis of glycerol carbonate from biodiesel by-product glycerol over calcined dolomite. *Journal of the Taiwan Institute of Chemical Engineers* 70: 179–187.

41 Zhao, S., Niu, S., Yu, H. et al. (2019). Experimental investigation on biodiesel production through transesterification promoted by the La-dolomite catalyst. *Fuel* 257: 116092.

42 Wilson, K., Hardacre, C., Lee, A.F. et al. (2008). The application of calcined natural dolomitic rock as a solid base catalyst in triglyceride transesterification for biodiesel synthesis. *Green Chemistry* 10 (6): 654–659.

43 Ilgen, O. (2011). Dolomite as a heterogeneous catalyst for transesterification of canola oil. *Fuel Processing Technology* 92 (3): 452–455.

44 Jindapon, W. and Ngamcharussrivichai, C. (2018). Heterogeneously catalyzed transesterification of palm oil with methanol to produce biodiesel over calcined dolomite: the role of magnesium oxide. *Energy Conversion and Management* 171: 1311–1321.

45 Shajaratun Nur, Z.A., Taufiq-Yap, Y.H., Rabiah Nizah, M.F. et al. (2014). Production of biodiesel from palm oil using modified Malaysian natural dolomites. *Energy Conversion and Management* 78: 738–744.

46 Niu, S., Zhang, X., Ning, Y. et al. (2020). Dolomite incorporated with cerium to enhance the stability in catalyzing transesterification for biodiesel production. *Renewable Energy* 154: 107–116.

47 Sánchez-Cantú, M., Pérez-Díaz, L.M., Rosales, R. et al. (2011). Commercial hydrated lime as a cost-effective solid base for the transesterification of wasted soybean oil with methanol for biodiesel production. *Energy & Fuels* 25 (7): 3275–3282.

48 Sánchez-Cantú, M., Pérez-Díaz, L.M., Pala-Rosas, I. et al. (2013). Hydrated lime as an effective heterogeneous catalyst for the transesterification of castor oil and methanol. *Fuel* 110: 54–62.

**49** Roschat, W., Siritanon, T., Yoosuk, B., and Promarak, V. (2016). Biodiesel production from palm oil using hydrated lime-derived CaO as a low-cost basic heterogeneous catalyst. *Energy Conversion and Management* 108: 459–467.

**50** Soares Dias, A.P., Puna, J., Neiva Correia, M.J. et al. (2013). Effect of the oil acidity on the methanolysis performances of lime catalyst biodiesel from waste frying oils (WFO). *Fuel Processing Technology* 116: 94–100.

**51** Boz, N., Degirmenbasi, N., and Kalyon, D.M. (2013). Transesterification of canola oil to biodiesel using calcium bentonite functionalized with K compounds. *Applied Catalysis B: Environmental* 138–139: 236–242.

**52** Soetaredjo, F.E., Ayucitra, A., Ismadji, S., and Maukar, A.L. (2011). KOH/bentonite catalysts for transesterification of palm oil to biodiesel. *Applied Clay Science* 53 (2): 341–346.

**53** do Nascimento, L.A.S., Angélica, R.S., da Costa, C.E.F. et al. (2011). Comparative study between catalysts for esterification prepared from kaolins. *Applied Clay Science* 51 (3): 267–273.

**54** Gaidoumi, A.E., Benabdallah, A.C., Bali, B.E., and Kherbeche, A. (2017). Synthesis and characterization of zeolite HS using natural pyrophyllite as new clay source. *Arabian Journal for Science and Engineering* 43 (1): 191–197.

**55** Belviso, C., Kharchenko, A., Agostinelli, E. et al. (2018). Red mud as aluminium source for the synthesis of magnetic zeolite. *Microporous and Mesoporous Materials* 270: 24–29.

**56** Zeng, S., Wang, R., Zhang, Z., and Qiu, S. (2016). Solventless green synthesis of sodalite zeolite using diatomite as silica source by a microwave heating technique. *Inorganic Chemistry Communications* 70: 168–171.

**57** Doyle, A.M., Alismaeel, Z.T., Albayati, T.M., and Abbas, A.S. (2017). High purity FAU-type zeolite catalysts from shale rock for biodiesel production. *Fuel* 199: 394–402.

**58** Mohadesi, M., Aghel, B., Maleki, M., and Ansari, A. (2020). The use of KOH/Clinoptilolite catalyst in pilot of microreactor for biodiesel production from waste cooking oil. *Fuel* 263: 116659.

**59** Dang, T.H., Chen, B.H., and Lee, D.J. (2013). Application of kaolin-based catalysts in biodiesel production via transesterification of vegetable oils in excess methanol. *Bioresource Technology* 145: 175–181.

**60** Abdelhady, H.H., Elazab, H.A., Ewais, E.M. et al. (2020). Efficient catalytic production of biodiesel using nano-sized sugar beet agro-industrial waste. *Fuel* 261: 116481.

**61** Saravanan Arumugamurthy, S., Sivanandi, P., Pandian, S. et al. (2019). Conversion of a low value industrial waste into biodiesel using a catalyst derived from brewery waste: an activation and deactivation kinetic study. *Waste Management* 100: 318–326.

**62** Cong, W.-J., Wang, Y.-T., Li, H. et al. (2020). Direct production of biodiesel from waste oils with a strong solid base from alkalized industrial clay ash. *Applied Energy* 264: 114735.

**63** Ngaosuwan, K., Goodwin, J.G., and Prasertdham, P. (2016). A green sulfonated carbon-based catalyst derived from coffee residue for esterification. *Renewable Energy* 86: 262–269.

**64** Galadima, A. and Muraza, O. (2020). Waste materials for production of biodiesel catalysts: technological status and prospects. *Journal of Cleaner Production* 263: 121358.

**65** Gohain, M., Bardhan, P., Laskar, K. et al. (2020). Rhodotorula mucilaginosa: a source of heterogeneous catalyst for biodiesel production from yeast single cell oil and waste cooking oil. *Renewable Energy* 160: 220–230.

**66** AaH, A.-M., Jamil, F., Al-Haj, L. et al. (2018). Biodiesel production over a catalyst prepared from biomass-derived waste date pits. *Biotechnology Reports* 20: e00284.

**67** Hangun-Balkir, Y. (2016). Green biodiesel synthesis using waste shells as sustainable catalysts with *Camelina sativa* Oil. *Journal of Chemistry* 2016: 6715232.

**68** Aghel, B., Mohadesi, M., Ansari, A., and Maleki, M. (2019). Pilot-scale production of biodiesel from waste cooking oil using kettle limescale as a heterogeneous catalyst. *Renewable Energy* 142: 207–214.

# 13

# Biodiesel Production Using Ionic Liquid-Based Catalysts

*B. Sangeetha and G. Baskar*

*Department of Biotechnology, St. Joseph's College of Engineering, Chennai, India*

## 13.1 Introduction

Ionic liquids (ILs) are organic salts that comprise of ions and exist as liquid at temperatures below 100 °C. Cations in ILs are organic, and anions can be organic or inorganic. The commonly used ILs are nitrogen-containing compounds such as alkylammonium, N,N′-dialkylimidazolium,N-alkylpyridinium,andN,N′-pyrrolidiniumorphosphorous-containing compounds such as alkylphosphonium [1]. The popular choice of anions includes $BF_4^-$, $PF_6^-$,$CH_3CO_2^-$,$CF_3CO_2^-$,$NO_3^-$,$T_{f2}N^-$,$[(CF_3SO_2)_2N]^-$, $[RSO_4]^-$, and $[R_2PO_4]^-$, where Tf and R stand for bis (trifluoromethylsulfonyl)imide anions and alkyl group, respectively.

The synthesis of ILs can be driven into two steps [2, 3]: (i) formation of preferred cation: the preferred cation can be synthesized by one of the three ways – by quaternization reactions of amine with a haloalkane, by the protonation of the amine by an acid, or by formation of zwitterion; (ii) direct combination or anion exchange: direct combination reaction can be carried out by reacting a zwitterion carrying the preferred cation with an acid carrying the preferred anion. Anion exchange reactions can be performed by one of the three ways – by acid–base neutralization, by anion metathesis, or by treatment of halide salts with Lewis acids to form Lewis acid-based ILs.

ILs are alternatives to hazardous and volatile materials, which owe promising physical and chemical properties such as broad liquid range, increased chemical and thermal stabilities, negligible vapor pressure that eliminates problem associated with volatility and flammability, high ionic conductivity, high thermal conductivity, broad electrochemical potential window, and are effective in dissolving a wide range of organic and inorganic compounds and simple recovery process [4]. Some ILs are believed to be "green" due to their nominal toxicity to humans and to the environment. In addition, they are frequently branded as "designer solvent," as their properties, such as polarity, hydrophobicity, solvating activity, density, melting point, viscosity, and others, can be tweaked to certain degree by modifying the cation, anion, or alkyl substituents on the cation [5]. This review highlights the usage of IL as a suitable catalyst for the greener synthesis of biodiesel amidst the conventional homogeneous catalysts.

*Biodiesel Production: Feedstocks, Catalysts, and Technologies*, First Edition. Edited by Samuel Lalthazuala Rokhum, Gopinath Halder, Suttichai Assabumrungrat, and Kanokwan Ngaosuwan.

## 13.2 Mechanism of IL-Catalyzed Biodiesel Production

The mechanism of biodiesel production catalyzed by ILs is different from that in their absence. Due to the temperature-sensitive polyether group, the hydrogen bond can be formed between alcohol and oxygen of polyether group at low temperature, enabling the IL to be hydrophilic. Upon heating, the hydrogen bond is broken, and the IL can become hydrophobic. Hence, the reaction mixture is homogeneous in the presence of ILs at high temperature, while the biphasic system is formed upon cooling [6].

The mechanism of basic IL-catalyzed transesterification reaction illustrated in the following steps: (i) deprotonation of methoxy group of MeOH by the negatively charged anion of ILs, (ii) confrontation of activated methoxy group by the electrophilic center (which is bonded next to two electronegative oxygen atoms) of the carbonyl's carbon of triglycerides, and (iii) translation of the intermediate to a diglyceride group that produces methyl ester or reversion to the starting material as each step in the transesterification reaction is reversible if the alcohol is in excess. The catalytic activity of basic IL depends on the competency of methoxy group activation by anion due to its basicity and the intermiscibility between the reactants [7, 8].

Similar to the mechanism of basic ILs, the mechanism of acidic IL-catalyzed transesterification reaction illustrated in the following steps: (i) protonation of the carbonyl group to become a more potent electrophile, (ii) formation of carbocation intermediate with the carbonyl group of triglycerides, and (iii) nucleophilic substitution reaction of carbocation intermediate with MeOH to form molecules of diglyceride, FAME, and a proton to catalyze the transesterification reaction. The anion in the IL plays a vital role in the Brønsted acidic ionic liquids (BAILs); the more acidic the conjugate acid of the anion (lower affinity of the anion), the stronger the IL's acidity. However, this does not necessarily result in a higher catalytic activity [8, 9].

## 13.3 Acidic and Basic Ionic Liquids (AILs/BILs) as Catalyst in Biodiesel Production

AILs are of three major categories: (i) Lewis acidic ILs (LAILs), which accept lone pair electrons from a Lewis base due to its Lewis acid site, (ii) BAILs, which possess one or multiple Brønsted acid sites to move a proton to a Brønsted base, and (iii) Brønsted–Lewis acidic ionic liquids (BLAILs), which possess both Brønsted and Lewis acidic sites working concurrently to promote the catalytic activity due to stronger acidity [10]. AILs can catalyze esterification of FFAs as well as transesterification of TAGs.

Basic ILs, alkaline in nature, are employed in transesterification reactions of biodiesel production to avoid soap formation when the feedstock contains relatively high FFA. BILs are used as an ancillary for traditional alkali catalysts such as KOH, NaOH, $K_2CO_3$, $NaHCO_3$, NaOAc, trimethylamine, and tetrabutylammonium acetate [11]. The commonly used BAILs are choline, amino acid, imidazolium, guanidine, and hydroxide-based ILs. These catalysts retain high thermal stability and low decomposition temperature and reveal high catalytic efficiency, ease of separation, and reusability. Theoretically, the catalytic activity is better when the IL has more basicity [12]. The literature summary on AILs and BILs as catalysts in esterification and transesterification of various feedstocks along with reaction conditions applied is depicted in Table 13.1.

Li et al. [13] prepared seven differential multi -$SO_3H$-functionalized AILs by using low cost less substituted amines, which contained multiple sites for the functionalization of sulfonic acid groups. The IL catalyst [EDA-PS][P-TSA] exhibited good catalytic performance with conversion of 97.58% at much shorter reaction time and at lowest dosage of catalyst loading. It also showed high generality for converting numerous free fatty acid feedstocks into biodiesel with substantial conversion within 93.59–94.33%. Ullah et al. [14] synthesized hydrophobic phosphonium-based stable AILs containing four specific fluorous anions by metathesis reaction. The hydrophobic ILs were preferred due to their capability to dissolve the fatty acids of oils, which causes the transesterification of fatty acid with methanol during biodiesel preparation. Among all, [TBP][$NTf_2$] showed good thermal stability, catalytic performance, and reusability. Wang et al. [17] reported that among three AILs, poly(1-(4-sulfonate)-butyl-3-vinylimidazolium) phosphotungstate exhibited good catalytic activity and reusability even after five recycles without significant deactivation when compared with poly(1-(4-sulfonate)-butyl-3-vinylimidazolium) hydrogen sulfate and poly(1-(4-sulfonate)-butyl-3-vinylimidazolium) trifluoromethanesulfonate in the esterification of oleic acid.

Zhang et al. [18] compared the basic IL catalyst, [Bmim]OH with potassium hydroxide, and tetrabutylammonium hydroxide during the transesterifiaction of castor oil. The results showed that the catalytic activity was good and stable for [Bmim]OH with good reusability than the latter two. The study conducted by Gholami et al. [19] reported that usage of basic IL, choline hydroxide, as a catalyst in microchannel reactor results in shorter reaction time and reduced energy consumption compared to stirred tank reactors. In contrast to other heterogeneous catalysts, it also lessens the amounts of catalyst and alcohol needed to achieve efficiencies higher than 90%. The results depicted that the relationship between reaction efficiency and catalyst concentration, alcohol-to-oil ratio, and reaction time was governed by a second order equation. The results also showed that the reaction efficiency was most affected by the reaction time.

## 13.4 Supported Ionic Liquids in Biodiesel Production

Instead of using single homogeneous catalyst, a blend of IL and other solid materials such as magnetic nanoparticles, metal–organic frameworks (MOFs), polymers, etc., which act as supports to AILs/BILs, can be used to enhance the catalytic activity and ease the catalyst separation. A literature summary of supported ILs along with reaction conditions was depicted in Table 13.2. Zhang et al. [20] reported on the use of magnetically recyclable acidic poly ionic liquids (PILs) immobilized on magnetic mesoporous carrier ($Fe_3O_4$@$SiO_2$@SBA-15), FnmS-PILs with the hydrophobic regulator catalyst through radical polymerization and anion exchange methods. The catalyst possessed uniform mesoporous structure, high number of acid sites, strong acid strength, strong hydrophobicity, large BET surface area, and strong magnetism, which allowed for easier separation from the reaction mixture. Also the catalyst exhibited strong catalytic activity with a biodiesel yield of 87.5% in transesterification reaction even after five cycles.

**Table 13.1** Acidic and basic ionic liquids in biodiesel production.

| Feedstock | Type of reaction | Type of reactor | Ionic liquid as catalyst | Methanol:oil molar ratio | Catalyst loading | Temperature | Time duration | Conversion or yield% | References |
|---|---|---|---|---|---|---|---|---|---|
| Oleic acid | Esterification | Laboratory flask with condenser and magnetic rotor | [EDA-PS][P-TSA] | 13 : 1 | 3 wt% | 70 °C | 1.8 h | 97.58% | [13] |
| Waste cooking oil | Transesterification | Batch reactor | $[TBP][NTf_2]$ | 18 : 1 | 4.5 wt% | 60 °C | 10 h | 81% (Y) | [14] |
| Oleic acid | Esterification | Round-bottom, three-necked flask equipped with a water-cooled reflux condenser and a magnetic stirrer | $[H_{NMP}]CH_3SO_3$ | 11.23 : 1 | 10.48 wt% | 52.86 °C | 2.81h | 97.35 | [15] |
| Oleic acid | Esterification | Reaction vessel under agitation | $[Hmim]HSO_4$ | 15 : 1 | 15 wt% | 110 ± 2 °C | 8 h | 95 | [16] |
| Palmitic acid | Esterification | RBF with reflux condenser | poly(1-(4-sulfonate)-butyl-3-vinylimidazolium) phosphotungstate PIL-3 | 6 | 3 wt% | 60 °C | 8 h | 91.6 | [17] |
| Castor oil | Transesterification | Three-port flask | [Bmim]OH | 6 : 1 | 1 wt% | 40 °C | 1 h | 95% | [18] |
| *Salvia leriifolia* oil | Transesterification | Microchannel reactor | (ChOH) | 0.37 | 6.11 wt% | — | 0.208 h | 96.8% efficiency | [19] |

**Table 13.2** Supported ionic liquids in biodiesel production.

| Feedstock | Type of reaction | Type of reactor | Ionic liquid as catalyst | Methanol:oil molar ratio | Catalyst loading | Temperature | Time duration | Conversion or yield% | References |
|---|---|---|---|---|---|---|---|---|---|
| Oleic acid | Esterification | Pressure tube | FnmS-PILs | 17 : 1 | 4 wt% | 75 °C | 3 h | 96.1% | [20] |
| *Euphorbia lathyris* L. crude oils | Transesterification | Pressure tube | FnmS-PILs | 17 : 1 | 4 wt% | 75 °C | 3 h | 91.7% | [20] |
| Soybean oil | Transesterification | RBF with water cooled reflux condenser and a magnetic stirrer | $Fe_3O_4$@HKUST-1-ABILs | 30 : 1 | 1.2 wt% | 65 °C | 3 h | 92.3% | [21] |
| *Jatropha curcas* L. | One-pot synthesis | Autoclave equipped with a magnetic stirrer | (β-CD-IL-$Fe_3O_4$) | 10 : 1 | 3 wt% | 130 °C | 3 h | m | [22] |
| Soybean oil | One-pot synthesis | Autoclave equipped with a mechanical stirring and electrical heating jacket | IL 1 – 1-vinyl-3-(3-sulfopropyl) imidazolium hydrogen sulfate<br>IL 2 – 1.4-butanediyl-3,3′-bis-1-vinylimidazoliu hydrogen sulfate | 30 : 1 | 8 wt% | - | 8 h | 94.2% | [23] |
| Soybean oil | One-pot synthesis | Cylindrical pressurized stainless reactor fitted with an electrical heating jacket and mechanical stirring | 1-Vinyl-3-(3-sulfopropyl) imidazolium hydrogen sulfate | 35 : 1 | 9 wt% | 120 °C | 6 h | 93.3% | [24] |
| Soybean oil | One-pot synthesis | Autoclave fitted with a temperature controller and mechanical stirring | $CoFe_2O_4$/MIL-88B(Fe)-$NH_2$/(Py-Ps) PMo | 30 : 1 | 8 wt% | 140 °C | 8 h | 95.6% | [25] |
| Oleic acid | Esterification | Batch reactor | [$(CH_2COOH)_2$IM]$HSO_4$@H-UiO-66 | 10.39 : 1 | 6.28 wt% | 80 °C | 5 h | 93.82% | [26] |
| Soybean oil | One-pot synthesis | Autoclave fitted with a temperature controller and magnetic stirring | AILs/HPW/UiO-66-2COOH | 35 : 1 | 10 wt% | 110 °C | 6 h | 95.8% | [27] |
| Soybean oil | One-pot synthesis | Autoclave reactor fitted with an electrical heating jacket | AIL/HPMo/MIL-100(Fe) | 30 : 1 | 9 wt% | 120 °C | 6 h | 92.3% | [28] |
| Soybean oil | Transesterification | Flask equipped with magnetic stirrer and reflux condenser | (BNPs-CCH) | 11 : 1 | 4.13 wt% | 60 °C | 4.4 h | 95.2% | [29] |
| Tung nut oil | Transesterification | Glass batch reactor | Solid acidic ionic liquid with vinyl imidazole and 1,3-propane sulfonate as monomer | 1.5 : 1 | 160 wt% | 65 °C | 6 h | 91% | [30] |
| Oleic acid | Esterification | Flask | [TTMPP-PS][$CF_3SO_3$] | 12 : 1 | 4 wt% | 70 °C | 3 h | 94.3% | [31] |
| Soybean oil | Transesterification | Hydrothermal reactor | $SO_4^{2-}$/$ZrO_2$-$SiO_2$(Et)-IL-i | 18 : 1 | 5 wt% | 150 °C | 3 h | 99% | [32] |
| Oleic acid | Esterification | Pressure tube | (MMFP-AIL) | 12 : 1 | 4 wt% | 90 °C | 3 h | 95% | [33] |

Chang and Zhou [22] reported on the use of BAIL supported on magnetic $Fe_3O_4$ [β-CD-6-Im-($CH_2$)3-$HSO_3$][$HSO_4$]-$Fe_3O_4$, which was easily recycled and exhibited excellent catalytic activity even after being reused for five times. Xie and Wan [21] designed a magnetically separable and efficient solid base catalyst, $Fe_3O_4$@HKUST-1-ABILs. The magnetic $Fe_3O_4$ nanoparticles were synthesized by a solvothermal method, then the core–shell-structured $Fe_3O_4$@HKUST-1 composites were prepared by a versatile layer-by-layer assembly method, and finally the amino-functionalized basic ILs (ABILs) were encapsulated on $Fe_3O_4$@ HKUST-1composites. The designed catalyst had good stability without a significant loss in catalytic activity and exhibited superparamagnetic behavior, which can be separated easily by external magnetic field. Xie and Wang [23] studied about the one-pot transformation of low-quality oils to biodiesel using magnetically separable dual Brönsted acid catalyst. Firstly, $Fe_3O_4$ nanoparticles were synthesized by chemical coprecipitation method, then silica-coated core–shell magnetic composites were prepared by using a sol–gel method, and finally the magnetic separable Brönsted acid catalyst was synthesized by grafting copolymerization of dual acidic IL monomer 1-vinyl-3-(3-sulfopropyl) imidazolium hydrogen sulfate [$SO_3H$-($CH_2$)3-IM-vinyl][$HSO_4$] and cross-linker 1,4-butanediyl-3,3′-bis-1-vinylimi dazoliu hydrogen sulfate on the magnetic support. The magnetic core permits magnetic separation of the solid catalyst, whereas the silica shell increases the support surface area and provides site for the vinyl functionalization, thereby providing good catalytic activity and stability for heterogeneous biodiesel production. Similarly, Xie and Wang [24] studied about the one-pot transformation of low-quality oils using polymeric Bronsted acidic IL 1-vinyl-3-(3-sulfopropyl) imidazolium hydrogen sulfate immobilized onto the magnetic support through radical grafting copolymerization.

Additionally, Xie and Wang [25] synthesized heterogenized polyoxometalate (POM)-based magnetically recyclable IL catalyst, $CoFe_2O_4$/MIL-88B(Fe)-NH/(Py-Ps)PMo, for the one-pot production of biodiesel from the low-quality oils. POM-based sulfonated ILs consisting of both Brönsted and Lewis acid sites, namely, (Py-Ps)PMo, were firstly primed by reaction of pyridine with 1,3-propanesultone, trailed by anion exchange with phosphomolybdic acid (HPMo). Thereafter, the so-prepared ILs were loaded on the amino-functionalized MOF ($CoFe_2O_4$/MIL-88B(Fe)- $NH_2$) to synthesize the recoverable solid catalyst.

Ye et al. [26] designed an efficient Brønsted–Lewis acidic catalyst [($CH_2$COOH)2IM]$HSO_4$@ H- UiO-66 via bidentate coordination between one -COO- group of [($CH_2$COOH)2IM]$HSO_4$ and two portions of unsaturated Zr ion defeats of the hierarchically porous Zr MOF (H-UiO-66). The catalyst was recovered by centrifugation and reused for which the biodiesel yield decreased from 93.82 to 90.95% after five runs. Xie and Wan [27] fabricated a solid catalyst, namely, AILs/POM/UiO-66-2COOH, which exhibited high surface area and increased acidity, owing to the merits of both Bronsted and Lewis acids together with the heterogeneous microenvironment of MOFs. The solid catalyst was prepared by functionalizing the UiO-66-2COOH MOFs with Keggin-type POM acids via the in situ preparation strategy, and sulfonated acidic ionic liquids (AILs) were then incorporated into the prepared POM/UiO-66-2COOH composites by pairing POM anions with functionalized IL cations. The catalyst served to be robust and stable even after five times of reuse without significant decrease in activity. Similarly, Xie and Wan [28] designed another recyclable solid catalyst, namely, AIL/HPMo/MIL-100(Fe) for the bioconversion of acidic oil feedstocks into biodiesel. The MOF, MIL-100(Fe), was modified with phosphomolybdic acid (HPMo) by hydrothermal approach, and finally the AIL, 1-(propyl-3-sulfonate) imidazolium hydrogen sulfate $[SO_3H\text{-}(CH_2)3\text{-}IM][HSO_4]$, was immobilized on the prepared HPMo/MIL-100(Fe) composite through ion-exchange method. The high stability for the solid catalyst was due to strong interaction between the heteropolyphosphomolybdenum anion and imidazolium cation.

Hosseini et al. [29] developed a novel catalyst BNPs-CCH, by stabilizing basic IL, chlorocholine hydroxide (CCH) on boehmite nanoparticles. The results demonstrated that enhanced catalytic performance in transesterification reaction is attributed due to the higher basicity of BNPs-CCH than CCH. Chen et al. [31] reported on the application of unusual self-solidifying TMPP containing ILs for biodiesel production via the catalytic esterification. These ILs were synthesized by reacting tris(2,4,6-trimethoxyphenyl) phosphine (TTMPP) with 1,3-propanesulfonate (PS) to form a TPPMP-PS zwitterion, trailed by acidification with different acids. The so-prepared solid-state ILs, [TTMPP-PS] $[CF_3SO_3]$, were soluble during the reaction but insoluble after the reaction. Thus, the catalyst exhibited combined advantages of both homogeneous and heteronomous catalysts with noticeable activity, exceptional stability, and recyclability. Fan et al. [32] designed supported ILs with both Brønsted and Lewis acid sites by immobilizing different dosage of acid IL sulfonic acid functionalized imidazolium hydrogen sulfate ($[Ps\text{-}im]HSO_4$) onto the acidic mesoporous material $SO_4^{2-}/ZrO_2\text{-}SiO_2(Et)$, which was developed by a one-step cocondensation technique combined with hydrothermal treatment. The BET surface area and pore volumes decreased, and average pore diameters increased with the increase of IL loading. They showed resistance toward acid and water and demonstrated satisfied catalytic performance until five runs. Pan et al. [33] synthesized an acidic IL, MMFP-IL, by quaternary ammonization of aminal groups and triazine rings in low-cost polymer, MMFP with 1,3-propanesultone, followed by treatment with $H_3PW_{12}O_{40}$ (HPW). The study demonstrated that MMFP-IL catalyst exhibited abundant mesoporous structures with large BET surface area ($283.0\,m^2\,g^{-1}$), strong acid strength, high acid concentration (2.2 mmol $g^{-1}$), sufficient distribution of acid sites, and good recyclability (up to four cycles) in contrast to other solid acids such as Amberlyst 15 and Nafion NR50.

## 13.5 IL Lipase Cocatalysts

The use of ILs as immobilizing agents on lipase-enzyme-catalyzed reactions is an emerging technique, and it is depicted in Table 13.3. Su et al. [34] synthesized biodiesel from Chinese tallow kernel oil using *Candida rugosa* lipase in $[Hmim][PF_6]$ IL. The study demonstrated that as the IL possessed a single mono-alkyl group, downstream processing was simple, and the biodiesel conversion was 95.4%. Fan et al. [35] employed dialkyl-based ILs, $[C_1C_3OHPyr][NTf_2]$, as a reaction media for *C. rugosa* lipase-catalyzed transesterification reaction for which the biodiesel yield was 82.4%. The study suggested that the yield can be increased by extending the reaction time to 6 h, and the ILs and lipase could be easily recycled by washing with water and acetone, respectively. Lozano et al. [36] used binary mixture of a hydrophobic $[C_{16}mim][NTf_2]$ and hydrophilic [Bmim][Cl] ILs in combination with Novozym 435 enzyme for algal oil extraction and biocatalysis to obtain 100% yield.

**Table 13.3** IL lipase co-catalysts in biodiesel production.

| Feedstock | Type of reaction | Type of reactor | IL | Enzyme | Optimal reaction conditions | Conversion (C) or yield (Y) | Recyclability | References |
|---|---|---|---|---|---|---|---|---|
| Chinese tallow kernel oil | Transesterification | Screw capped flask | $[Hmim][PF_6]$ | *Candida rugosa* lipase | IL:oil: 1 : 1 mg:g temperature: 40 °C time: 24 h molar ratio of methanol:oil: 4 : 1 enzyme: 20 wt% | 95.4 (Y) | — | [34] |
| Soybean oil | Transesterification | Screw capped vial | $[C_1C_3OHPyr]NTf_2$ | *Candida rugosa* lipase | IL: 1.0 g temperature: 25 °C time: 2 h molar ratio of methanol:oil: 2.5 : 1 enzyme: 30 mg | 82.4 (Y) | — | [35] |
| *Chlorella vulgaris* or *Chlorella protothecoides* | One-pot synthesis | Screw capped vial | $[C_{16}mim][NTf_2]$ and [Bmim][Cl] (95 : 5 (w/w)) | Candida Antarctica lipase B | IL:microalgae: 1.0 g: 0.5 g temperature: 60 °C time: 2 h molar ratio of methanol:oil: 6 : 1 enzyme: 18 wt% | 100.0 (Y) | 5 | [36] |

$[NTf_2]$ anion in hydrophobic IL served as an outstanding reaction medium for biodiesel synthesis, and [Bmim][Cl] in hydrophilic IL served as a suitable solvent for dissolution of cellulosic biomass. The separation of biodiesel was simple and clean via iterative centrifugation, paving way for the recovery and reuse of binary mixture of ILs enzyme catalytic system. In summary, anion of ILs added a significant effect on biodiesel yield than cation counterparts of ILs [37]. Hydrophobic IL was preferred as an ideal medium for different lipase-catalyzed reactions since it accounted for high enzyme activity, high solubility of substrate, and high solvent viscosity [38].

## 13.6 Optimization and Novel Biodiesel Production Technologies Using ILs

In order to increase the process efficiency and to make the biodiesel product cheaper, the application of the optimal reaction conditions for the selected feedstock, alcohol, and catalyst system is desired. In developing and optimizing the biodiesel production process, a statistical design of experiments is important, because it allows the simultaneous analysis of the effects of many process variables and their interactions. The response surface methodology (RSM) is commonly used in defining the optimal reaction conditions for achieving the maximum esters yield.

The application of microwave-assisted heating system and ultrasound-assisted IL catalytic system for biodiesel production lessens reaction times, enhances catalytic activities, and improves the selectivity and yield with operational simplicity and environmentally benign characteristics. In microwave-assisted heating, the samples are directly heated through the interaction between microwave energy and molecular dipole moments of the materials [39], which increases the rate of mass transfer and results in the rupture of cell wall due to localized high temperature and pressure gradients [40]. The ILs, which are thermally stable, absorb microwave energy and enhance the efficiency of IL-catalyzed reactions [12]. In ultrasonication, low frequency but high intensity ultrasonic waves cause collisions between the reactants during vibration and turbulence resulting from ultrasonic cavitation effect [41]. Ultrasonication offers mechanical energy for mixing, reduces the viscosity of IL, and increases the mass transfer rate, thereby increasing the rate of transesterification compared to mechanical agitation. Due to consumption of high energy and power during ultrasonication, it is thought provoking when applied in industrial-scale operation [42]. The review of latest investigations of microwave- and ultrasound-assisted biodiesel production using IL as catalysts is depicted in Table 13.4.

The study investigated by Yan et al. [43] indicated that microwave heater consumed 70.24% if the total energy whereas ultrasound consumed only 29.76% of the total energy. The combined assistance of ultrasonic microwave decreased the reaction time from 180 to 15 min, and hence the study declared that usage of ultrasonic microwave industrial reactor for biodiesel production is energy efficient and time saving. The study conducted by Ding et al. [44] concluded that the reaction rate is enhanced by microwave application, which saved 44% of total energy consumption. The results specified that biodiesel conversion of >90% can be attained within 4 h rather than 8 h using a conventional heating system with a power of 150 W. The study performed by Handayani et al. [46] found that mixed usage of

**Table 13.4** Novel technologies of biodiesel production and optimization using ionic liquids.

| Feedstock | Type of reaction | Type of reactor | Technology adopted | Catalyst | Optimization reaction conditions | Optimization method | Conversion or yield% | References |
|---|---|---|---|---|---|---|---|---|
| Oleic acid | Esterification | Ultrasonic microwave reactor | Ultrasonic microwave combined intensification | [Bsmim]$HSO_4$ | MW power: 117.52 W, ultrasonic power: 50 W reaction time: 17.24 min, M/O molar ratio: 22, ILs/O mass ratio: 11.15 | Response surface methodology | 97.92% | [43] |
| Palm oil | Transesterification | Microwave reactor | Microwave irradiation | ([$HSO_3$-Bmim] $HSO_4$) | Methanol:oil: 11 : 1 catalyst loading: 9.17 wt%<br>MW power: 168 W<br>reaction time: 6.43 h | Response surface methodology | 98.93% (Y) | [44] |
| Wet microalgal *Nannochloropsis* sp. biomass | Transesterification | Microwave reactor | Microwave irradiation | [Emim] [MeSO4] | Wet algae to methanol (wt/vol): 1 : 4, methanol:IL: 1 : 0.5, reaction time: 14.05 min | Response surface methodology | 40.9% (Y) | [45] |
| *Calophyllum inophyllum* oil | Transesterification | Microwave-assisted batch reactor | Microwave irradiation | (Bmim$HSO_4$) | Methanol:oil: 40 : 1 catalyst loading: 15 wt%<br>temperature: 60 °C<br>time: 120 min | — | 93.99 (Y) | [46] |
| Palm oil | Transesterification | Microwave heated continuous flow reactor | Microwave irradiation | ChOH | Methanol to oil: 13.2 : 1, flow rate: 20 ml $min^{-1}$, MWpower: 800 W, catalyst loading: 6 wt% | Design of experiments and response surface methodology | 89.72% | [47] |
| Oleic acid | Esterification | Double-jacketed glass flask | Ultrasonic irradiation of 20 kHz | [DABCODBS] $[CF_3SO_3]_2$ | Methanol:oil: catalyst loading: 0.64 mol%<br>temperature: 59 °C<br>time: 83 min<br>amplitude: 60% in continuous mode | Response surface methodology | 95.43% | [48] |
| Oleic acid | Esterification | Double-jacketed glass flask | Ultrasonic irradiation of 20 kHz | [DABCODBS] $[CF_3SO_3]_2$ | Methanol:oil: 9 : 1 catalyst loading: 0.4 mol%<br>temperature: 60 °C<br>time: 1 h | — | 89% | [49] |

IL [$BmimHSO_4$] and NAOH (0.5 wt% ($BmimHSO_4$) + 0.5 wt% NaOH) with microwave heating enhanced the rate of reaction and increased the yield of biodiesel production to 93.99%. Also Wahidin et al. [45] reported that synergetic effect of combining IL-methanol solvent and microwave heating at optimized reaction conditions leads to the production of high biodiesel yield per dry biomass of 42.22%. The study conducted by Masri et al. [48] depicted that there was a significant decrease in reaction time from 420 to 83 min using ultrasonication with the biodiesel conversion value of 93.20%. Increasing microwave power increases the temperature, which in turn increases the methyl ester content. Molecules are more active at high temperature than at low temperature. However, at high temperature methanol evaporation becomes a concern in biodiesel production using microwave technology [50]. Masri et al. [49] reported that the conversion of biodiesel using $SO_3H$-functionalized dicationic IL by mechanical stirring was 59%, which is comparatively lower than by ultrasonication-assisted method of biodiesel production with a conversion of 89%.

## 13.7 Recyclability of the Ionic Liquids on Biodiesel Production

The recyclability of catalysts used in the biodiesel production process is of much concern due to the increased cost of biodiesel production. Since ILs are usually 2–100 times more affluent than organic solvents [51], their recovering and reusing will sway the overall process cost significantly [52]. Moreover, dumping of waste catalyst may have an adverse effect on the environment. So recycling of ILs decreases the cost of biodiesel production and lessens the disposal of ILs as waste, thus minimizing its environmental impacts. The recyclability of ILs in esterification and transesterification reactions was investigated by almost all researchers in their testified studies.

Recyclability can encompass separation, purification, and reuse of ILs in sequential stages. At first, ILs are recovered from reactants and products, then refined if needed, and finally it is recycled. The process of recycling is governed by the type of ILs. In chemical reactions, hydrophobic ILs are more advantageous, since it results in a multiphase system at the end of the reaction, which can be recovered by membrane technology more easily than hydrophilic ones. Addition of potassium salt or sucrose [53] to the hydrophilic ILs enables easier phase separation and recycling.

### 13.7.1 Recovery of ILs

Since ILs are partially miscible in alkyl esters, its recovery is much easier than other heterogeneous catalysts. The techniques such as distillation, super critical $CO_2$ extraction, membrane separation technology, crystallization, and sorption are used for recovering ILs. Heat-based methods can also be employed for recovering ILs due to its high thermal stability and low vapor pressure. ILs can be recovered after abstraction of volatile components by distillation. Nonetheless, recovery of IL by distillation is not promising, when IL should be recovered from the components with a very high boiling point. In few cases, ILs can be recovered by $scCO_2$ extraction as ILs insoluble in $scCO_2$, whereas organic compounds are soluble. During crystallization, crystal formation results from merely cooling the mixture below certain temperature. The temperature at which an IL solidifies is noteworthy for crystallization after which it can be separated from the liquid mixture by filtration under vacuum. Crystallization of IL can also be executed with addition of an antisolvent. The IL, ChCl, is recovered from glycerol by adding n-butanol as an antisolvent followed by cooling the mixture to −20 °C [54]. Due to variation in size and charge of ILs from other components, ILs can either be permeated or retained using suitable membrane technology. Dicationic ILs exhibit higher retention values since the size of the cation determines the effectiveness of the method [55]. In addition, as dicationic ILs are in solid form, they were easily recovered at the end of the reaction by centrifugation, contrary to conventional acids and monocationic ILs system, which were dissolved in this same reaction mixture. On the other hand, after ultrasound-assisted esterification reaction, the reaction mixture was on homogeneous form with the IL catalysts severely dispersed in the phase. Therefore, fatty acid methyl ester, reactants, and catalyst were separated via liquid–liquid extraction by addition of water into the system, tailed by centrifugation [49].

Physical adsorption is a desirable option as it is simple with the regeneration of the adsorbent, which lowers the cost of operation [56]. Activated carbon is employed for recovering of imidazolium-based ILs from an aqueous solution [57]. The hydrophobicity and size of IL as well as the surface chemistry of adsorbent influence the adsorbent capacity. Among numerous sorption methods, the ion-exchange resin (Amberlite IRN-150) and activated carbon are the most competent in the sorption of [Bmim] cation and in the recovery of [Bmim][Cl] [58].

### 13.7.2 Reuse of Ionic Liquids

There was no significant decrease in the catalytic activity of the most recycled ILs even after being used more than a few times. In general, ILs are highly miscible with organic solvents [59], and some impurities inevitably hoard in the IL catalyst during the recycling process, thus dropping the IL concentration [60]. Moreover, some inhibiting compounds may be formed in the production process and poison the catalyst, thus lowering its activity [61]. IL is unglued from glycerol and few other raw materials by rotary evaporator (−0.10 Mpa, 80 °C) for reuse [44]. The probable reason for the decrease in the oil conversion within five cycles was mostly owing to the accumulation of triglycerides or other impurities on the solid catalyst surface. The sharp decrease in the catalytic activity after recycling run could be ascribed to major leaching of AIL from the MOF support [28]. The reusability test performed by Phrompthilak et al. indicated that the ChOH IL catalyst may not be appropriate to be reused several times in continuous flow microwave irradiation system because of possible contamination and decomposition (Phromphithak et al. [47]). The latest review on catalytic activity of recycled ILs is showed in Table 13.5, where the number of recycling and the conversion/ester yield in the first and final steps is depicted.

## 13.8 Kinetics of IL-Catalyzed Biodiesel Production

The information about transesterification reaction kinetics is a prerequisite for the production process development, design, operation, scale-up, analysis, and control. Roman et al. [16] evaluated the kinetic parameters upon esterification of oleic acid using 1-methylimidazolium hydrogen sulfate, [HMIM]$HSO_4$ IL, and found that the reaction was

**Table 13.5** Reusability of ILs in biodiesel production.

| Feedstock | Type of reaction | Name of ionic liquid | No. of cycles | First cycle conversion or yield% | Final cycle conversion or yield% | References |
|---|---|---|---|---|---|---|
| Oleic acid | Esterification | $[Bsmim]HSO_4$ | 5 | 97.92% | >95% | [43] |
| Palm oil | Transesterification | $([HSO_3\text{-}BMIM]HSO_4)$ | 6 | 98.93% | 87.65% | [44] |
| Oleic acid | Esterification | MMFP-AIL | 4 | 95% | 88% | [33] |
| Oleic acid | Esterification | $[DABCODBS][CF_3SO_3]_2$ | 5 | 93.11% | 89.03 | [48] |
| Oleic acid | Esterification | $[(CH_2COOH)_2IM]$ $HSO_4$@H-UiO-66 | 5 | 93.82 | 90.95 | [26] |
| Soybean oil | One-pot synthesis | AIL/HPMo/ MIL-100(Fe) | 5 | 95.8 | 90.3 | [28] |
| Soybean oil | Transesterification | CCH | 5 | 95.2 | 83% | [29] |
| Palm oil | Transesterification | ChOH | 5 | 89.72% | 49.3% | [47] |

governed by third order kinetics with an activation energy of 6.8 kJ $mol^{-1}$, which depicted that the employed IL catalyst can be a replacement of traditional acidic catalyst. Ding et al. [44] conducted a kinetic research on transesterification of palm oil catalyzed by $[HSO_3\text{-}BMIM]HSO_4$ under microwave irradiation and found that the reaction followed pseudo-first order kinetic model with activation energy of 56.12 kJ $mol^{-1}$. Panchal et al. [30] designed a novel solid acidic IL polymer PIL for transesterification of tung nut oil and reported that the transesterification reaction is governed by first order rate kinetics and activation energy was found to be 72.81 kJ $mol^{-1}$. Casiello et al. [62] performed a kinetic study on transesterification of soybean oil using methanol in the presence of binary IL catalyst ZnO/TBAI (TBAI = tetrabutylammonium iodide). The results showed that the reaction followed pseudo-first order kinetics with an activation energy of 48.46 kJ $mol^{-1}$, which is in agreement with the reports for transesterification activation energy values ranging between 26 and 115 kJ $mol^{-1}$. Pan et al. [33] investigated the esterification of oleic acid using an acidic IL (IL)-functionalized mesoporous melamine formaldehyde polymer (MMFP-IL) and reported that the reaction was governed by first order and activation energy ($Ea = 35.3$ kJ $mol^{-1}$) was relatively low, leading to its high catalytic activity. Thermodynamic studies revealed that reaction system was unspontaneous, endothermic, and endergonic in nature. Interestingly, the MMFP-IL catalyst was heterogeneous, stable, and recyclable.

## 13.9 Techno-Economic Analysis and Environmental Impact Analysis of Biodiesel Production Using Ionic Liquid as Catalyst

Several techno-economic analyses on biodiesel production have been investigated in the literature. However, techno-economic and environmental impact analysis of biodiesel production using ILs as catalysts was not reported in the literature. Naveenkumar and Baskar [63] executed the techno-economic analysis of 21 million kg $yr^{-1}$ biodiesel production from *Calophyllum inophyllum* oil using Zn-doped CaO nanocatalyst and reported that the annual biodiesel revenue was 15224000 $ yr$^{-1}$, and the payback period was about 1.15 years. Sae-Ngae et al. [64] conducted techno-economic and environmental impact analysis on biovalorization of agro-industrial wastes for biodiesel feedstocks by oleaginous yeasts. The study reported that highest profitability and lowest environmental impact per kilogram of biodiesel is possible when glycerol waste is used as a biodiesel feedstock, resulting in zero-waste process and circular economy. The greenhouse gas (GHG) emission was considered as a vital parameter in environmental impact analysis study, which was quantified through the metric of global warming potential (GWP). The study suggested that combined process of acid hydrolysis with sterilization, use of non-sterile process, coproduct utilization, and optimization might potentially reduce GHG emissions. The pooled use of techno-economic and environmental impact assessment could be useful to offer information for policy makers and producers in finding the possible bottlenecks and selecting appropriate enactments.

## 13.10 Conclusion

The use of ILs as promising and environmentally benign catalyst for biodiesel production in the recent years was investigated extensively. Despite its advantages, there are few shortcomings in using ILs as a catalyst for cost-effective biodiesel production. Further, imminent investigations on techno-economic analysis, life cycle assessment, environmental impact assessment, and scale-up technologies would be essential to commercialize biodiesel when ILs are employed as catalysts.
Abbreviations

| | |
|---|---|
| ($[HSO_3$-BMIM$]HSO_4$) | 1-Butyl-3-methylimidazolium hydrogen sulfate |
| ([TBP]$[NTf_2]$) | Tetrabutylphosphonium bis(trifluoromethylsulfonyl) imide |
| ($BMIMHSO_4$) | 1-Butyl-3-methylimidazolium hydrogen sulfate |
| (BNPs-CCH) | Boehmite nanoparticles-chlorocholine hydroxide |
| (MMFP-AIL) | Mesoporous melamine-formaldehyde polymer-acidic ionic liquids |
| (β-CD-IL-$Fe_3O_4$) | [β-Cyclodextrin-6-Im-$(CH_2)$3-$HSO_3$]$[HSO_4]$-$Fe_3O_4$ |
| [Bmim][Cl] | 1-Butyl-3-methylimidazolium chloride |
| [Bmim][Cl] | 1-Butyl-3-methylimidazolium chloride |
| [Bmim]OH | 1-Butyl-3-methylimidazolium hydroxide |
| [Bsmim]$HSO_4$ | 1-(4-Sulfobutyl)-3-methylmidazolium hydrosulfate |
| $[C_{16}$mim]$[NTf_2]$ | 1-Hexadecyl-3-methylimidazolium bis9trifluoromethylsulfonyl)imide |
| $[C_1C_3$OHPyr]$NTf_2$ | N-methyl-N-propanolpyrrolidinium bis(trifluoromethanesulfonyl)imide |
| [DABCODBS]$[CF_3SO_3]_2$ | Di-(4-sulfonic acid) dibutyl-triethylene diammonium di(trifluoromethanesulfonate) |
| [EMIM][MeSO4] | 1-Ethyl-3-methylimmidazolium methyl sulfate |
| [Hmim]$[PF_6]$ | 1-Hexyl-3-methyl-imidazolium-hexafluorophosphate |

| | |
|---|---|
| $[HMIM]HSO_4$ | 1-Methylimidazolium hydrogen sulfate |
| $[H_{NMP}]CH_3SO_3$ | N-methyl-2-pyrrolidonium methyl sulfonate |
| $[TTMPP-PS][CF_3SO_3]$ | Tris(2,4,6-trimethoxyphenyl)phosphine-1,3-propanesulfonate trifluoromethanesulfonate |
| CaO | Calcium oxide |
| ChCl | Choline chloride |
| ChOH | Choline hydroxide |
| $Fe_3O_4$@HKUST-1-ABILs | $Fe_3O_4$@HKUST-1amino-functionalized basic ionic liquids |
| FnmS-PILs | ($Fe_3O_4$@SiO2@SBA-15)-poly ionic liquids |
| ILs | Ionic liquids |
| wt% | Weight % |

## References


1 Zhao, H. and Baker, G.A. (2013). Ionic liquids and deep eutectic solvents for biodiesel synthesis: a review. *J. Chem. Technol. Biotechnol.* **88** (1): 3–12.

2 Ratti, R. (2014). Ionic liquids: synthesis and applications in catalysis. *Adv. Chem.* **2014**: 1–16.

3 Payagala, T., Huang, J., Breitbach, Z.S. et al. (2007). Unsymmetrical dicationic ionic liquids: manipulation of physicochemical properties using specific structural architectures. *Chem. Mater.* **19** (24): 5848–5850.

4 Welton, T. (1999). Room-temperature ionic liquids. Solvents for synthesis and catalysis. *Chem. Rev.* **99** (8): 2071–2083.

5 Bovio, S., Podestà, A., and Milani, P. (2010). *Ionic Liquids: From Knowledge to Application.* American Chemical Society.

6 Wu, Q., Wan, H., Li, H. et al. (2013). Bifunctional temperature-sensitive amphiphilic acidic ionic liquids for preparation of biodiesel. *Catal. Today* **200** (1): 74–79.

7 Fan, M., Huang, J., Yang, J., and Zhang, P. (2013). Biodiesel production by transesterification catalyzed by an efficient choline ionic liquid catalyst. *Appl. Energy* **108**: 333–339.

8 Ishak, Z.I., Sairi, N.A., Alias, Y. et al. (2017). A review of ionic liquids as catalysts for transesterification reactions of biodiesel and glycerol carbonate production. *Catal. Rev. Sci. Eng.* **59** (1): 44–93.

9 He, L., Qin, S., Chang, T. et al. (2013). Biodiesel synthesis from the esterification of free fatty acids and alcohol catalyzed by long-chain Brønsted acid ionic liquid. *Catal. Sci. Technol.* **3** (4): 1102–1107.

10 Tiong, Y.W., Yap, C.L., Gan, S., and Yap, W.S.P. (2018). Conversion of biomass and its derivatives to levulinic acid and levulinate esters via ionic liquids. *Ind. Eng. Chem. Res.* **57** (14): 4749–4766.

11 Hajipour, A.R. and Rafiee, F. (2009). Basic ionic liquids. A short review. *J. Iran. Chem. Soc.* **6** (4): 647–678.

12 Troter, D.Z., Todorović, Z.B., Dokić-Stojanović, D.R. et al. (2016). Application of ionic liquids and deep eutectic solvents in biodiesel production: a review. *Renew. Sustain. Energy Rev.* **61**: 473–500.

**13** Li, M., Chen, J., Li, L. et al. (2021). Novel multi -$SO_3H$ functionalized ionic liquids as highly efficient catalyst for synthesis of biodiesel. *Green Energy Environ.* 6: 271–282. https://doi.org/10.1016/j.gee.2020.05.004.

**14** Ullah, Z., Bustam, M.A., Man, Z. et al. (2019). Phosphonium-based hydrophobic ionic liquids with fluorous anions for biodiesel production from waste cooking oil. *Int. J. Environ. Sci. Technol.* **16** (3): 1269–1276.

**15** Han, B., Yin, F., Liu, S. et al. (2019). Optimization of oleic acid esterification for biodiesel production using brønsted acidic ionic liquid as a catalyst. *Chiang Mai J. Sci.* **46** (4): 714–726.

**16** Roman, F.F., Ribeiro, A.E., Queiroz, A. et al. (2019). Optimization and kinetic study of biodiesel production through esterification of oleic acid applying ionic liquids as catalysts. *Fuel* **239** (November 2018): 1231–1239.

**17** Wang, Y., Zhao, D., Chen, G. et al. (2019). Preparation of phosphotungstic acid based poly(ionic liquid) and its application to esterification of palmitic acid. *Renew. Energy* **133**: 317–324.

**18** Zhang, A., Wang, Q., He, Y. et al. (2020). Preparation of biodiesel based on alkaline ionic liquid [Bmim]OH catalyzed castor oil. *IOP Conf. Ser. Mater. Sci. Eng.* **729** (1): 012048. https://doi.org/10.1088/1757-899X/729/1/012048.

**19** Gholami, A., Pourfayaz, F., Hajinezhad, A., and Mohadesi, M. (2019). Biodiesel production from Norouzak (*Salvia leriifolia*) oil using choline hydroxide catalyst in a microchannel reactor. *Renew. Energy* **136**: 993–1001.

**20** Zhang, H., Li, H., Pan, H. et al. (2018). Magnetically recyclable acidic polymeric ionic liquids decorated with hydrophobic regulators as highly efficient and stable catalysts for biodiesel production. *Appl. Energy* **223** (January): 416–429.

**21** Xie, W. and Wan, F. (2018). Basic ionic liquid functionalized magnetically responsive $Fe_3O_4$@HKUST-1 composites used for biodiesel production. *Fuel* **220** (January): 248–256.

**22** Chang, F. and Zhou, Q. (2020). Facile preparation of β-cyclodextrin-$Fe_3O_4$ magnetic-Brønsted acidic ionic liquid for biodiesel production. *Waste Biomass Valori.* **11** (5): 1983–1988. https://doi.org/10.1007/S12649-018-0506-4.

**23** Xie, W. and Wang, H. (2021). Grafting copolymerization of dual acidic ionic liquid on core-shell structured magnetic silica: a magnetically recyclable Brönsted acid catalyst for biodiesel production by one-pot transformation of low-quality oils. *Fuel* **283**: 118893.

**24** Xie, W. and Wang, H. (2020). Immobilized polymeric sulfonated ionic liquid on core-shell structured $Fe_3O_4/SiO_2$ composites: a magnetically recyclable catalyst for simultaneous transesterification and esterifications of low-cost oils to biodiesel. *Renew. Energy* **145**: 1709–1719.

**25** Xie, W. and Wang, H. (2020). Synthesis of heterogenized polyoxometalate-based ionic liquids with Brönsted-Lewis acid sites: a magnetically recyclable catalyst for biodiesel production from low-quality oils. *J. Ind. Eng. Chem.* **87**: 162–172.

**26** Ye, C., Qi, Z., Cai, D., and Qiu, T. (2019). Design and synthesis of ionic liquid supported hierarchically porous Zr metal-organic framework as a novel Brønsted-Lewis acidic catalyst in biodiesel synthesis. *Ind. Eng. Chem. Res.* **58** (3): 1123–1132.

**27** Xie, W. and Wan, F. (2019). Immobilization of polyoxometalate-based sulfonated ionic liquids on UiO-66-2COOH metal-organic frameworks for biodiesel production via one-pot transesterification-esterification of acidic vegetable oils. *Chem. Eng. J.* **365** (January): 40–50.

**28** Xie, W. and Wan, F. (2019). Biodiesel production from acidic oils using polyoxometalate-based sulfonated ionic liquids functionalized metal–organic frameworks. *Catal. Lett.* **149** (10): 2916–2929.

**29** Hosseini, S., Moradi, G.R., and Bahrami, K. (2019). Synthesis of a novel stabilized basic ionic liquid through immobilization on boehmite nanoparticles: a robust nanocatalyst for biodiesel production from soybean oil. *Renew. Energy* **138**: 70–78.

**30** Panchal, B., Chang, T., Qin, S. et al. (2019). Optimization and kinetics of tung nut oil transesterification with methanol using novel solid acidic ionic liquid polymer as catalyst for methyl ester synthesis. *Renew. Energy* **151**: 796–804.

**31** Chen, J., Li, M., Li, M. et al. (2020). Self-solidifying quaternary phosphonium-containing ionic liquids as efficient and reusable catalysts for biodiesel production. *ACS Sustain. Chem. Eng.* **8** (18): 6956–6963.

**32** Fan, M., Liu, H., and Zhang, P. (2018). Ionic liquid on the acidic organic-inorganic hybrid mesoporous material with good acid-water resistance for biodiesel production. *Fuel* **215** (November 2017): 541–550.

**33** Pan, H., Li, H., Zhang, H. et al. (2019). Acidic ionic liquid-functionalized mesoporous melamine-formaldehyde polymer as heterogeneous catalyst for biodiesel production. *Fuel* **239** (November 2018): 886–895.

**34** Su, F., Peng, C., Li, G.L. et al. (2016). Biodiesel production from woody oil catalyzed by Candida rugosa lipase in ionic liquid. *Renew. Energy* **90**: 329–335.

**35** Fan, Y., Wang, X., Zhang, L. et al. (2018). Lipase-catalyzed synthesis of biodiesel in a hydroxyl-functionalized ionic liquid. *Chem. Eng. Res. Des.* **132**: 199–207.

**36** Lozano, P., Bernal, J.M., Gómez, C. et al. (2020). Green biocatalytic synthesis of biodiesel from microalgae in one-pot systems based on sponge-like ionic liquids. *Catal. Today* **346**: 87–92.

**37** Liu, Y., Chen, D., Yan, Y. et al. (2011). Biodiesel synthesis and conformation of lipase from Burkholderia cepacia in room temperature ionic liquids and organic solvents. *Bioresour. Technol.* **102** (22): 10414–10418.

**38** Yang, Z. and Pan, W. (2005). Ionic liquids: green solvents for nonaqueous biocatalysis. *Enzyme Microb. Technol.* **37** (1): 19–28.

**39** Wahidin, S., Idris, A., and Shaleh, S.R.M. (2016). Ionic liquid as a promising biobased green solvent in combination with microwave irradiation for direct biodiesel production. *Bioresour. Technol.* **206**: 150–154.

**40** Martinez-Guerra, E., Howlader, M.S., Shields-Menard, S. et al. (2018). Optimization of wet microalgal FAME production from *Nannochloropsis* sp. under the synergistic microwave and ultrasound effect. *Int. J. Energy Res.* **42** (5): 1934–1949.

**41** Gole, V.L. and Gogate, P.R. (2012). Intensification of synthesis of biodiesel from nonedible oils using sonochemical reactors. *Ind. Eng. Chem. Res.* **51** (37): 11866–11874.

**42** Patle, D.S., Sharma, S., Gadhamsetti, A.P. et al. (2018). Ultrasonication-assisted and benzimidazolium-based Brønsted acid ionic liquid-catalyzed transesterification of castor oil. *ACS Omega* **3** (11): 15455–15463.

**43** Yan, J., Zhao, Y., Li, K. et al. (2020). Efficient production of biodiesel from ionic liquid catalyzed esterification using ultrasonic-microwave combined intensification. *Chem. Eng. Process. Process Intensif.* **149**: 107870.

**44** Ding, H., Ye, W., Wang, Y. et al. (2018). Process intensification of transesterification for biodiesel production from palm oil: microwave irradiation on transesterification reaction catalyzed by acidic imidazolium ionic liquids. *Energy* **144** (2018): 957–967.

45 Wahidin, S., Idris, A., Yusof, N.M. et al. (2018). Optimization of the ionic liquid-microwave assisted one-step biodiesel production process from wet microalgal biomass. *Energy Convers. Manag.* **171** (March): 1397–1404.

46 Handayani, P.A., Abdullah, A., and Hadiyanto, H. (2017). Biodiesel production from Nyamplung (*Calophyllum inophyllum*) oil using ionic liquid as a catalyst and microwave heating system. *Bull. Chem. React. Eng. Catal.* **12** (2): 293–298.

47 Phromphithak, S., Meepowpan, P., Shimpalee, S., and Tippayawong, N. (2020). Transesterification of palm oil into biodiesel using ChOH ionic liquid in a microwave heated continuous flow reactor. *Renew. Energy* **154**: 925–936.

48 Masri, A.N., Abdul Mutalib, M.I., Yahya, W.Z.N. et al. (2020). Rapid esterification of fatty acid using dicationic acidic ionic liquid catalyst via ultrasonic-assisted method. *Ultrason. Sonochem.* **60** (November 2018): 104732.

49 Masri, A.N., Abdul Mutalib, M.I., Aminuddin, N.F., and Lévêque, J.M. (2018). Novel $SO_3H$-functionalized dicationic ionic liquids – a comparative study for esterification reaction by ultrasound cavitation and mechanical stirring for biodiesel production. *Sep. Purif. Technol.* **196** (August 2017): 106–114.

50 Zhang, H., Ding, J., and Zhao, Z. (2012). Microwave assisted esterification of acidified oil from waste cooking oil by CERP/PES catalytic membrane for biodiesel production. *Bioresour. Technol.* **123**: 72–77.

51 Plechkova, N.V. and Seddon, K.R. (2008). Applications of ionic liquids in the chemical industry. *Chem. Soc. Rev.* **37** (1): 123–150.

52 Wu, B., Liu, W.W., Zhang, Y.M., and Wang, H.P. (2009). Do we understand the recyclability of ionic liquids? *Chem. Eur. J.* **15** (8): 1804–1810.

53 Wu, B., Zhang, Y.M., and Wang, H.P. (2008). Aqueous biphasic systems of hydrophilic ionic liquids + sucrose for separation. *J. Chem. Eng. Data* **53** (4): 983–985.

54 Hayyan, M., Mjalli, F.S., Hashim, M.A., and AlNashef, I.M. (2010). A novel technique for separating glycerine from palm oil-based biodiesel using ionic liquids. *Fuel Process. Technol.* **91** (1): 116–120.

55 Mai, N.L., Ahn, K., and Koo, Y.M. (2014). Methods for recovery of ionic liquids – a review. *Process Biochem.* **49** (5): 872–881.

56 Francisco, M., Mlinar, A.N., Yoo, B. et al. (2011). Recovery of glucose from an aqueous ionic liquid by adsorption onto a zeolite-based solid. *Chem. Eng. J.* **172** (1): 184–190.

57 Palomar, J., Lemus, J., Gilarranz, M.A., and Rodriguez, J.J. (2009). Adsorption of ionic liquids from aqueous effluents by activated carbon. *Carbon* **47** (7): 1846–1856.

58 Vijayaraghavan, K., Pham, T.P.T., Cho, C.W. et al. (2009). An assessment on the interaction of a hydrophilic ionic liquid with different sorbents. *Ind. Eng. Chem. Res.* **48** (15): 7283–7288.

59 Zhang, P., Sun, Y., Zhang, Q. et al. (2018). Upgrading of pyrolysis biofuel via esterification of acetic acid with benzyl alcohol catalyzed by Brønsted acidic ionic liquid functionalized ethyl-bridged organosilica hollow nanospheres. *Fuel* **228**: 175–186.

60 Huang, K., Wu, R., Cao, Y. et al. (2013). Recycling and reuse of ionic liquid in homogeneous cellulose acetylation. *Chinese J. Chem. Eng.* **21** (5): 577–584.

61 Seddon, K.R., Stark, A., and Torres, M.J. (2000). Influence of chloride, water, and organic solvents on the physical properties of ionic liquids. *Pure Appl. Chem.* **72** (12): 2275–2287.

62 Casiello, M., Catucci, L., Fracassi, F. et al. (2019). ZnO/ionic liquid catalyzed biodiesel production from renewable and waste lipids as feedstocks. *Catalysts* **9** (1): 1–14.

**63** Naveenkumar, R. and Baskar, G. (2020). Optimization and techno-economic analysis of biodiesel production from *Calophyllum inophyllum* oil using heterogeneous nanocatalyst. *Bioresour. Technol.* **315**: 123852.

**64** Sae-Ngae, S., Cheirsilp, B., Louhasakul, Y. et al. (2020). Techno-economic analysis and environmental impact of biovalorization of agro-industrial wastes for biodiesel feedstocks by oleaginous yeasts. *Sustainable Environ. Res.* **30** (1): 11.

# 14

# Metal–Organic Frameworks (MOFs) as Versatile Catalysts for Biodiesel Synthesis

*Vasudeva Rao Bakuru*[1], *Marilyn Esclance DMello*[1], *and Suresh Babu Kalidindi*[2]

[1] *Materials Science and Catalysis Division, Poornaprajna Institute of Scientific Research, Bangalore Rural, India*
[2] *Department of Inorganic and Analytical Chemistry, School of Chemistry, Andhra University, Visakhapatnam, India*

## 14.1 Introduction

Metal–organic frameworks (MOFs) are coordinated networks that are crystalline and porous in nature. MOFs are constructed by using two types of building blocks: (i) metal ions or clusters that are positively charged and (ii) the organic ligands that are anionic in nature. The strong and directional coordinate covalent bonds between these two building blocks form well-defined frameworks with diverse chemistry and stability [1]. For example, MOF-5 ($Zn_4O(BDC)_3$, $BDC^{2-}$ = benzene dicarboxylate) is formed by linking of $Zn_4O$ positively charged secondary building units (SBU)s with $BDC^{2-}$ anion. MOF-5 consists of octahedral $[Zn_4O]^{6+}$ clusters connected by benzene dicarboxylic acid (BDC) linker to form a porous cubic framework [2]. MOF crystal nucleation and growth could be achieved on slow deprotonation of the carboxylic acids in presence of bases such as dimethylformamide under specific conditions. Some common metal clusters are shown in Figure 14.1a. A wide range of MOFs (large number of topologies) with surface areas up to 10 000 $m^2\,g^{-1}$ (measured by Brunauer–Emmett–Teller (BET)) are synthesized by rational variation in metal ions/cluster and judicious selection of ditopic /polytopic/polydentate organic ligands [3]. Figure 14.1b shows some organic linkers used for the construction of MOFs.

Since 1990s, MOF constructions and applications have been established by various research groups (Kitagawa et al.; Yaghi et al.; and Ferey et al. to name a few), and today more than 20 000 MOF structures have been so far reported [4]. The thermodynamic stability of MOFs majorly depends on the metal–ligand bond strengths. High-valent metal ions possessing high charge density forms stronger coordination bonds. Hence, carboxylate-based ligands together with high-valent metal ions, such as $Zr^{4+}$, $Cr^{3+}$, $Ti^{4+}$, and $Al^{3+}$ based on UiO (University of Oslo) and Materials Institute Lavoisier (MIL) series, have exhibited remarkable stability in water and acidic conditions (for example, MIL-101, MIL-53, and UiO-66) [5–8]. In addition, nitrogen-containing imidazole, triazole, and pyrazolate-based

*Biodiesel Production: Feedstocks, Catalysts, and Technologies*, First Edition. Edited by Samuel Lalthazuala Rokhum, Gopinath Halder, Suttichai Assabumrungrat, and Kanokwan Ngaosuwan.

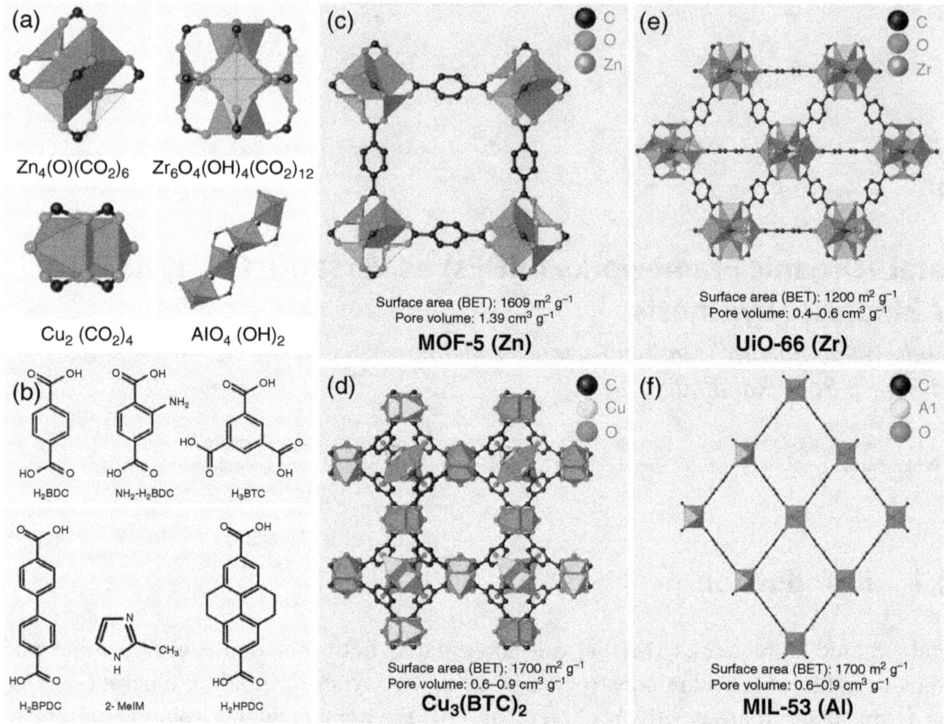


**Figure 14.1** Common examples of (a) metal clusters (secondary building units) in metal–organic frameworks, (b) organic linkers used for synthesis of metal–organic frameworks and structure and properties of (c) MOF-5 {$Zn_4O(BDC)_3$, $BDC^{2-}$ = benzene dicarboxylate}, (d) $Cu_3(BTC)_2$ {$BTC^{3-}$ = 1,3,5-benzenetricarboxylate}, (e) UiO-66, UiO-University of Oslo {$Zr_6O_4(OH)_4(BDC)_{12}$}, and (f) MIL-53, MIL-Materials Institute Lavoisier {$AlO_4(OH)_2(BDC)_2$, $BDC^{2-}$ = benzene dicarboxylate}.

MOFs assembled by using soft divalent metal ions including $Zn^{2+}$, $Ni^{2+}$, $Cu^{2+}$, and $Mn^{2+}$ have also shown exceptional stability in alkaline environments (for example, ZIF-8) [9]. Most common examples of MOFs are given in Figure 14.1c–f. The control over the geometry, connectivity, length ratio, and functional group of a linker can dictate the size, shape, and internal surface topology, hence tuning the structure of the overall MOF assembly for a targeted application [10]. Usually, the long-ranged ordered structure in MOFs is confirmed by powder X-ray diffraction (PXRD) patterns that characterizes the crystallinity of the sample and BET surface area derived from $N_2$ sorption studies confirms the porosity.

MOFs have been studied for a wide range of applications including gas storage, energy conversion, chemical and gas sensing, drug delivery, proton conductivity, and catalysis [10]. Ton scales of some MOFs are produced by Badische Anilin und Soda Fabrik (BASF), and few are available at Merck-Sigma Aldrich, MOF technologies, and with other chemical suppliers. Furthermore, MOF composition and tunable pore structures can commend them as catalysts. They offer attractive opportunities for synthesis with complementary catalytically active groups, numerous functional groups, and bifunctional catalysis/tandem reactions [11, 12]. The tailorable pore structures offer rapid transport of reactant and product molecules and extensive opportunities as shape-selective catalysts [11]. MOFs have

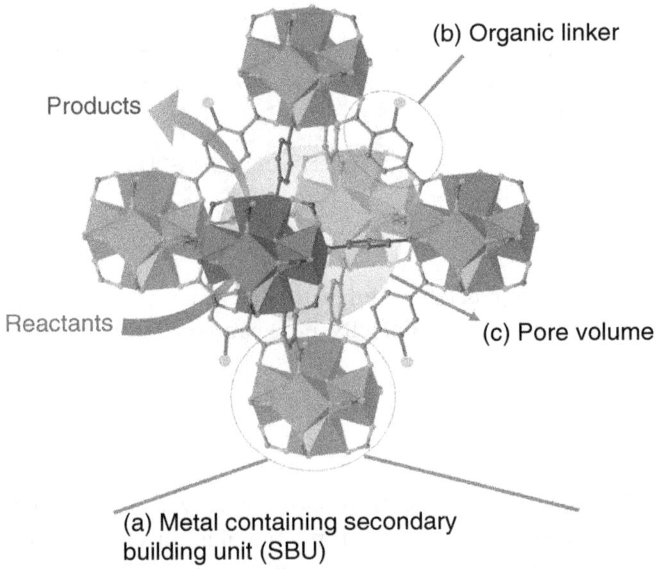


**Figure 14.2** Tunable active sites present in metal–organic frameworks: (a) metal-containing secondary building unit (SBUs), (b) organic linker, and (c) pore volume.

been widely explored for heterogeneous catalysts as they contain well-defined and tunable active sites at (i) secondary building units, (ii) the organic ligands, and (iii) pore volume (by encapsulation of active catalytic active species) (Figure 14.2).

### 14.1.1 Metal-Containing Secondary Building Units

In MOFs, SBUs mostly made up of transitional metal ions coordinate with electron-rich organic linkers to construct frameworks with uniform pores. Though the metal ions present in SBUs of many MOFs are coordinately saturated, at the same time significant number of MOFs in which SBUs containing open metal sites are also reported. In fact, as-synthesized MOFs with open metal SBUs will have weakly bound water or solvent molecules satisfying the coordination. These weakly bound molecules in general can be removed by heating MOFs at elevated temperatures under vacuum, giving rise to coordinately unsaturated open metal centers [13]. The open metal centers with unique coordination chemistry can act as Lewis acidic sites for catalysis. Also, Lewis acidic sites can stem from the defects present in MOFs [14]. It has been established that majority of MOFs contains noticeable amount of defect sites [15]. In MOFs such as UiO-66, missing linkers are present up to 10%, effectively reducing the coordination of SBU to ~11 instead of 12(Scheme 14.1) [16]. In UiO-66, each benzenedicarboxylic linker binds to four Zr(IV) ions on both ends. In case of a missing linker defect, to balance the charge, two of these Zr(IV) ion sites need to be saturated with $Cl^-$ or $OH^-$. As a result, two open metal centers (Lewis acid sites) are created for one missing linker defect. Further $\mu_3$-OH groups on $M_6$- SBU can exhibit Brønsted acidity depending on the metal with which it is coordinated. Unlike other materials, the Lewis acid centers in MOFs are present in a tunable porous environment.

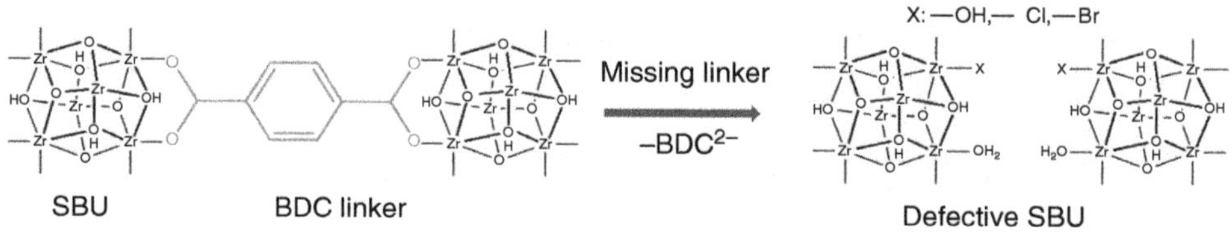


**Scheme 14.1** Schematic representation of creation of missing linker defects in UiO-66 MOF.

Due to this the sites are readily accessible to reactants, and further confinement effects and pore chemistry (such as hydrophilicity and hydrophobicity, etc.) can be additional riders for catalysis.

### 14.1.2 Organic Linker

Pore surface functionalization via organic linker modification has been widely used strategy to anchor desired active sites in MOFs. Introduction of functional groups ($-NH_2$, -OH, etc.) on linkers can turn an innocent MOF into catalytically active one. For example, by replacing the terephthalic acid by the 2-aminoterephthalic acid in Al-MIL-101, the MOF has been turned into active catalyst for the Knoevenagel condensation of benzaldehyde with malononitrile and with ethyl cyanoacetate at 80 °C in toluene [17, 18]. Furthermore, functional groups on linkers can be used in tandem with Lewis active sites on SBU for synergetic catalysis. For example, $NH_2$-MIL-101(Al) was used as a bifunctional catalyst (tandem transformation), site-isolated Lewis acidity arising from $Al^{3+}$ centers and Brønsted basic sites of amino terephthalate linkers for Meinwald rearrangement–Knoevenagel condensation reaction at 60 °C in acetonitrile [19].

### 14.1.3 Pore Volume

Controlled growth/encapsulation of catalytically active species in sufficiently large MOF pores not only immobilizes them in a solid matrix but also offers control over catalytic properties through pore chemistry. Many reactive species (metal oxide and metal nanoparticles, clusters, quantum dots, transition metal complexes, polyoxometalates [POMs], and organic molecules) have been immobilized/ encapsulated uniformly using MOF as hosts [20]. Synergy between the active sites of guests in the pores and active sites of the MOF led to the development of efficient catalysts for various important chemical transformations [21].

Biodiesel has turned out to be the most common biofuels having resemblances to petroleum-derived diesel in its main characteristics including octane number (45–67), energy content (37.27 MJ $kg^{-1}$), viscosity (3.6–5.0 cSt at 313 K), and phase changes. Commonly, biodiesel is produced from transesterification of vegetable oils or animal fats with the addition of methanol. Hence it is also known as fatty acid methyl ester (FAME). Although most commercial biodiesel is produced by using alkali homogenous catalysts (for example, KOH, NaOH, etc.) due to the high activating and corrosion resistant properties, they can be very sensitive in the presence of water and free fatty acids present in plant oils.

This will reduce the yield of biodiesel product and increase the cost of production. Furthermore, a smart choice of heterogeneous catalysts in the process of batch or continuous reaction may directly influence the large commercial production of biodiesel [22]. Despite the available choice of many heterogeneous solid-state catalysts like alkali earth metal oxides and transition metal oxides for producing biodiesel, they can be resistant to mass transfer and may render inefficient. This limitation can be minimized by using catalyst supports and effective catalysts with high surface area through existence of pores synergistically tunable with available active catalytic centers, preventing sintering in reaction medium.

Lately, the ever-expanding research in development of stable MOFs as new kinds of high-performance catalysts offers a wide playground for chemical transformations related to biodiesel synthesis. Overall, MOFs fall under unique class of heterogeneous catalysts in which all its constituents are tunable up to atomic level. This allows much desired structure–property correlation to be established.

## 14.2 Biodiesel Synthesis Over MOF Catalysts

Biodiesel is widely produced by two types of chemical transformations, namely, transesterification and esterification (Scheme 14.2). These two transformations yield fatty acid alkyl esters, which are commonly known as biodiesel.

Both transformations are catalyzed by either acid or base catalysts. MOFs (as acid/base catalysts) have been well studied for transesterification and esterification reactions. A good number of reports exist in the literature, and important studies have been summarized in this chapter.

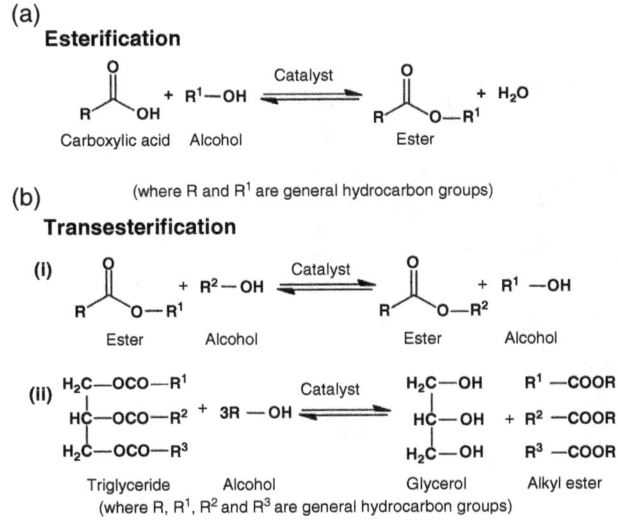


**Scheme 14.2** (a) General esterification of carboxylic acids and (b) transesterification of esters and triglycerides with alcohols.

### 14.2.1 Transesterification Reaction

Tranesterification is the most commonly used process for the large-scale production of biodiesel and involves the treatment of triglycerides (from oils) with short-chain alcohols(methanol/ethanol) to give fatty acid alkyl esters and glycerol(Scheme 14.2b). Glycerol is a valuable side product of this reaction and is often converted to value-added fine chemicals such as lactic acid, acrolein, and 1,3-propanediol.

#### 14.2.1.1 Transesterification at SBUs of MOFs

Nicolas et al. exploited defects present on the external surface (particle surface) of zeolite imidazolate framework (ZIF-8) for the transesterification (Figure 14.1a) of vegetable oil into biodiesel [23]. The ZIF-8 [Zn-(MeIm)$_2$)] (Figure 14.3a) is one of the interesting families of MOFs and is fabricated from zinc precursor and 2-methylimidazole (MeIm). The structure of ZIF-8 resembles the sodalite structure of zeolites. The ZIF-8 shows a high surface area (~1400 $m^2$ $g^{-1}$), a pore diameter of about 11 Å, and thermal stability up to 420 °C. Transesterification of vegetable oil has been carried out with different alcohols, and among all, methanol showed the highest conversions. At 473 K, the ZIF-8 catalyst exhibited >90% conversion of vegetable oil into monoglycerides in 2 h. Under the same reaction conditions, commercial $ZnAl_2O_4$ yielded ~55.0% of vegetable oil conversion. In ZIF-8, the window openings of the pores are about 3.3 Å, which is very small when compared to the sizes of triglycerides. Therefore, the activity of ZIF-8 has been ascribed to the active sites (acidic–basic sites) that are located at the external surface, but not in the microporosity of ZIF-8. The kind of active sites present on the external surface is established using FTIR (CO-adsorption) technique and density functional theory(DFT) calculations. Though a great variety of sites are present on the surface of ZIF-8, strong Zn (II) Lewis acid sites in combination with N– moieties and OH groups (basic ones) had played a key role in the combined activation of alcohols and esters during the reaction(Figure 14.3b).

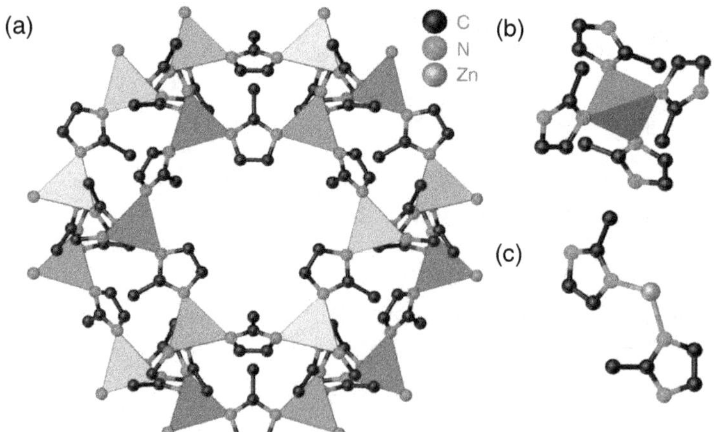


**Figure 14.3** (a) Structure of ZIF-8 MOF, (b) secondary building unit (SBU)-Zn(MeIM)$_2$ of ZIF-8, and (c) open metal site ($Zn^{2+}$) of ZIF-8 metal cluster.

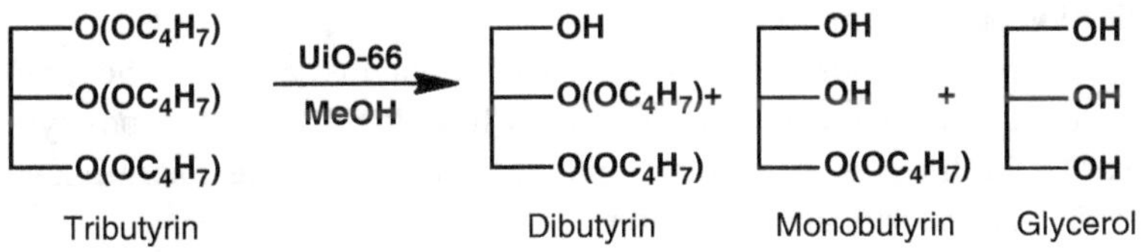


**Scheme 14.3** Transesterification of tributyrin with methanol.

Zirconium-based UiO-66 MOFs ($\{Zr_6O_4(OH)_4[(O_2C)-C_6H_4-(CO_2)]_6\}$) with missing linker defects are found to effectively catalyze transesterification of soybean oil with methanol(Scheme 14.3). The UiO-66 MOF was synthesized from Zr(IV)-precursor and (BDC) linker. It has two types of pores: octahedral pores (12 Å) and tetrahedral (7.5 Å) with pore windows of ~7 Å [24]. This MOF is thermally stable up to 550 °C and mechanically stable (10 tons $cm^{-2}$). The SBU($Zr_6O_4(OH)_4$) in UiO-66 can be linked with 12 BDC linkers. However, in this class of MOFs, missing linker defects (Lewis acidic sites) are very common, and rarely coordination number close to 12 has been achieved for $Zr_6$-SBUs in UiO-66. It is possible to control the number of defects in UiO-66 by manipulating the reaction conditions. Li et al. synthesized a series of UiO-66 (UiO-66-100-1, UiO-66-100-2, UiO-66-160-2, and UiO-66-220-2) MOFs with different amounts of defects by varying temperature and BDC to Zr (IV) (terephthalic acid/$ZrCl_4$) ratios (Table 14.1). Among all, UiO-66-100-1 showed the highest number of missing linker defects of 2.2 per SBU. The Lewis acidity produced out of these defects was analyzed using $NH_3$ temperature-programmed desorption (TPD) technique and found that four samples have both strong and weak Lewis acidic sites. The total amount of acidity is in correlation with the number of missing linker defects present. Owing to many acidic sites created by missing linker defects, UiO-66-100-1 exhibited the significant conversion of soybean oil (>98.5%, at 140 °C for 5 h) with methyl oleate as the main product. The performance of UiO-66-100-1 is much better when compared with $WO_3/ZrO_2$, $ZrO_2/Al_2O_3$, $SO_4^{2-}/SnO_2$, [25] La/zeolite beta, [26] and ETS-10 catalysts, which required much high temperature or longer reaction times to achieve such conversions.

**Table 14.1** The transesterification results over different catalysts.

| Catalyst | Zr/BDC molar ratio and synthesis temp. (°C) | No. of defects | Conversion (%) | Yield (%) | | |
|---|---|---|---|---|---|---|
| | | | | Dibutyrin | Monobutyrin | Glycerol |
| UiO-66-100-1 | 1.0, (100 °C) | 2.2 | 99.3 | 13.5 | 18.4 | 67.4 |
| UiO-66-100-2 | 0.5, (100 °C) | 1.7 | 81.9 | 38.9 | 10.7 | 32.3 |
| UiO-66-160-2 | 0.5, (160 °C) | 0.6 | 42.2 | 16.3 | 6.2 | 19.8 |
| UiO-66-220-2 | 0.5, (220 °C) | 0.2 | 38.8 | 20.5 | 8.6 | 9.7 |
| Blank | — | — | 6.3 | 6.3 | 0.0 | 0.0 |

***Reaction conditions:*** 120 °C, 1.3 mmol of tributyrin, 52 mmol methanol, 5 h.

#### 14.2.1.2 Transesterification at Linker Active Sites

Functionalized linkers are greatly explored in MOFs for heterogeneous catalysis. The functionalization of the linkers can be achieved through direct functional linkers or postsynthetic modification (PSM). So this class of tunable materials has been widely scrutinized for biodiesel synthesis via transesterification reactions.

Recently, Farrusseng and coworkers reported a bifunctional DMOF-$NH_2$ [$Zn_2$ (2-aminoterephthalate)$_2$(dabco)] (dabco = 1,4-diazabicyclo[2.2.2]octane) for the transesterification of ethyldecanoate [27]. The MOF contains $Zn_2$ paddle wheel SBUs and is connected with 1,4-bdc linker to form a distorted 2D square-grid {$Zn_2$(1,4-bdc)$_2$}, and further, the axial sites of SBUs are linked by dabco, which then act as pillars to form a 3D structure. PSM strategy through "click chemistry" was employed for linker modification within MOF. First, the amino groups (DMOF-$NH_2$-**1**) at the linker(amino-BDC) of the MOF are converted into their analogous azido (DMOF-$N_3$-**1a**). Further, click reaction was carried out at azided functionality using phenylacetylene and propargylamine to give **1b**(1,2,3-triazolyl) and **1c**(1,2,3-triazolyl with tertiary amine at the 4-position), respectively. MOF **1d** containing both these linker functionalities was also synthesized. (Figure 14.4). All the materials preserved their crystallinity even after PSM as confirmed by powder XRD. The **1b** shows moderate basicity due to the presence of triazolyle group (pKb ≈ 9.4) as well as strong lipophilicity. On the other hand, **1c** exhibits stronger basic nature (pKb ≈ 3 for trialkylamines) and lower lipophilicity. Transesterification of ethyldecanoate with MeOH was carried out over **1b**, **1c**, and **1d** base catalysts, and **1d** exhibited the highest conversion of 84% at 130 °C for 20 h. The optimized balance between basicity and lipophilicity in MOF **1d** led to better performance.

Recently, Zheng et al. reported UiO-66(Zr)-based solid superacids for transesterification of sunflower oil with methanol [27–29]. Typically, super acidic UiO-66-[$C_3NH_2$][$SO_4H$] and UiO-66-[$C_3NH_2$][$SO_3CF_3$] were synthesized in two-step process. In the first step, the $NH_2$-UiO-66(Zr) MOF was quaternized with 1,3-propane sultone, and in the next step, the quaternized MOFs were ion-exchanged with $H_2SO_4$ or $HSO_3CF_3$ [30]. The grafted solid acid catalysts maintained their crystallinity, which was confirmed from the reflections in

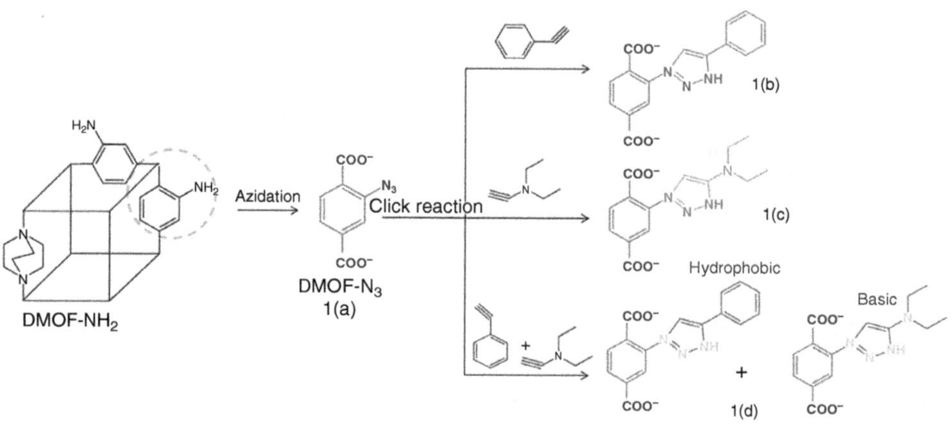


**Figure 14.4** Synthesis of functional DMOF-1 catalysts through "click chemistry. *Source:* Adopted from Ref. [27]."

PXRD patterns [31]. The type and strength of acidic sites were assigned using $^{31}P$ solid-state nuclear magnetic resonance (NMR) with trimethylolpropane (TMP), triethylphosphine oxide (TMPO) as probe molecules, and the data revealed that both Brønsted and Lewis acidic sites were present in the MOFs. The Lewis acid sites present in pristine MOFs are mainly due to structural defects (open metal sites), which are minimized after grafting with sulfonic acidic groups($HSO_3CF_3$) [32]. The Hammett indicator method confirmed super acidity ($H_0 < -12.4$) of the UiO-66-[$C_3NH_2$][$SO_3CF_3$] [33]. All the samples were investigated for transesterification of sunflower oil. A biodiesel yield of ~98.4% was observed over UiO-66-[$C_3NH_2$][$SO_3CF_3$]. This activity of UiO-66-[$C_3NH_2$][$SO_3CF_3$] is much superior when compared to Amberlyst-15 (55.7%), SBA-15-$SO_3$H-0.2 (61.4%), Nafion NR50 (62.9%), and the homogeneous acid ionic liquid (IL) of aniline-[$C_3NH_2$][$SO_3CF_3$] (90.5%) and $H_2SO_4$ (93.7%). The UiO-66-[$C_3NH_2$][$SO_3CF_3$] is recyclable up to five cycles. Various attractive structural features such as high acid concentrations, super acidity, and selective activation of methanol inside the micropores of MOF are responsible for the observed activity of UiO-66-[$C_3NH_2$][$SO_3CF_3$].

### 14.2.2 Esterification of Carboxylic Acids

Levulinate esters are attractive fuel additives due to their high octane number, high lubricity, and flashpoint. Both diesel and gasoline can be blended with levulinate esters. Thus, esterification of lignocellulosic biomass-derived levulinic acid has been actively investigated. The esterification of levulinic acid is generally carried out using mineral acids ($H_2SO_4$ and p-toluenesulfonic acid). However, the use of solid acid catalysts is an attractive prospect as it makes handling and separation much easier. Heteropolyacids, sulfated metal oxides, and zeolites have been reported with some good activities.

#### 14.2.2.1 Esterification of Carboxylic Acids at SBUs of MOFs

UiO-66 MOFs are attractive catalysts for esterification as they contain versatile SBUs. As mentioned earlier, this class of MOF possess missing linker defects. (Scheme 14.1). Defective UiO-66(Zr) acted as a versatile catalyst for the esterification of levulinic acid with ethanol. Over UiO-66(Zr), for EtOH: LA molar ratio of 15 : 1, an ethyl levulinate yield of 94% was achieved in 8 h at 78 °C. The catalytic activity of UiO-66(Zr) was in general superior when compared with other solid acid catalysts such as zeolites. A careful analysis of catalytic data obtained from the different batches of UiO-66(Zr) revealed that the catalytic activity is greatly dependent on the missing linker defects. The presence of linker defects was confirmed by thermal gravimetric analysis (TGA), and defect concentration varies from 2.5 to 13.2% depending on the synthesis conditions. UiO-66(Zr) MOF with 13.2% linker defects exhibited the highest TOF of 117 $h^{-1}$, which is very much comparable to p-toluene sulfonic acid (117 $h^{-1}$). The catalyst was stable and reusable up to three cycles. Suggesting the heterogeneous nature of catalysis, a Zr leaching of only 0.35% of the total Zr in the MOF was observed. Further, hydrated and dehydrated SBUs of defective UiO-66(Zr) exhibited quite different activities for esterification reaction. In the hydrated SBU, at the missing linker defect, physisorbed water is linked to the open Zr(IV) site next to Zr-OH site. This water molecule can act as both Brønsted base and acid. As suggested by DFT calculations, levulinic acid can replace this water molecule and react with methanol previously

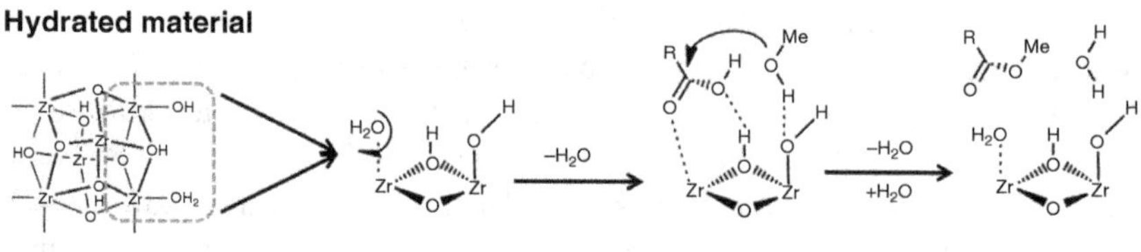


**Scheme 14.4** Schematic representation of mechanism for the esterification of propionic acid with methanol over a hydrated UiO-66 (Zr) material. *Source:* Adopted from Ref. [34], Copyright 2015, Elsevier.

stabilized on the neighboring Zr-OH site (Scheme 14.4). The synergy aroused out these sites is witnessed experimentally [34].

Kalidindi et al. explored Zr, Ce, and Hf-based isostructural UiO-66 MOFs for esterification of levulinic acid with methanol/ethanol [35]. The $\mu_3$-OH groups present on the triangle face of $M_6$- SBU(M = Zr(IV), Ce(IV) have a different degree of Brønsted acidity depending on the metal ion. The acidity order was found to be UiO-66(Zr)<UiO-66(Ce) < UiO-66(Hf), which is consistent with oxophilicity of the metal ion(Figure 14.5a) [36]. Among three, UiO-66(Hf) exhibited the highest activity, the yields of ethyllevulinate(EL) were found to be 84.1, 50.4, 15.7, and 8.3% over UiO-66(Hf), UiO-66(Ce), UiO-66(Zr), and blank, respectively, for 4 h at 80 °C(Figure 14.5b).

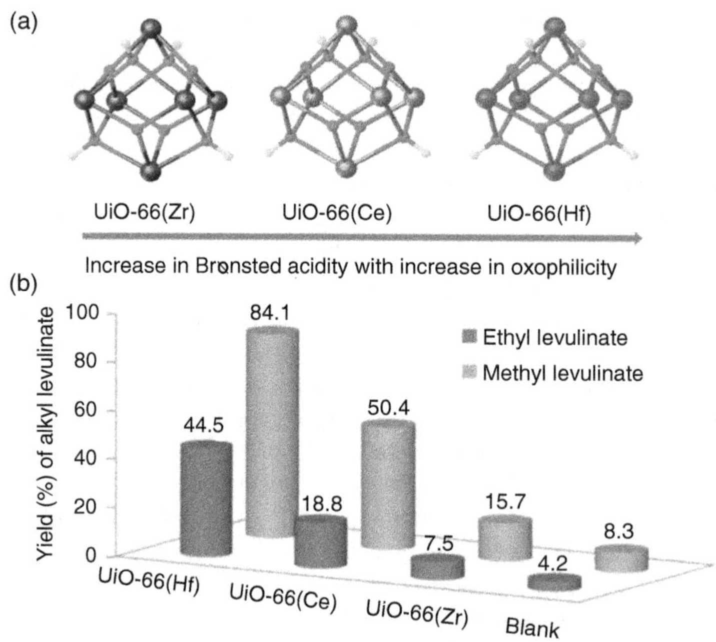


**Figure 14.5** (a) Secondary building units (SBU) corresponding to UiO-66(Zr), UiO-66(Ce), and UiO-66(Hf) MOF, respectively, the $\mu_3$-OH sites (Brønsted acidic) represented in cyan color, (b) plot of yield (%) versus catalysts. *Reaction conditions:* levulinic acid (1 mmol), alcohol (15 mmol), catalyst amount 20 mg, and reaction carried for 4 h at 80 °C. *Source:* Reproduced with permission Ref. [35], Copyright 2019, Royal Society of Chemistry.

#### 14.2.2.2 Esterification of Carboxylic Acids at Linker Active Sites

Linker modification with Brønsted acidic functional groups was effectively used for acceleration esterification reactions. Use of 1,2,4,5-benzenetetracarboxylic acid instead of 1,4-benzenedicarboxylic acid for the synthesis of UiO-66(Zr) gave a pore environment that contains two free carboxylic acid groups (Figure 14.6a) [37]. The PXRD pattern of UiO-66-$(COOH)_2$ contained an additional peak at $2\theta = 4.2°$, which is attributed to the missing linkers at SBU [38]. Both UiO-66 and UiO-66-$(COOH)_2$ exhibited Type-I $N_2$ isotherms with surface areas of $1110\,m^2\,g^{-1}$ and $880\,m^2\,g^{-1}$, respectively. The lower surface area of UiO-66-$(COOH)_2$ is mainly due to partial filling of pore volume by free carboxylic acid groups. The acidity of MOFs was characterized by pyridine-FTIR technique. UiO-66-$(COOH)_2$ showed two characteristic peaks at $1536\,cm^{-1}$ and $1490\,cm^{-1}$. The former peak is due to pyridinium ion formation (due to Brønsted acid sites) and later one because of both Lewis and Brønsted acid sites. On the other hand, UiO-66 MOFs displayed only single peak at $1450\,cm^{-1}$ for pyridine on Lewis acid sites (Figure 14.6b). In UiO-66-$(COOH)_2$ MOF, the free acid groups acted in synergy with Lewis acid sites (the defect sites) on the SBU of MOF to give excellent ethyl levulinate yields. Under optimized reaction conditions (78 °C, 0.39 mol.% of catalyst,1 : 20 molar ratio b/w LA/EtOH for 24 h), the UiO-66-$(COOH)_2$ catalyst gave 98.0% ethyllevulinate yield, whereas UiO-66 catalyst gave a yield of 25.6% under same conditions.

Mannepalli et al. reported Hf- and Zr- based UiO-66 MOFs with different functional groups {UiO-66(Hf), UiO-66(Hf)-$NH_2$, UiO-66(Hf)-$SO_3H$, and UiO-66(Zr)-$SO_3H$)} for esterification of levulinic acid [39]. Among all, the UiO-66(Hf)-$SO_3H$ displayed highest Brønsted acidity of 14.3 $[H^+]$/mol $mol^{-1}$ as confirmed by acid–base titrations. Both $-SO_3H$ groups on the linker and hydroxyl groups (Hf-$\mu_3$-OH) at metal clusters contributed to the acidity of MOF. The yields of isopropyl levulinate were 88.0, 52.0, 8.0, and 0.0% over UiO-66(Hf)-$SO_3H$, UiO-66(Zr)-$SO_3H$, UiO-66(Hf), and UiO-66(Hf)-$NH_2$, respectively. The UiO-66(Hf)-$SO_3H$ was recycled for four cycles with a decrease of ~4% in the yield of isopropyl levulinate. The PXRD data of recycled catalyst confirmed structural stability of MOF catalyst.

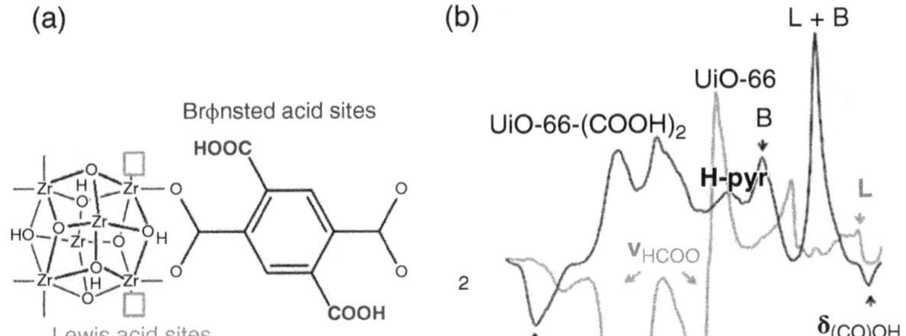


**Figure 14.6** (a) Representation of synergistic active centers within the structure of UiO-66-$(COOH)_2$. (b) Infrared spectra of pyridine absorbed UiO-66 and UiO-66-$(COOH)_2$. *Source:* Reproduced with permission Ref. [37], Copyright 2019, American Chemical Society.

#### 14.2.2.3 Esterification at Pore Volume (Guest Incorporation)

Often, catalytic active species are incorporated inside MOF pores or embedded inside MOF matrix to bring synergetic catalysis between the guest active sites and framework active sites. Active species such as POMs and ILs are hybridized with MOFs for effective synthesis of biodiesel.

$H_3PW_{12}O_{40}$ POM has been combined with Cu-BTC MOF to bring catalytic synergy between Brønsted acid sites of POM and unsaturated sites (Lewis acid sites) of MOF. The hybrid NENU-3a $\{Cu_{12}(BTC)_8\}$-$[H_3PW_{12}O_{40}]\}$ was formed spontaneously upon mixing of $Cu(NO_3)_2$, $H_3PW_{12}O_{40}$, and $H_3BTC$ ligand (Figure 14.7a). The final shape of the particles was octahedral (o- NENU-3a). When this synthesis was carried out in the presence of a modulator p-toluic acid (pTA), cubic-shaped NENU-3a was realized (Figure 14.7b) [40]. The morphology of the crystals had profound effect on the catalytic activity of NENU-3a for esterification reaction, especially for sterically demanding substrates. Cubic-NENU-3a showed conversions close to 90% for the transformation of long-chain fatty acids ($C_{12}-C_{22}$) into corresponding monoalkyl esters in comparison to <22% using octahedral-NENU-3a. Hindering accessibility to substrates in octahedral-NENU-3, HPW units are covered by the BTC ligands in {111} facets (Figure 14.7c and d). In cubic-NENU-3, HPW sites are much more accessible through {100} facets. This technique is useful for sterically demanding substrates that are hard to diffuse into the MOF pores.

In another study, ILs were incorporated into the porous MOF catalyst by approximate ligand substitution. The construction of $[(CH_2COOH)_2IM]$@H-UiO-66 catalyst was conducted via bidentate coordination between one -COO- group of $[(CH_2COOH)_2IM]HSO_4$ and two portions of unsaturated Zr(IV) defects of the hierarchically porous Zr MOF (H-UiO-66) [41]. The developed catalyst was studied for esterification of oleic acid with methanol, which yielded 93.82% of biodiesel. The observed novel activity is attributed hierarchical structure of UiO-66 and simultaneous presence of acid sites (Lewis and Brønsted acid sites).

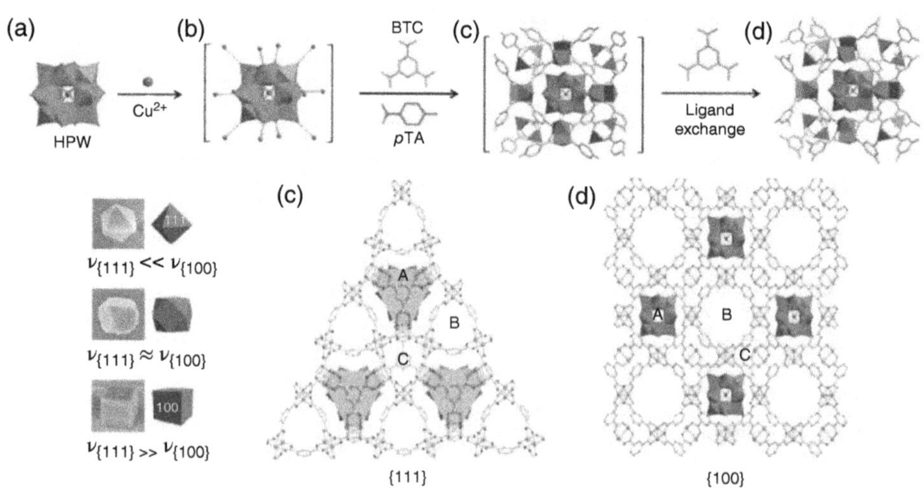


**Figure 14.7** (a) Schematic representation of formation of NENU-3a $\{Cu_{12}(BTC)_8\}$-$[H_3PW_{12}O_{40}]\}$ material. (b) Various shapes of NENU-3a $\{Cu_{12}(BTC)_8\}$-$[H_3PW_{12}O_{40}]\}$, (c) NENU-3a along {111} direction, and (d) NENU-3a along {100} direction. *Source:* Reproduced with permission Ref. [40], Copyright 2015, American Chemical Society.

## 14.3 Conclusion

MOFs with tunable active sites at inorganic secondary building unit, organic linker, and pore volume (through guest incorporation) are versatile heterogeneous catalysts for the synthesis of biodiesel. The examples discussed in this chapter clearly demonstrate various advantageous of MOFs for catalysis such as presence of well-defined and tunable active sites, large porosity coupled with wide window openings, and confinement effects. MOFs are effective acid/base catalysts for both transesterification and esterification reactions for the production of biodiesel. The activities of some of MOF catalysts discussed here surpassed activities of traditional acid catalysts such as zeolites. Further, cooperation between the active sites present at different parts of the frameworks afforded impressive biodiesel yields via catalytic synergy. Though so far the studies are encouraging, most of the studies discussed here are carried out in batch reactors and in small scale. MOF catalysts need to be tested under conditions that are relevant for practical applications such as in continuous flow reactors, etc. Further hurdles for commercialization of MOF-based catalysts for biodiesel production include issues related production and regeneration of the catalyst.

## References


**1** Zhou, H.-C., Long, J.R., and Yaghi, O.M. (2012). Introduction to metal-organic frameworks. *Chem. Rev.* 112: 673–674.

**2** Zheng, C., Greer, H.F., Chiang, C.Y., and Zhou, W. (2014). Microstructural study of the formation mechanism of metal-organic framework MOF-5. *CrystEngComm.* 16 (6): 1064–1070.

**3** Chen, T.H., Popov, I., Kaveevivitchai, W., and Miljanić, O.Š. (2014). Metal-organic frameworks: rise of the ligands. *Chem. Mater.* 26 (15): 4322–4325.

**4** Furukawa, H., Cordova, K.E., O'Keeffe, M., and Yaghi, O.M. (2013). The chemistry and applications of metal-organic frameworks. *Science* 341: 6149.

**5** Férey, G., Mellot-Draznieks, C., Serre, C. et al. (2005). A chromium terephthalate-based solid with unusually large pore volumes and surface area. *Science* 309 (5743): 2008–2009.

**6** Cavka, J.H., Jakobsen, S., Olsbye, U. et al. (2008). A new zirconium inorganic building brick forming metal organic frameworks with exceptional stability. *J. Am. Chem. Soc.* 130 (42): 13850–13851.

**7** Serre, C., Millange, F., Thouvenot, C. et al. (2002). Very large breathing effect in the first nanoporous chromium(III)-based solids: MIL-53 or CrIII(OH)·$\{O_2C\text{-}C_6H_4\text{-}CO_2\}\cdot\{HO_2C\text{-}C_6H_4\text{-}CO_2H\}x\cdot H_2Oy$. *J. Am. Chem. Soc.* 124 (45): 13519–13526.

**8** Kalidindi, S.B., Nayak, S., Briggs, M.E. et al. (2015). Chemical and structural stability of zirconium-based metal-organic frameworks with large three-dimensional pores by linker engineering. *Angew. Chemie. Int. Ed.* 54 (1): 221–226.

**9** Demessence, A., Alessandro, D.M.D., Foo, M.L., and Long, J.R. (2009). Strong $CO_2$ binding in a waterstable, triazolate-bridged metal-organic framework functionalized with ethylenediamine. *J. Am. Chem. Soc.* 131: 8784–8786.

**10** Yuan, S., Feng, L., Wang, K. et al. (2018). Stable metal–organic frameworks: design, synthesis, and applications. *Adv. Mater.* 30 (37): 1–35.

**11** Yang, D. and Gates, B.C. (2019). Catalysis by metal organic frameworks: perspective and suggestions for future research. *ACS Catal.* 9 (3): 1779–1798.

**12** Bakuru, V.R., DMello, M.E., and Kalidindi, S.B. (2019). Metal-organic frameworks for hydrogen energy applications: advances and challenges. *ChemPhysChem.* 20 (10): 1177–1215.

**13** Kalmutzki, M.J., Hanikel, N., and Yaghi, O.M. (2018). Secondary building units as the turning point in the development of the reticular chemistry of MOFs. *Sci. Adv.* 4 (10): 9180.

**14** Hu, Z. and Zhao, D. (2017). Metal-organic frameworks with Lewis acidity: synthesis, characterization, and catalytic applications. *CrystEngComm.* 19 (29): 4066–4081.

**15** Wang, J., Liu, L., Chen, C. et al. (2020). Engineering effective structural defects of metal-organic frameworks to enhance their catalytic performances. *J. Mater. Chem. A* 8 (8): 4464–4472.

**16** Wu, H., Chua, Y.S., Krungleviciute, V. et al. (2013). Unusual and highly tunable missing-linker defects in zirconium metal-organic framework UiO-66 and their important effects on gas adsorption. *J. Am. Chem. Soc.* 135 (28): 10525–10532.

**17** Hartmann, M. and Fischer, M. (2012). Amino-functionalized basic catalysts with MIL-101 structure. *Microporous Mesoporous Mater.* 164: 38–43.

**18** Serra-Crespo, P., Ramos-Fernandez, E.V., Gascon, J., and Kapteijn, F. (2011). Synthesis and characterization of an amino functionalized MIL-101(Al): separation and catalytic properties. *Chem. Mater.* 23 (10): 2565–2572.

**19** Srirambalaji, R., Hong, S., Natarajan, R. et al. (2012). Tandem catalysis with a bifunctional site-isolated Lewis acid–Brønsted base metal–organic framework, NH2-MIL-101(Al). *Chem. Commun.* 48 (95): 11650–11652.

**20** Zhang, X., Chen, Z., Liu, X. et al. (2020). A historical overview of the activation and porosity of metal-organic frameworks. *Chem. Soc. Rev.* 49 (20): 7406–7427.

**21** Huang, Y.B., Liang, J., Wang, X.S., and Cao, R. (2017). Multifunctional metal-organic framework catalysts: synergistic catalysis and tandem reactions. *Chem. Soc. Rev.* 46 (1): 126–157.

**22** Lin, L., Cunshan, Z., Vittayapadung, S. et al. (2011). Opportunities and challenges for biodiesel fuel. *Appl. Energy* 88 (4): 1020–1031.

**23** Lazare, S., Bazer-bachi, D., Bonnier, F. et al. (2010). Catalysis by a non-functionalized MOF in transesterification: acido-basicity at the external surface of ZIF-8 probed by FTIR and ab initio calculations. *J. Am. Chem. Soc.* 132 (35): 12365–12377.

**24** Bakuru, V.R., Velaga, B., Peela, N.R., and Kalidindi, S.B. (2018). Hybridization of Pd nanoparticles with UiO-66(Hf) metal-organic framework and the effect of nanostructure on the catalytic properties. *Chem. A Eur. J.* 24 (60): 15978–15982.

**25** Furuta, S., Matsuhashi, H., and Arata, K. (2004). Biodiesel fuel production with solid superacid catalysis in fixed bed reactor under atmospheric pressure. *Catal. Commun.* 5 (12): 721–723.

**26** Shu, Q., Yang, B., Yuan, H. et al. (2007). Synthesis of biodiesel from soybean oil and methanol catalyzed by zeolite beta modified with $La^{3+}$. *Catal. Commun.* 8 (12): 2159–2165.

**27** Savonnet, M., Camarata, A., Canivet, J. et al. (2012). Tailoring metal-organic framework catalysts by click chemistry. *Dalt. Trans.* 41 (14): 3945–3948.

**28** Hall, J.N. and Bollini, P. (2020). Metal-organic framework MIL-100 catalyzed acetalization of benzaldehyde with methanol: Lewis or Brønsted acid catalysis? *ACS Catal.* 10: 3750–3763.

**29** Trickett, C.A., Osborn Popp, T.M., Su, J. et al. (2019). Identification of the strong Brønsted acid site in a metal–organic framework solid acid catalyst. *Nat. Chem.* 11 (2): 170–176.

**30** Peng, W.L., Mi, J., Liu, F. et al. (2020). Accelerating biodiesel catalytic production by confined activation of methanol over high-concentration ionic liquid-grafted UiO-66 solid superacids. *ACS Catal.* 10 (20): 11848–11856.

**31** Kandiah, M., Nilsen, M.H., Usseglio, S. et al. (2010). Synthesis and stability of tagged UiO-66 Zr-MOFs. *Chem. Mater.* 22 (24): 6632–6640.

**32** Zheng, A., Liu, S.-B., and Deng, F. (2017). $^{31}P$ NMR chemical shifts of phosphorus probes as reliable and practical acidity scales for solid and liquid catalysts. *Chem. Rev.* 117 (19): 12475–12531.

**33** Zhang, X., Zhao, Y., Xu, S. et al. (2014). Polystyrene sulphonic acid resins with enhanced acid strength via macromolecular self-assembly within confined nanospace. *Nat. Commun.* 5: 1–9.

**34** Cirujano, F.G., Corma, A., and Llabrés i Xamena, F.X. (2015). Conversion of levulinic acid into chemicals: synthesis of biomass derived levulinate esters over Zr-containing MOFs. *Chem. Eng. Sci.* 124: 52–60.

**35** Bakuru, V.R., Churipard, S.R., Maradur, S.P., and Kalidindi, S.B. (2019). Exploring the Brønsted acidity of UiO-66 (Zr, Ce, Hf) metal-organic frameworks for efficient solketal synthesis from glycerol acetalization. *Dalt. Trans.* 48 (3): 843–847.

**36** Hu, Z., Nalaparaju, A., Peng, Y. et al. (2016). Modulated hydrothermal synthesis of UiO-66(Hf)-type metal-organic frameworks for optimal carbon dioxide separation. *Inorg. Chem.* 55 (3): 1134–1141.

**37** Wang, F., Chen, Z., Chen, H. et al. (2019). Interplay of Lewis and Brønsted acid sites in Zr-based metal-organic frameworks for efficient esterification of biomass-derived levulinic acid. *ACS Appl. Mater. Interfaces* 11 (35): 32090–32096.

**38** Chen, Z., Wang, X., Noh, H. et al. (2019). Scalable, room temperature, and water-based synthesis of functionalized zirconium-based metal-organic frameworks for toxic chemical removal. *CrystEngComm.* 21 (14): 2409–2415.

**39** Gupta, S.S.R. and Kantam, M.L. (2019). Catalytic conversion of furfuryl alcohol or levulinic acid into alkyl levulinates using a sulfonic acid-functionalized hafnium-based MOF. *Catal. Commun.* 124: 62–66.

**40** Liu, Y., Liu, S., He, D. et al. (2015). Crystal facets make a profound difference in polyoxometalate-containing metal-organic frameworks as catalysts for biodiesel production. *J. Am. Chem. Soc.* 137 (39): 12697–12703.

**41** Ye, C., Qi, Z., Cai, D., and Qiu, T. (2019). Design and synthesis of ionic liquid supported hierarchically porous Zr metal-organic framework as a novel Brønsted-Lewis acidic catalyst in biodiesel synthesis. *Ind. Eng. Chem. Res.* 58 (3): 1123–1132.

# Part 3

# Technologies, By-product Valorization and Prospects of Biodiesel Production

# 15

# Upstream Strategies (Waste Oil Feedstocks, Nonedible Oils, and Unicellular Oil Feedstocks like Microalgae)

*Aleksandra Sander and Ana Petračić*

*Department of Mechanical and Thermal Process Engineering, University of Zagreb, Faculty of Chemical Engineering and Technology, Zagreb, Croatia*

## 15.1 Introduction

Sustainability of feedstocks should be considered while defining upstream and/or downstream strategies for biodiesel production. Feedstocks availability, production technology, economics, competition, and greenhouse gas (GHG) savings must be taken into account [1–5]. Upstream strategies include the use of advanced feedstocks like waste cooking oil (WCO), waste animal fats (WAFs), nonedible oils, or genetically engineered oils, as well as selection of appropriate catalyst and reaction conditions [6]. On the other hand, downstream strategies deal with development of new methods for separation, purification, recovery of alcohol, catalyst, and production of pure glycerol [7].

### 15.1.1 Classification of Biodiesel

The literature offers different ways of classifying biofuels. According to one, biofuels can be categorized based on the type of feedstock used and technology used as first-, second-, third-, and fourth-generation biofuels [8]. Regarding biodiesel, first generation is produced from edible oil crops and fats, second generation is produced from nonfood oil crops and wastes, while third generation refers to production of biodiesel from microalgae [9]. Fourth generation of biofuels was introduced as biofuels that are produced from genetically engineered microalgae with the general purpose of enhancing the quality and productivity of microalgae [10]. First, second and third generation of biodiesel are produced by chemical transesterification, while fourth-generation biodiesel utilizes thermochemical technologies. The key difference, according to which the classifications differ, is that according to some researchers WCO and WAFs are placed in the third generation, along with microalgae [11]. Recently, additional fifth generation is

*Biodiesel Production: Feedstocks, Catalysts, and Technologies*, First Edition. Edited by Samuel Lalthazuala Rokhum, Gopinath Halder, Suttichai Assabumrungrat, and Kanokwan Ngaosuwan.

added [12]. According to this classification, microalgae are divided from WCO and WAFs and placed to fourth generation, while fifth generation belongs to genetically engineered biofuels.

Biofuels are also classified as conventional and advanced biofuels. Conventional biodiesel is commercially produced from vegetable oils and animal fats and is characterized by its blending ability with petroleum diesel. Advanced biodiesel is produced from nonfood crops and waste materials as well as microalgae. A complete list of advanced feedstocks is given in the Annex IX of the Renewable Energy Directive (RED II, [13]).

According to the Renewable Fuels Standard of the Energy Independence and Security Act of 2007, "The term 'advanced biofuel' means renewable fuel, other than ethanol derived from corn starch, that has lifecycle greenhouse gas emissions, as determined by the Administrator, after notice and opportunity for comment, that are at least 50% less than baseline lifecycle greenhouse gas emissions." Among other advanced fuels listed, biomass-based diesel produced from "low nutrient input/high per acre yield crops; agricultural or forestry waste; or other sustainable biomass feedstocks including algae" is also considered as advanced biofuel. Similar definition of the advanced biofuels is given by the Renewable Energy Directive on the promotion of the use of energy from renewable sources (RED II, [13]).

### 15.1.2 Commercial Production of Biodiesel

Commercial production of biodiesel from the advanced feedstocks is limited due to the existing technologies used for conventional biodiesel produced from first-generation feedstocks. However, during the last two decades, a significant progress in development of new and modification of the existing technologies, which includes adoption of different types of waste feedstocks, can be observed. Recently, a significant number of investigations were focused on microalgae, and the obtained results indicate that the process is technically feasible. A simplified diagram describing the biodiesel production from advanced feedstocks is given in Figure 15.1.

Biodiesel is produced on the industrial level globally in a large number of factories mainly from unused vegetable oils. Utilization of advanced feedstocks requires additional steps in the production line, so sustainable compromise between cheap feedstock and new technology must be found in order to develop the method that will be economically feasible. Feasibility of the biodiesel production technology is influenced by the cost of all steps in the production line, from production/collecting of feedstocks to purification of crude biodiesel. According to the literature, the price of feedstock represents about 70–75% of the overall biodiesel production cost, and it could be reduced by selecting a cheaper feedstock. [14]. To produce biodiesel that will compete with the production of fossil diesel, development of technology that will produce biodiesel from low quality advanced feedstocks is mandatory. With using wastes as feedstocks for biodiesel production, competition with food as well as the cost of disposal or treatment of waste materials is significantly reduced.

According to the Union zur Förderung von Oel- und Proteinpflanzen e.V. (UFOP) reports for 2009 and 2019, a share of second-generation feedstocks did not change significantly.

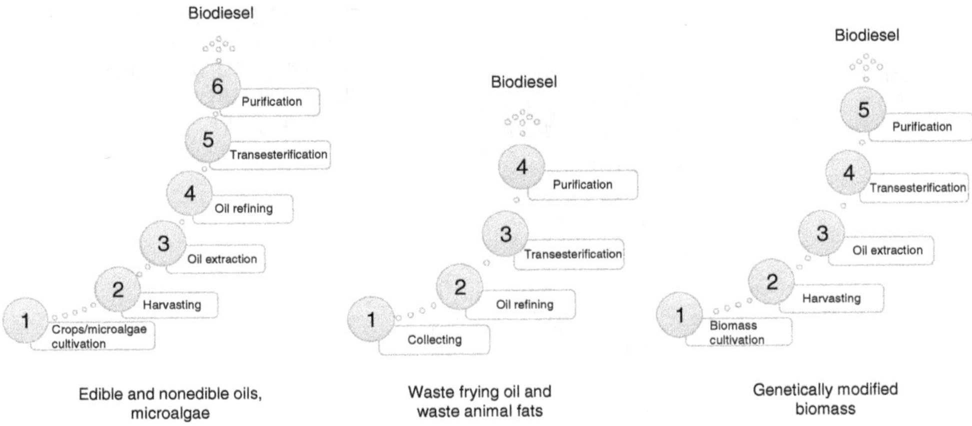


**Figure 15.1** Steps in production of advanced biodiesel from waste cooking oil (WCO) and waste animal fats (WAF).

Most used feedstocks for biodiesel production remained palm, soybean, and rapeseed oil (see Figure 15.2).

Leading producers of biodiesel are European Union (rapeseed, palm, soybean, sunflower, WCO, WAF, other), Indonesia (crude palm oil), the United States (soybean, rapeseed, WCO, WAF, other), Brazil (crude soybean oil, WAF), and Argentina (soybean), and this counts for about 84% of total biodiesel production in 2019 (see Figure 15.3). Small number of plants produce biodiesel from nonedible feedstocks (mostly WCO, WAF) and HVO.

Constant increase in fuel consumption in the transportation section should be the main driving force for production of biodiesel. Based on the OECD-FAO agricultural outlook 2019–2028, the production of biofuels will increase but slower than during previous decades. The major feedstocks will remain the same due to decreased economical support for development of new technologies for biodiesel production. Additional issues like national policies, capacities of the agricultural sector for production of feedstocks, automobile industry, and achievement of feedstocks sustainability criteria remain uncertain.

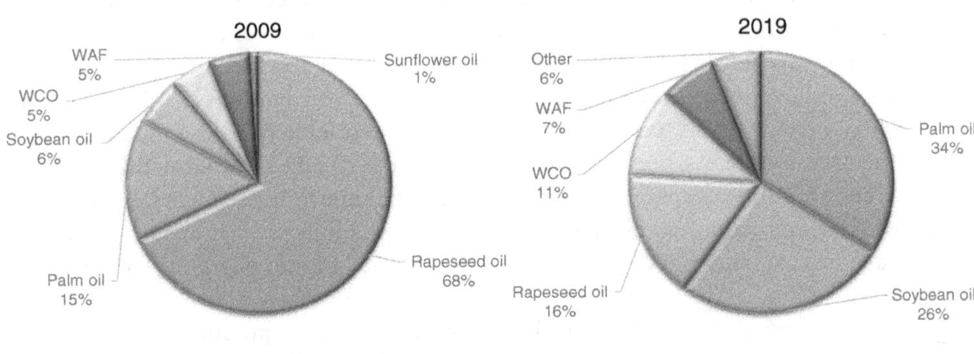


**Figure 15.2** Global share of feedstocks.

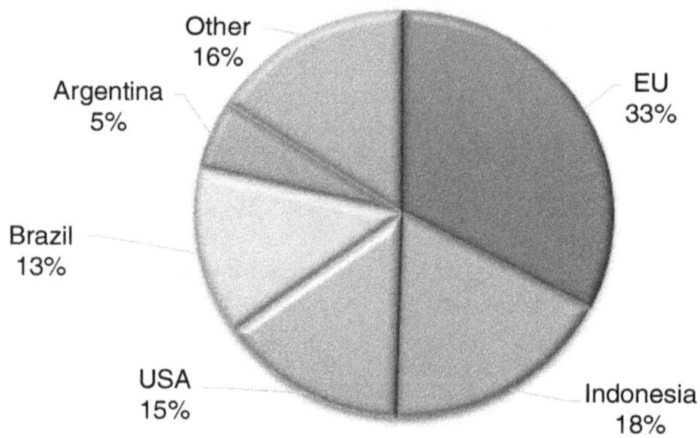


**Figure 15.3** Biodiesel production by countries. *Sources:* Barros [15]; Kenneth [16]; Rahmanulloh [17]; Flach et al. [18]; OECD Publishing [19].

## 15.2 Biodiesel Feedstocks

The major sources of triglycerides are edible and nonedible oil and animal fats. After extraction, crude oil besides glycerides (mono-, di-, and tri-) contains different types of impurities that require refining. The selection of the appropriate refining method is strongly influenced by the variable composition of crude oil due to its origin (type, environmental conditions of the feedstock cultivation and/or collecting, and method used for extraction) [20].

The feasibility of a great number of nonedible oil seeds that can be cultivated in different regions around the world as a potential feedstock for biodiesel production has been investigated. Some of these tree-based seeds can be cultivated under different climate conditions and soil types and can produce seeds for a long time period. For instance, *Jatropha curcas* can grow on a low quality soil and can produce seeds for 50 years [21].

Seed and oil processing, geography and climate region, type and nutrition content of the soil, and handling and storage conditions strongly influence the properties of oils, which will further determine the yield and quality of the produced biodiesel. Since a large number of oils extracted from nonedible crops contain a high concentration of free fatty acids (FFAs), a simple base-catalyzed transesterification cannot be performed without pretreatment.

Generally, to obtain oil from seeds, seeds have to be processed. After harvesting, water needs to be removed from seeds by drying, usually by direct exposing of seeds to sunlight for a sufficiently long period of time. This stage is crucial since improper handling can result in seed deterioration and increased water content. Additional steps include husking, cleaning, and grinding. Oil is then extracted from the obtained particles by the appropriate method. Prior to the conversion of oil to biodiesel, oil is refined. Through the refining process crude oil is degummed, neutralized, bleached, and deodorized. After extraction, solid material is dried and subjected to further processing to produce some kind of valuable by-product, like animal feed or soil nutrient (fertilizers).

Properties of most commonly used feedstocks for biodiesel production are given in Table 15.1, based on Scopus literature search (November 2020).

**Table 15.1** Oil content, total acid number (TAN) in different types of feedstocks, and biodiesel yield (with number of references per feedstock given in brackets).

| Edible oils | Oil content, % | TAN, mg KOH $g^{-1}$ | Maximum biodiesel yield, % |
|---|---|---|---|
| Soybean (2817) | 15–22 | 0.37–0.78 | 98,03 [22] |
| Palm (2764) | 30–60 | 6.9–50.8 | 92,2 [22] |
| Rapeseed/canola (2297) | 38–46 | 0.6–0.8 | 83,34 [22] |
| Sunflower (1124) | 25–35 | 0.2–0.5 | 98,0 [22] |
| Corn (909) | 20 | 0.1–5.72 | 99,48 [22] |
| Coconut (494) | 63–65 | 1.99–12.8 | 98,7 [22] |
| Peanut (165) | 45–55 | 2.36 | 89,0 [22] |
| Linseed (142) | 40–45 | 0.8 | 97,1 [23] |
| Safflower (121) | 35–50 | 0.35 | 98 [24] |
| Nonedible oils | | | |
| *Jatropha* oil (2819) | 40–60 | 13.5–22.6 | 99,87 [22] |
| Castor (701) | 50 | 10.8 | 97,8 [25] |
| Cotton seed oil (424) | 17–25 | 0.6–2.87 | 96 [22] |
| Karanja oil (392) | 27–39 | 14.7 | 97–98 [26] |
| Neem oil (262) | 20–30 | 16.6 | 96 [27] |
| Mahua oil (257) | 50–61 | 21 | 98 [26] |
| Rubber seed oil (214) | 50–60 | 0.52 | 99,5 [28] |
| Mustard oil (156) | 25–30 | 8,6 | 91 [22] |
| Camelina oil (150) | 30–35 | 0.63–4.61 | 96 [29] |
| WCG oil (108) | 15–28 | Up to 15.4 | up to 98 [30] |
| Jojoba oil (72) | 45–55 | 0.43–0.57 | 96 [31] |
| Tobacco oil (65) | 35–49 | 1.34–37 | 91 [26] |
| Some advanced feedstocks | | | |
| Microalgae (1852) | | | [32] |
| *Nannochloropsis gaditana* | 12.6 | 23.8–24.6 | 94.2 |
| *Nannochloropsis oculata* | 22–29 | 83.4 | 11.4 |
| *Dunaliella tertiolecta* | 11–16 | 167.3 | 82.3 |
| *Scenedesmus* | 17–24 | 20 | 97 |
| *Chlorella vulgaris* | 41–58 | 3.8 | 97 |

### 15.2.1 Edible Oils as Feedstock for Biodiesel Production

Edible oils belong to first-generation feedstocks. Biodiesel is still produced mostly from these types of oils meaning that palm, rapeseed (canola), and soybean oil are main feedstocks. Oils can be extracted from seeds, tree fruits, and kernels and can be classified as saturated oils (palm oil) and unsaturated oil (rapeseed, sunflower, and soybean) [33]. Palm and coconut oils contain the highest amount of oil; the majority of oils can be easily converted to biodiesel due to its low total acid number (TAN) by the use of alkali catalyst, and sufficiently high biodiesel yield can be obtained (see Table 15.1). The main advantages of these feedstocks are crops availability, simple conversion with high yield, capacity to produce high purity biodiesel, and lower GHG emissions. The utilization of first-generation feedstocks should be minimized in order to reduce high production cost and avoid competition with food industry.

### 15.2.2 Nonedible Oils as Feedstocks for Biodiesel Production

To overcome disadvantages of first-generation feedstocks, many nonedible oil plants have been considered as feedstock for biodiesel production. Nonedible oils are of various purity and composition. In contrast to edible oils, composed mostly of palmitic, stearic, oleic, and linoleic fatty acid, some nonedible oils are composed of different fatty acids as well as other compounds, which can influence the properties of biodiesel. High concentration of saturated fatty acids contributes to poor cold flow characteristics. Too high concentration of unsaturated fatty acids will reduce oxidation stability and cetane number. Presence of polyunsaturated fatty acids will increase the risk of polymerization and carbon residue deposition [34]. The length of fatty acids chain positively influences the cetane number and heating value (saturated) [24]. Density and viscosity are highly dependent on the length and saturation level of the fatty acid chains [35]. Additionally, whole biodiesel production technology may be influenced by the composition of this type of oils [36–38]. Some oils have a high concentration of FFAs and/or a high degree of saturation, which makes the production of biodiesel more complex (see Table 15.1). However, by optimization of reaction conditions, high yield of biodiesel can be obtained.

Besides the oil content and the yield per hectare, the analysis of nonedible crops feasibility should include land availability, possibility of cultivation, energy consumption, impact on air quality, greenhouse emissions, impact on soil (fertilizers, pesticides, and erosion) and biodiversity, water requirements, employment, and complete economic evaluation (feedstocks and processing technology) [9].

### 15.2.3 Waste Feedstocks (Waste Cooking Oils, Waste Animal Fats, Waste Coffee Ground Oil, Olive Pomace)

Given the fact that biodiesel produced from oil used in the food industry is not competitive with fossil-based diesel (price of raw material contributes to more than 70% of the total production cost), it is necessary to use cheaper raw materials for its production. Therefore, waste edible oil, by-products of the manufacturing process of edible oils, inedible oils, and WAF are considered the economically acceptable raw materials [39, 40]. The use of waste oils and fats is very important because it prevents the formation and accumulation of new waste and reduces the cost of waste treatment [41]. Another advantage is the fact that WCO and WAF are raw materials that are readily available. The steady increase in the production of meat and meat products results in an increase in the amount of waste suitable for the production of biodiesel. On the other hand, a large number of restaurants and industrial production of food products continuously generate spent cooking oil, which is also cheap raw material, easily available in large quantities for the production of biodiesel [42]. Earlier studies have demonstrated that the waste edible oils and animal fats are reliable, available, and environmentally friendly raw materials for biodiesel production [43]. The use of animal fat as a raw material for the production of biodiesel has economic and environmental advantages over the previously used, mainly vegetable oils [40, 41]. Waste coffee grounds show great potential for biodiesel production, containing around 10–15% of oil (w/w). Coffee is a very popular beverage worldwide, making the feedstock available and affordable [44, 45]. Olive pomace is another potential waste oily feedstock that can be used for biodiesel production [46]. The main disadvantage of these advanced raw materials is too high proportion of FFAs that prevents easy conversion of triglycerides into fatty acid esters [47, 48]. As for biodiesel produced from nonedible oils, high yield of biodiesel can be achieved [49, 50]. Waste coffee grounds and olive pomace usually require an additional drying step before oil extraction, though there are methods for direct transesterification from wet biomass [32, 51, 52].

### 15.2.4 Unicellular Oil Feedstocks (Microalgae, Yeasts, Cyanobacteria)

Microalgal oil has attracted a lot of attention during the past decade. Depending on the species, microalgae can contain up to 60% of oil. Unlike terrestrial plants, microalgae can be grown and harvested frequently, and do not require arable land to grow, excluding the risk of land use change. They can be grown in closed photobioreactors or in open ponds. Their need for nutrients can partly be satisfied from wastewater and industrial carbon dioxide, but the risk from pathogens that could kill microalgal culture has to be considered [53, 54]. Genetic modification of microalgae is the latest trend in improving biodiesel production efficiency from microalgal oil. The focus of modifications is on increasing microalgal photosynthetic efficiency and light penetration as well as enhancing the lipid and carbohydrate content [55]. Biodiesel production from microalgae consists of several steps, starting from harvesting, optional drying, oil extraction, optional oil purification, transesterification, and biodiesel purification. Researchers also explored direct or *in situ* transesterification of microalgal biomass. The main obstacles in microalgal biodiesel production are related to drying of biomass since it is the most energy intensive step [32].

Cyanobacteria potentially have better energy balance than microalgae because they can generate reactive or fixed nitrogen from the surrounding air, thus removing the need for fertilizers. They were mostly used for production of gasoline substitutes such as ethanol and isobutanol [56].

Yeasts and molds can use waste materials as substrates since they are not photosynthetic organisms, accumulating lipids from agricultural waste yielding from 5 to 19% w/w [57].

## 15.3 Composition of Oils and Fats

Edible fats and vegetable oils are mainly composed of glycerides (mono-, di-, and triglycerides) of different fatty acid profiles, which directly influences the selection of the refining technology. Apart from glycerides, oils and fats also contain different types of compounds like phospholipids, FFAs, metals, tocopherols, squalenes, sterols, moisture, and oxidized products. Tocopherols, sterols, and squalenes improve oxidative stability of oils, so during pretreatment stage the loss of these components should be minimal. Phospholipids as well as FFAs reduce the yield of biodiesel during homogeneous and heterogeneous transesterification of vegetable oils [58]. Additionally, microalgal oils contain glycolipids and sulfolipids that also need to be removed. It was also reported that phospholipids reduce the catalytic activity of methoxide and inhibit the catalytic performance of enzymes. Besides that, phospholipids are responsible for too high phosphorus content in biodiesel and can damage the diesel engine parts (accumulation, damage of catalytic converter). Waste coffee ground oil contains caffeine, kahweol, cholesterol, and avenasterol [59]. A great number of vegetable oils are characterized by some presence of toxic compounds. These oils are classified as nonedible oils due to their unsuitability for human food but have a great potential as feedstock for biodiesel production. Chemical composition and fatty acids profile of nonedible vegetable oils may vary significantly [60].

Composition of WCO is highly influenced by the composition of oil used for frying. However, during frying, the composition changes as a result of different chemical and physical transformation of food subjected to high temperatures and air [61]. As a result of frying, during which oxidation, hydrolysis, polymerization, isomerization, and decomposition of oil may occur, various compounds are formed [62]. Alongside with glycerides and FFA, WCO contains heterocycles, products of Maillard reaction, metal traces, spices, and a large number of organic compounds as well as solid food particles [63]. In order to produce a high quality biodiesel, pretreatment of WCO is mandatory.

Chemical composition and fatty acids profile of WAFs are influenced by the fat origin. Due to the high amount of saturated fatty acids, WAFs are at room temperature in solid state [39]. This type of fats is also characterized by a high amount of FFA and presence of other impurities like water, metals, and other trace elements and suspended solids, which may adversely affect biodiesel properties [64, 65].

## 15.4 Methods for Oil Extraction

Methods used for oil extraction are mechanical, solvent, and enzymatic extraction. Commercially, oil is extracted by mechanical pressing and solvent extraction [60]. Before extraction, seeds need to be processed. Processing includes cleaning, removal of law quality, in some cases removal of husk, and crushing. To improve extraction efficiency, it is often necessary to remove excess water from fresh material by means of drying. In some rural areas, traditional methods for extraction of oils, like hot water flotation, ghanis, and manual presses, are still used. The extraction efficiency of these methods is low.

### 15.4.1 Mechanical Extraction

The most commonly used method for oil extraction is mechanical pressing. Oil is extracted by means of applied pressure on the fresh material. By applying mechanical force on seeds, cell wall fractures and oil are squeezed out. Presses can be driven manually (ram press) or by an engine. Engine-driven screw presses are more efficient. They can extract from 58 to 80% of oil in contrast to manually driven presses that extract 60–65%. To improve the extraction efficiency, it is sometimes necessary to boil seeds. In that way, yield may be increased up to 89% [26]. Mechanical presses can be classified as hydraulic and screw presses. Extraction could be performed as batch or continuous process. Batch processes are in general more efficient since the larger amount of oil can be extracted. As for all other batch processes, after one batch is processed, the extractor needs to be completely cleaned prior to feeding with another batch. In continuously operated extractors, the seeds are continuously fed into the extractor, and oil is pressed out. Efficiency of continuously operated mechanical extractors can be increased by increasing the pressure on the seed or the process duration [66]. The mayor advantage of mechanical extraction is that this method produces high quality oil. Other advantages include low equipment cost, maintenance, repair and operational skills, and low power requirement. However, there are some drawbacks related to this type of extractors, like extraction efficiency (high oil content in the residual material) and necessity of purification (filtration, degumming) of crude oils. Extraction efficiency is influenced by the type of seed since conventional mechanical presses are usually designed to extract oil for a defined type of seed. Residual oil in the processed material can be extracted by utilization of chemical or enzymatic extraction.

### 15.4.2 Solvent Extraction

In order to improve extraction efficiency of mechanical extraction methods, solvent extraction methods have been developed. Solvent extraction refers to extraction of oil from solid material by means of the appropriate solvent. During solvent extraction, oil diffuses through the cell walls into the selective solvent. Physical properties of both phases play a huge role in the rate of mass transfer between the solid material and solvent. Extraction efficiency is influenced by the solvent used for extraction, particle size, temperature, and hydrodynamic conditions. The crucial step is selection of the solvent that will selectively extract oil from solid material. Besides the demands regarding selectivity, cost, toxicity, chemical and thermal stability, and availability, solvent should be of low viscosity to enhance the mass transfer. Smaller particle size is also favorable, considering the larger interfacial area available for mass transfer as well as reducing the extraction time. Even though one of the major advantages of solvent extraction is that it can be conducted at room conditions, performing extraction at elevated temperatures is sometimes beneficial since the mass transfer is enhanced due to the increased solubility of solute. It is a well-known fact that higher mixing intensities will result in enhanced mass transfer as a consequence of lower hydrodynamic and mass transfer resistances. Solvent extraction can be performed in batch and continuous mode.

For the chemical extraction of oils from plant material, an organic solvent is used. Plant materials consist of a wide variety of natural compound, which can be extracted simultaneously with oil and will influence its properties. Extraction of oils with hot water is a traditional method in some nondeveloped countries. Even though this method is environmentally friendly, processing of oil at elevated temperatures may lead to some undesirable reactions, like oxidation. Although different solvents have been investigated for this purpose, *n*-hexane remains the major solvent used commercially due to the high solubility of oils and fats and its low boiling point (easily recoverable). In order to separate oil from solution, additional processing step is necessary. Consequently, solvent extraction consumes large amount of energy since regeneration step (distillation) must be taken into consideration. However, both extract (oil) and the insoluble solid material can be valuable products, so this can make the process economically more feasible. Advanced technologies like ultrasonic and microwave extraction, or supercritical fluid extraction, prevail limitations of the conventional extraction. Utilization of ultrasound or microwaves enhances the rate of mass transfer, so the extraction time can be significantly reduced. However, when plant material is exposed to ultrasonic or microwave irradiation, decomposition may occur due to additional heating. Increased efficiency of ultrasonic-assisted extraction was attributed to the propagation of waves, which resulted in the acoustic cavitation phenomena responsible for the disruption of cell wall and promotes penetration of solvent. In a way, hydrodynamic conditions are improved [67]. Microwave-assisted extraction process is characterized by a lower solvent consumption and can be considered as environmentally friendly method since water and alcohols are used as solvents. Microwaves interact with the compound with the highest dielectric constant (in case of plant material – water) and generate heat inside cell. Water molecules evaporate, and pressure inside cell is increased, which will result in rupturing the cell walls. Since the material is exposed to microwaves for a short period of time, degradation is minimized [68].

Supercritical fluid extraction is more effective than conventional solvent extraction method. Enhanced mass transfer resulted from improved selectivity and better properties (lower viscosity, higher diffusion coefficients). Solvation capacity can be significantly influenced by the change of pressure and temperature [69]. The most commonly used solvent is supercritical $CO_2$. The major advantages are low cost, high selectivity to nonpolar compounds, and environmentally friendly solvent. Due to the high equipment and operation cost, this method is not used commercially.

Due to the fact that solvent extraction generates organic waste, the solvent should be reusable and easily regenerable. Commercially used solvents are volatile organic compounds (VOCs), which are flammable and are of varying toxicity, so the major task of many researchers is to find a way of replacing VOCs with environmentally friendly, so-called *green solvents*.

### 15.4.3 Enzymatic Extraction

A novel, simple, low-cost, and environmentally friendly method for extraction of oil from plant material is aqueous enzymatic extraction. As mentioned previously, the major disadvantage of hot water extraction is low efficiency since it takes long time for water to penetrate the cell walls and allow oil to be transferred out from the cell. The extraction efficiency can be significantly improved with addition of enzymes, which will degrade cell wall, resulting in facilitated extraction of oil. Since the lipids are insoluble in water, they will form emulsion that can easily be demulsified by changing temperature or by the application of enzymes. Phospholipids are also removed from oil by this method, so degumming step can be avoided. Oil yield can be additionally improved if a combination of different enzymes is used. The selection of enzymes depends on the structure of plant material and cell wall. The efficiency of aqueous enzymatic extraction is influenced by the size of plant material, temperature, pH, hydrodynamic conditions, water-to-substrate ratio, and enzyme-to-substrate ratio. Generally speaking, smaller particles are favorable because they exhibit higher mass transfer area. However, for plant material with high oil content, too small particles tend to stick together, resulting in significantly lower area in contact with solvent, and consequently lower extraction efficiency will be established [70]. Enzymes are sensitive to pH and temperature. Each enzyme has its optimum pH value and temperature at which it must be used for high oil yield. Concentration of enzymes and water quantity should also be determined carefully in order to achieve the maximum possible yield [71].

## 15.5 Purification of Oils and Fats

After extraction, oil is usually further processed by some refining method [25]. Methods used for refining of vegetable food oils are filtering, deacidification, winterizing, bleaching, degumming, and deodorization. However, not all steps are necessary for purification of oils, which will be used for production of biodiesel. Filtering is necessary for oils obtained by mechanical and enzymatic extraction. WCOs and WAFs after rendering also require removal of solids by means of filtration. Deacidification is the most commonly used step for purification of oils prior to base-catalyzed chemical transesterification since some oils and fats, especially waste oils and fats, contain high amounts of FFAs. Winterization of some oils is also necessary step because by this method saturated fatty acid content is removed, and low temperature fuel properties are improved [72, 73]. Both oil and biodiesel can be winterized. Another process that is advisable for purification of oils is degumming since phospholipids reduce the yield of biodiesel by deactivation of catalyst. Additionally, separation of glycerol from biodiesel is inhibited. By degumming, phospholipids or gums are removed from the crude oil or fat [74]. Bleaching and deodorization of oils to be used as feedstock for biodiesel production are not essential steps. Deodorization step involves removal of volatile compounds (ketones and aldehydes) as well as some other compounds (pesticides and PAH) and is usually performed by steam distillation. Usually, after distillation, citric acid can be added for removal of metal traces due to the formation of metal complexes. The major disadvantage of deodorization is that during this process different chemical reactions can occur that will destroy the structure of triglycerides; in other words the quality of oil will be lowered. Bleaching is used for removal of trace metals, soaps, and pigments. However, residual oils recovered from wastes generated at vegetable oils refineries after bleaching and deodorization can be used as feedstocks for biodiesel production [75, 76]. Waste frying oil and animal fats contain too high concentration of metals that needs to be reduced in order to obtain high quality feedstock for biodiesel production. Metals are commercially removed by adsorption, but based on some researchers, during deacidification with deep eutectic solvents, the concentration of metals was also reduced [64].

### 15.5.1 Deacidification

The main disadvantage of low quality feedstocks is too high proportion of FFAs that prevents easy conversion of triglycerides into fatty acid esters [47, 48]. Today, there are different technologies on the market for pretreatment of vegetable oils and animal fats, which are available for commercial production [77]. The FFAs content, in addition to economic aspects, influences the selection of appropriate pretreatment methods. The most commonly used methods of pretreatment are the neutralization of acids or their esterification with acid catalysts, although vacuum steam distillation is sometimes used as well. Some alternative methods of pretreatment of raw materials have been investigated, such as transesterification by glycerol, biological pretreatment by microorganisms or enzymes, and adsorption, liquid–liquid extraction, or extraction by supercritical fluids [39]. Each method has its advantages and disadvantages, but in most cases the economic aspect dictates the final decision regarding the selection of pretreatment method. Along with the existing methods such as esterification and alkaline stripping, liquid–liquid extraction exhibits a great potential as well because of its mild operating conditions. As for the application of the eutectic solvent for the feedstock pretreatment, according to available literature data, eutectic mixtures have not been used as selective solvents in the extraction of FFAs from raw materials of different origin. However, their application is investigated in a pretreatment process of a feedstock with the increased level of FFAs by esterification, wherein the eutectic solvent acted as a catalyst. Eutectic mixtures were generally used as catalysts in the process of esterification of palmitic acid [78] and esterification of palm oil [79, 80].

### 15.5.2 Winterization

Some oils and fats contain waxes or saturated triglycerides. This high temperature melting compounds are removed by winterization. During winterization, oil is subjected to low temperatures for a defined period of time. As a result of cooling, saturated acids components crystallize. Crystals are further separated by centrifugation, and the oil can be decanted. After winterization the concentration of FFAs can be increased, so it is advisable to perform winterization before deacidification [81]. Winterization step can be performed during purification of biodiesels with improper cold flow properties [82]. Winterization is a simple economically feasible method used to improve oils and biodiesels cold flow properties.

### 15.5.3 Demetallization

Traces of metals in feedstocks usually come from plant/animal growth and from processing equipment. Plants, while growing, absorb trace elements from the surrounding ground, and animals absorb them through plants and other food [83]. Reduction of metal content in biodiesel feedstocks is usually performed by adsorption or by caustic refining followed by water washing steps [84]. Commercially available adsorbents for removal of heavy metals have been developed. Granulometric properties of adsorbent (pore size, pore volume, and specific surface area) and the composition of oil/fat influence adsorption efficiency. Special care must be taken when dealing with feedstock with high acid number, since it was shown that bentonite types of adsorbents are soluble in acidic feedstocks. Traces of metals are also removed from oils and fats during oil refining steps: degumming, deodorization, and bleaching. Recently, a novel method for reduction of metal content in oils and fats by means of liquid–liquid extraction with deep eutectic solvents was proposed [64].

### 15.5.4 Degumming

Some vegetable oils, WCOs, and WAFs contain too high concentration of phospholipids. Phospholipids will settle on the bottom of the tank due to their higher density after a long time period. To enhance the separation of phospholipids from oils and fats, different methods have been investigated [74]. Selection of the appropriate method depends on the type of phospholipids. Phospholipids can be hydratable and nonhydratable. Hydratable compounds can easily be removed by filtration or centrifugation after simple treatment of oil with small amount of water. Process can be performed in batch or continuous mode. The process is mostly affected by temperature and water content. Degumming efficiency is increased at higher concentration of water and at elevated temperatures. Nonhydratable phospholipids can be separated from oils and fats by chemical, physical, and biological degumming methods. In chemical degumming processes, chemical reagent is added to oil. For instance, addition of phosphoric or citric acid resulted with formation of Ca and Mg salts (metal complexes) that can easily be removed from suspension. Chemical processes are generally performed with addition of acid or ethylenediaminetetraacetic acid (EDTA). Chemical processes consume large amount of energy and chemicals. Membrane degumming methods are based on the selective separation of molecules of different sizes. Ceramic and polymeric membranes have been investigated. The major disadvantages of membranes are high price, the use of expensive solvents, poor permeate flux rate, and possibility of contamination. For the biological degumming process, phospholipase enzymes are used. The process is environmentally acceptable and highly effective. However, due to the high cost of enzymes, applicability to limited concentration of gums, and long duration, this method is not economically feasible.

## 15.6 Production of Biodiesel

The idea of using vegetable oils as fuel for diesel engine was first introduced by Diesel at the end of the nineteenth century when he stated that: "Even though the use of vegetable oils for engine fuels may seem insignificant today, such oils may become in the course of time as important as the petroleum and coal tar products of the present time" (1912). However, today it is a well-known fact that oils without pretreatment cannot be used directly due to the several disadvantages like high viscosity, low cetane number, low flash point, and presence of FFAs and natural gums (phosphatides). Because of their improper properties and composition, vegetable oils can cause problems to the engine like formation of deposits (carbon, heavy gum, and waxes), corrosion, poor cold weather start-up, plugging and gumming of filters, lines and injectors, improper injection, incomplete combustion, and so on [85]. Properties of oils can be altered by several methods: blending, pyrolysis, microemulsification, and transesterification [14]. The simplest method for reduction of viscosity and in some cases improvement in emissions is blending of oils with fossil diesel, biodiesel, or alcohols. However, problem due to the lubricating oil thickening will still remain (polyunsaturated fatty acids) since the molecular structure of oil in blend is the same. Density and viscosity of oils can be modified by catalytic thermal cracking or pyrolysis [86]. This is a promising technology since the process is similar to conventional hydro processing of fossil fuels [87]. Microemulsification is another possible way to reduce the oil viscosity by formation of microemulsions; stable mixtures of polar and nonpolar phases, from oil, water, and/or alcohol; and surfactant [88]. Conventional method of production of biodiesel is chemical transesterification of oil with alcohol (methanol or ethanol) in the presence of a catalyst. Originally biodiesel was synthesized from high quality vegetable oils. The feasibility of a large number of vegetable and animal originated materials to be used as feedstocks for the production of biodiesel have been investigated so far [89–92]. In recent times, the application of ionic liquids, deep eutectic solvents, or nanomaterials as catalysts, in the process of transesterification, is investigated. For low quality feedstocks noncatalytic reaction under supercritical conditions or membrane reactors can be used.

### 15.6.1 Catalysts for Biodiesel Production

Biodiesel is produced by a transesterification reaction during which the oil or fat reacts with an alcohol in the presence of a catalyst, and the corresponding alkyl esters of the fatty acids present in the feed are obtained. The reaction may be carried out with homogeneous or heterogeneous catalysts, wherein the homogeneous catalyst may be alkaline or acidic and heterogeneous catalysts may be solid catalysts or enzymes [93]. However, such a reaction mechanism is very sensitive to the purity of raw materials or the presence of water and FFAs. The use of high quality edible oils as feedstock for biodiesel production results in ethical doubts as well as in an increase in the price of biodiesel. The increase in global demands for the production of biodiesel results in increasing interest for the use of waste materials of plant and animal origin or nonedible oils [48]. The main problems associated with the use of waste materials are a high proportion of FFAs that must be reduced to below 1% and the frequent presence of water. Namely, the presence of FFAs and water results in the undesired reactions of saponification and hydrolysis, which hinder transesterification reaction and substantially reduce conversion. In the production of biodiesel from lower quality feedstock, it is necessary to pretreat the raw material or, if this does not guarantee satisfactorily low concentrations of FFAs, to carefully select a catalyst that will provide conversion of FFAs into high quality oil.

### 15.6.2 Homogeneous Catalysts

The usual process of industrial synthesis of biodiesel is based on the reaction, which is usually carried out in the presence of an alkaline catalyst and includes the starting material of high quality, i.e. vegetable oil with small amounts of FFAs [77, 94]. The result of the reaction is the formation of a mixture of alkyl (usually methyl or ethyl) esters of fatty acids, i.e. biodiesel and glycerol as the by-product [48]. Most commonly used catalysts are sodium methoxide, sodium hydroxide, potassium hydroxide, and potassium methoxide. The advantages of synthesis of biodiesel in the presence of alkaline catalysts are low-cost catalysts and their accessibility, high activity (high conversion of about 98%), and mild operating conditions of the process (low pressure; moderate temperature (25–70 °C); short reaction time (30 min to 2 h)) [25]. However, if raw materials of low purity are used, the saponification occurs, and the resulting soaps reduce the efficiency of the process, making the separation difficult and favoring the formation of microemulsions [62]. Besides the quality of feedstocks, the efficiency of alkali-catalyzed transesterification is influenced by the type and amount of catalyst used, methanol-to-oil mass ratio, temperature, and reaction time. Efficiency can be improved by the application of ultrasound, microwaves, or addition of cosolvent. The major drawbacks are expensive, time consuming, and often complicated downstream processes like disposing and treatment of high amount of alkaline wastewater, purification of glycerol, alcohol recovery, and removal of inorganic salts [32, 95].

If the feedstock contains too high concentrations of FFAs or water, transesterification may be performed in the presence of acid catalysts such as sulfonic acid, sulfuric acid, phosphoric acid, hydrochloric acid, or an organic sulfonic acid [14, 91]. This way high conversion can be achieved, but because of disadvantages such as the corrosivity of the catalyst, long reaction times, high concentration of catalyst and/or alcohol, and elevated operating temperatures, this method is not widely used. Even though no soap is produced, biodiesel must be purified (neutralization, removal of salt deposits), which in turn increases the production cost. Production of biodiesel from low quality feedstocks can be performed in two steps: acidic esterification followed by alkali transesterification. That way high conversion can be achieved [25]. The efficiency may be increased by proper selection of reaction conditions and utilization of ultrasound or microwaves.

### 15.6.3 Heterogeneous Catalysts

For transesterification of low quality feedstocks, heterogeneous catalysts can be used without purification step, and high yield of biodiesel is easily achieved [87, 96–99]. By employment of heterogeneous catalysts, no wastewaters are produced since biodiesel of high purity can be produced without further needs for purification. Furthermore, catalysts can be regenerated and reused and may be of low cost. However, commercial heterogeneous catalysts are usually expensive, so homogeneous transesterification is preferred. Some of these solid catalysts can simultaneously perform esterification and transesterification, resulting in shorter reaction times and consequently lower energy consumption. Novel catalysts, which have both alkali and acidic characteristics (zeolites, mesoporous silica, and heteropolyacids), can simultaneously reduce the high acid and water content [100]. Heterogeneous catalysts can also be alkaline and acidic. The major advantages of heterogeneous catalysts, which are high yield, medium catalyst cost, high reaction rate, reusability, can be used in continuous process. Examples of alkaline catalysts are metal oxides (TiO2–MgO, CaO, mixed Ca and Zr oxides, titanium calcium, and potassium impregnated zinc oxide), alkaline zeolites and clays, snail shells, boilers ash from agricultural waste, and biochar [62]. Major drawbacks of these types of catalyst are high energy requirement, exhaustive catalyst preparation, catalyst leaching, and low surface area. Examples of acidic catalysts are zinc stearate immobilized on silica gel, $SO_4^{2-}/ZrO_2$ superacid catalyst, ($SO_4^{2-}$/$SnO_2$), $Zn_{1.2}H_{0.6}PW_{12}O_{40}$ nanotubes with Lewis and Brønsted acid sites, $SiO_2$, $SO_4^{2-}/SnO_2$-$SiO_2$, nanoparticles of sulfated zirconia doped with manganese sulfatepolymeric materials, and ion exchange resins. Low concentration of active sites, high cost of the materials for preparation of catalyst, exhaustive catalyst preparation, and catalyst leaching are major disadvantages of heterogeneous acidic catalysts. Special group of heterogeneous catalysts belongs to enzymes (biological catalysis). Free and immobilized enzymes as well as some microbial cells can be used for transesterification. Different types (sorts) of lipases have been investigated. Generally speaking, reactions with enzymes are environmentally friendly; no by products are formed, and reaction can be performed at moderate operating conditions, which is together with no need for washing and easy separation of biodiesel from glycerol, and reduces energy consumption. Enzymes can be used for transesterification of oils and fats of different quality. The presence of water (above 0.2%), phospholipids, and chlorophylls can inhibit enzymatic activity. However, high cost of enzymes and the fact that they can be easily deactivated by methanol are some of their drawbacks. Lipases are of different activity, so by combination of two enzymes, they can be used for simultaneous esterification of FFA and transesterification of triglycerides.

## 15.7 Future Prospects

Evaluation of feedstock's feasibility through political, socioeconomic, and environmental aspects should be done for sustainable production of advanced biodiesel. Besides cultivation of fast growing nonedible crops or microorganisms on low quality lands or wastes, it is also necessary to provide efficient waste feedstocks supply chain (by increasing the number of suppliers form households, restaurants, industry; adequate infrastructure for collecting and storage of wastes; transportation; imports).

Best methods for extraction, reaction, and purification should be developed for specific advanced feedstocks and/or implementation of process intensification methods like ultrasound. New catalysts should be developed (especially from waste materials like eggshells).

The biorefinery concept that enables the production of various valuable products, like biopolymers, fine chemicals, and biomaterials, along with biodiesel from biomass is one of the most promising concepts that assure sustainability [90, 101]. The facility should be flexible to handle different types of feedstocks. Incorporation of the additional steps (to assure pretreatment of WCO and WAF) in the existing factories is one option. It is true that some initial equipment investments are necessary; however, in time this will positively influence the price of biodiesel because that way waste materials are used for production of a valuable product.

Continuous process is also an option for improvement of economic feasibility of biodiesel production process since continuously operated systems may be designed for higher mass transfer rates and increased product quality, resulting with shorter production time and lower energy consumption and costs.

## References

1 Ajala, O.E., Aberuagba, F., Odetoye, T.E., and Ajala, A.M. (2015). Biodiesel: sustainable energy replacement to petroleum-based diesel fuel – a review. *ChemBioEng Reviews 2* (3): 145–156. https://doi.org/10.1002/cben.201400024.

2 Azad, A.K., Rasul, M.G., Khan, M.M.K. et al. (2016). A review on socio-economic aspects of sustainable biofuels. *International Journal of Global Warming 10* (1): 32–54. https://doi.org/10.1504/IJGW.2016.077903.

3 Gissi, E., Gaglio, M., Aschonitis, V.G. et al. (2018). Soil-related ecosystem services trade-off analysis for sustainable biodiesel production. *Biomass and Bioenergy 114* (2017): 83–99. https://doi.org/10.1016/j.biombioe.2017.08.028.

4 Ziolkowska, J.R. (2013). Evaluating sustainability of biofuels feedstocks: a multi-objective framework for supporting decision making. *Biomass and Bioenergy 59*: 425–440. https://doi.org/10.1016/j.biombioe.2013.09.008.

5 Živković, S.B., Veljković, M.V., Banković-Ilić, I.B. et al. (2017). Technological, technical, economic, environmental, social, human health risk, toxicological and policy considerations of biodiesel production and use. *Renewable and Sustainable Energy Reviews 79* (February): 222–247. https://doi.org/10.1016/j.rser.2017.05.048.

6 Tabatabaei, M., Karimi, K., Horváth, I.S., and Kumar, R. (2015). Recent trends in biodiesel production. *Biofuel Research Journal 2* (3): 258–267. https://doi.org/10.18331/BRJ2015.2.3.4.

7 Hasheminejad, M., Tabatabaei, M., Mansourpanah, Y. et al. (2011). Upstream and downstream strategies to economize biodiesel production. *Bioresource Technology 102* (2): 461–468. https://doi.org/10.1016/j.biortech.2010.09.094.

8 Mat Aron, N.S., Khoo, K.S., Chew, K.W. et al. (2020). Sustainability of the four generations of biofuels – a review. *International Journal of Energy Research 44* (12): 9266–9282. https://doi.org/10.1002/er.5557.

9 Alalwan, H.A., Alminshid, A.H., and Aljaafari, H.A.S. (2019). Promising evolution of biofuel generations. Subject review. *Renewable Energy Focus 28* (March): 127–139. https://doi.org/10.1016/j.ref.2018.12.006.

10 Pandey, R. and Kumar, G. (2017). A comprehensive review on generations of biofuels: current trends, development and scope. *International Journal on Emerging Technologies (Special Issue NCETST-2017) 8* (1): 561–565. www.researchtrend.net.

11 Singh, D., Sharma, D., Soni, S.L. et al. (2020). A review on feedstocks, production processes, and yield for different generations of biodiesel. *Fuel 262* (July): https://doi.org/10.1016/j.fuel.2019.116553.

12 Hajjari, M., Tabatabaei, M., Aghbashlo, M., and Ghanavati, H. (2017). A review on the prospects of sustainable biodiesel production: a global scenario with an emphasis on waste-oil biodiesel utilization. *Renewable and Sustainable Energy Reviews 72* (October 2015): 445–464. https://doi.org/10.1016/j.rser.2017.01.034.

13 European Commission (2018). *Directive (EU) 2018/2001 of the european parliament and of the council of 11 December 2018 on the promotion of the use of energy from renewable sources (recast)*. Official Journal of The European Union. https://eur-lex.europa.eu/eli/dir/2018/2001/oj.

14 Atabani, A.E., Silitonga, A.S., Badruddin, I.A. et al. (2012). A comprehensive review on biodiesel as an alternative energy resource and its characteristics. *Renewable and Sustainable Energy Reviews 16* (4): 2070–2093. https://doi.org/10.1016/j.rser.2012.01.003.

15 Barros, S. (2020). Brazil biofuels annual. *USDA Foreign Agricultural Service.*

16 Kenneth, J. (2020). Argentina biofuels annual. *USDA Foreign Agricultural Service.*

17 Rahmanulloh, A. (2020). Indonesia biofuels annual. *USDA Foreign Agricultural Service.*

18 Flach, B., Lieberz, S., and Bolla, S. (2019). EU biofuels annual. *USDA Foreign Agricultural Service Monthly Biodiesel Production Report, US Energy Information Administration.*

19 Publishing, O.E.C.D. (2019). *OECD-FAO Agricultural Outlook 2019–2028*. Organisation for Economic Co-operation and Development OECD.

20 Vaisali, C., Charanyaa, S., Belur, P.D., and Regupathi, I. (2015). Refining of edible oils: a critical appraisal of current and potential technologies. *International Journal of Food Science and Technology 50* (1): 13–23. https://doi.org/10.1111/ijfs.12657.

21 Bojan, S.G. and Durairaj, S.K. (2012). Producing biodiesel from high free fatty acid *Jatropha curcas* oil by a two step method-an Indian case study. *Journal of Sustainable Energy & Environment 3*: 63–66.

22 Anwar, M. (2021). Biodiesel feedstocks selection strategies based on economic, technical, and sustainable aspects. *Fuel 283* (October 2020): 119204. https://doi.org/10.1016/j.fuel.2020.119204.

23 Ullah, F., Bano, A., and Ali, S. (2013). Optimization of protocol for biodiesel production of linseed (*Linum usitatissimum* L.) oil. *Polish Journal of Chemical Technology 15* (1): 74–77. https://doi.org/10.2478/pjct-2013-0013.

**24** Yesilyurt, M.K., Cesur, C., Aslan, V., and Yilbasi, Z. (2020). The production of biodiesel from safflower (*Carthamus tinctorius* L.) oil as a potential feedstock and its usage in compression ignition engine: a comprehensive review. *Renewable and Sustainable Energy Reviews 119* (November): https://doi.org/10.1016/j.rser.2019.109574.

**25** Osorio-González, C.S., Gómez-Falcon, N., Sandoval-Salas, F. et al. (2020). Production of biodiesel from castor oil: a review. *Energies 13* (10): 1–22. https://doi.org/10.3390/en13102467.

**26** Aransiola, E.F., Ehinmitola, E.O., Adebimpe, A.I. et al. (2019). Prospects of biodiesel feedstock as an effective ecofuel source and their challenges. In: *Advances in Eco-Fuels for a Sustainable Environment (Issue x)* (ed. K. Azad). Elsevier Ltd. https://doi.org/10.1016/b978-0-08-102728-8.00003-6.

**27** Abbah, E.C., Nwandikom, G.I., Egwuonwu, C.C., and Nwakuba, N.R. (2016). Effect of reaction temperature on the yield of biodiesel from neem seed oil. *American Journal of Energy Science 3* (3): 16–20.

**28** Ulfah, M., Mulyazmi, S.T., Burmawi, B. et al. (2018). Biodiesel production methods of rubber seed oil: a review. *IOP Conference Series: Materials Science and Engineering 334*: 012006. https://doi.org/10.1088/1757-899X/334/1/012006.

**29** Yang, J., Caldwell, C., Corscadden, K. et al. (2016). An evaluation of biodiesel production from *Camelina sativa* grown in Nova Scotia. *Industrial Crops and Products 81*: 162–168. https://doi.org/10.1016/j.indcrop.2015.11.073.

**30** Sander, A., Petračić, A., Vuković, J.P., and Husinec, L. (2020). From coffee to biodiesel – deep eutectic solvents for feedstock and biodiesel purification. *Separations* 7 (2): 22. https://doi.org/10.3390/SEPARATIONS7020022.

**31** Sandouqa, A. and Al-Hamamre, Z. (2019). Energy analysis of biodiesel production from jojoba seed oil. *Renewable Energy 130*: 831–842. https://doi.org/10.1016/j.renene.2018.07.015.

**32** Sitepu, E.K., Heimann, K., Raston, C.L., and Zhang, W. (2020). Critical evaluation of process parameters for direct biodiesel production from diverse feedstock. *Renewable and Sustainable Energy Reviews 123* (March 2019): 109762. https://doi.org/10.1016/j.rser.2020.109762.

**33** Sawangkeaw, R. and Ngamprasertsith, S. (2013). A review of lipid-based biomasses as feedstocks for biofuels production. *Renewable and Sustainable Energy Reviews 25*: 97–108. https://doi.org/10.1016/j.rser.2013.04.007.

**34** Atabani, A.E., Mahlia, T.M.I., Masjuki, H.H. et al. (2013). A comparative evaluation of physical and chemical properties of biodiesel synthesized from edible and non-edible oils and study on the effect of biodiesel blending. *Energy 58*: 296–304. https://doi.org/10.1016/j.energy.2013.05.040.

**35** Sander, A., Antonije Košćak, M., Kosir, D. et al. (2018). The influence of animal fat type and purification conditions on biodiesel quality. *Renewable Energy 118*: 752–760. https://doi.org/10.1016/j.renene.2017.11.068.

**36** Ahorsu, R., Medina, F., and Constantí, M. (2018). Significance and challenges of biomass as a suitable feedstock for bioenergy and biochemical production: a review. *Energies 11* (12): https://doi.org/10.3390/en11123366.

**37** Kim, D.S., Hanifzadeh, M., and Kumar, A. (2018). Trend of biodiesel feedstock and its impact on biodiesel emission characteristics. *Environmental Progress and Sustainable Energy 37* (1): 7–19. https://doi.org/10.1002/ep.12800.

**38** Takase, M., Zhao, T., Zhang, M. et al. (2015). An expatiate review of neem, jatropha, rubber and karanja as multipurpose non-edible biodiesel resources and comparison of their fuel, engine and emission properties. *Renewable and Sustainable Energy Reviews* 43: 495–520. https://doi.org/10.1016/j.rser.2014.11.049.

**39** Banković-Ilić, I.B., Stojković, I.J., Stamenković, O.S. et al. (2014). Waste animal fats as feedstocks for biodiesel production. *Renewable and Sustainable Energy Reviews 32*: 238–254. https://doi.org/10.1016/j.rser.2014.01.038.

**40** Chakraborty, R., Gupta, A.K., and Chowdhury, R. (2014). Conversion of slaughterhouse and poultry farm animal fats and wastes to biodiesel: parametric sensitivity and fuel quality assessment. *Renewable and Sustainable Energy Reviews* https://doi.org/10.1016/j.rser.2013.08.082.

**41** Behçet, R., Oktay, H., Çakmak, A., and Aydin, H. (2015). Comparison of exhaust emissions of biodiesel–diesel fuel blends produced from animal fats. *Renewable and Sustainable Energy Reviews 46*: 157–165. https://doi.org/10.1016/j.rser.2015.02.015.

**42** Adewale, P., Dumont, M.-J., and Ngadi, M. (2015). Recent trends of biodiesel production from animal fat wastes and associated production techniques. *Renewable and Sustainable Energy Reviews 45*: 574–588. https://doi.org/10.1016/j.rser.2015.02.039.

**43** Dufour, J. and Iribarren, D. (2012). Life cycle assessment of biodiesel production from free fatty acid-rich wastes. *Renewable Energy* https://doi.org/10.1016/j.renene.2011.07.016.

**44** Atabani, A.E., Al-Muhtaseb, A.H., Kumar, G. et al. (2019). Valorization of spent coffee grounds into biofuels and value-added products: pathway towards integrated bio-refinery. *Fuel 254* (February): 115640. https://doi.org/10.1016/j.fuel.2019.115640.

**45** Kondamudi, N., Mohapatra, S.K., and Misra, M. (2008). Spent coffee grounds as a versatile source of green energy. *Journal of Agricultural and Food Chemistry 56* (24): 11757–11760. https://doi.org/10.1021/jf802487s.

**46** Lama-Muñoz, A., Álvarez-Mateos, P., Rodríguez-Gutiérrez, G. et al. (2014). Biodiesel production from olive-pomace oil of steam-treated alperujo. *Biomass and Bioenergy 67* (August): 443–450. https://doi.org/10.1016/j.biombioe.2014.05.023.

**47** Chai, M., Tu, Q., Lu, M., and Yang, Y.J. (2014). Esterification pretreatment of free fatty acid in biodiesel production, from laboratory to industry. *Fuel Processing Technology 125*: 106–113. https://doi.org/10.1016/j.fuproc.2014.03.025.

**48** Santacesaria, E., Tesser, R., Di Serio, M. et al. (2007). Kinetics and mass transfer of free fatty acids esterification with methanol in a tubular packed bed reactor: a key pretreatment in biodiesel production. *Industrial & Engineering Chemistry Research 46* (15): 5113–5121. https://doi.org/10.1021/ie061642j.

**49** Chhetri, A., Watts, K., and Islam, M. (2008). Waste cooking oil as an alternate feedstock for biodiesel production. *Energies 1* (1): 3–18. https://doi.org/10.3390/en1010003.

**50** Encinar, J.M., Sánchez, N., Martínez, G., and García, L. (2011). Study of biodiesel production from animal fats with high free fatty acid content. *Bioresource Technology 102* (23): 10907–10914. https://doi.org/10.1016/j.biortech.2011.09.068.

**51** Nguyen, H.C., Nguyen, M.L., Wang, F.M. et al. (2020). Biodiesel production by direct transesterification of wet spent coffee grounds using switchable solvent as a catalyst and solvent. *Bioresource Technology 296* (October 2019): 122334. https://doi.org/10.1016/j.biortech.2019.122334.

**52** Tongcumpou, C., Usapein, P., and Tuntiwiwattanapun, N. (2019). Complete utilization of wet spent coffee grounds waste as a novel feedstock for antioxidant, biodiesel, and bio-char production. *Industrial Crops and Products 138*: https://doi.org/10.1016/j.indcrop.2019.111484.

**53** Knothe, G. and Razon, L.F. (2017). Biodiesel fuels. *Progress in Energy and Combustion Science 58*: 36–59. https://doi.org/10.1016/j.pecs.2016.08.001.

**54** Lardon, L., Hélias, A., Sialve, B. et al. (2009). Life-cycle assessment of biodiesel production from microalgae. *Environmental Science and Technology 43* (17): 6475–6481. https://doi.org/10.1021/es900705j.

**55** Chen, C.-Y., Yeh, K.-L., Aisyah, R. et al. (2011). Cultivation, photobioreactor design and harvesting of microalgae for biodiesel production: a critical review. *Bioresource Technology 102* (1): 71–81. https://doi.org/10.1016/j.biortech.2010.06.159.

**56** Nozzi, N.E., Oliver, J.W.K., and Atsumi, S. (2013). Cyanobacteria as a platform for biofuel production. *Frontiers in Bioengineering and Biotechnology 1*: https://doi.org/10.3389/fbioe.2013.00007.

**57** Meeuwse, P., Sanders, J.P.M., Tramper, J., and Rinzema, A. (2013). Lipids from yeasts and fungi: tomorrow's source of biodiesel? *Biofuels, Bioproducts and Biorefining 7* (5): 512–524. https://doi.org/10.1002/bbb.1410.

**58** Iyer, R. (2016). The issue of reducing or removing phospholipids from total lipids of a microalgae and an oleaginous fungus for preparing biodiesel. *Biofuels 7* (1): 55–72. https://doi.org/10.1080/17597269.2015.1118778.

**59** Defernez, M., Wren, E., Watson, A.D. et al. (2017). Low-field $^{1}$H NMR spectroscopy for distinguishing between arabica and robusta ground roast coffees. *Food Chemistry 216*: 106–113. https://doi.org/10.1016/j.foodchem.2016.08.028.

**60** Atabani, A.E., Silitonga, A.S., Ong, H.C. et al. (2013). Non-edible vegetable oils: a critical evaluation of oil extraction, fatty acid compositions, biodiesel production, characteristics, engine performance and emissions production. *Renewable and Sustainable Energy Reviews 18*: 211–245. https://doi.org/10.1016/j.rser.2012.10.013.

**61** Mannu, A., Vlahopoulou, G., Urgeghe, P. et al. (2019). Variation of the chemical composition of waste cooking oils upon bentonite filtration. *Resources 8* (2): 1–15. https://doi.org/10.3390/resources8020108.

**62** Fonseca, J.M., Teleken, J.G., de Cinque Almeida, V., and da Silva, C. (2019). Biodiesel from waste frying oils: methods of production and purification. *Energy Conversion and Management* 184: 205–218. Elsevier Ltd. https://doi.org/10.1016/j.enconman.2019.01.061.

**63** Mannu, A., Garroni, S., Porras, J.I., and Mele, A. (2020). Available technologies and materials for waste cooking oil recycling. *Processes 8* (3): 1–13. https://doi.org/10.3390/PR8030366.

**64** Petračić, A., Sander, A., and Cvetnić, M. (2019). A novel approach for the removal of trace elements from waste fats and oils. *Separation Science and Technology* 1–15: https://doi.org/10.1080/01496395.2019.1706575.

**65** Toldrá-Reig, F., Mora, L., and Toldrá, F. (2020). Trends in biodiesel production from animal fatwaste. *Applied Sciences (Switzerland) 10* (10): https://doi.org/10.3390/app10103644.

**66** Bhargavi, G., Nageswara Rao, P., and Renganathan, S. (2018). Review on the extraction methods of crude oil from all generation biofuels in last few decades. *IOP Conference Series: Materials Science and Engineering 330* (1): https://doi.org/10.1088/1757-899X/330/1/012024.

**67** Fuad, F.M. and Don, M.M. (2016). Ultrasonic-assisted extraction of oil from *Calophyllum inophyllum* seeds: optimization of process parameters. *Jurnal Teknologi 78* (10): 199–206. https://doi.org/10.11113/jt.v78.4946.

**68** Kumar, R.C., Benal, M.M., Prasad, B.D. et al. (2018). Microwave assisted extraction of oil from pongamia pinnata seeds. *Materials Today: Proceedings 5* (1): 2960–2964. https://doi.org/10.1016/j.matpr.2018.01.094.

**69** Sovovà, H. and Stateva, R.P. (2011). Supercritical fluid extraction from vegetable materials. *Reviews in Chemical Engineering 27* (3–4): 79–156. https://doi.org/10.1515/REVCE.2011.002.

**70** de Mariano, R.G., B., Couri, S., and Freitas, S.P. (2009). Enzymatic technology to improve oil extraction from *Caryocar brasiliense* camb. (Pequi) Pulp. *Revista Brasileira de Fruticultura 31* (3): 637–643. https://doi.org/10.1590/S0100-29452009000300003.

**71** Mwaurah, P.W., Kumar, S., Kumar, N. et al. (2020). Novel oil extraction technologies: process conditions, quality parameters, and optimization. *Comprehensive Reviews in Food Science and Food Safety 19* (1): 3–20. https://doi.org/10.1111/1541-4337.12507.

**72** González Gómez, M., Howard-Hildige, R., Leahy, J., and Rice, B. (2002). Winterisation of waste cooking oil methyl ester to improve cold temperature fuel properties. *Fuel 81* (1): 33–39. https://doi.org/10.1016/S0016-2361(01)00117-X.

**73** Zhong, H., Watanabe, M., Enomoto, H. et al. (2016). Winterization of vegetable oil blends for biodiesel fuels and correlation based on initial saturated fatty acid constituents. *Energy and Fuels 30* (6): 4841–4847. https://doi.org/10.1021/acs.energyfuels.6b00310.

**74** Sharma, Y.C., Yadav, M., and Upadhyay, S.N. (2019). Latest advances in degumming feedstock oils for large-scale biodiesel production. *Biofuels, Bioproducts and Biorefining 13* (1): 174–191. https://doi.org/10.1002/bbb.1937.

**75** Huang, Y.P. and Chang, J.I. (2010). Biodiesel production from residual oils recovered from spent bleaching earth. *Renewable Energy 35* (1): 269–274. https://doi.org/10.1016/j.renene.2009.07.014.

**76** Piloto-Rodríguez, R., Melo, E.A., Goyos-Pérez, L., and Verhelst, S. (2014). Conversion of by-products from the vegetable oil industry into biodiesel and its use in internal combustion engines: a review. *Brazilian Journal of Chemical Engineering 31* (2): 287–301. https://doi.org/10.1590/0104-6632.20140312s00002763.

**77** Bhosle, B.M. and Subramanian, R. (2005). New approaches in deacidification of edible oils – a review. *Journal of Food Engineering 69* (4): 481–494. https://doi.org/10.1016/j.jfoodeng.2004.09.003.

**78** Tang, B., Lee, Y.J., Park, H.E., and Row, K.H. (2014). Pretreatment of biodiesel by esterification of palmitic acid in Brønsted-Lowry acid based deep eutectic solvents. *Analytical Letters 47* (14): 2443–2450. https://doi.org/10.1080/00032719.2014.908386.

**79** Hayyan, A., Hashim, M.A., Hayyan, M. et al. (2014). A new processing route for cleaner production of biodiesel fuel using a choline chloride based deep eutectic solvent. *Journal of Cleaner Production 65*: 246–251. https://doi.org/10.1016/j.jclepro.2013.08.031.

**80** Hayyan, A., Hashim, M.A., Hayyan, M. et al. (2013). A novel ammonium based eutectic solvent for the treatment of free fatty acid and synthesis of biodiesel fuel. *Industrial Crops and Products 46*: 392–398. https://doi.org/10.1016/j.indcrop.2013.01.033.

**81** Günç Ergönül, P. and Nergiz, C. (2015). The effect of different filter aid materials and winterization periods on the oxidative stability of sunflower and corn oils. *CyTA – Journal of Food 13* (2): 174–180. https://doi.org/10.1080/19476337.2014.931889.

**82** Vijayan, S.K., Naveena Victor, M., Sudharsanam, A. et al. (2018). Winterization studies of different vegetable oil biodiesel. *Bioresource Technology Reports 1*: 50–55. https://doi.org/10.1016/j.biteb.2018.02.005.

**83** Ansari, R., Kazi, T.G., Jamali, M.K. et al. (2009). Variation in accumulation of heavy metals in different verities of sunflower seed oil with the aid of multivariate technique. *Food Chemistry 115* (1): 318–323. https://doi.org/10.1016/j.foodchem.2008.11.051.

**84** Vera, C., Busto, M., Yori, J. et al. (2011). Adsorption in biodiesel refining – a review. *Biodiesel – Quality, Emissions and By-Products* 427–458: https://doi.org/10.5772/1094.

**85** Mahmudul, H.M., Hagos, F.Y., Mamat, R. et al. (2017). Production, characterization and performance of biodiesel as an alternative fuel in diesel engines – a review. *Renewable and Sustainable Energy Reviews 72* (April 2016): 497–509. https://doi.org/10.1016/j.rser.2017.01.001.

**86** Rajalingam, A., Jani, S.P., Kumar, A.S., and Khan, M.A. (2016). Production methods of biodiesel. *Journal of Chemical and Pharmaceutical Research 8* (3): 170–173. www.jocpr.com.

**87** Abbaszaadeh, A., Ghobadian, B., Omidkhah, M.R., and Najafi, G. (2012). Current biodiesel production technologies: a comparative review. *Energy Conversion and Management 63*: 138–148. https://doi.org/10.1016/j.enconman.2012.02.027.

**88** Murray, R., King, G., and Wyse-Mason, R. (2019). Micro-emulsification vs. transesterification: an investigation of the efficacy of methanol use in improving vegetable oil engine performance. *Biofuels* 1–10. https://doi.org/10.1080/17597269.2019.1598316.

**89** Fatih Demirbas, M. (2009). Biorefineries for biofuel upgrading: a critical review. *Applied Energy 86* (SUPPL. 1): S151–S161. https://doi.org/10.1016/j.apenergy.2009.04.043.

**90** Groom, M.J., Gray, E.M., and Townsend, P.A. (2008). Biofuels and biodiversity: principles for creating better policies for biofuel production. *Conservation Biology 22* (3): 602–609. https://doi.org/10.1111/j.1523-1739.2007.00879.x.

**91** Gutiérrez, C.D.B., Serna, D.L.R., and Alzate, C.A.C. (2017). A comprehensive review on the implementation of the biorefinery concept in biodiesel production plants. *Biofuel Research Journal 4* (3): 691–703. https://doi.org/10.18331/BRJ2017.4.3.6.

**92** Liew, W.H., Hassim, M.H., and Ng, D.K.S. (2014). Review of evolution, technology and sustainability assessments of biofuel production. *Journal of Cleaner Production 71*: 11–29. https://doi.org/10.1016/j.jclepro.2014.01.006.

**93** Atadashi, I.M., Aroua, M.K., Abdul Aziz, A.R., and Sulaiman, N.M.N. (2013). The effects of catalysts in biodiesel production: a review. *Journal of Industrial and Engineering Chemistry* 19 (1): 14–26. https://doi.org/10.1016/j.jiec.2012.07.009.

**94** Soltani, S., Rashid, U., Yunus, R., and Taufiq-Yap, Y.H. (2015). Synthesis of biodiesel through catalytic transesterification of various feedstocks using fast solvothermal technology: a critical review. *Catalysis Reviews – Science and Engineering 57* (4): 407–435. https://doi.org/10.1080/01614940.2015.1066640.

**95** Hama, S. and Kondo, A. (2013). Enzymatic biodiesel production: an overview of potential feedstocks and process development. *Bioresource Technology 135*: 386–395. https://doi.org/10.1016/j.biortech.2012.08.014.

**96** Endalew, A.K., Kiros, Y., and Zanzi, R. (2011). Heterogeneous catalysis for biodiesel production from *Jatropha curcas* oil (JCO). *Energy 36* (5): 2693–2700. https://doi.org/10.1016/j.energy.2011.02.010.

**97** Lee, A.F. (2014). Catalysing sustainable fuel and chemical synthesis. *Applied Petrochemical Research 4* (1): 11–31. https://doi.org/10.1007/s13203-014-0056-z.

**98** Santacesaria, E., Vicente, G.M., Di Serio, M., and Tesser, R. (2012). Main technologies in biodiesel production: state of the art and future challenges. *Catalysis Today 195* (1): 2–13. https://doi.org/10.1016/j.cattod.2012.04.057.

**99** Sharma, Y.C., Singh, B., and Korstad, J. (2011). Latest developments on application of heterogenous basic catalysts for an efficient and eco friendly synthesis of biodiesel: a review. *Fuel 90* (4): 1309–1324. https://doi.org/10.1016/j.fuel.2010.10.015.

**100** Abomohra, A.E.F., Elsayed, M., Esakkimuthu, S. et al. (2020). Potential of fat, oil and grease (FOG) for biodiesel production: a critical review on the recent progress and future perspectives. *Progress in Energy and Combustion Science 81*: 100868. https://doi.org/10.1016/j.pecs.2020.100868.

**101** Venkata Mohan, S., Nikhil, G.N., Chiranjeevi, P. et al. (2016). Waste biorefinery models towards sustainable circular bioeconomy: critical review and future perspectives. *Bioresource Technology 215* (December 2018): 2–12. https://doi.org/10.1016/j.biortech.2016.03.130.

# 16

# Mainstream Strategies for Biodiesel Production

*Narita Chanthon[1], Nattawat Petchsoongsakul[1], Kanokwan Ngaosuwan[2], Worapon Kiatkittipong[3], Doonyapong Wongsawaeng[4], Weerinda Appamana[5], and Suttichai Assabumrungrat[1,6]*

[1] *Center of Excellence on Catalysis and Catalytic Reaction Engineering, Department of Engineering, Faculty of Engineering, Chulalongkorn University, Bangkok, Thailand*
[2] *Division of Chemical Engineering, Faculty of Engineering, Rajamangala University of Technology Krungthep, Bangkok, Thailand*
[3] *Department of Chemical Engineering, Faculty of Engineering and Industrial Technology, Silpakorn University, Nakhon Pathom, Thailand*
[4] *Department of Nuclear Engineering, Faculty of Engineering, Chulalongkorn University, Bangkok, Thailand*
[5] *Department of Chemical and Materials Engineering, Faculty of Engineering, Rajamangala University of Technology Thanyaburi, Pathum Thani, Thailand*
[6] *Bio-Circular-Green-economy Technology & Engineering Center, BCGeTEC, Department of Chemical Engineering, Faculty of Engineering, Chulalongkorn University, Bangkok, Thailand*

## 16.1 Introduction

A mainstream process is used to convert renewable feedstocks (triglyceride sources) including refined edible oils/nonedible oils/algae, microalgae, and waste oils/advance solar oil (referred to as first to fourth generation) to biodiesel. There are many possible reaction routes such as pyrolysis, microemulsion, and transesterification [1, 2]. Transesterification has been revealed in numerous published articles to be the most adopted method for biodiesel synthesis, which can be performed with acid, base, and enzyme catalysts [3]. A base catalyst exhibits superior performances in term of high transesterification rate at mild conditions in a short reaction time as compared with acid and enzyme catalysts [4]. Conventional homogenous base catalysts (NaOH, KOH, $CH_3ONa$, and $CH_3OK$) can produce economical biodiesel using feedstocks with low fatty acid (refined edible oil), especially on the commercial scale [5]. However, using homogeneous base-catalyzed transesterification requires an expensive downstream purification process, producing a large amount of alkaline wastewater and environmental toxicity from the high basicity catalyst to obtain biodiesel with properties conforming to the standard [6]. Several pieces of research have focused on the design and development of heterogeneous catalysts for transesterification using various feedstocks [7]. The important factors that can hinder the transesterification efficiency are the low solubility of the two reactants (oils and alcohol) and the reversible reaction in nature.

*Biodiesel Production: Feedstocks, Catalysts, and Technologies*, First Edition. Edited by Samuel Lalthazuala Rokhum, Gopinath Halder, Suttichai Assabumrungrat, and Kanokwan Ngaosuwan.

Using the higher alcohol-to-oil ratio greater than the theoretical molar ratio of 3 : 1 can help shift the reaction forward. However, the mass transfer limitation problem could be solved by changing the reaction conditions such as alcohol-to-oil molar ratio and reaction temperature as related to the solubility of the reaction mixture [8]. Addition of a cosolvent could minimize the limiting mass transfer occurring from the two-phase mixture, but this method also extensively needs downstream purification as well as a homogeneous catalyst [9]. Supercritical transesterification can generate biodiesel in the absence of a catalyst; however, this process requires high capital and operating cost using a highly resistant vessel with high pressure (10–25 MPa) and high temperature (200–400 °C) to stimulate the homogeneous phase of oil and alcohol in a critical condition [10]. These two methods might not be suitable for large-scale biodiesel production. Therefore, the advancement of the process technology as intensified reactors enhances not only the mixing degree of nonpolar oil and polar alcohol but also heat transfer to provide a higher transesterification rate and biodiesel production yield [11]. The intensified reactors can be categorized into six types of biodiesel reactors including rotating reactor, tubular flow reactor, cavitation reactor, microwave reactor, multifunctional reactor, and other intensified reactors [6].

This chapter emphasizes the fundamental concepts, the operating parameters and the operating conditions of using intensified reactors to produce biodiesel as a mainstream process. In addition, the operation mode of batch and continuous processes is reviewed, which is dependent on the biodiesel production capacity. The prospects and challenges of the mainstream strategies are also explored to gain more insight and to select the appropriate technology using both homogeneous and heterogeneous catalysts based on economic and environmental considerations for sustainable biodiesel production.

## 16.2 Mainstream Strategies and Technology for Biodiesel Production

### 16.2.1 Current Mainstream Operation

#### 16.2.1.1 Batch Mode

The batch process as a simple process can carry out a chemical reaction by initial adding and mixing all components for the desired reaction time. Generally, a chemical reaction in a laboratory is mostly studied in a batch process because it requires smaller amounts of reactants, lower construction or setup cost, and can be easily controlled. To enhance the efficiency of biodiesel synthesis, the design and development of suitable catalysts and intensification are mainly carried out in a batch reactor due to the remarkable batch characteristics mentioned previously. Up to date, the worldwide bigger share of biodiesel production is still in a batch process because of the available local feedstocks suitable to produce on a small and medium scale [8]. However, the limitation of the batch process is the production capacity, which depends on the reactor volume, the rather low efficiency owing to start-up and shutdown, and the higher manpower requirement, which might not be proper for large-scale biodiesel production.

#### 16.2.1.2 Continuous Mode

The continuous process is mainly proposed for commercial and industrial scales where the reactants are continuously fed, and the products are continually removed at the same time. This continuous mode offers the small reactor volume and area requirement, as well as the low operating cost to obtain the higher production capacity with uniform product quality. Conventional continuous stirred tank reactor (CSTR) and tubular reactor have been used to produce biodiesel via homogenous-catalyzed transesterification, which is an obstacle for the subsequent separation and purification unit of biodiesel production [12]. Continuous biodiesel production using a heterogeneous system, mostly carried out in the laboratory and pilot scale, is comprehensively needed to break through the integration of feedstocks selection, catalyst design, and the choice of intensified reactor to achieve the economic and environmental aspect of biodiesel synthesis.

## 16.2.2 Process Mainstream for Biodiesel Production Based on the Reactor Types

### 16.2.2.1 Rotating Reactor

A rotating reactor is a type of an intensified reactor based on the mechanical agitation method. The main target of the rotating reactor is the production of the centrifugal force to induce turbulence generation and increase the heat/mass transfer rate. Both spinning disk reactors (SDRs) and rotating tube reactors (RTRs) are mainly remarkable in this type due to their enhancement of heat and mass transfer.

***16.2.2.1.1 Spinning Disk Reactor (SDR)*** An SDR is a promising intensified continuous reactor, which composes of a rotating disk inside a stationary chamber (**Figure 16.1a**). The reactor core is defined as a feed-in at the upper disk and a feed-out at the lower disc. Moreover, the SDR can be modified into a rotor–stator spinning disk reactor (RSDR), which comprises two flat discs separated by a narrow gap (<0.4 mm). The upper disk is rotated, while the lower disk remains immobile. The improvement of mixing inside SDR and RSDR derives from liquid dispersion, which is introduced at the center of a rotating disc, splashing on the discs and flowing outward radially. The rotation of the spinning disk mainly produces the centrifugal force, resulting in thin film formation with high surface shear. These high shear forces can provide rapid mixing, which reduces residence time and enhances overall heat/mass transfer [14].

Biodiesel production in a continuous system has been developed by using the SDR. The effects of gap width, rotational speed, feed flow rate, disk surface characteristic (smooth disk vs. grooved disc), feed ratio, reaction temperature, and catalyst concentration were investigated. Generally, vegetable oil and methanol are mostly utilized as feedstocks to synthesize biodiesel via transesterification. Both heterogeneous and homogeneous catalysts can be selected for catalyzing transesterification in SDRs [6].

As part of the effects related to reactor performance, the decrease of gap width leads to the increase in the shear stress between the two-disk interface, resulting in high mixing intensity. Increasing the rotational speed enhances the mixing intensity, which occurs in the turbulent regime inside the reactor. On the contrary, increasing the feed flow rate reduces the conversion, which relates to the reduction of residence time. The reactor rotational speed influences the radius and thickness of the liquid film on the disk surface. The radial velocity grows because of the increase in rotational speed and liquid flow rate. At a high rotational speed operating condition, efficient biodiesel production is allowed due to excellent fluid dynamics. Finally, the disk surface is a significant factor on conversion with changing the residence time distribution and back mixing inside the reactor to improve mixing intensity [15].

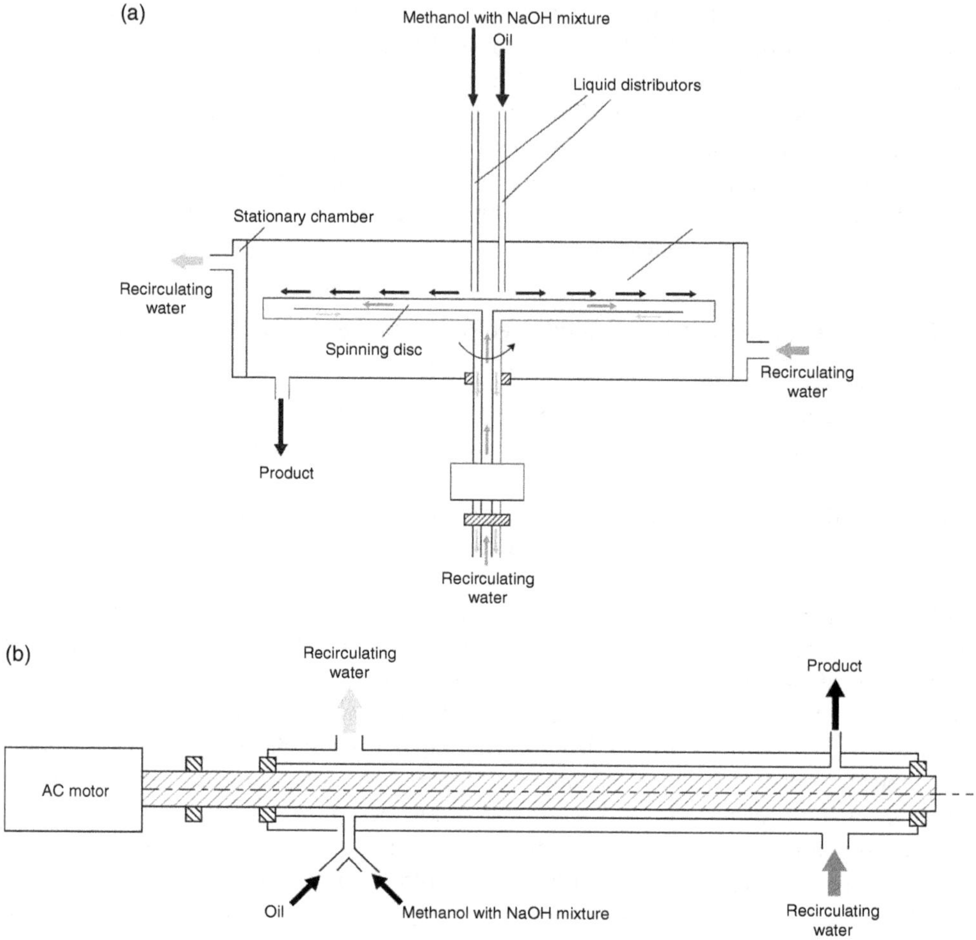


**Figure 16.1** Rotating reactor-assisted biodiesel production system (a) spinning disk reactor and (b) rotating tube reactor. *Source:* Adapted from Ref. [13].

***16.2.2.1.2 Rotating Tube Reactor (RTR)*** An RTR is a novel reactor that increases the intensity of mixing through centrifugal force generated on the surface of the reactor. The RTR mainly consists of three concentric tubes including the rotating inner tube, the stationary middle tube, and the outer tube. **Figure 16.1b** shows the system of the RTR. The inner tube or the rotor, which connects to the motor, is rotated at high speed to generate a thin film, resulting in augmentation of the interfacial surface area and reduction of heat mass transfer resistance. The inside of the RTR is divided into two sections: the narrow gap between the inner and middle tubes specifying the reaction zone and the wider gap between the middle and outer tubes specifying the heating/cooling zone.

To enhance biodiesel production using the RTR in the transesterification process, rotational speed, residence time, flow rate, mixing intensity, temperature, and catalyst concentration are the key parameters affecting biodiesel yield in the RTR [16]. The rotational speed is a crucial parameter for operating the RTR because it affects the hydrodynamic regime. Mostly, the rotational speed lower than 1000 rpm is used in the RTR. The hydrodynamic regime of the liquid mixture (oil and methanol phases) is converted from pool flow to annular flow when increasing the rotational speed, resulting in reduced residence time, increased heat mass transfer rate, and increased biodiesel yields. Moreover, the centrifugal force is induced by the high rotational speed, which assists glycerol separation (a byproduct of biodiesel production).

#### 16.2.2.2 Tubular Flow Reactor

A tubular reactor is a flow reactor type that is usually used on the industrial scale because of the proper structure and simple operating system with continuous production. The four main reactors in the tubular flow reactor section are packed bed reactor (PBR), oscillatory flow reactor (OFR), microchannel reactor, and static mixer reactor, as shown in **Figure 16.2**.

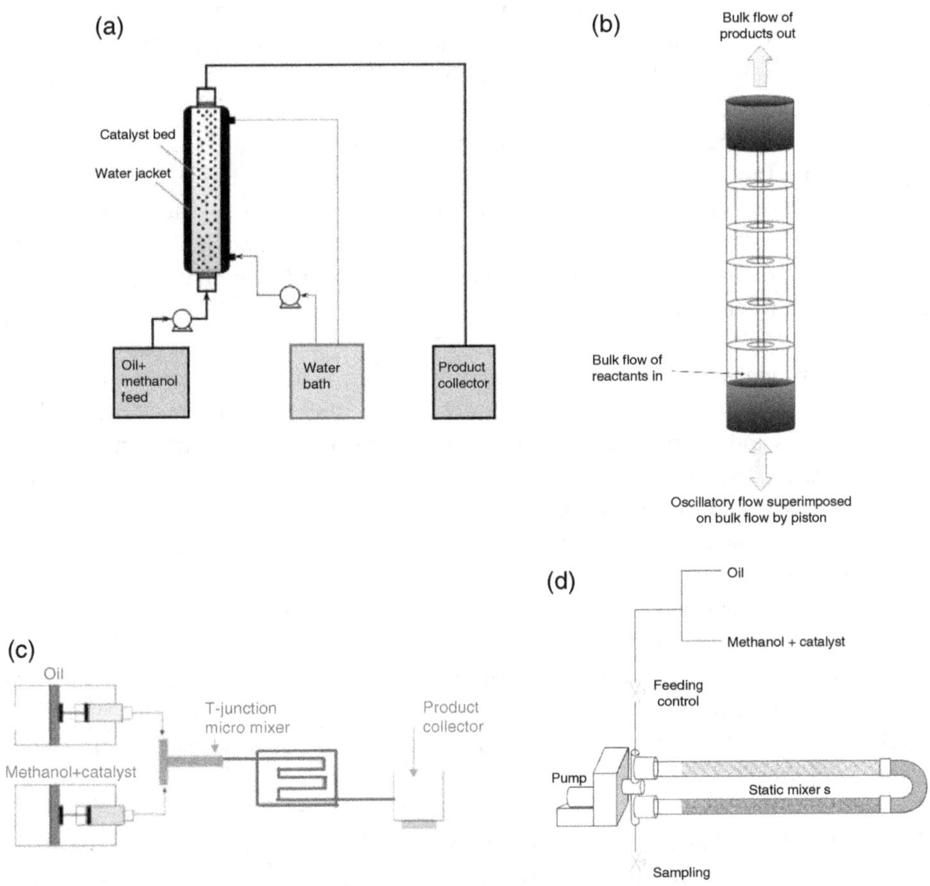


**Figure 16.2** Tubular flow reactor (a) packed bed reactor, (b) oscillatory flow reactor, (c) microchannel reactor, and (d) static mixer reactor. *Source:* Adapted from Ref. [13, 17–19].

***16.2.2.2.1 Packed Bed Reactor*** A PBR is mostly used for the continuous transesterification process using a heterogeneous catalyst or an enzymatic catalyst [9]. The fundamental operation of the PBR is the reaction medium flowing continuously through the reactor (or recirculation), which contains the packed catalyst bed and controls the operating condition. The PBR is divided into two types, including a fixed bed reactor and a rotating PBR. The former is the simplest PBR, which consists of solid catalyst particles packed into the bed, while the latter is the improved reactor by centrifugal force, which is induced by high rotational speed.

Basic parameters of biodiesel production using the PBR include the effect of methanol-to-oil molar ratio, concentration of catalyst, residence time, and temperature. Recently, biodiesel synthesis was studied in a PBR using calcium oxide-nanocrystal cellulose (NCC) supported on polyvinyl alcohol. The highest conversion was obtained at 98.4% at the optimal condition of 0.5 wt% NaOH catalyst concentration and 6 : 1 methanol-to-oil molar ratio at 65 °C [20].

***16.2.2.2.2 Oscillatory Flow Reactor*** An OFR is the intensified continuous flow reactor. The OFR includes a tube containing baffles, which operates with a pulsed flow rate. Turbulent is created while the piston moves forward and reverses the direction. Furthermore, the oscillation of the liquid medium creates a recirculating flow and eddy motions within the OFR, leading to increased efficiency of mixing and heat mass transfer based on long residence time.

Generally, the parameters of biodiesel production using the OFR are oscillation frequency, methanol-to-oil molar ratio, catalyst type, residence time, and reaction temperature. The most proper conditions were found at methanol-to-waste cooking oil molar ratio of 6 : 1, catalyst concentration of 1 wt% NaOH, oscillation frequency of 0.67 Hz, and 30 min of reaction time. The highest biodiesel yield based on waste cooking oil of 72.5% could be achieved when using the OFR [21].

***16.2.2.2.3 Microchannel Reactor*** A microchannel reactor is a miniaturized reactor with dimensions of 10–1000 μm. The main attributes of the microchannel reactor are short diffusion distances and large surface area-to-volume ratio, resulting in increasing heat and mass transfer rates. The use of the microchannel reactor offers a short transesterification time compared with a conventional mechanical stirred reactor.

Parameters affecting biodiesel synthesis in microchannel reactors are catalyst type, cosolvent addition, methanol-to-oil molar ratio, catalyst concentration, and reaction temperature [22]. Moreover, various microchannel structures such as T-shape mixer, T-junction microchannel, or zigzag microchannel extremely affect mixing intensity [23–26]. The optimal conditions for biodiesel production using the microchannel reactor were obtained at methanol-to-oil molar ratio of 8.5 : 1 and catalyst loading of 1.2 wt% KOH at 59 °C. The biodiesel yield under this condition was obtained at 99.5% in 14.9 s using a zigzag microchannel mixer [26].

***16.2.2.2.4 Static Mixer Reactor*** A static mixer reactor is the combination of a microchannel reactor with static elements. The static elements that enter inside the microchannel can increase the contact area and contact time, improving the mixing performance and reducing the molecular diffusion path. In the static mixer reactor, vortex formation results in enhanced fluid mixing, higher heat mass transfer, and higher yield. A fatty acid ethyl ester (FAEE) yield of 99.5% was achieved at ethanol-to-oil molar ratio of 9 and catalyst concentration of 1.0 wt% NaOH at 50 °C [17].

#### 16.2.2.3 Cavitational Reactor

A cavitational reactor is a reactor type that focuses on the enhancement of mixing intensity by generating cavitation in the medium for reducing the mass transfer resistance. Because alcohol and oil are immiscible, the generation of cavitation to achieve intense turbulence and streaming, which provide smaller bubbles and a higher interfacial area, is very crucial. Moreover, the acceleration of heat from the attraction and collapse of bubbles generates sufficient heat, leading to intensified operation. Basically, the cavitational effect can be generated through two routes including acoustic cavitation and hydrodynamic cavitation (HC).

***16.2.2.3.1 Acoustic Cavitation Reactor/Ultrasonic Cavitation Reactor*** For two decades, low-frequency ultrasound (20–40 kHz) is a remarkable technique for improving the mixing process of a two-phase liquid [27]. An ultrasonic cavitation reactor (US) is a reactor that is modified by applying low-frequency ultrasound into the reaction medium zone. Ultrasound generates cavitation bubbles in the liquid medium followed by their collapse [28], which disrupts the phase boundary of the immiscible mixture, resulting in the emulsification of the medium and reduction of heat mass transfer resistance.

There are many advantages of the ultrasonic reactor including (i) time saving: the formation of bubbles increases the interfacial surface area of the liquid mixture, leading to reduced reaction time; (ii) lower energy consumption: the collapse of bubbles releases heat to the mixture resulting in less addition of heat from the external source; and (iii) covering homogeneous and heterogeneous systems: the acceleration of low-frequency ultrasound is effective in both catalytic systems. The comparison of a US reactor operating on dual-frequency with a mechanical stirrer was investigated in the previous study [29]. The results indicated that the US reactor achieved a high biodiesel yield at ca. 5 min, while the mechanical stirrer required 60 min. Moreover, the US reactor could enhance the heterogeneous catalytic transesterification by hindering the catalyst deactivation [30]. Although the operation of ultrasonic achieves high effectiveness in small-scale production, the scale-up of the biodiesel production process via the ultrasonic cavitation reactor could be a great improvement in the future.

***16.2.2.3.2 Hydrodynamic Cavitation Reactor*** HC is the cavity phenomenon generated from the passage of a rapidly moving liquid through a constriction like an orifice plate, venturi, or throttling valve. When the reaction medium flows through a constriction, the velocity increases, and the pressure in the liquid decreases below the vapor pressure generating a vapor cloud. These cavities may be driven to collapse within the reaction medium, which increases temperature and pressure.

The developments of biodiesel production via interesterification using HC have been explored under several operating conditions including the effect of upstream pressure, molar ratio, catalyst concentration, and plate geometry. A previous study confirmed that the optimal yield obtained with oil-to-methyl acetate ratio of 1 : 12, catalyst concentration of 1 wt%, and inlet pressure of 3 bar using a slit venturi as a constriction was 89.2% [31].

***16.2.2.3.3 Shockwave Power Reactor*** A shockwave power reactor (SPR), namely, a rotor–stator reactor, is a novel reactor the HC reactor type. The operating principle of the SPR is the creation of cavitation by the high-speed rotation of the rotor, which rotates inside the stator. When the reaction mixture enters the cavitation zone, the fast formation and collapse of bubbles occurred, resulting in shockwave generation, intense mixing, and heat release, respectively.

The major design parameter is the rotor–stator diameter ratio, which extremely impacts the interacting surface of reactants, leading to an enhanced biodiesel yield. The other design parameters of the shockwave reactor are such as ratio of the cavity diameter to the rotor diameter, ratio of cavity depth to the gap between the rotor and the stator, and other operating parameters such as rotational speed and residence time [32]. The optimal parameters of this SPR for biodiesel production were reported as the rotor–stator diameter ratio of 0.73, the ratio of the cavity diameter to the rotor diameter of 0.06, and the ratio of cavity depth to the gap between the rotor and the stator of 0.5 at the rotational speed of 25 510 rpm and a residence time of 30.10 s, which can produce an fatty acid methyl ester (FAME) yield of 96.6%.

#### 16.2.2.4 Microwave Reactor

Microwave radiation is part of the electromagnetic spectrum with the frequency in the range of 0.3–300 GHz. Commercial household microwave ovens operate at 2450 MHz to avoid cellular phone and telecommunication interferences. Industrial applications rely on the frequencies of 915 and 2450 MHz, and most of the microwave-assisted chemistry researches are performed at 2450 MHz (wavelength of 12.24 cm) because it is worldwide approved and is commercially utilized in microwave chemistry apparatus. Microwave reactors use microwave radiation to impart energy into reaction mixtures to accelerate the reaction rate. The conversion can be achieved rapidly compared with conventional reactors.

Many researchers studied microwave reactors for biodiesel production. As the main components of biodiesel production are vegetable oil and methanol (transesterification), both components are polar and ionic. Microwave irradiation assisted in mixing these components to achieve high conversion. Moreover, microwave irradiation in the production of biodiesel can be divided according to various types of feedstocks. Moreover, for the algae or microalgae compound, the solubility of the material in the solvent strongly influences the effectiveness of microwave radiation on component extraction. In terms of transesterification, the microwave could enhance transesterification through the thermal effect and methanol evaporation as the microwave strongly interacts with it [33, 34]. During microwave irradiation, triglyceride and methanol experience a substantial activation energy reduction from increased dipolar polarization [35]. This is because microwave radiation interaction with the reaction mixture at the molecular level is related to ionic conduction and dipolar rotation [36]. The extent of the activation energy reduction depends on the medium and the mechanism of the reaction. In general, a hydroxyl group (-OH) attached to a big molecule as though it was anchored to a stationary raft with more localized rotation causes localized superheating and pushes the reaction forward faster [37].

Microwave irradiation is a green and preferred technique for biodiesel synthesis because of several advantages: substantially reduced reaction duration and solvent, lower energy need, improved selectivity, and conversions, generating smaller amounts of by-products. Nonetheless, microwave reactors face limitations as reported by many experts due to the difficulty in penetration on a large scale, making scaling up of the microwave technology from a laboratory evaluation to an industrial production a challenge. Other limitations of the microwave conventional processes include difficulty in temperature measurement and control, nonuniform temperature distribution in large reactors or as the reaction mixture volume increases, and the required radiation intensity. Therefore, continuous microwave-assisted biodiesel production has been proposed as illustrated in **Figure 16.3**. This system can use both serpentine plug-type and helical-type reactor, which can control the reaction temperature and irradiation time for homogeneous biodiesel production. However, a clear understanding of the microwave effects on the extraction of oil feedstocks as well as transesterification would help improve the reactor design. Likewise, a clear understanding of the effects of microwave radiation on various process parameters (catalyst, reaction mixture, or solvents) is also important to develop a safe reactor.

#### 16.2.2.5 Multifunctional Reactor (Reactive Distillation, Membrane, Centrifugal Reactors)

***16.2.2.5.1 Reactive Distillation*** Typically, distillation is a separation unit to control the qualities of biodiesel production. A distillation column was employed for the separation of the methanol–water–glycerol ternary mixture. Reactive distillation (RD), a combined reactor and distillation column, was developed in order to offer a smaller unit and overcome chemical reaction equilibrium. Boon-anuwat et al. [40] investigated four process designs including RD and conventional processes catalyzed by homogeneous and heterogeneous catalysts. The results showed energy requirements are ordered as Conv. Hetero > Conv. Homo > RD Homo > RD Hetero. Most importantly, employing RD can decrease alcohol-to-oil feed molar ratio from ca. 10–15 to only 4. Recently, to enable using waste cooking oil in

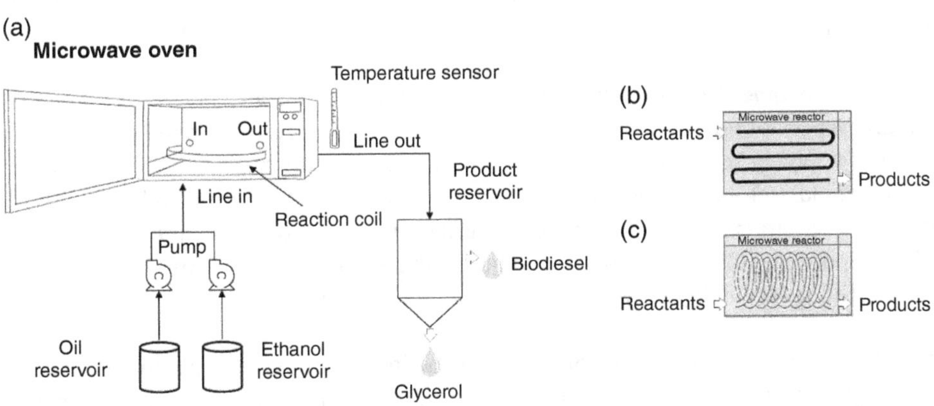


**Figure 16.3** Continuous microwave-assisted biodiesel production system (a), serpentine plug-type reactor (b), and helical coil-type reactor (c). *Source:* Adapted from Ref. [38, 39].

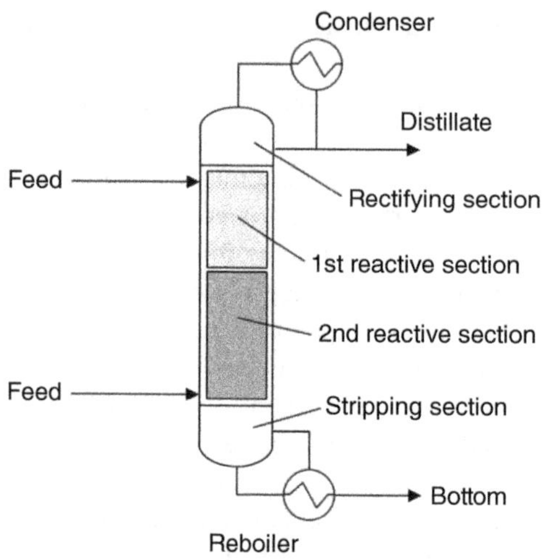


**Figure 16.4** Schematic diagram of hybridized reactive distillation.

a single RD, a hybrid design of ester- and transesterification processes in a novel RD column has been proposed as expressed in **Figure 16.4** [41].

The fully thermally coupled distillation column, commonly known as the Petlyuk column, used interconnected vapor–liquid streams to reach heat transfer through direct contact. Petlyuk column exhibits energy saving up to 50% compared to a conventional distillation train, as it does not present the remixing effect. Gomez-Castro et al. [42] purposed the feasibility of using the Petlyuk column to synthesize biodiesel under high-pressure condition. This configuration was carried out for esterification and product purification in the same shell. This alternative thermally coupled systems show a substantial improvement in energy consumptions compared to a conventional process and RD column.

Rather complex RD having a dividing wall column called "reactive dividing wall column" combines two columns into one by adding a dividing wall into the column where the side stream product can be obtained between the distillate and bottom products. This would reduce capital and operating costs, as well as the required installation space. In the study of the reactive dividing wall column by Kiss et al. [43], only ~15% excess methanol is required for complete conversion of the fatty acids feedstock. Biodiesel and water could be obtained as pure bottom product and side stream, respectively, while the methanol excess as pure distillate product could be easily recycled. Moreover, this configuration offers several key benefits over the conventional process, e.g. no requirement on neutralization of the alkali catalyst, no water washing, and no wastewater treatment. This provides up to 25% energy savings. However, in order to achieve high biodiesel yield and purity, using RD requires highly precise process control due to the process nonlinearity [8].

***16.2.2.5.2 Membrane Reactor*** A membrane reactor can be an alternative of combined reactor and separator for biodiesel production **(Figure 16.5)**. Several researchers show the potential of applying the membrane reactor for biodiesel production [44]. Using a carbon membrane, biodiesel, methanol, and glycerol can permeate, while canola oil cannot [45]. In the case of commercial polysulfone, triglycerides and glycerol can be rejected with the membrane pore size of 0.2 μm, while biodiesel and methanol can permeate. However, methanol still has much higher permeability than biodiesel [46].

***16.2.2.5.3 Centrifugal Reactor*** A continuous centrifugal contactor separator (CCCS) is another highly intensified method for biodiesel production **(Figure 16.6)**. The rotor in this reactor is rapidly rotating within a stationary vessel. Due to the high centrifugal force, biodiesel can be separated *in situ* from the glycerol layer. Both homogeneous catalysts, e.g.

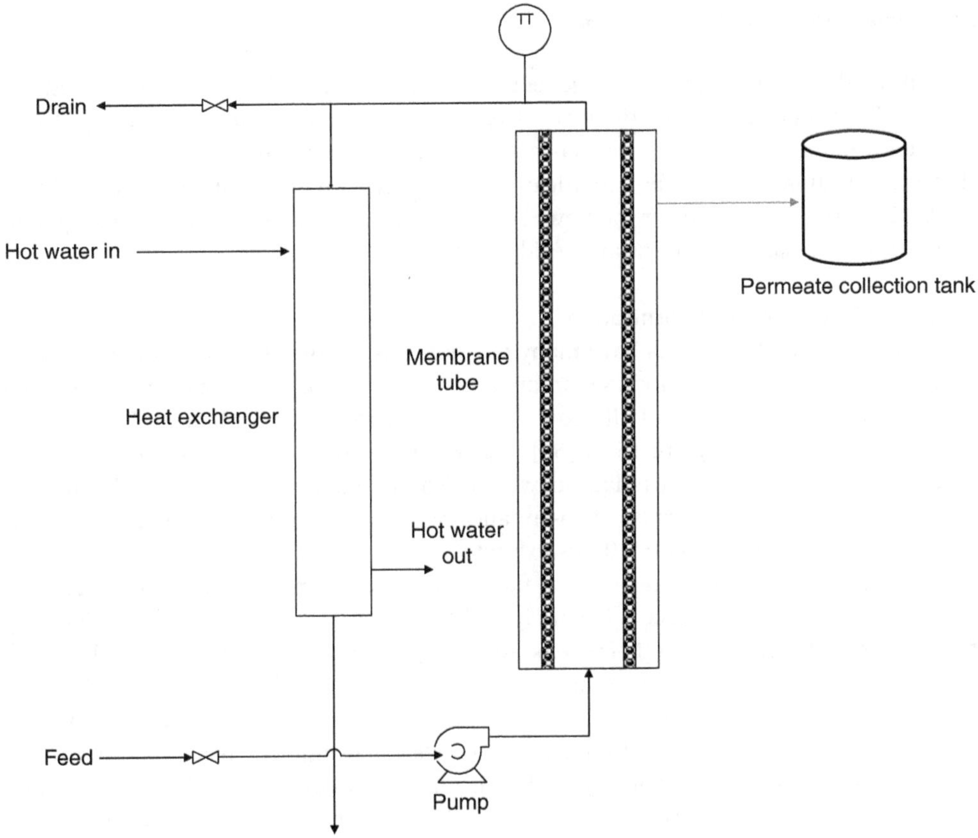


**Figure 16.5** Schematic diagram of separated membrane reactor.

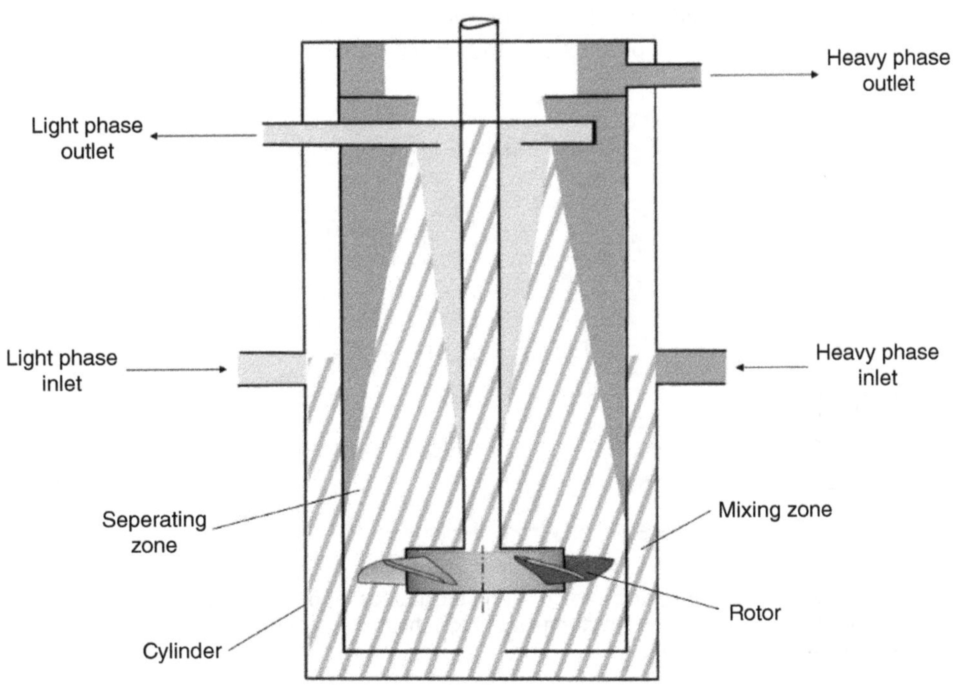


**Figure 16.6** Schematic cross-sectional view of the centrifuge contactor reactor. *Source:* Adapted from Ref. [8].

sodium methoxide [47], and heterogeneous catalysts, e.g. CaO [48], could be effectively applied. Promising high productivity of 638 kg biodiesel/($m^3$ reactor-h) catalyzed by CaO is recently reported [48]. The residence time (<10 s) could be an obstacle to the transesterification process. The modification of a delay loop to operate a half capacity of the CCCS was applied to lower the flow rate and increase the holdup volume, resulting in enhancing the small droplet size and residence time [6].

#### 16.2.2.6 Other Process Intensification

Nowadays, biodiesel processes have many technologies to pursue. Process intensification significantly helps improve biodiesel production. Some researchers developed a sequential combination of microwave and ultrasonic technologies to reduce the biodiesel production time as can be seen in **Figure 16.7a** [50]. **Figure 16.7b** exhibits the integration of a high shear mixer with microwave irradiation, which also successfully produced a high biodiesel yield of 99.8% with 5 min because of co-enhancement of mass and heat transfer in the flow reactor [49]. These combined methods exhibited reduced reaction time, less catalyst, and less energy consumption. There are a few publications on plasma as an alternative technology. The plasma-assisted reactor is anticipated to reduce the reaction time considerably from the role of highly energetic free electrons [51]. The separation process also becomes easy because of no glycerol formation during the process. To achieve higher energy electrons, the plasma should be energized with a voltage higher than 1 kV and up to 8 kV [52]. Moreover, the dielectric barrier discharge (DBD) plasma technology has also been used to enhance biodiesel oxidation stability via hydrogenation by conversion of polyunsaturated FAMEs (C18 : 2 and C18 : 3) to C18 : 1 and C18 : 0 at low temperature and one atmosphere in the absence of a catalyst [53].

Conventional biodiesel production processes utilize high-power heating and magnetic stirring. Electromagnetic induction heating developed for induction heating technology can also be applied in biodiesel production. Induction heating generates intense heat in

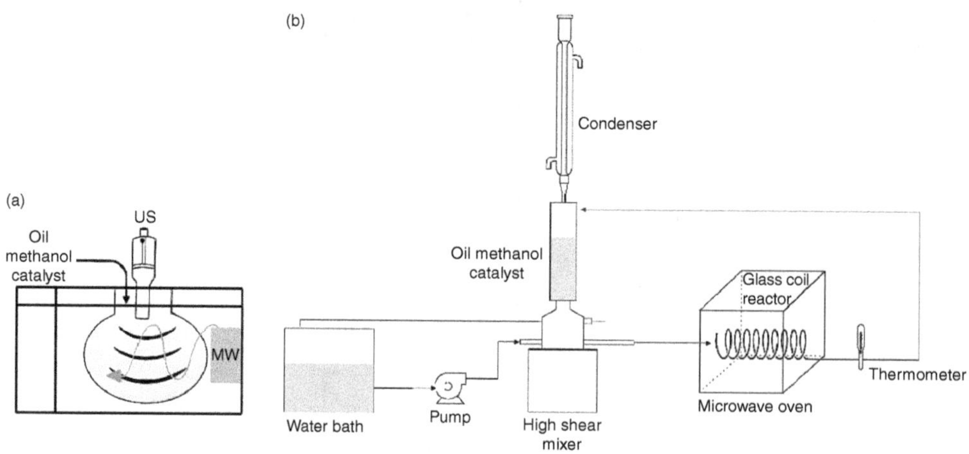


**Figure 16.7** Integration of different intensification techniques for biodiesel production (a) microwave-ultrasonic combined reactor and (b) high shear mixing-microwave combined reactor. *Source:* Adapted from Ref. [38, 49].

electrically conductive objects (normally metal) via electromagnetic induction. The electromagnetic induction principle is similar to that occurring in a transformer. When alternating current flows through a conductor (an induction coil in this case), it produces a time-varying magnetic field around the coil, which induces an eddy current in the metal inside the coil. The inherent electrical resistance in the metal to the flow of eddy current generates heat through joule heating. An induction heater (IH) comprises three primary elements of electromagnetic induction, skin effects, and heat transfer [54]. This process generates a substantially rapid heating rate. This advanced biodiesel synthesis technology can significantly enhance the production rate with the potential to reduce production costs and time.

In addition, a 1200 W commercial high-power household fruit blender has been successfully explored as a simple and effective technology for biodiesel synthesis in a batch system employing a homogeneous catalyst [55]. The fruit blender created process intensification from the cavitational effect, resulting from the rapidly spinning impeller, causing a pressure reduction in the liquid generating numerous microbubbles, enhancing the mixing efficiency, and increasing the reaction kinetics. Likewise, a 700 W commercial household hand blender has been successfully explored to produce biodiesel in a continuous system using a homogeneous catalyst [56].

The overview of each reactor type for biodiesel production, which includes the catalysis system, current status, operating condition, key parameter, and merit and demerit are indicated in Table 16.1.

## 16.3 Future Prospects and Challenges

The mainstream process is a core of biodiesel production connecting with upstream selection and purification of feedstocks and downstream separation and purification of biodiesel, which is facilitated by a novel reactor or process intensification to obtain a high efficiency, economical, and sustainable biodiesel process. The existing commercial biodiesel productions are carried out in batch, CSTR PBR, rotating tube, and shockwave reactors. The pilot scales of biodiesel production include membrane reactor, RD, and centrifugal reactor. Apart from that, the SDR, microtubular, microchannel, ultrasonic reactor, microwave, plasma, electromagnetic induction, and combined reactors are being pursued on a laboratory scale. Basically, the selection of the novel reactor strongly depends on the feedstocks and catalyst types. Moreover, the choice of the operation mode in a batch or continuous mode is related to biodiesel production capacity. For instance, the bubble column reactor and RD are compatible with various feedstocks, but these reactors might be limited due to the complicated design and operation control for the industrial scale. The efficiency of microwave and cavitation reactors is defined by the intensity of their radiations, which can be a demerit for the higher biodiesel production capacity. The RTR and OFR can operate in a continuous mode, but these reactors are more compatible with the homogeneous-catalyzed transesterification process. The centrifugal contractor as one of the continuous modes of the intensified reactor not only can be used for both homogeneous and heterogeneous systems but also can simultaneously produce and separate biodiesel from the reaction mixture. In addition, the design and development of the heterogeneous catalyst for the continuous biodiesel synthesis to attain high activity, high stability, and low cost is crucial

**Table 16.1** Overview of intensified reactors for biodiesel production.

| Reactor | Catalysis system/ operating mode | Current status | Operating condition | | | | Key parameters | Advantages | Limitations | References |
|---|---|---|---|---|---|---|---|---|---|---|
| | | | Temp (°C) | Time | Specific condition | Yield/ conversion (%) | | | | |
| Spinning disk reactor | Homogeneous/ continuous | Laboratory scale | 40–60 | 2–3 s | Rotational speed 500–2000 rpm<br>small gap 0.1–0.4 mm | 90–97 | Gap thickness, rotational speed, and flow rate | • High yield efficiency with respect to yield, energy, and reaction time<br>• Reduced reactor size<br>• Simple structure and easy maintenance | • Might not be suitable for heterogeneous system | [14] |
| Rotating tube reactor | Homogeneous/ continuous | Laboratory, pilot, and industrial scale | 25–75 | <1 min | Film thickness 1.1 cm Rotational speed 250–1000 rpm | 80–97 | Film thickness and rotational speed | • Operation at mild conditions<br>• Easy separation<br>• High yield efficiency with respect to yield, energy, and reaction time<br>• Potentially used on industrial scale | Might not be suitable for heterogeneous system | [8, 16] |
| Packed bed reactor | Homogenous, heterogeneous, and enzyme/batch and continuous | Laboratory and pilot scale | 30–350 | 30 min–48 h | Flow rate 0.06–8.31 $min^{-1}$ | 72–98 (conv) | Residence time and suitable catalyst desired | • Low cost and environmentally friendly<br>• High biodiesel yield<br>• Reduce mechanical damage | • Require knowledge on catalyst pelletization<br>• Obtain high pressure drop<br>• Difficult to measure and control temperature | [6, 8] |
| Oscillatory flow reactor | Homogeneous acid and base/ batch | Laboratory and pilot scale | 25–60 | 2–30 min | Oscillation frequency 0.67–4.1 Hz | 72–>95 | Oscillatory frequency and residence time | • Low energy consumption<br>• Potentially developed in a continuous mode<br>• Large variation in residence time | • Might not be suitable for heterogeneous system | [21] |
| Static mixer reactor | Homogenous/ continuous | Laboratory scale | 40–60 | 6–10 min | Flow rate 0.1–72 l $h^{-1}$ | 95–99 | High vigorous mixing depended on structure of motionless element in a tube | • Reduce energy consumption<br>• Low operating and maintenance cost<br>• Can be used for separation unit<br>• High yield | • Might not be suitable for heterogeneous system | [9, 57] |

| | | | | | | | | | |
|---|---|---|---|---|---|---|---|---|---|
| Microchannel (T and Y shape and zigzag) | Homogeneous acid and base/continuous | Laboratory scale | 55–70 | 3–180 s | Diameter 0.24–2 mm<br>Flow rate 0.13–3.8 ml $min^{-1}$ | 91–100 | Ratio of surface area to volume of microchannel | • Accurate temperature control<br>• Low energy consumption<br>• Easy scale up by increasing number of microchannel | • Reaction mixture clogging and corrosion<br>• Low biodiesel production capacity | [57] |
| Ultrasonic reactor | Homogenous, heterogeneous, and enzyme/batch and continuous | Laboratory, pilot, and industrial scale | 25–60 | 4–240 min | Frequency 20–50 kHz | 64–100 | Frequency amplitude and duty cycle | • Enhance heat and mass transfer<br>• Reduce reaction time<br>• Reduce catalyst and alcohol for transesterification | • Difficult to control reaction temperature<br>• Nonuniform ultrasonic intensity with large reactor volume | [9] |
| Hydrodynamic cavitation reactor | Homogenous and heterogeneous/batch and continuous | Laboratory, pilot, and industrial scale | 30–100 | 3–10 min | ΔP 17–38 bar | 92–99.9 | Plate geometry and cavitation number | • Enhance heat and mass transfer<br>• Highest yield efficiency with short reaction time and high capacity<br>• Simple design and easy to scale up | • Require low flow rate to obtain cavitation phenomena<br>• Might deteriorate heterogeneous catalyst | [57, 58] |
| Shockwave power reactor | Homogenous/continuous | Laboratory and industrial scale | N/A | 31 s | Rotational speed 25510 rpm<br>Power generator 4 kW | 96–99 | Ratio of<br>• Rotor and stator<br>• Cavity to rotor | • Similar to hydrodynamic cavitation | • Similar to hydrodynamic cavitation | [8, 32] |
| Microwave reactor | Homogenous and heterogeneous/batch and continuous | Laboratory scale | 50–140 | 10 s–60 min | Power 25–1200 W | 74–99.8 | Microwave power | • Uniform temperature<br>• Less energy consumption<br>• Shorter reaction time | • Difficult to scale up<br>• Require internal mixing | [9] |
| Reactive distillation | Homogenous and heterogeneous/batch and continuous | Laboratory and pilot scale | 50–150 | 5 min–4 h | Alcohol-to-oil ratio 4-8 : 1 | 72–99 | Reboiler duty and reflux ratio | • Simultaneous reaction and separation in one unit<br>• Minimization alcohol-to-oil ratio | • Require precise temperature control of feed and reboiler | [9] |
| Membrane reactors | Homogenous, heterogeneous, and enzyme/batch and continuous | Laboratory and pilot scale | 55–70 | 5 min–4 h | Pressure 0.01–0.5 MPa,<br>Pore size 0.05–1.4 μm | 64–98 | Types, thickness, and pore size of membrane | • Similar to reactive distillation | • Expensive material and easy fouling membrane<br>• Accuracy control of operating parameter to avoid fouling | [8, 9] |

*(Continued)*

**Table 16.1** (Continued)

| Reactor | Catalysis system/ operating mode | Current status | Operating condition | | | | Key parameters | Advantages | Limitations | References |
|---|---|---|---|---|---|---|---|---|---|---|
| | | | Temp (°C) | Time | Specific condition | Yield/ conversion (%) | | | | |
| Centrifugal contact separator reactor | Homogenous, heterogeneous, and enzyme/batch and continuous | Laboratory and pilot scale | 60–80 | 2–5 min | Flow rate 3–11 ml $min^{-1}$<br>Rotation speed 2100–3600 rpm | 95–98 | Rotational speed and feed flow rate | • Low-cost equipment<br>• Low operating cost<br>• Potentially used on commercial scale | • Extremely low retention time | [6] |
| Combination of microwave with high mixing reactor | Homogenous/ batch and continuous | Laboratory scale | 60–80 | 5 min | Power 300–600 W | 96–99 | Microwave power | • Provide excellent heat and mass transfer<br>• Reduce reaction time | • Difficult to scale up as well as microwave reactor | [49] |
| Hybrid catalytic-plasma reactor | Heterogeneous/ batch | Laboratory scale | 65 | WHSV 1.85 $h^{-1}$ | Applied voltage 7 kV DC | 92 | High voltage to generate energetic electron | • High conversion in a short reaction time | • Require further study of plasma reactor configuration | [59] |
| Electromagnetic induction reactor | Homogenous/ batch | Laboratory scale | 65 | 0.43 min | DC voltage 30 V<br>Power input 200 W<br>Current output 1.2 A | 66 | Input power | • Faster reaction time | • Require further study of induction material configuration | [60] |
| High-power household fruit blender | Homogenous/ batch and continuous | Laboratory scale | 50–65 | 2–5 min | Stirring speed 10 000–12 000 rpm | 80–97 | Input power | • Simple and low-cost system | • Limited to small scale production | [55, 56] |

knowledge that should be explored together with the scaling up of the intensification reactor for commercialization. However, these intensification technologies require a huge investment cost to develop to the commercialization scale because these technologies are in the developing stage needing the new connecting unit operations and labor skills. The government policy and social acceptance are two issues for driving biodiesel production to an economic and sustainable process. The government should not only support by subsidizing taxes but should also provide the availability of low-cost and food security feedstocks and infrastructure for setting up the new biodiesel plant. Social acceptance can be viewed as the rising demand for the biodiesel market by making people aware of the energy security and environmental friendliness, which is a characteristic of biodiesel. Furthermore, the integration of a glycerol conversion unit to a high-value chemical can make the biodiesel plant a circular economic and interesting investment.

## Acknowledgment

The authors would like to acknowledge the supports from the "Research Chair Grant" National Science and Technology Development Agency (NSTDA).

## References

1 Syafiuddin, A., Chong, J.H., Yuniarto, A., and Hadibarata, T. (2020). The current scenario and challenges of biodiesel production in Asian countries: a review. *Bioresource Technology Reports* 12: 100608.

2 Singh, D., Sharma, D., Soni, S.L. et al. (2020). A review on feedstocks, production processes, and yield for different generations of biodiesel. *Fuel* 262: 116553.

3 Kant Bhatia, S., Kant Bhatia, R., Jeon, J.-M. et al. (2021). An overview on advancements in biobased transesterification methods for biodiesel production: oil resources, extraction, biocatalysts, and process intensification technologies. *Fuel* 285: 119117.

4 Lotero, E., Liu, Y., Lopez, D.E. et al. (2005). Synthesis of biodiesel via acid catalysis. *Industrial & Engineering Chemistry Research* 44 (14): 5353–5363.

5 Aransiola, E.F., Ojumu, T.V., Oyekola, O.O. et al. (2014). A review of current technology for biodiesel production: state of the art. *Biomass and Bioenergy* 61: 276–297.

6 Raheem, I., Mohiddin, M.N.B., Tan, Y.H. et al. (2020). A review on influence of reactor technologies and kinetic studies for biodiesel application. *Journal of Industrial and Engineering Chemistry* 91: 54–68.

7 Alagumalai, A., Mahian, O., Hollmann, F., and Zhang, W. (2021). Environmentally benign solid catalysts for sustainable biodiesel production: a critical review. *Science of The Total Environment* 768: 144856.

8 Tabatabaei, M., Aghbashlo, M., Dehhaghi, M. et al. (2019). Reactor technologies for biodiesel production and processing: a review. *Progress in Energy and Combustion Science* 74: 239–303.

9 Athar, M. and Zaidi, S. (2020). A review of the feedstocks, catalysts, and intensification techniques for sustainable biodiesel production. *Journal of Environmental Chemical Engineering* 8 (6): 104523.

10 Lee, J.-S. and Saka, S. (2010). Biodiesel production by heterogeneous catalysts and supercritical technologies. *Bioresource Technology* 101 (19): 7191–7200.

11 Chuah, L.F., Klemeš, J.J., Yusup, S. et al. (2017). A review of cleaner intensification technologies in biodiesel production. *Journal of Cleaner Production* 146: 181–193.

12 Gupta, J., Agarwal, M., and Dalai, A.K. (2020). An overview on the recent advancements of sustainable heterogeneous catalysts and prominent continuous reactor for biodiesel production. *Journal of Industrial and Engineering Chemistry* 88: 58–77.

13 Qiu, Z., Zhao, L., and Weatherley, L. (2010). Process intensification technologies in continuous biodiesel production. *Chemical Engineering and Processing: Process Intensification* 49 (4): 323–330.

14 Appamana, W., Sukjarern, P., Ngaosuwan, K., and Assabumrungrat, S. (2019). Intensification of continuous biodiesel production using a spinning disc reactor. *Journal of Chemical Engineering of Japan* 52 (6): 545–553.

15 Qiu, Z., Petera, J., and Weatherley, L.R. (2012). Biodiesel synthesis in an intensified spinning disk reactor. *Chemical Engineering Journal* 210: 597–609.

16 Lodha, H., Jachuck, R., and Singaram, S.S. (2012). Intensified biodiesel production using a rotating tube reactor. *Energy & Fuels* 26: 7037–7040.

17 Santana, H.S., Tortola, D.S., Silva, J.L., and Taranto, O.P. (2017). Biodiesel synthesis in micromixer with static elements. *Energy Conversion and Management* 141: 28–39.

18 Natarajan, Y., Nabera, A., Salike, S. et al. (2019). An overview on the process intensification of microchannel reactors for biodiesel production. *Chemical Engineering and Processing: Process Intensification* 136: 163–176.

19 Feng, Y., Zhang, A., Li, J., and He, B. (2011). A continuous process for biodiesel production in a fixed bed reactor packed with cation-exchange resin as heterogeneous catalyst. *Bioresource Technology* 102 (3): 3607–3609.

20 Zik, N.A.F.A., Sulaiman, S., and Jamal, P. (2020). Biodiesel production from waste cooking oil using calcium oxide/nanocrystal cellulose/polyvinyl alcohol catalyst in a packed bed reactor. *Renewable Energy* 155: 267–277.

21 García-Martín, J.F., Barrios, C.C., Alés-Álvarez, F.-J. et al. (2018). Biodiesel production from waste cooking oil in an oscillatory flow reactor. Performance as a fuel on a TDI diesel engine. *Renewable Energy* 125: 546–556.

22 Tiwari, A., Rajesh, V.M., and Yadav, S. (2018). Biodiesel production in micro-reactors: a review. *Energy for Sustainable Development* 43: 143–161.

23 Jachuck, R., Pherwani, G., and Gorton, S.M. (2009). Green engineering: continuous production of biodiesel using an alkaline catalyst in an intensified narrow channel reactor. *Journal of Environmental Monitoring* 11 (3): 642–647.

24 Guan, G., Kusakabe, K., Moriyama, K., and Sakurai, N. (2009). Transesterification of sunflower oil with methanol in a microtube reactor. *Industrial & Engineering Chemistry Research* 48: 1357–1363.

25 Guan, G., Kusakabe, K., Moriyama, K., and Sakurai, N. (2008). Continuous production of biodiesel using a microtube reactor. *Chemical Engineering Transactions* 14: 237–242.

26 Dai, J.-Y., Li, D.-Y., Zhao, Y.-C., and Xiu, Z.-L. (2014). Statistical optimization for biodiesel production from soybean oil in a microchannel reactor. *Industrial & Engineering Chemistry Research* 53 (22): 9325–9330.

**27** Carmen Stavarache, M.V., Nishimura, R., and Maeda, Y. (2005). Fatty acids methyl esters from vegetable oil by means of ultrasonic energy. *Ultrasonics Sonochemistry* 12: 367–372.

**28** Choedkiatsakul, I., Ngaosuwan, K., and Assabumrungrat, S. (2013). Application of heterogeneous catalysts for transesterification of refined palm oil in ultrasound-assisted reactor. *Fuel Processing Technology* 111: 22–28.

**29** Choedkiatsakul, I., Ngaosuwan, K., Cravotto, G., and Assabumrungrat, S. (2014). Biodiesel production from palm oil using combined mechanical stirred and ultrasonic reactor. *Ultrasonics Sonochemistry* 21 (4): 1585–1591.

**30** Poosumas, J., Ngaosuwan, K., Quitain, A.T., and Assabumrungrat, S. (2016). Role of ultrasonic irradiation on transesterification of palm oil using calcium oxide as a solid base catalyst. *Energy Conversion and Management* 120: 62–70.

**31** Maddikeri, G.L., Gogate, P.R., and Pandit, A.B. (2014). Intensified synthesis of biodiesel using hydrodynamic cavitation reactors based on the interesterification of waste cooking oil. *Fuel* 137: 285–292.

**32** Abbaszadeh-Mayvan, A., Ghobadian, B., Najafi, G., and Yusaf, T. (2018). Intensification of continuous biodiesel production from waste cooking oils using shockwave power reactor: process evaluation and optimization through response surface methodology (RSM). *Energies* 11: 2845.

**33** Loupy, A., Petit, A., Ramdani, M. et al. (1993). The synthesis of esters under microwave irradiation using dry-media conditions. *Canadian Journal of Chemistry* 71 (1): 90–95.

**34** Yuan, H., Yang, B., and Zhu, G. (2009). Synthesis of biodiesel using microwave absorption catalysts. *Energy & Fuels* 23 (1): 548–552.

**35** Perreux, L. and Loupy, A. (2001). A tentative rationalization of microwave effects in organic synthesis according to the reaction medium, and mechanistic considerations. *Tetrahedron* 57 (45): 9199–9223.

**36** Choedkiatsakul, I., Ngaosuwan, K., Assabumrungrat, S. et al. (2015). Biodiesel production in a novel continuous flow microwave reactor. *Renewable Energy* 83: 25–29.

**37** Tierney, J.P. and Lidström, P. (2005). *Microwave Assisted Organic Synthesis*. Wiley Online Library.

**38** Khedri, B., Mostafaei, M., and Safieddin Ardebili, S.M. (2019). A review on microwave-assisted biodiesel production. *Energy Sources, Part A: Recovery, Utilization, and Environmental Effects* 41 (19): 2377–2395.

**39** Gude, V.G., Patil, P., Martinez-Guerra, E. et al. (2013). Microwave energy potential for biodiesel production. *Sustainable Chemical Processes* 1 (1): 1–31.

**40** Boon-anuwat, N.-n., Kiatkittipong, W., Aiouache, F., and Assabumrungrat, S. (2015). Process design of continuous biodiesel production by reactive distillation: comparison between homogeneous and heterogeneous catalysts. *Chemical Engineering and Processing: Process Intensification* 92: 33–44.

**41** Petchsoongsakul, N., Ngaosuwan, K., Kiatkittipong, W. et al. (2017). Process design of biodiesel production: hybridization of ester-and transesterification in a single reactive distillation. *Energy Conversion and Management* 153: 493–503.

**42** Gomez-Castro, F.I., Rico-Ramirez, V., Segovia-Hernandez, J.G., and Hernandez, S. (2010). Feasibility study of a thermally coupled reactive distillation process for biodiesel production. *Chemical Engineering and Processing: Process Intensification* 49 (3): 262–269.

43 Kiss, A.A. and Ignat, R.M. (2012). Enhanced methanol recovery and glycerol separation in biodiesel production – DWC makes it happen. *Applied Energy* 99: 146–153.

44 Shuit, S.H., Ong, Y.T., Lee, K.T. et al. (2012). Membrane technology as a promising alternative in biodiesel production: a review. *Biotechnology Advances* 30 (6): 1364–1380.

45 Dubé, M., Tremblay, A., and Liu, J. (2007). Biodiesel production using a membrane reactor. *Bioresource Technology* 98 (3): 639–647.

46 Hapońska, M., Nurra, C., Abelló, S. et al. (2019). Membrane reactors for biodiesel production with strontium oxide as a heterogeneous catalyst. *Fuel Processing Technology* 185: 1–7.

47 Kraai, G., Schuur, B., Van Zwol, F. et al. (2009). Novel highly integrated biodiesel production technology in a centrifugal contactor separator device. *Chemical Engineering Journal* 154 (1–3): 384–389.

48 Fayyazi, E., Ghobadian, B., van de Bovenkamp, H.H. et al. (2018). Optimization of biodiesel production over chicken eggshell-derived CaO catalyst in a continuous centrifugal contactor separator. *Industrial & Engineering Chemistry Research* 57 (38): 12742–12755.

49 Choedkiatsakul, I., Ngaosuwan, K., Assabumrungrat, S. et al. (2015). Integrated flow reactor that combines high-shear mixing and microwave irradiation for biodiesel production. *Biomass and Bioenergy* 77: 186–191.

50 Gole, V.L. and Gogate, P.R. (2013). Intensification of synthesis of biodiesel from non-edible oil using sequential combination of microwave and ultrasound. *Fuel Processing Technology* 106: 62–69.

51 Lieberman, M.A. and Lichtenberg, A.J. (2005). *Principles of Plasma Discharges and Materials Processing*. Wiley.

52 Istadi, I., Anggoro, D.D., Marwoto, P. et al. (2009). Biodiesel production from vegetable oil over plasma reactor: optimization of biodiesel yield using response surface methodology. *Bulletin of Chemical Reaction Engineering & Catalysis* 4 (1): 23–31.

53 Kongprawes, G., Wongsawaeng, D., Hosemann, P. et al. (2021). Improvement of oxidation stability of fatty acid methyl esters derived from soybean oil via partial hydrogenation using dielectric barrier discharge plasma. *International Journal of Energy Research* 45 (3): 4519–4533.

54 El Sherbiny, S.A., Refaat, A.A., and El Sheltawy, S.T. (2010). Production of biodiesel using the microwave technique. *Journal of Advanced Research* 1 (4): 309–314.

55 Wongsawaeng, D., Ngaosuwan, K., Kiatkittipong, W. et al. (2019). Simple and effective technology for sustainable biodiesel production using high-power household fruit blender. *Journal of Cleaner Production* 237: 117842.

56 Kamjam, M., Wongjaikham, W., Wongsawaeng, D. et al. (2021). Continuous biodiesel production based on hand blender technology for sustainable household utilization. *Journal of Cleaner Production* 297: 126737.

57 Mazubert, A., Poux, M., and Aubin, J. (2013). Intensified processes for FAME production from waste cooking oil: a technological review. *Chemical Engineering Journal* 233: 201–223.

58 Laosuttiwong, T., Ngaosuwan, K., Kiatkittipong, W. et al. (2018). Performance comparison of different cavitation reactors for biodiesel production via transesterification of palm oil. *Journal of Cleaner Production* 205: 1094–1101.

59 Buchori, L., Istadi, I., Purwanto, P. et al. (2016). Preliminary testing of hybrid catalytic-plasma reactor for biodiesel production using modified-carbon catalyst. *Bulletin of Chemical Reaction Engineering & Catalysis* 11 (1): 59–65.

60 Kurniati, S., Soeparman, S., Setyo Yuwono, S. et al. (2019). A novel process for production of calophyllum inophyllum biodiesel with electromagnetic induction. *Energies* 12 (3): 383.

# 17

# Downstream Strategies for Separation, Washing, Purification, and Alcohol Recovery in Biodiesel Production

*Ramón Piloto-Rodríguez[1] and Yosvany Díaz-Domínguez[2]*

[1] *Center for the Studies of Renewable Energies, Universidad Tecnológica de La Habana, Havana, Cuba*
[2] *Faculty of Chemical Engineering, Universidad Tecnológica de La Habana, Havana, Cuba*

## 17.1 Introduction

Biodiesel is a biofuel comprised of monoalkyl esters of long-chain fatty acids derived from vegetable oils or animal fats. A monoalkyl ester is the product of the reaction of a straight-chain alcohol, such as methanol or ethanol, with a triglyceride (TG) to form glycerol and esters of long-chain fatty acids. When the TG reacts with alcohol in presence of a catalyst, the transesterification reaction occurs, where these are converted subsequently to diglyceride (DG), monoglyceride (MG), and finally to glycerol. Through these reversible reactions, the bounded fatty acids are released and converted to fatty acids methyl or ethyl esters (biodiesel).

Various methods are used for production of biodiesel; the most reported are homogeneous and heterogeneous (alkali or acid) and enzymatic catalysis. During the transesterification there are side reactions; one of them is the saponification reaction. If the oil contains free fatty acids (FFAs) in less than 2%, they will react with the base catalyst to form soap and water. This reaction consumes catalyzer and difficult the separation process. Additionally, the formation of water in the reaction also inhibits the direct reaction. A second side reaction is hydrolysis where water generated from vegetable oil or formed during saponification hydrolyzes the TG to form more FFA and glycerol. To reduce the possibility of side reactions, the vegetable oil is pretreated with acid catalyst, which esterifies FFA to form esters (esterification reaction). The esterification reaction with methanol (assumed always in this chapter as the alcohol used) produces as by-product water to be removed.

Biodiesel produced by catalyzed transesterification contains several impurities: unreacted alcohol, unreacted glycerides, residual catalyst, FFA, glycerol, water, and others. One reason increasing production cost of biodiesel is the expenses involved in biodiesel separation and purification. Biodiesel produced should be purified according to ASTM D6751, EN 14214, or any local biodiesel standard. Separation and purification are main issues since impurities have a negative effect on the engine performance and exhaust emissions. Before its final use

*Biodiesel Production: Feedstocks, Catalysts, and Technologies*, First Edition. Edited by Samuel Lalthazuala Rokhum, Gopinath Halder, Suttichai Assabumrungrat, and Kanokwan Ngaosuwan.

in transportation, biodiesel is blended with diesel fuel according to the international standards and regulations, but its use as neat biodiesel (B100) is also reported, mainly in rural areas.

### 17.1.1 Factors Affecting Biodiesel Yield

The water content in neat oil or waste cooking oil (WCO) accelerates hydrolysis and simultaneously reduces biodiesel yield. Water content should not exceed 0.5% to obtain high yield. It is more critical for an acid-catalyzed reaction than base-catalyzed reaction. WCO are preheated at around 105 °C to evaporate water. As was previously mentioned, the FFA content is a critical issue. If FFA content exceeds 2%, transesterification reaction will barely proceed by homogenous base catalyst.

In most cases, methanol is used for the production of biodiesel because recovery of methanol from the final product is much easier, the biodiesel yield is higher than using other alcohols, viscosity of biodiesel obtained using methanol is lesser, and lower production cost, but higher toxicity in the case of methanol. Ethanol is less toxic than methanol and can be obtained from renewable sources, but beside the previously mentioned drawbacks, compared with methanol, ethanol tends to form azeotrope with water, which makes difficult the separation of water from alcohol during distillation process [1].

### 17.1.2 Transesterification Reaction Conditions

Methanol for transesterification with NaOH catalyst is more than reported at 6 : 1 molar ratio of alcohol to oil reaching high biodiesel yields. It is also established that 1.5% of catalyst concentration is most effective for biodiesel production [2]. Concerning stirring speed, optimization is necessary for different raw materials based on the difference in physical properties. The reaction temperature should not exceed the boiling point of the alcohol to avoid vaporization, but if the reaction temperature is kept below 50 °C, then the viscosity of biodiesel will increase [3]. The time between 1 and 3 h is recommended for the reaction, but it will depend on conversion technology, mechanism, feedstock, and biodiesel yield.

The fatty acid methyl ester (FAME) composition in biodiesel depends on material feedstock [4]. For instance, FAME obtained from soybean has a tendency to easily oxidize, the one produced from fish oils easily produces sludge, while the ones produced from palm oil easily form wax. The obtaining and refining process also influences the quality of the FAME. Because biodiesel properties are strongly related to the feedstock, the FAME profile is wide among different products, and also the technological production and strategies must be adapted to these particularities.

The following ranges of FFA are found in feedstocks: refined vegetable oils <0.05%; crude vegetable oil 0.3–0.7%; animal fat 5–30%; and grease 40–100%. When working with feedstocks that contain higher than 5% of FFA, it is important to convert them or biodiesel yield will be very low. The four main techniques for converting the FFA to biodiesel are acid catalysis or acid catalysis followed by alkali catalysis, enzymatic methods, and glycerolysis.

### 17.1.3 Separation After FAME Conversion

Regardless the used catalysts alkali or acidic, it has to be removed from the final products (biodiesel and glycerol). Basic catalysts are more difficult to remove, implying several washing steps with hot distilled water.

In a batch process, since agitation is stopped in the reactor, there are different procedures: (i) the reaction mixture is settled in the reactor for an initial separation of FAME and glycerol; (ii) the mixture is pumped into a settling vessel; and (iii) the mixture is separated in a centrifuge. Methanol must be removed from both the glycerol and FAME streams using an evaporator or a flash unit; esters are neutralized, washed using warm acid water to remove residual methanol and salts, and then are dried. Biodiesel is storage, and the glycerol is neutralized and washed with soft water and send to refining.

When the feedstock is yellow grease or animal fats, the system is slightly modified, adding an acid esterification step. The sulfuric acid and methanol mixture is added, and the system is stirred. If a two-step acid treatment is used, the stirring is suspended until methanol separates and is then removed. Once the conversion to FAME has reached the equilibrium, the methanol/water/acid mixture is removed by a centrifuge.

High FFA feedstock will react with the catalyst and form soaps if they are fed to a base-catalyzed system. The acid esterification is in this case applied, and the resulting soaps are separated using a centrifuge. Some TGs are lost with the soaps during this process.

Heterogeneous catalyzers are good alternatives for using in raw materials with relative high FFA content (0.1–0.5%) [5]. Similar to homogenous catalysts, heterogeneous catalysts are basic or acidic. These include porous zirconia, titanium based, alumina, ion-exchange resins, and acid zeolites, among others. In heterogeneous catalysis, biodiesel yield is in a range of 90–96%. Moreover heterogeneous-catalyzed biodiesel does not have the drawback of soap formation and consequently is less complicated than the further purification steps, reducing costs of the whole process. Nevertheless, the cost of heterogeneous catalyzer is considerable higher, but a recovery of around 80% of catalyzer is possible after reaction, compensating in part its cost. The major disadvantage of heterogeneous catalyst is the lower yield compared to homogeneous catalysis and longer reaction time due to limiting mass transfer [6]. Heterogeneous-catalyzed reaction is complex because it occurs in a three-phase system consisting of a solid and two immiscible liquids (oil and methanol) [7].

The separation of compounds in the output mixture is by exploiting the differences in boiling points between them. If the boiling points are enough different, such as between water and biodiesel, an evaporator or flash vaporizer can be used for the separation. In the evaporator, the liquid is heated to a temperature in which only the more volatile species will vaporize. In the evaporator, the separation is accomplished by supplying heat, while the mixture is held at a fixed pressure. On the contrary, flash vaporizer first heats the liquid at an elevated pressure and then liquid is sent through a flash valve that decreases the pressure. This decrease in pressure causes the more volatile portion of the liquid mixture to vaporize.

For miscible fluids with similar boiling points such as methanol and water, the distillation column is more effective. In the distillation column, the concentration of the more volatile component (methanol) is enriched above the feed point, and the less volatile (water) is enriched below the feed point. The vaporization in the column is driven by heat supplied in the reboiler, which is subsequently removed in the condenser. Trayed distillation columns are the most common where contacting occurs between the vapor stream and the liquid stream. The liquid enters the feed tray through the downcomer from the above tray, flows across the tray where interacts with vapor, spills over, and finally flows through the downcomer to the next tray.

In the case of algae feedstock, it is often applied in situ transesterification, where the first step is to separate biomass from the transesterified mixture by filtration or centrifugation. The reported studies for biodiesel purification using in situ transesterified mixture are mostly done by wet washing [8]. Ultrasonication-assisted in situ transesterification is performed for algal biomass, and then biomass is filtered followed by settling phase separation of the filtrate, wet washing of filtrate using water, and finally drying using anhydrous $Na_2SO_4$ [8, 9].

Enzymes (lipases) are a different heterogeneous catalyst type, which have gained attention recently due to the fact that the process has simple separation and purification steps [6]. The immobilized lipases have found application in continuous biodiesel production. The excess of methanol and glycerol after transesterification influences on lipases, making them inactive. The transesterification with immobilized lipase in supercritical $CO_2$ contributes in reducing the interaction between methanol and enzyme, reducing its toxicity with a fast separation of $CO_2$ from the product [10].

### 17.1.4 Washing

The primary purpose of the ester washing is the removal of any soap formed during the transesterification. If the washing process is insufficient, impurities like methanol, glycerol, and catalyzer will remain in the FAME phase affecting biodiesel quality, with a negative impact on engines performance, exhaust emissions, and increasing maintenance costs. Remaining glycerol implies fuel tank bottom deposits, injector fouling, storage problem, settling, and engine durability problems. In the case of water, it reduces the heat of combustion, increases corrosion, failure of fuel pump, and enhances bacteriological growth; likewise methanol decreases flash point and viscosity and causes corrosion. Rests of soap decrease oxidation stability, add corrosion problems, damage injectors, and plug filters [6].

Water provides a medium for addition of acid to neutralize the remaining catalyst and to remove the product salts. Wet washing of crude biodiesel could remove 90% of catalyst. The residual methanol should be removed before the washing step to prevent the addition of methanol to the wastewater effluent. However, some processes remove the methanol with the wash water, removing it later from the wash water. The use of warm water (60 °C) prevents precipitation of saturated FAME and delays the formation of emulsions during the washing. Softened water (slightly acidic) eliminates Ca and Mg and neutralizes the remaining base catalysts. The water washing must be carefully controlled to avoid emulsion formation. Wet washing conventionally is being done using hot distilled water and various solvents, and acidified water through water washing is a most widely used purification technique. Water washing processes are not recommended for purification of crude biodiesel from calcium-based heterogeneous processes as it might form very stable calcium soaps dissolved into crude biodiesel.

The phase separation between esters and water is typically complete. However, the equilibrium solubility of water in esters is higher than the specified water content for B100. Therefore, after the washing step always remains a content of water that may be removed by vacuum drier that can either be batch or continuous, or by molecular sieves, silica gels, and even also by a heating step at 105 °C until boiling ends. The last one is widely used for water removal and is quite simple.

A general scheme of flow processes for wet and dry washing is shown in Figure 17.1. In the wet washing the glycerol free biodiesel is washed with water/acid (HCl)/organic solvents followed by phase separation for 10–60 min. Two phases are obtained after separation; the washed water/solvent/acid with impurities and the treated biodiesel. Nevertheless,

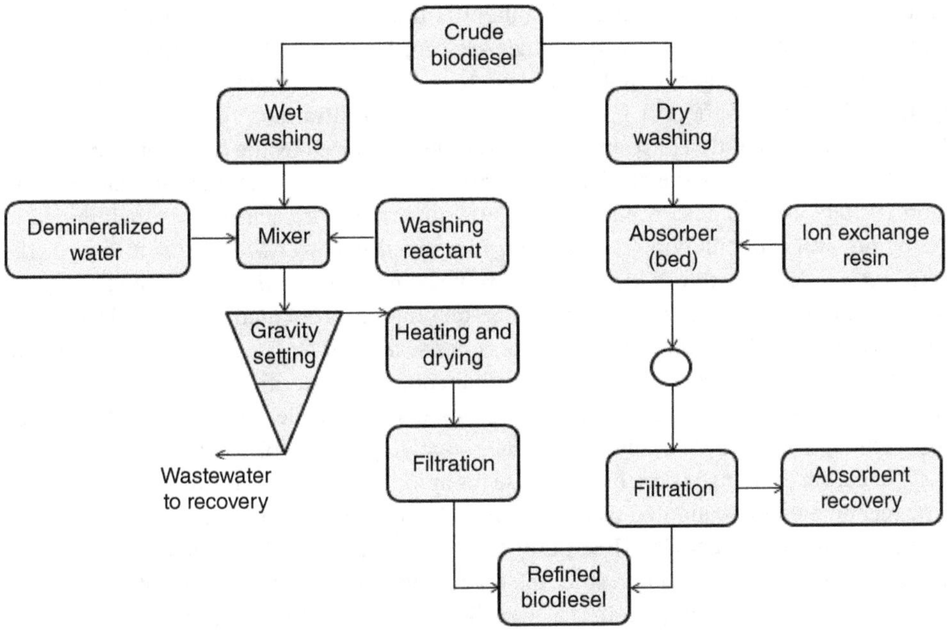


**Figure 17.1** General scheme of flow processes for wet and dry washing.

wet washing has several drawbacks such as significant amount of wastewater produced; use of centrifugation for phase separation, making the process more expensive and more energy intensive; 2–3 times washing increasing the operation time; and the requirement of an additional drying step. Besides it, using water and acid in wet washing tends to increase water content and acidity in biodiesel.

The presence of methanol in one or both phases affects the solubility of ester in glycerol and glycerol in ester. The ester washing step is used to neutralize any residual catalyst, to remove any soaps formed during the esterification reaction, and to remove residual free glycerol and methanol. FAME drying is required to reduce the amount of water present in biodiesel according to standards.

Dry washing does not use solvents for biodiesel purification. In dry washing, the use of ion-exchange resins is generalized. These resins having negatively charged sulfonate groups ($SO_3^{2-}$) are able to establish bonds with positively charged impurities like water, glycerol, metal ions, acid ions, and soap leading to purified biodiesel. The temperature strongly influences the adsorption capacity of the resin, since at room temperature the resin is surrounded by water layers, leaving no space available for the binding of other impurities, while at high temperature the water is removed from the surface of the resins, which can bind with other impurities.

Dry washing using industrial or agricultural wastes brings the advantages of no wastewater discharge, but wastes are unable to remove all the impurities (water and acidity) from crude biodiesel, needing further treatment and increasing the cost. Some reports are based on the use of solid wastes from ceramic industry and raw sugarcane bagasse [8].

The use of inorganic chemicals is an excellent strategy for purification of crude biodiesel produced using alkaline transesterification. The use of 1% magnesol and 2% silica at 65 °C purifies crude biodiesel, matching the standards limits [6]. Silica is reported to remove impurities from crude biodiesel with a single washing stage at 0.2 bar vacuum pressure and 90 °C, contact times longer than 45 min, and 1.1% of $SiO_2$ adsorbent dosage. Magnesol is reported to have better effect at 1.5–3% followed by microfiltration.

The use of a combination of resins in dry washing is reported by the cation exchange resin Diaion PK208LH actin as the catalyst for transesterification and the anion-exchange resin Diaion PA306S used to adsorb impurities. The temperature of each resin is maintained at 50 °C by hot water circulation. The reaction is stopped when FAME concentration in the output starts to decrease due to loss of anion exchange resins catalytic activity. Biodiesel purified using anion exchange resin Diaion PA306S met the EN14214 quality standard values for the impurities and FAME content. The process has several advantages as no requirement of upstream refining the crude biodiesel, simultaneous transesterification, and purification reducing the process time and increasing the productivity, while the glycerin adsorbed by anion-exchange resin can be easily recovered by supplying methanol. No decrease in catalytic activity of cation exchange resin, but decrease for anion exchange resin was observed, which can be regenerated by sequential pass of methanol to recover glycerin, acetic acid in methanol to displace fatty acids, NaOH solution to displace acetic acid ion, deionized water to remove NaOH, and methanol to restore the resin. The large number of solvents required to regenerate the anion-exchange resin is a clear disadvantage of this technique [8].

## 17.2 Glycerol Separation and Refining

Glycerol separation is usually a gravity-settling process, and the settling duration is 18–24 h. Free glycerol is presented with the ester after the transesterification; its presence after washing is result of noneffective water washing or other causes. The ester/glycerol separation is typically the first step of recovery in many biodiesel processes. The separation process is based on the facts that FAME and glycerol are barely soluble, but there is a significant difference in density between them. Although gravitational settling is applied in batch process, centrifugation is employed for continuous process where the oil is continuously fed into the transesterification reaction and continuous purification takes place. Separation of aqueous impurities and metal ions takes place simultaneously.

The rate of separation is affected by several factors. Most biodiesel processes use relatively intense mixing, at least at the beginning of the reaction, to incorporate the low soluble alcohol into the oil phase. If this mixing continues for the entire reaction, the glycerol can be dispersed in very fine droplets throughout the mixture. This dispersion requires from 1 h to several hours to allow droplets to coalesce into the glycerol phase. For this reason, mixing is generally slowed down during the progress of the reaction to reduce time required for phase separation. The more nearly neutral the pH, the quicker the glycerol phase will coalesce. This is one reason to minimize the total catalyst use. In some batch systems the reaction mixture is neutralized at the beginning of the glycerol/ester phase separation step.

Filtering is an essential part of biodiesel production. Feedstocks should be filtered to at least 100 μm, and the obtained biodiesel to at least 5 μm to ensure no contaminants is carried with the fuel that could damage engines. The fuel could be cooled before filtering to capture some of the saturated esters as they crystallize and thereby lower the cloud point of the fuel. Once biodiesel is separated from the glycerol, it still contains soap, metal ions, water, acid ions, catalyst, residual alcohol, and bound glycerol in form of MG, DG, and TG. Glycerol is partially refined as a coproduct from biodiesel production; about 10% by weight of the input reactants are converted to glycerol in the transesterification. Partially refined glycerol is worth more than the biodiesel product.

The recovered glycerol from the transesterification reaction contains residual methanol and catalyst, fat/oil, and some esters. It may also contain phosphatides, sulfur compounds, proteins, aldehydes, ketones, and insolubles. The catalyst is neutralized in a chemical refining; the neutralization step leads to the precipitation of salts. Also the soaps produced in the esterification must be removed by coagulation and precipitation with aluminum sulfate or ferric chloride. Control of pH is very important because low pH leads to dehydration of glycerol, and high pH leads to polymerization. Glycerol may then be bleached using activated carbon or clay. There is a physical refining with a first step of removing fatty, insoluble, or precipitated solids by filtration and/or centrifugation. This removal may require pH adjustment; then water is removed by evaporation. The final purification of glycerol is completed using vacuum distillation with steam injection, followed by activated carbon bleaching. The advantages of this approach are that this is a well-established technology. The primary disadvantage is that the process is highly energy consuming.

The glycerol separation by microfiltration membrane has taken interest and has been researching using nonreactive media [11]. However, due to methanol excess used in transesterification, the solubilization of glycerides in the nonpolar FAME phase is still a problem in the membrane refining.

Electrostatic coalescence is reported for the separation of biodiesel from glycerol. The applied electric field polarizes drops, and each drop acts as a dipole [12]. This promotes glycerin coalescence in biodiesel stream and is developed for heterogeneous-catalyzed transesterification. As the mixture enters the electric field established by the electrode surfaces, glycerol droplets coalesce. Once the droplets reach a size that overcomes gravity, the droplets fall to the glycerol phase at the vessel bottom.

Ion exchange purification of glycerol is an attractive alternative to vacuum distillation for smaller capacity plants. The ion-exchange system uses cation, anion, and mixed bed exchangers to remove catalyst and other impurities. The glycerol is first diluted with soft water to 15–35% of glycerol-in-water solution. The ion exchange is followed by vacuum distillation or flash drying for water removal, often leading to 85% of partially refined glycerol.

There are a few separation technologies that can be used for ultrapurification of glycerin such as distillation units, supercritical fluid extraction followed by fractionation steps, and membrane separation technology (MST) by a combination of nanofiltration, microfiltration, and electrodialysis [13]. MST has been proven to be the most efficient among others. Glycerin ultrapurification by using MST could be very promising in reducing biodiesel production cost. Several studies highlight the importance of membrane technologies, based on economical and operating advantages in biodiesel industry [5, 14].

## 17.3 Membrane Reactors

Process integration by combining chemical reaction and membrane separation in a single unit offers numerous benefits compared with conventional processes [5]. The operational principle is that the membrane acts as selective barrier to regulate the mass transfer rates of substances by permeability across the membrane. This kind of reactors establishes a reaction and a membrane-based separation simultaneously in the same physical enclosure. Through this, it is possible to separate the FAME/glycerin in methanol from unreacted TG by shifting the reaction equilibrium in favor of the biodiesel product.

Cao et al. [14] applied the membrane reactor for the recycling of methanol, glycerin, and NaOH in a biodiesel production process at room temperature. The result was the reduction of methanol-to-oil ratio and the reuse of NaOH in biodiesel produced. Nevertheless, in order to comply with standards of biodiesel, further purification of the FAME-rich phase must be conducted to remove methanol, MG, DG, and catalysts, but no additional devices are necessary to separate glycerol, reducing the operation time. The separation principle is related to the immiscibility of glycerol and FAME, where free glycerol forms droplets joining to the hydrophilic ends of the soap. These droplets can be separated of FAME by size exclusion using membranes.

Membrane technology for biodiesel purification had showed the ability to remove free glycerol particles from biodiesel without using a water washing step. Some reported membrane types are polyacrylonitrile and $\alpha$-$Al_2O_3/TiO_2$ ceramic [6, 11]. The membrane technology has main advantages in a way that it increases yield; purified biodiesel meets the standards; the formation of emulsions is avoided; operation is easy; and water use is very low compared to wet washing. As main drawbacks cleaning of membrane requires ionized water, cost of membranes is relative high, and scalable application must be developed. Solvent-resistant polymeric membranes are reported for biodiesel purification [15] by synthesized nanofiltration membranes composed of vinylidene fluoride as a support and poly(dimethylsiloxane) as coating material. A basic scheme of membrane separation unit for biodiesel purification is shown in Figure 17.2.

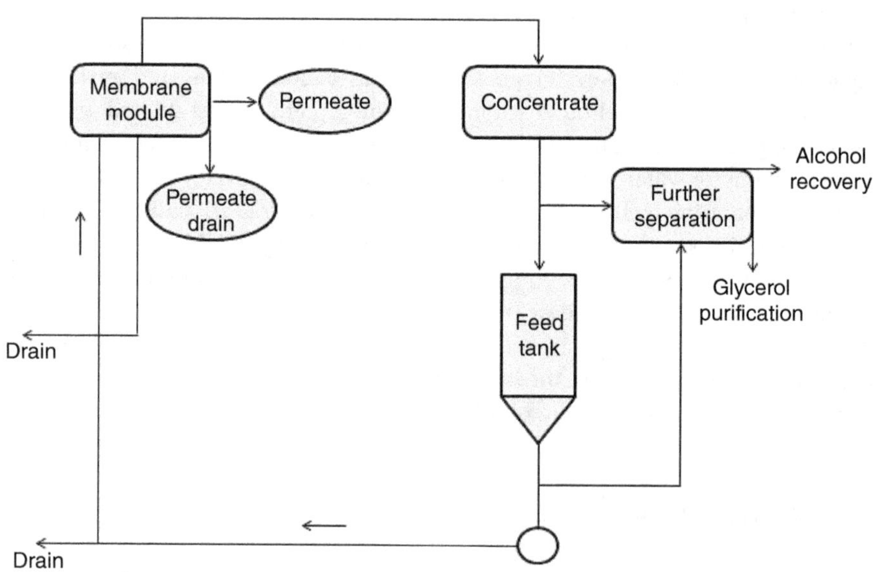


**Figure 17.2** Scheme of membrane separation unit for biodiesel purification.

## 17.4 Methanol Recovery

Methanol aggressively corrodes certain metals. In biodiesel no methanol shall be detected; therefore, it is important to eliminate it after the transesterification reaction. Content of 0.2% or less (EN14110) must be assured. The recovery of the residual methanol (excess) in transesterification saves input costs for the process and eliminates the emission of methanol. The emissions reduction is needed because methanol is highly flammable and toxic. The majority of the biodiesel production has adopted distillation as a separation unit for excess methanol, and through this more than 90% of alcohol is recovered. The recovery of methanol using distillation column and operating under vacuum prevents degradation of biodiesel and glycerol.

Methanol is fully miscible with water and with glycerol. However, it has a low solubility in fats and oils. The solubility in glycerol and water means that methanol will prefer these phases when there is a two-phase system present. The low solubility in fats and oils is the reason for the solubility-limited phase of the overall transesterification reaction. It is advantageous to remove the methanol before phase separation.

The main approach for methanol recovery is to separate it from the two phases. The conventional operation used for the recovery is distillation, which is an energy intensive process, either conventional or vacuum, or partially recovered in a single stage flash, and the use of multiple columns for the same purpose might increase operating as well as capital cost. Methanol recovery prior to separation of glycerol-rich and biodiesel-rich phases is also reported [6]. An alternative to distillation is a falling film evaporator. Residual methanol in the FAME phase can be removed in the water wash step in ester post-processing.

The use of cosolvent is designed to overcome slow reaction time caused by the low solubility of the alcohol in the TG phase. One approach is the Biox process. This process uses a cosolvent, tetrahydrofuran (THF), to solubilize the methanol. The cosolvent helps in forming a homogeneous phase facilitating transesterification. The formation of the homogeneous phase reduces the need for extensive stirring, use of multiple batch reactors and extensive reaction time, and improves the yield and product quality. The result is a fast reaction, on the order of 5–90 min, and no catalyst residues in either the ester or the glycerol phase. After the reaction is complete, the excess methanol and the THF are recovered in a single step. This is a continuous process designed for high FFA feedstocks that operates at ambient temperature. Other cosolvents, such as methyl tert-butyl ether (MTBE), have been investigated. The cosolvent must be completely removed from the glycerin as well as the biodiesel.

## 17.5 Additization

Additization is the addition of materials that have a specific functionality that modifies one or more biodiesel fuel properties. Examples include additives for cetane number improvement, cloud point, pour point, antioxidants, or other stability enhancing agents.

Biodiesel due to the large number of molecules with double bonds it contains is much less oxidatively stable than petroleum-based diesel fuel. One problem related to FAME properties is corrosion, which is observed in the fuel tank and engine part in contact with biodiesel or its blends. The main cause of corrosion is a lack of oxidation stability, producing some corrosive acids. Initially the oxidation generates peroxides, which subsequently form acids. Oxidation stability of biodiesel is improved by adding around 400 ppm of antioxidant [16]. To address this, stability-enhancing additive technology is well developed in the food industry and many of these additives can be used to stabilizing biodiesel. Tert-butyl hydroquinone (TBHQ) and butylated hydroxytoluene (BHT) are effective in oxidation inhibition. The oxidation stability of biodiesel depends not only on feedstock, processing condition, and contaminants but also storage conditions, storage duration, temperature, exposure to air and light, and presence of metals and impurities. Various additives are regularly used; using around 50–250 ppm of additive, the induction time (parameter based on Rancimat technique) of biodiesel can be increased in more than 10 h [16]. Various synthetic antioxidants are reported such as pyrogallol, propylgallate, and 3-tert-butyl-4-hydroxyanisole (BHA), but its effectiveness depends on the feedstocks characteristics and biodiesel processing.

The recommendation of the highest amount of 500 ppm of antioxidants such as Kerobit 3627, Chimec CH 4636, Irgastab BD 100, or Baynox Plus can increase the oxidation stability of B10 to 8 h, but too high dosage of antioxidant additive may deteriorate other fuel properties. With the use of 1000 mg $kg^{-1}$ of Vulkanox ZKF antioxidant, only acid value may be affected, while the viscosity, density, pour point, carbon residue, and sulfated ash content remain unaffected. However, if Vulkanox ZKF is additized into a biodiesel of high acid value, its acid value will increase beyond the standard limit of 0.5 mg KOH $g^{-1}$. With the use of 250 mg $kg^{-1}$ of antioxidants, all parameters including acid value remain unchanged [16].

Oxidative stability of biodiesel is improved by adding natural and synthetic antioxidants at varying concentrations between 250 and 1000 ppm. The various natural and synthetic antioxidants α-tocopherol, BHA, butyl-4-methylphenol, TBHQ, 2,5-di-tert-butyl-hydroquinone, ionol BF200, propylgallate, and pyrogallol improve the oxidative stability of biodiesel from soybean, cottonseed, poultry fat, and yellow grease [17]. It may also be improved by partial hydrogenation, increasing stability but with an opposite relationship with cold flow properties; therefore, for any kind of FAME to be enhanced by this technique, a strategy of hydrogenation degree must be established.

Concerning emissions NO is decreased by adding antioxidant additives such as BHA, BHT, and TBHQ. These additives are individually dissolved at varying concentrations in B10 and B20. Both BHA and TBHQ are effective in lowering NO emissions. Also significant reductions of NOx are obtained by adding ascorbic acid, α-tocopherol acetate, butylated hydroxytoluene, p-phenylenediamine, and ethylenediamine [6].

Diethyl ether (DEE) is another oxygenated additive, which can be used in biodiesel blends to suppress the NOx emissions. DEE is an excellent ignition enhancer and has a low auto ignition temperature. It is also good enhancing cold starting. Up to 20% of DEE in the blend may reduce 15% of NOx and smoke opacity.

Oxygenated additives such as methanol, ethanol, and n-butanol are also used in biodiesel, mainly for enhancing engines performance and reducing exhaust emissions. It is reported the use of ethyl levulinate as additive, with significant improvement in low temperature flow properties of blend, lowering the cloud point, pour point, and cold filter plugging point [18]. Ethanol is more than reported in biodiesel–ethanol blends or biodiesel–diesel–ethanol blends. Also reported is the use of isopropanol and butanol [6]. These blends can modify fuel viscosity, but failure of the flash point and increases in moisture content must be attended. Also phase or component separation is something to pay special attention in these systems.

Re-additization with BHT for long term storage biodiesel blends (B20) is reported [19]. The amount of BHT was varied up to 600 ppm. The authors concluded that 6 h minimum induction period are expected to have a potential storage life of greater than six months and likely greater than one year. Re-additizing is more effective and requires less antioxidant additive when applied at higher induction periods. Once the induction time is short, it is probable that the fuel will have no oxidation reserve remaining, producing acids soon; therefore, re-additization will no longer be applicable.

The cold flow performance of biodiesel is derived from its FAME composition varying among feedstocks. It requires high fraction of unsaturated fatty acids. However, it is inevitable to use biodiesel with inferior cold flow properties in cold atmosphere. Two ways to solve this problem are additization or winterization. Additization is only effective when biodiesel has low saturation content. Winterization removes high melting point saturated esters and the remaining lower saturated fatty acids, improving the cold flow performance. Many cold flow additives are available for diesel fuel; most have limited effectiveness on B100 but work with varying degrees of effectiveness with B20 [4].

Normally in biodiesel obtained from fats when mixing with diesel fuel, low temperature flow ability worsens. In order to use biodiesel during winter or in a cold region, a low temperature flow ability improver is added. The low temperature performance of the blend depends not only on the low temperature performance of FAME but also on the low temperature blend characteristics, type of improvers, and the amount of low temperature flow ability additive used. As the amount of saturated FAME increases, the required amounts of cold flow improver are increased due to the relationship between cold flow properties and the FAME saturated/unsaturated ratio. The cold flow improver should work even in base diesel fuel since the final use is a blend diesel fuel–biodiesel. The cold flow improver should be selected depending on the kind of FAME and the amount of blending. Biodiesel blends up to B5 are sometimes possible to treat with existing cold flow improvers, but higher blending ratio may request the use of further advanced additive technologies [16].

The effective use of additives such as Tween-80, dihydroxy fatty acid (DHFA), acrylated polyester prepolymer, palm-based polyol, additive synthesized using DHFA + ethyl hexanol (DHFAEH), and castor oil ricinoleate is reported. All the additives used showed significant reductions of pour point and cloud point for palm oil biodiesel. The biggest reduction of pour point was about 7.5 °C by adding 1% of DHFA, while the biggest reduction of the cloud point was about 10.5 °C by the addition of 1% DHFA + 1% polyester prepolymer [20].

Bio-oil obtained from fast pyrolysis of biomass pinewood is reported as antioxidant additive for biodiesel. A ten-solvent system is used to extract the natural additives from bio-oil, and the mixture of biodiesel and extracted bio-oil compounds is stirred and centrifuged. The insoluble fraction of bio-oil is settled to the bottom, while the homogeneous upper phase composed of biodiesel and soluble bio-oil compounds is separated and used as the additive to be incorporated to neat biodiesel. Direct blending of a small concentration of bio-oil (<4%) and biodiesel increased the biodiesel oxidation stability by 175% [21]. Hydrothermal treatment of bio-oil derived from the pyrolysis of pinewood is used for obtaining a product rich in phenolic compounds, being applied as additive in sunflower oil biodiesel. The oxidation stability of biodiesel is considerably improved by incorporating these type of additives at less than 2%, with improvement of 400% in oxidation stability compared with neat biodiesel [22].

Another approach is the transformation of glycerol into oxygenated compounds. The products obtained from the esterification of glycerol can be used as oxygenated additives for engines. Monoacetylglycerol, diacetylglycerol, and triacetylglycerol are consecutively obtained in a three-step reaction. Among them, triacetylglycerol is the most effective oxygenated additive, improving biodiesel viscosity and cold properties, cetane number, and reducing NOx emissions. Acetic anhydride is a strong acylating agent that renders the esterification reaction more efficiently than carboxylic acids. The use of acetic anhydride as reagent results in favoring the production of triacetylglycerol [23]. There are various heterogeneous, acid catalysts for glycerol esterification, mainly ion-exchange organic resins, such as Amberlyst-15, zeolites, and silica-supported heteropoly acids. The additives from glycerol have the potential to replace the existing petroleum-based additives such as MTBE and ethyl tert-butyl ether (ETBE) [24].

Transesterification of methyl acetate with glycerol was also developed to produce acetin simultaneously with biodiesel using Ca and Sn mixed metal hydroxides as catalyst [25]. $CaSn(OH)_6$ catalyst performed much better compared with other hydroxy stannates and metal oxides. A high glycerol conversion (78.2%) was attributed to higher basicity exhibited by $CaSn(OH)_6$ catalyst. Other novel catalysts have been explored in glycerol acetylation. Okoye [26] used glycerol as a carbon precursor to synthesize solid acid catalyst via partial carbonization and sulfonation to catalyze glycerol acetylation producing a mixture of diacetin and triacetin.

Cetane, combustion, and lubricity improvers are also widely used [13]. Regardless of biodiesel generally has higher cetane number than diesel fuel, it is frequently used. Nevertheless most of biodiesel from different feedstocks are by definition cetane improvers. There are also a lot of reports about the enhance on lubricity just by adding biodiesel to diesel fuel, and biodiesel in many studies has shown better combustion performance in blends concerning brake specific fuel consumption and thermal efficiency, but the use of additional additives is more related to comply with the standards and regulations of the final product.

## 17.6 Conclusion

Several downstream strategies concerning biodiesel separation and purification are established, and others are cutting-edge technologies. Dry and wet washing are well known and have pros and cons. The most active in innovations and new technologies is the dry washing technique. Glycerol separation and refining is a main issue in order to reduce biodiesel costs but to have additional industrial output. Glycerol is also a feed for production of antioxidant additives for biodiesel. Methanol recovery is a well-established technology, but it has potential for innovation yet such as the Biox process. Membrane technology is a way to integrate processes and contributes to better separation and enhances quality of products. Additization is an extremely important end process before commercialization and use of biodiesel. The number of additives is vast, and these have different functions. Several additives developed for food industry are also applied to biodiesel. The oxidation stability and exhaust emissions are significantly improved by adding natural and synthetic antioxidants. The obtaining of antioxidant additives from bio-oil obtained by biomass pyrolysis is a promising method, since it is produced by renewable energy technologies and is a good strategy for integration of bioenergy systems. In this respect oxygenated additives such as methanol, ethanol, and n-butanol are also used as biodiesel additives, which can be obtained from biomass, enhancing engines performance and reducing exhaust emissions.

## References


1 Van Gerpen, J.H. (2005). *Biodiesel processing and production. Fuel Processing Technology* **86** (10): 1097–1107.

2 Chen, L., Yin, P., and Liu, L. (2011). *Biodiesel production over copper vanadium phosphate. Energy* **36** (1): 175–180.

3 Gnanaprakasam, A., Sivakumar, V.M., Surendhar, A., and Thirumarimurugan, M. (2013). Recent strategy of biodiesel production from waste cooking oil and process influencing parameters: a review. *Journal of Energy* 2013: http://dx.doi.org/10.1155/2013/926392.

4 Alleman, T.A. and McCormick, R.L. (2016). *Biodiesel Handling and Use Guide*, 5e). DOE/GO-102016-4875. U.S. Department of Energy.

5 Hasheminejad, M., Tabatabaei, M., Mansourpanah, Y. et al. (2011). *Upstream and downstream strategies to economize biodiesel production. Bioenergy Resources* **102**: 461–468.

6 Suthar, K., Dwivedi, A., and Joshipura, M. (2019). *A review on separation and purification techniques for biodiesel production with special emphasis on Jatropha oil as a feedstock. Asia-Pacific Journal of Chemical Engineering* 14: **e2361**.

7 Rizwanul, I.M., Ong, H.C., Mahlia, T.M.I. et al. (2020). *State of the arts of catalysts for biodiesel production. Frontiers in Energy Research* **8**: 1–17.

8 Yellapu, S.K., Kaur, R., Kumar, L.R. et al. (2018). Recent developments of downstream processing for microbial lipids and conversion to biodiesel. *Bioresource Technology* https://doi.org/10.1016/j.biortech.2018.01.129.

9 Suganya, T., Kasirajan, R., and Renganathan, S. (2014). Ultrasound-enhanced rapid in situ transesterification of marine macroalgae *Enteromorpha compressa* for biodiesel production. *Bioresource Technology* **156**: 283–290.

10 Toldrá-Reig, F., Mora, L., and Toldrá, F. (2020). *Trends in biodiesel production from animal fatwaste. Appied Sciences* **10**: 3644.

11 Sérgi, M.C., Curvelo, N., and Davantel, S.T. (2010). *Separation of biodiesel and glycerol using ceramic membranes. Journal of Membrane Science* **352**: 271–276.

12 Sams, G.W. W. A. Summers, S.S. Randhava, and H.G. Wallace (2008). Removal of glycerin from biodiesel using an electrostatic process. US Patent 20100108523 A1.

13 Tabatabaei, M., Karimi, K., Horváth, I.S., and Kumar, R. (2015). *Recent trends in biodiesel production. Biofuel Research Journal* **7**: 258–267.

14 Cao, P., Dube, M.A., and Temblay, A.Y. (2008). *Methanol recycling in the production of biodiesel in a membrane reactor. Fuel* **87**: 825–833.

**15** Torres, J.J., Toledo, J., Ochoa, N.A. et al. (2018). *Biodiesel purification using polymeric nanofiltration composite membranes highly resistant to hard conditions. Chemical Engineering and Technology* **41** (2): 253–260.

**16** Goto, S., Oguma, M., and Chollacoop, N. (2010). Current status of biodiesel fuel in East-Asia and ASEAN countries. In: *EAS-ERIA Biodiesel Fuel Trade Handbook* (ed. S. Goto). Jakarta: ERIA Research Project.

**17** Ali, O.M., Mamat, R., Rosli Abdullah, N. et al. (2016). *Effects of different chemical additives on biodiesel fuel properties and engine performance. A comparison review. MATEC Web of Conferences* **38**: 03002.

**18** Madiwale, S., Karthikeyan, A., and Bhojwani, V. (2017). *A comprehensive review of effect of biodiesel additives on properties, performance, and emission. IOP Conference Series: Materials Science and Engineering* **197**: 012015.

**19** Christensen, E.D., Alleman, T., and McCormick, R.L. (2018). *Re-additization of commercial biodiesel blends during long-term storage. Fuel Processing Technology* **177**: 56–65.

**20** Ming, T.C., Ramli, N., Lye, O.T. et al. (2005). *Strategies for decreasing the pour point and cloud point of palm oil products. European Journal of Lipid Science and Technology* **107** (7–8): https://doi.org/10.1002/ejlt.200400944.

**21** García, M., Botella, L., Gil-Lalaguna, N. et al. (2016). *Antioxidants for biodiesel: additives prepared from extracted fractions of bio-oil. Fuel Processing Technology* https://doi.org/10.1016/j.fuproc.2016.10.001.

**22** Gil-Lalaguna, N., Bautista, A., Gonzalo, A., and Luis Sánchez, J. (2017). *Obtaining biodiesel antioxidant additives by hydrothermal treatment of lignocellulosic bio-oil. Fuel Processing Technology* https://doi.org/10.1016/j.fuproc.2017.05.020.

**23** García-Martín, J.F., Alés-Álvarez, F.J., Torres-Gárcia, M. et al. (2019). Production of oxygenated fuel additives from residual glycerine using biocatalysts obtained from heavy-metal-contaminated *Jatropha curcas* L. roots. *Energies* **12** (740): en12040740.

**24** Nda-Umar, U.I., Ramli, I., Taufiq-Yap, Y.H., and Muhamad, E.N. (2019). *An overview of recent research in the conversion of glycerol into biofuels, fuel additives and other bio-based chemicals. Catalysts* **9** (15): 9010015.

**25** Sandesh, S., Kumar, P., Manjunathan, P. et al. (2016). *Synthesis of biodiesel and acetins by transesterification reactions using novel CaSn(OH)6 heterogeneous base catalyst. Applied Catalysis A: General* **523**: 1–11.

**26** Okoye, P., Abdullah, A., and Hameed, B. (2017). *Synthesis of oxygenated fuel additives via glycerol esterification with acetic acid over bio-derived carbon catalyst. Fuel* **209**: 538–544.

# 18

# Heterogeneous Catalytic Routes for Bio-glycerol-Based Acrylic Acid Synthesis

*Nittan Singh*[1,2], *Pavan Narayan Kalbande*[1,2], *and Putla Sudarsanam*[1,2]

[1] *Catalysis and Inorganic Chemistry Division, CSIR-National Chemical Laboratory, Pune, India*
[2] *Academy of Scientific and Innovative Research (AcSIR), CSIR-National Chemical Laboratory, Pune, India*

## 18.1 Introduction

Acrylic acid is a vital carboxylic acid for the chemical and polymer industries. It contains two important functional groups, namely, vinyl and carboxylic acid [1, 2]. Hence, it can be used for the synthesis of a number of useful chemicals and polymers. Primarily, it is used for the synthesis of acrylic esters. Self-polymerization of acrylic acid and its esters gives very useful polymers. Acrylic esters can also be combined with other monomers (e.g. styrene, butadiene, acrylamides, acrylonitrile, and vinyl compounds) to give commercially valuable copolymers. Because of this huge demand especially in developing countries, the global market value of acrylic acid is expected to grow from $13.60 million in 2015 to $18.32 billion by 2022, with an annual growth rate of 5.14% over the 2016–2022 period [3]. To meet this demand, efficient and sustainable production pathways are needed for acrylic acid production, particularly using renewable feedstocks, such as biomass [4, 5].

In industry, acrylic acid is produced from propylene oxidation as shown in Scheme 18.1. It is a two-step reaction: first, the catalytic oxidation of propylene gives acrolein, which is subsequently oxidized in the presence of a catalyst to acrylic acid in the second step [1]. Propylene is a fossil fuel-derived chemical, and it is produced via refining of gasoline as well as from splitting, cracking, and reforming of the hydrocarbon mixture. To control atmospheric greenhouse gas emissions (e.g. $CO_2$) as well as to meet the increasing energy demand, using renewable energy sources, especially biomass (green and sustainable carbon feedstock), for the synthesis of fuels and chemicals including acrylic acid has received great attention over the last two decades (Figure 18.1) [1, 2]. In this chapter, the synthesis of acrylic acid from bio-glycerol using various heterogeneous solid catalysts is summarized. Glycerol is largely obtained as a waste by-product from the biodiesel industry via esterification of lipids in the presence of a base catalyst [6–12]. For comparison, the synthesis of acrylic acid from propylene was initially discussed.

*Biodiesel Production: Feedstocks, Catalysts, and Technologies*, First Edition. Edited by Samuel Lalthazuala Rokhum, Gopinath Halder, Suttichai Assabumrungrat, and Kanokwan Ngaosuwan.

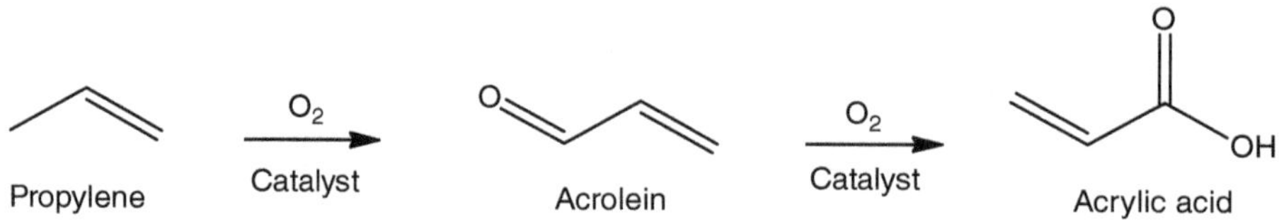


**Scheme 18.1** Commercial synthesis of acrylic acid via propylene oxidation.

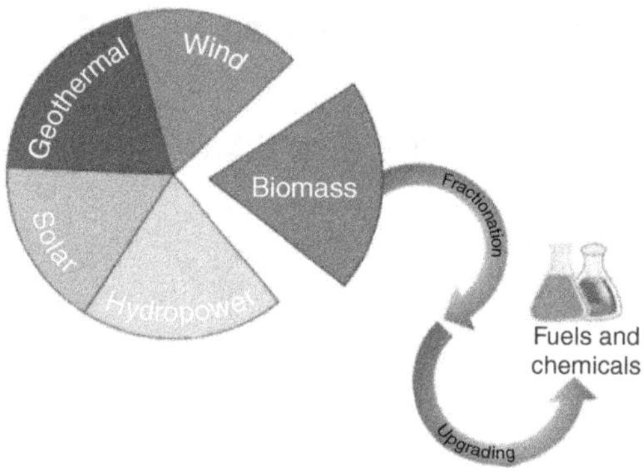


**Figure 18.1** Renewable energy sources and biomass valorization to fuels and chemicals. *Source:* Ref. [4].

## 18.2 Acrylic Acid Synthesis from Propylene

Propylene ($C_3H_6$) is a colorless gas with a pungent smell. It is highly flammable and commercially produced via gasoline refining. For the synthesis of acrylic acid from propylene, the selective oxidation of propylene was conducted using two reactors under gas phase conditions. In the first reactor, propylene is oxidized to acrolein in the range of 300–370 °C temperature and 1–2 bar pressure using a catalyst [1, 2, 13–15]. The subsequent catalytic oxidation of produced acrolein in the second reactor gives acrylic acid in the range of 250–300 °C temperature and 0.3–0.8 bar pressure. Mo and Bi-based mixed oxide catalysts are typically used for the first step, i.e. propylene to acrolein. For this, promoters, such as Al, P, K, Sn, Co, Fe, V, Ni, Sn, Si, Sb, etc., are used to achieve high reaction rates with superior product selectivity. For the second step, i.e. acrolein to acrylic acid, V and Mo-based mixed oxide catalysts with Ni, Fe, Sn, Si, Sb, Co, Cu, W, etc., as promoters are used. Both reaction steps need often a few seconds to avoid over-oxidation, thus achieving higher yields of acrylic acid [1, 2, 13–15]. Steam is often used to control these steps to eventually get improved yields of acrylic acid.

## 18.3 Acrylic Acid Synthesis from Glycerol

Processing of triglyceride feedstocks (oils and fats) results in approximately 100 g of glycerol by-product for every liter of biodiesel produced via base-catalyzed transesterification reaction with simple alcohols, like methanol (Figure 18.2) [5]. With the current annual world biodiesel production of just over 33 billion liters and projections for this to exceed 41

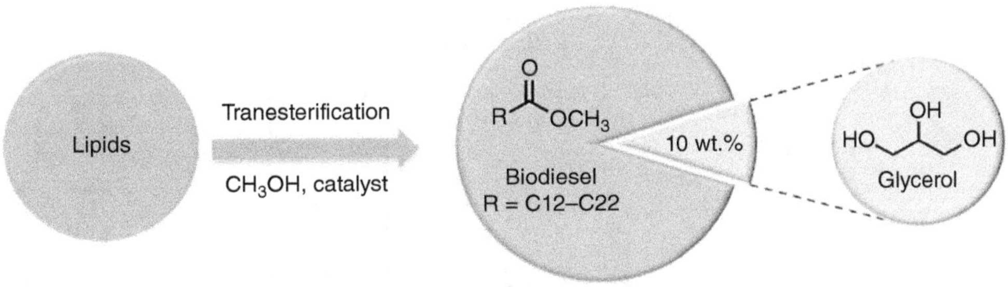


**Figure 18.2** Glycerol is a by-product from biodiesel synthesis. *Source:* Adapted from Ref. [5].

billion liters by 2025, significant use of glycerol is necessary to both improve the economics of biodiesel production and avoid landfill dumping of this growing waste stream [16]. Glycerol is a truly versatile chemical with the ability to give a variety of drop-in chemicals and fuel-grade molecules via catalytic oxidation, hydrogenolysis, dehydration, transesterification, etherification, polymerization, and carboxylation routes [4, 5]. The drawback with most of these pathways is that they provide an untried alternative to existing commercial technologies that are satisfying the current demand for these products. Pragmatically, glycerol needs to be converted into a high value product that could be used to enhance or improve the efficiency of biodiesel. One such product from glycerol is acrylic acid, a valuable precursor for the polymer industry. The efficient conversion of glycerol to acrylic acid at industrially feasible conditions can boost the polymer industry because glycerol is a waste product from the biodiesel industry.

The conversion of glycerol to acrylic acid can be done either by one-step (oxidative dehydration) or two-step route (Figure 18.3) [2]. The two-step method involves different types of reactions (dehydration, oxidation, deoxydehydration, hydrolysis, and ammoxidation), giving different intermediates (acrolein, propylene, lactic acid, allyl alcohol, 3-hydroxypropanoic acid, and acrylonitrile). For both reaction routes, we need catalysts to enhance their reaction rates. Hence, the selective conversion of glycerol to acrylic acid is highly dependent on the nature of the catalyst. This chapter discusses the synthesis of acrylic acid from glycerol via acrolein route using heterogeneous solid catalysts. Homogeneous catalysts can also be used for glycerol to acrolein and then to acrylic acid. These catalysts perform well but require complicated workups to separate the catalyst from the reaction mixture and/or corrode equipment. Heterogeneous catalysts, by nature of their simple separation from the reaction mixture, would seem to overcome these limitations. Many research papers have been written comparing heterogeneous alternatives for the glycerol-to-acrylic acid conversion.

### 18.3.1 Glycerol Dehydration to Acrolein

In this section, the catalytic synthesis of acrolein from glycerol via dehydration route is discussed, which will give valuable insights for developing a suitable catalyst for the efficient synthesis of acrylic acid from glycerol. Selective oxidation of propylene is a traditionally used route for acrolein synthesis, and it is an important intermediate for the chemical industry. To develop a sustainable method, the production of acrolein from glycerol has

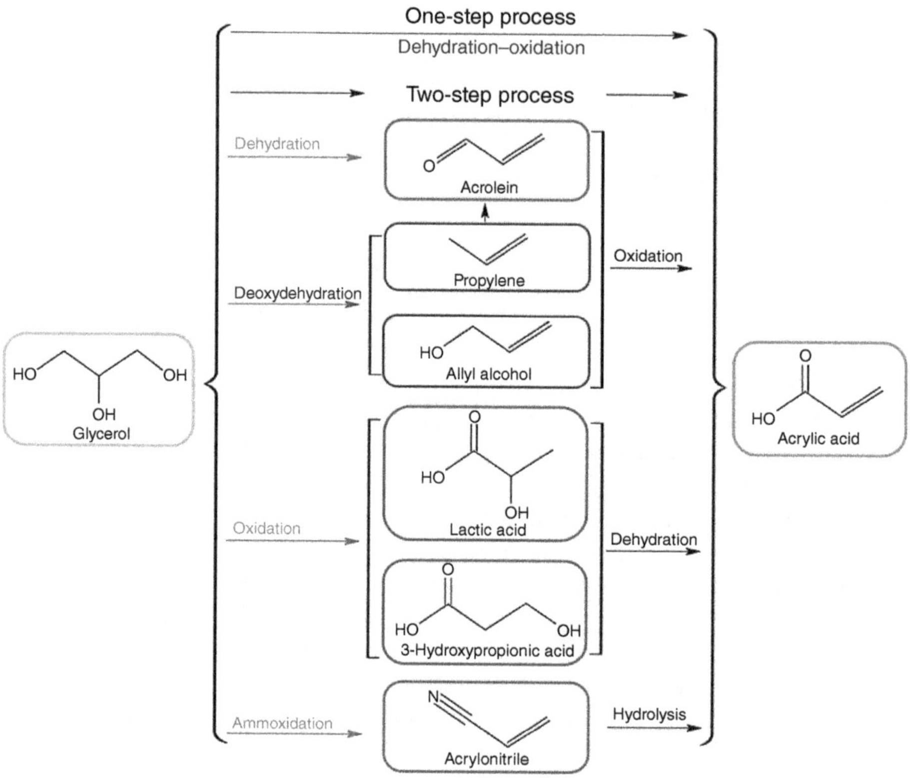


**Figure 18.3** Conversion of glycerol to acrylic acid via different routes. *Source:* Ref. [2].

received great attention over the last two decades [1, 2]. Dehydration is an acid-catalyzed reaction, and many solid acids are developed for the dehydration of glycerol to acrolein. The typically used solid acids are heteropoly acids, zeolites, bare and functionalized metal oxides, etc. As shown in Figure 18.4, the formation of acrolein from glycerol is favored on the Brønsted acid sites, whereas acetol formation is possible in the case of solid catalysts having Lewis acid and/or basic sites [17]. For acrolein synthesis, the Brønsted acid sites initiate the protonation of the secondary hydroxyl group of glycerol and then the release of $H_3O^+$ to give 1,3-dihydroxypropene. It undergoes a keto-enol rearrangement to 3-hydroxypropionaldehyde, and the subsequent dehydration gives acrolein (Figure 18.4a). It is obvious that the catalyst with a higher ratio of Brønsted/(Brønsted + Lewis) acid sites can give improved yields of acrolein from glycerol dehydration. The catalyst should also exhibit high hydrothermal stability because water is a key by-product, and it can cause the leaching of acid sites, thus rapidly decreased catalytic activity. Therefore, developing a water-tolerant solid acid catalyst with optimized Brønsted acid sites is essential for sustainable acrolein production from glycerol dehydration reaction.

The glycerol dehydration reaction is mostly carried out in the gas phase in a tubular reactor using a solid acid catalyst(s). Till now, a heteropoly acid ($Cs_{2.5}H_{0.5}PW_{12}O_{40}$) was found to exhibit the highest acrolein selectivity of 98% with complete conversion of glycerol conversion at 275 °C, 1 bar pressure, and 1 h time on stream (TOS) [17]. Although excellent

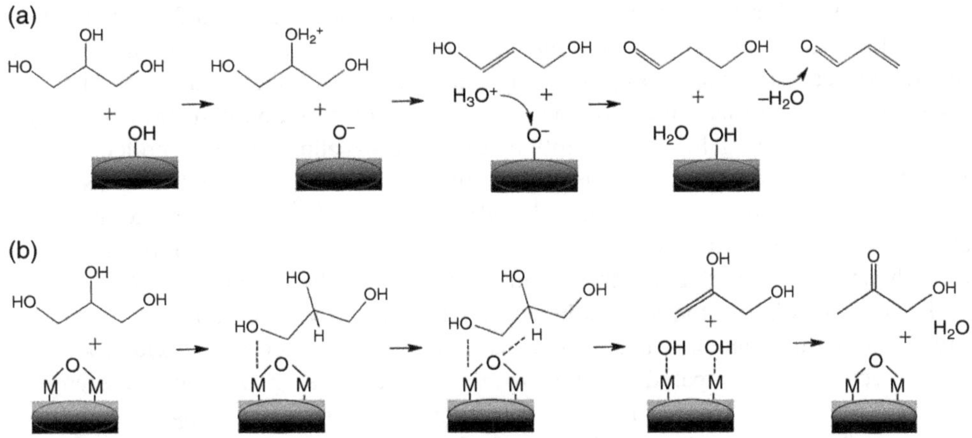

**Figure 18.4** The reaction mechanism of glycerol dehydration over (a) Brønsted acid sites and (b) Lewis acid sites. *Source:* Adapted from Ref. [17].

catalytic activities are achieved in the vapor-phase reaction of glycerol dehydration, it must be noted that these are achieved under short TOS and low space velocity conditions. This is mainly due to the formation of coke on the catalyst's surface, which will block the active sites, thus resulting in rapid catalyst deactivation after a few hours of reaction [2]. Coke formation generally occurs via polycondensation of glycerol and/or acrolein. Several potential strategies are used to prevent coke formation/catalyst deactivation during glycerol dehydration reaction. For instance, co-feeding of $O_2$/air and using a precious metal-modified solid acid are found to improve the catalyst stability in the dehydration of glycerol under gas-phase reactions.

Alternatively, in recent times, considerable efforts have been made to develop liquid-phase dehydration routes for glycerol conversion into acrolein. For example, Shen et al. [18] reported various Brønsted acidic ionic liquids (ILs) for glycerol dehydration using a semibatch liquid-phase reactor. About 57.4 and 50.8% acrolein yields are obtained over [Bmim]$H_2PO_4$ and [Bmim]$HSO_4$ ILs, respectively, at 270 °C reaction temperature. The presence of moderate acid strength in the Brønsted acidic IL catalysts favored the dehydration of glycerol to acrolein in liquid phase [18, 19]. Other catalysts, such as heteropoly acid (e.g. silicotungstic, phosphotungstic, and phosphomolybdic acids), metal sulfates, metal phosphates, metal oxides (alumina), sulfonic acid, and zeolites are also used for glycerol dehydration in liquid phase. But, acrolein yield is below 50% in liquid-phase glycerol dehydration using heterogeneous catalytic routes. Hence, the gas phase reaction is continued to be a potential route for acrolein synthesis from glycerol, but developing thermally stable catalysts with a strong coke-resistance capacity remains a great challenge for vapor phase glycerol dehydration reaction.

### 18.3.2 Acrylic Acid Synthesis from Glycerol

Cascade reaction methods play an important role for the sustainable chemical industry (Figure 18.5) [4]. Improved atom economy (negligible atom loss) and reduced E-factor (low amounts of waste generation) can be substantially improved in the case of cascade reactions as they limit the number of steps needed for the isolation and purification of intermediates and products as well as for the recovery of the catalysts. Especially, the cascade route is crucial for biomass processing as it involves a number of steps to obtain the drop-in chemical. In this context, one-pot transformation (oxidative dehydration) of glycerol to acrylic acid has gain lots of attention as this process eliminates the acrolein synthesis step [1, 2]. Since this process involves both dehydration and oxidation reactions, a multifunctional catalyst with both acid (dehydration) and redox (oxidation) sites is needed for the efficient synthesis of acrylic acid from glycerol. Although a considerable number of research articles is reported for acrylic acid synthesis using two-bed reactors (two-step or single-step), we discuss the application of single-bed catalytic reactors for glycerol-to-acrylic acid in this chapter. Using a single-bed catalytic reactor for acrylic acid production is highly economical compared with the process carried out using two-step two reactors or a two-bed one reactor. The reason is that the design and operation of a single-bed catalytic reactor are facile, and it enhances efficiency and reduces the process investment.

Figure 18.6 shows the schematic representation of a one-bed catalytic system for glycerol oxydehydration to acrylic acid [1]. Vanadium (V)-based multifunctional catalysts with

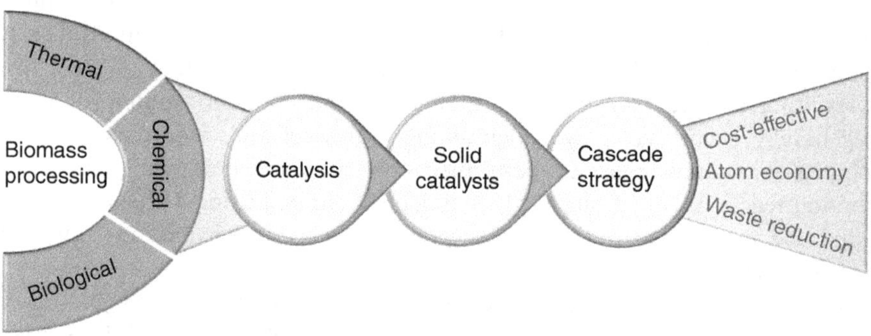


**Figure 18.5** Concept of sustainable biomass processing. *Source:* Ref. [4].

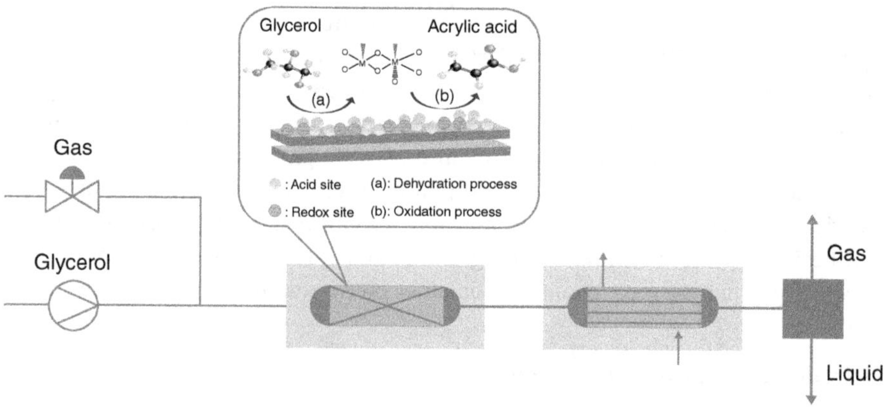


**Figure 18.6** Glycerol oxydehydration to acrylic acid in one-bed catalytic system. The catalyst must contain both acid sites (for dehydration of glycerol to acrolein) and redox sites (for oxidation of acrolein to acrylic acid). *Source:* Ref. [1].

molybdenum (Mo) and tungsten (W) were found to be efficient for one-pot synthesis of acrylic acid from glycerol [1, 2]. Mo- or W-based oxides catalyze dehydration of glycerol and vanadium oxide is useful for acrolein oxidation, thus the combination of $VO_x$ with $MoO_x$ and/or $WO_x$ has received great attention for acrylic acid synthesis from glycerol. For instance, the V-Mo mixed oxide ($MoV_2O_8$) showed 3.5 times higher catalytic efficiency than the bare metal oxides for acrylic acid production [20].

In recent years, developing ternary Mo–V–W–O [21] and Nb–V–W–O [22] oxides as multifunctional catalysts is a primary focus for the oxydehydration of glycerol to acrylic acid under gas phase conditions. In these ternary oxides, W can act both as a promoter and as an acidic material. For instance, Chieregato et al. [22] developed a W–V–Nb mixed oxides catalyst for oxidehydration of glycerol into acrylic acid. The catalyst contains a hexagonal tungsten bronze structure. Various reaction parameters were studied to achieve higher reaction rates and improved acrylic acid yields. It was demonstrated that the acrolein oxidation step must be faster than the glycerol dehydration step to prevent the formation of heavy compounds, like ketals and oligomers, thus achieving higher yields of acrylic acid. To achieve this, the use of higher oxygen partial pressures was recommended, which induces the formation of oxidizing $V^{5+}$ sites for acrolein oxidation into acrylic acid during the reaction. About 50.5% acrylic acid yield with a complete conversion of glycerol was achieved at optimized reaction conditions. The combination of a heteropoly acid ($H_3SiW_{12}O_{40}$) with vanadium oxide and a zeolite (HZSM-5) was also studied for acrylic acid synthesis [23]. But this reaction was carried out in the liquid phase because of the low thermal stability of heteropoly acid. For instance, Thanasilp et al. [23] developed various metal-doped $H_3SiW_{12}O_{40}$/HZSM-5 catalysts (metal = Co, Ce, Ni, and V) for liquid-phase oxydehydration of glycerol to acrylic acid using $H_2O_2$ as the oxidizing agent at 90 °C reaction temperature. Among those catalysts, the V-doped $H_3SiW_{12}O_{40}$/HZSM-5 catalyst (6 wt% V loading) showed the best activity with glycerol conversion of 99.67% and acrylic acid yield of 36.23% at 90 °C and 6 h reaction time, due to the improved mass transport properties of the catalyst.

The redox pair $V^{5+}/V^{4+}$ was found to play a key role for acrylic acid synthesis from acrolein [1, 24, 25]. As shown in Figure 18.7b, the oxidation of acrolein to acrylic acid on V-based catalysts follows the Mars–van Krevelen mechanism [24]. As per this mechanism, the oxygen species of $V_2O_5$ involve in the oxidation of acrolein to acrylic acid, with consequent reduction of $V^{5+}$ to $V^{4+}$, giving an oxygen vacancy. Next, the oxygen species from gas phase replenishes the oxygen vacancy while favoring the oxidation of $V^{4+}$–$V^{5+}$ on the catalyst surface. Although significant research has been done toward one-pot oxydehydration of glycerol to acrylic acid, achieving higher yields of acrylic acid with low coke formation remains a great challenge. Thus, future studies should focus on developing efficient multifunctional solid catalysts with durable activity and high coke-resistance ability.

## 18.4 Conclusion

The synthesis of biomass-based acrylic acid represents a sustainable route to overcome the challenges associated with the petrochemical route, i.e. oxidation of propylene or propane. In this context, the dehydration–oxidation of glycerol (a waste by-product from the

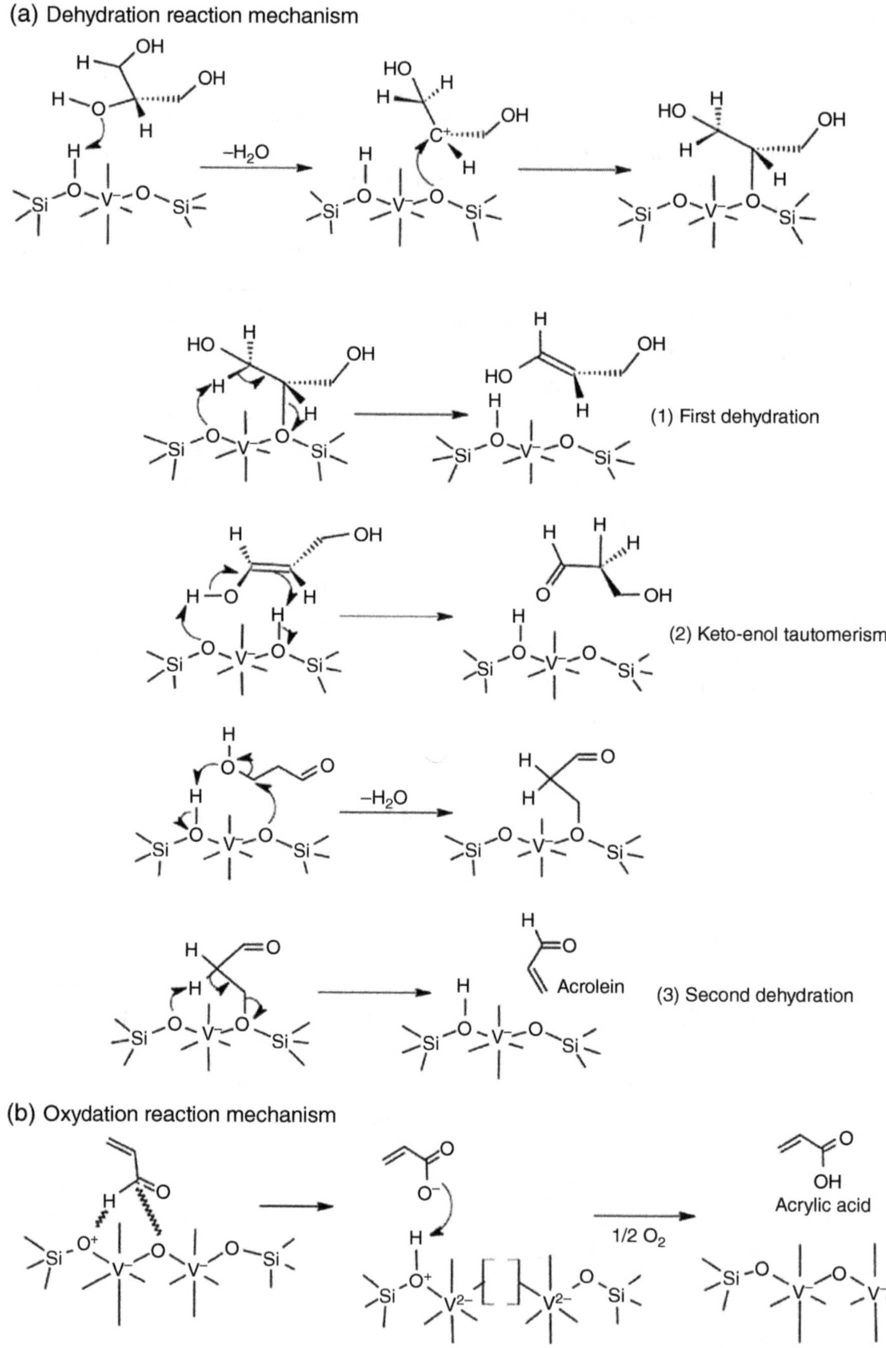


**Figure 18.7** Reaction mechanisms for (a) dehydration of glycerol into acrolein and (b) oxidation of acrolein into acrylic acid (it follows the Mars–van Krevelen mechanism). *Source:* Ref. [24].

biodiesel industry) offers a new promising route to produce acrylic acid in a single step. For efficient one-pot dehydration–oxidation of glycerol to acrylic acid, developing multifunctional catalysts with optimum amounts of both acid sites (for dehydration of glycerol to acrolein) and redox sites (for the oxidation of acrolein to acrylic acid) is crucial. The research revealed that V-based mixed oxide catalysts (W–V–O, Mo–V–O, W–V–Nb–O, and Mo–V–W–O) are found to be the most active and selective catalysts for efficient synthesis of acrylic acid in vapor phase state. It must be noted that these reactions were conducted on a laboratory scale. Rapid deactivation of the catalyst due to coke formation on the catalyst surface is a primary barrier for commercializing the process for bio-glycerol-based acrylic acid synthesis. Significant research should be done toward developing potential strategies for inhibiting the coke formation and enhancing catalytic stability. These strategies include doping of noble metals in the structure of the catalysts, tuning acidity and porosity, optimizing water concentration, feeding oxygen, and reaction temperature. Importantly, understanding the composition of coke, its formation, and mechanism during the reaction is important to develop next-generation catalysts for efficient acrylic acid synthesis from glycerol.

## Acknowledgments

NS and PNK thank the Council of Scientific and Industrial Research (CSIR), New Delhi, for research fellowships.

## References

**1** Wu, S.T., She, Q.M., Tesser, R. et al. (2020). *Catalysis Reviews* 62: 481–523.

**2** Sun, D., Yamada, Y., Sato, S., and Ueda, W. (2017). *Green Chemistry* 19: 3186–3213.

**3** Future, M.R. (2018). Acrylic Acid Market Research Report- Forecast to 2022. India. p. 166.

**4** Sudarsanam, P., Zhong, R., den Bosch, S.V. et al. (2018). *Chemical Society Reviews* 47: 8349–8402.

**5** Sudarsanam, P., Peeters, E., Makshina, E.V. et al. (2019). *Chemical Society Reviews* 48: 2366–2421.

**6** Mallesham, B., Sudarsanam, P., and Reddy, B.M. (2014). *Industrial & Engineering Chemistry Research* 54: 18775–18785.

**7** Mallesham, B., Sudarsanam, P., and Reddy, B.M. (2014). *Catalysis Science & Technology* 4: 803–813.

**8** Sudarsanam, P., Mallesham, B., Nagaprasad, A. et al. (2013). *Fuel Processing Technology* 106: 539–545.

**9** Mallesham, B., Sudarsanam, P., Gangadhara, R., and Reddy, B.M. (2013). *Green Chemistry* 15: 478–489.

**10** Reddy, P.S., Sudarsanam, P., Raju, G., and Reddy, B.M. (2012). *Journal of Industrial and Engineering Chemistry* 18: 648–654.

**11** Reddy, P.S., Sudarsanam, P., Mallesham, B. et al. (2011). *Journal of Industrial and Engineering Chemistry* 17: 377–381.

12 Reddy, P.S., Sudarsanam, P., Raju, G., and Reddy, B.M. (2010). *Catalysis Communications* 11: 1224–1228.

13 Cannizzaro, M., Forzatti, P., Pasquon, I., and Villa, P.L. (1981). Oxidation of propylene to acrylic acid on a Fe–Mo–Bi oxide system. *Reaction Kinetics and Catalysis Letters* **17** (1): 57–62.

14 Shimizu, N., I. Yanagisawa, M. Takata, T. Sato (1979). Process for producing acrylic acid from propylene. US Patent, Editor. Nippon Shokubai Kagaku Kogyo Co. Ltd., Osaka, Japan.

15 Engelbach, H. R. Krabetz, G. Duembgen, C.-H. Willersinn, U. Lebert, F. Thiessen (1977). Manufacture of acrylic acid by oxidation of propylene with oxygen-containing gases in two separate catalyst stages. BASF Aktiengesellschaft (Ludwigshafen (Rhine), DT), United States.

16 Suo, X., Zhang, H., Ye, Q. et al. (2015). Design and control of an improved acrylic acid process. *Chemical Engineering Research and Design* **104**: 346–356.

17 Alhanash, A., Kozhevnikova, E.F., and Kozhevnikov, I.V. (2010). *Applied Catalysis A* 378: 11–18.

18 Shen, L., Yin, H., Wang, A. et al. (2014). Liquid phase catalytic dehydration of glycerol to acrolein over Brønsted acidic ionic liquid catalysts. *Journal of Industrial and Engineering Chemistry* **20** (3): 759–766.

19 Shen, L., Yin, H., Wang, A. et al. (2012). Liquid phase dehydration of glycerol to acrolein catalyzed by silicotungstic, phosphotungstic, and phosphomolybdic acids. *Chemical Engineering Journal* **180**: 277–283.

20 Possato, L.G., Cassinelli, W.H., Meyer, C.I. et al. (2017). *Applied Catalysis A* 532: 1–11.

21 Yun, Y.S., Lee, K.R., Park, H. et al. (2015). *ACS Catalysis* 5: 82–94.

22 Chieregato, A., Soriano, M.D., Basile, F. et al. (2014). One-pot glycerol oxidehydration to acrylic acid on multifunctional catalysts: focus on the influence of the reaction parameters in respect to the catalytic performance. *Applied Catalysis B: Environmental* **150–151**: 37–46.

23 Thanasilp, S., Schwank, J.W., Meeyoo, V. et al. (2015). One-pot oxydehydration of glycerol to value-added compounds over metal-doped SiW/HZSM-5 catalysts: effect of metal type and loading. *Chemical Engineering Journal* **275**: 113–124.

24 Paula, A.S., Possato, L.G., Ratero, D.R. et al. (2016). *Microporous and Mesoporous Materials* 232: 151–160.

25 Rasteiro, L.F., Vieira, L.H., Possato, L.G. et al. (2017). Hydrothermal synthesis of Mo-V mixed oxides possessing several crystalline phases and their performance in the catalytic oxydehydration of glycerol to acrylic acid. *Catalysis Today* **296**: 10–18.

# 19

# Sustainability, Commercialization, and Future Prospects of Biodiesel Production

*Pothiappan Vairaprakash*[1] *and Arumugam Arumugam*[2]

[1] *Department of Chemistry, School of Chemical and Biotechnology, Center for Bioenergy, SASTRA Deemed to Be University, Thanjavur, India*
[2] *Department of Chemical Engineering, School of Chemical and Biotechnology, Center for Bioenergy, SASTRA Deemed to Be University, Thanjavur, India*

## 19.1 Introduction

Sustainability means the ability to sustain forever. In the biosphere, living systems prepare various bio-macromolecules and building blocks utilizing elements (C, H, N, O, Ca, and P) obtained from the lithosphere and dispose of these elements back into the lithosphere, following the law of conservation of mass. These nature-driven processes undergo forever and have been sustained over a millennium. In the battleground of these elements, the carbon cycle and the oxygen cycle are linked through the compound $CO_2$. Change in the amount of $CO_2$ would reshape the atmosphere [1–3]. While the Earth condensed from a ball of gas around 5 billion years ago, the Earth's atmosphere was consisting mainly of nitrogen, $CO_2$, and water vapor. Then the first life evolved on the Earth's primordial atmosphere. Later, about 2.5 billion years ago, the plants evolved as photoautotrophs and developed the ability to create glucose from $CO_2$ and water in the presence of sunlight. As plants consumed $CO_2$, the $CO_2$ level in the atmosphere reduced to 260–280 ppm. After the evolution of herbivores around 20 million years ago, the $CO_2$ level remained around 260–280 ppm in the atmosphere, till the industrial revolution lasted from the middle of the eighteenth century [4]. Since the 1980s, the $CO_2$ level increases by 20 ppm every decade, and by 2019, the $CO_2$ level in our atmosphere was ~ 410 ppm [5].

The increase in the $CO_2$ level is due to the overexploitation of carbon sources present in the lithosphere. These carbon sources include forest-derived biomass and fossil fuels. The forests are live carbon sinks and convert the $CO_2$ into carbon-based biomasses in the process of photosynthesis. While burning the forest to clear land for other activities, the live carbon sink is being destroyed, and the stored carbon is released as $CO_2$ to the atmosphere. This is an example of profit-targeted anthropological activities, leading to the increase in $CO_2$ levels. Another prominent cause for the increase in the $CO_2$ level is the burning of

*Biodiesel Production: Feedstocks, Catalysts, and Technologies*, First Edition. Edited by Samuel Lalthazuala Rokhum, Gopinath Halder, Suttichai Assabumrungrat, and Kanokwan Ngaosuwan.

fossil fuels. The forest burnt or cut for an economic value can be replenished over a while by afforestation activities. But the fossil fuel burnt is always burnt. It cannot be replenished as fossil fuels are formed in an anaerobic process that took place over a millennium [6]. For the development of society and life, humankind is completely dependent on these fossil resources. These fossil resources, which include oil, gas, and coal, are nonrenewable resources. Any development built with the base of fossil resources is not a sustainable development. To proceed toward sustainability, there is a long history in the development of biodiesel as a sustainable energy carrier.

## 19.2 Biodiesel as a Promising Renewable Energy Carrier

Under the circumstance of the rapid depletion of fossil fuels, biodiesel has emerged as a renewable energy carrier [7, 8]. Biodiesel is a general term used for all the transesterified products obtained from plant/animal oil. The history of biodiesel started along with the invention of the diesel engine by Rudolph Diesel in the 1890s [9]. But the popularity of biodiesel increased during 2005; it is the period the world started thinking about the scarcity of oil resources (Chart 19.1).

During the initial period of the development of the internal combustion (IC) engine, Diesel himself carried out few experiments on vegetable oil fuels to explore the possibility of using the diesel engine for agriculture in remote areas having limited access to

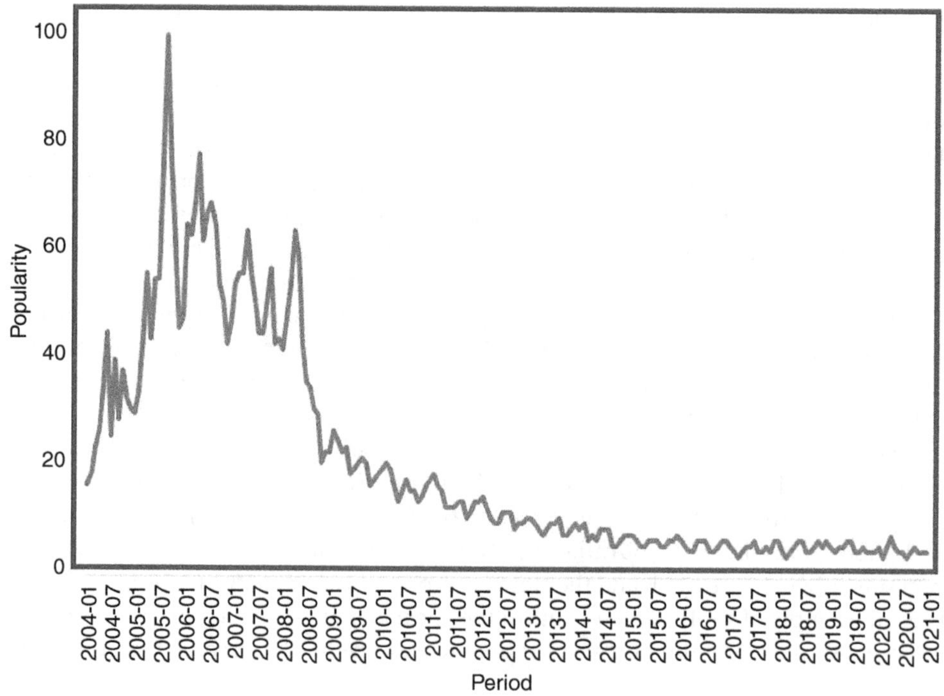


**Chart 19.1** Popularity of biodiesel (Google trends).

petroleum fuel. There were many continuous efforts by Rudolph Diesel to use vegetable oil in the early diesel engines [9]. After Rudolph Diesel, the research on the use of vegetable oil fuels in the IC engines did not attract much attention due to the mass availability of petroleum at less price. The focus of development was toward increasing the efficiencies of IC engines rather than using vegetable oil as a fuel in IC engines. The periods that attained more interest in using vegetable oil as fuel to IC engines were the times of high oil prices and scarcity of oil. Another problem associated with vegetable oil is its high viscosity (Figure 19.1). The IC engines failed to work with the high viscous vegetable oils [10, 11].

Blending with diesel [12–15] and emulsification [16–19] was the different attempts made to reduce the viscosity of the traditional vegetable oils to make IC engine compatible biodiesel. Due to the failures in these attempts and the low cost of fossil fuels, there was little attention paid toward vegetable oil-based alternate fuels. The first success in the reduction of viscosity was reported in a Belgian patent 422 877 filed by G. Chavanne [20] in the year 1937 via the transesterification process (Figure 19.2). The transesterified products obtained from various plant and animal oils are called biodiesel in general. Transesterification is a process of conversion of one ester into another one. In the transesterification of vegetable/animal oil, the methyl ester of fatty acid was obtained from the triglycerides, the glycerol ester of fatty acids. The methyl ester of fatty acid is a smaller molecule, having less viscosity than the oil and can be easily burnt in IC engines.

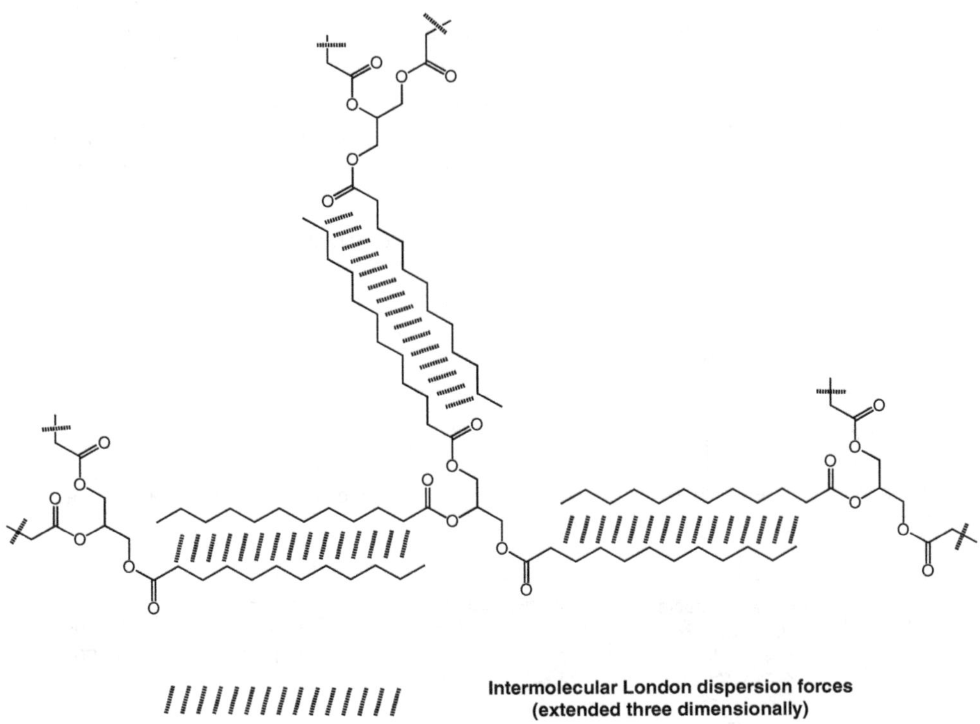


**Figure 19.1** Viscosity implied by intermolecular forces in vegetable oil.

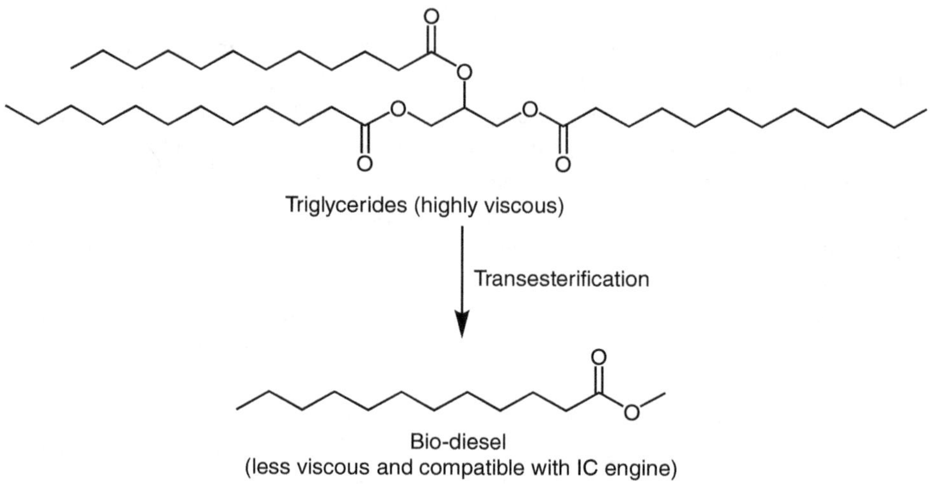


**Figure 19.2** Representative process in the transesterification of vegetable/animal oil.

## 19.3 Overview of the Biodiesel Production Process

Various catalysts have been developed for the effective transesterification of vegetable/animal oils [21–24]. These catalysts include chemical and biological catalysts (Figure 19.3). In the catalyst development, several aspects of sustainability have been introduced, which include the use of biological catalyst, use of simple inorganic chemicals as catalysts in the transesterification process, immobilizing the catalyst on inert support for the easy recovery of the catalysts after the completion of the transesterification process, etc. Lipase is the biological catalyst used in the transesterification process [25–31]. In the biodiesel preparation mediated by the lipase, both the free lipase and immobilized lipase have been utilized.

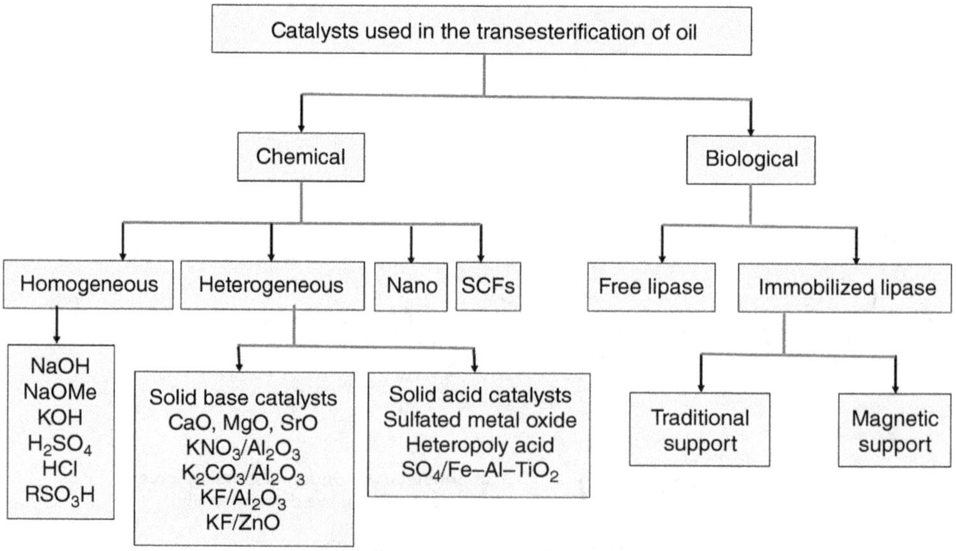


**Figure 19.3** Catalysts developed for the transesterification.

In the immobilization of lipase, both the traditional and magnetic support have been explored. The magnetic support had provided an additional advantage of the easy recovery of catalyst from the produced biodiesel [32]. In chemical catalysts, various types of catalysts have been developed by research groups around the globe. These chemical catalysts can be categorized as (i) acidic catalyst and basic catalyst, (ii) homogeneous catalyst and heterogeneous catalysts, and (iii) nanocatalysts. Supercritical fluids (SCFs) have been used as an effective solvent system in the transesterification of triglycerides into biodiesel under catalyst-free conditions [33–38].

## 19.4 Evolution in the Feedstocks Used for the Sustainable Production of Biodiesel

In the initial attempt of biodiesel production using the transesterification process, the main focus was to replace the fossil fuel in the IC engines. Hence, various edible oils were used in the initial development. This edible oil-based biodiesel was competitive in the food chain; hence, it could not replace fossil fuel completely. To bring sustainability in the production of biodiesel, nonedible oils were used in the transesterification processes [39]. Recently, vegetable oil, animal fat, algae, and microbial oil are being used as feedstock (Figure 19.4). The selection is based on the availability, regions, and economic aspects of the country.

## 19.5 First-Generation Biodiesel and the Challenges in Its Sustainability

The biodiesel produced from the edible oils belonged to the category first-generation biodiesel. The commonly used edible vegetable oils are rapeseed oil (*Brassica napus* L.) [40–43], sunflower oil (*Helianthus annuus*) [44–46], soybean oil (*Glycine*

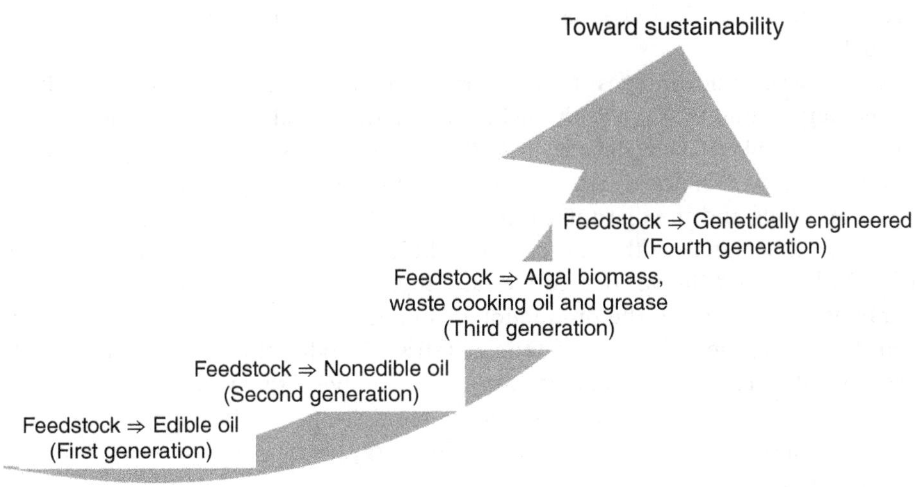


**Figure 19.4** Generational evolution in biodiesel feedstock towards sustainability.

*max*) [47–50], peanut oil (*Archis hypogaea*) [51–53], palm oil, and palm kernel oil (*Elaeis guineensis*) [54–56]. Potential edible oils used in the production of biodiesel are rice bran oil (*Oryza sativum*) [36, 57–59], coconut oil (*Cocos nucifera*) [11, 60], sesame oil (*Sesamum indicum* L.) [61, 62], cottonseed oil (*Gossypium hirsutum*) [63–66], moringa oil (*Moringa oleifera*) [67–69], and corn oil (*Zea mays*) [70, 71]. Commodities having comparable time scales of production and consumption are considered to be sustainable. The formation of fossil resources took millions of years, and fossil fuels are being consumed at a much faster rate. Hence, fossil fuels are not sustainable resources. In this regard, the first-generation biodiesel was considered sustainable. The edible oils used in the production of first-generation biodiesel are obtained from the crops belonging into three categories; annual crops, biennial crops, and perennial crops. The period of annual crops lies between few months to one year. The period of biennial crops lies up to two years; during the first planting season, the plants will germinate, grow, and survive, and the perennial crops need two planting seasons and sometimes more than that to complete the life cycle. Unlike fossil fuels, first-generation biodiesel has a comparable timescale of production and consumption. Hence they can be considered as a sustainable and renewable energy carriers.

The first-generation biodiesel had a very low sulfur content compared with petroleum products [72]. Hence, the first-generation biodiesel caused much low pollution and other deteriorating effects to the environment. The first-generation biodiesel can undergo decomposition naturally as they have unsaturation. In terms of unsaturation, the edible feedstock can be classified as high monounsaturation oils, oils with high poly unsaturation values and oils with high saturation values. Among these, the oils with high monounsaturation in fatty acids are the best feedstock for biodiesel production. Availability of crops and easy conversion process are the greatest advantages of first-generation feedstocks. In the production of biodiesel, fossil energy might be utilized in various processes including the cultivation of crops, harvesting, and the conversion of the crop into biodiesel. Considering the total energy input and the energy released while burning, the first-generation biodiesel production results in the positive net energy balance (NEB) indicating that the biodiesel energy content exceeds the fossil energy input [73]. Hence the biodiesel with the positive NEB value would reduce the $CO_2$ emission to the atmosphere, in comparison with the use of fossil fuel.

Despite the several advantages, first-generation biodiesel is also associated with a few disadvantages (Figure 19.5). With the increase in the global population, the demand for food might double in a few decades [74, 75]. The feedstocks used in the preparation of the first-generation biodiesel are competitive to the food crops. Limitations in the production of crops and the disruption in the food chain and food web are the major disadvantages associated with the use of edible oil for the production of biodiesel [76–80]. Among the various edible oils, soybean oil and corn oil are the most commonly used edible oil in the production of biodiesel. Hill et al. suggested in their study on the environmental and economic benefits of biodiesel that even if the whole of the corn yield and soybean yield is converted into biodiesel that would hardly replace only 9 and 11% of the diesel and gasoline demand, respectively [73]. Hence, first-generation biodiesel was considered a renewable energy carrier, but not the sustainable one (Figure 19.5).

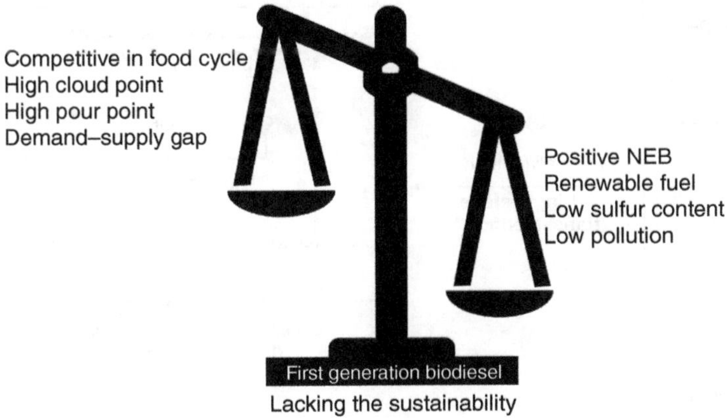


**Figure 19.5** Pros and cons of first-generation biodiesel.

## 19.6 Development of Second-Generation Biodiesel to Address the Sustainability

In the plant kingdom, the plants lacking a self-immune system produce secondary metabolites to protect themselves from invading organisms and herbivores [81]. The oils containing a large amount of these secondary metabolites are nonedible oils. These nonedible oils were identified to rectify the drawbacks of edible oil usage in biodiesel production. The biodiesel produced from the nonedible feedstock is termed second-generation biodiesel [82–85]. The commonly used nonedible oils are *Jatropha curcas* [86, 87], neem (*Azadirachta indica*) [88, 89], jojoba oil (*Simmondsia chinensis*) [22, 90, 91], pongamia oil (*Pongamia pinnata*) [92, 93], and mahua oil (*Madhuca indica*) [94–96], and the potential nonedible oils were castor oil (*Ricinus communis*) [97, 98], pinnai oil (*Calophyllum inophyllum*) [26, 27, 99, 100], algae oil (*Cyanobacteria*) [73, 101–103], tall oil (*Carnegiea gigantea*) [104–106], and coffee ground (*Coffea arabica*) [107–109]. Nonedible oils are considered to be more potent than edible oil and animal fats. The use of nonedible oils does not interfere with the global food demand. The cultivation of nonedible oils does not require quality land for its sustained growth. They can be cultivated in poor areas and wastelands with minimum efforts and low agricultural inputs including fertilizers, pesticides, and energy. In general, crops producing the nonedible oil for biodiesel are disease resistant and pest resistant. The other main advantages of nonedible oils include renewability, liquid nature, lower sulfur content, higher heat content, biodegradability, lower aromatic content, and ready availability [110].

Due to the presence of secondary metabolites and the free fatty acids (FFAs) in large amounts, the transesterification of nonedible oils requires a little harsh condition and a large amount of alcohol [111]. The presence of FFAs in large amounts leads to the formation of a large amount of soap in the alkali-mediated transesterification process. The layer separation between the glycerol and biodiesel is disrupted by the presence of soap. To address this difficulty caused by the formed soap, transesterification of nonedible oils is carried out in two stages. In the first stage, the FFAs are converted into the corresponding methyl ester via sulfuric acid-mediated esterification. The second stage is the conversion of triglycerides into biodiesel via alkali mediated transesterification in methanol (Figure 19.6). In the double-stage

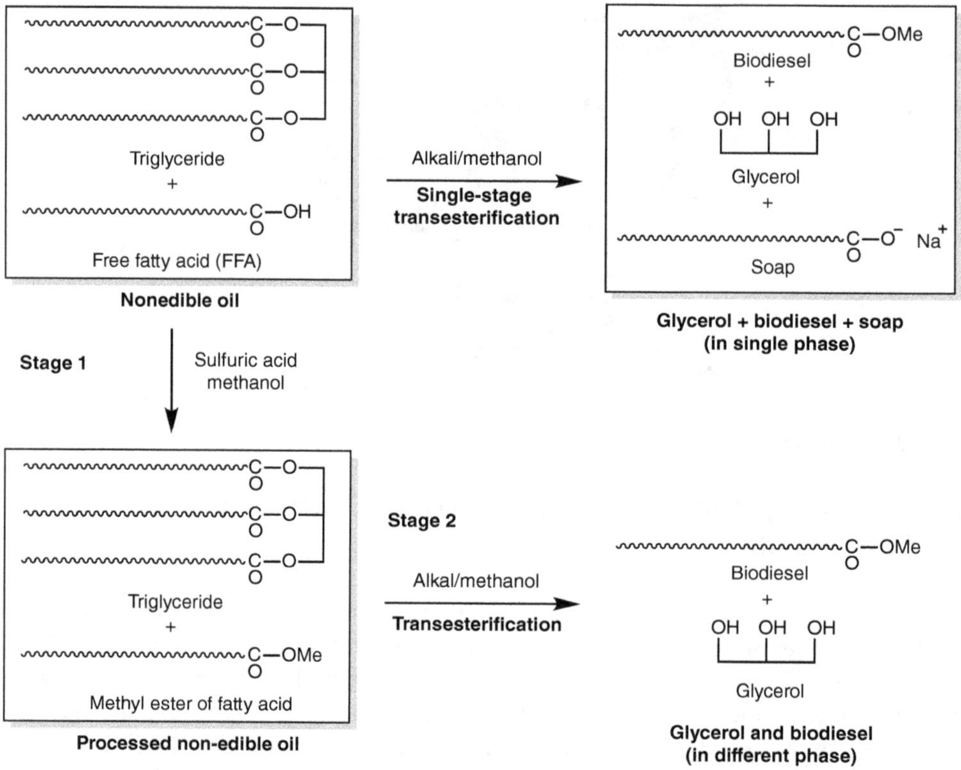


**Figure 19.6** Single-stage vs double-stage transesterification of nonedible oil.

conversion of nonedible oil into biodiesel, the optimum methanol-to-oil ratio is in the range from 8 : 1 to 30 : 1 at different experimental conditions [112, 113]. Despite these drawbacks, nonedible oils are the best sustainable energy carriers among the others. The other resources used in the preparation of biodiesel are algae oil and waste cooking oil.

## 19.7 Algae-Based Biodiesel

The biodiesel produced using microalgae as feedstock is termed third-generation biodiesel [39, 114–116]. The algae is the aquatic plant having one cell, which consists of more lipids. As the algae contain a large amount of lipids, they are the most promising resources for the oil, which can be transesterified to make biodiesel [101–103, 114]. For example, *Chlorella protothecoides* [117–119], *Chlorella zofingiensis* [120–122], and *Schizochytrium limacinum* [123–125] were good sources having more than 50% oil of their dry weight. The algae are cultivated in either an open pond system or a closed pond system [126, 127] and harvested by centrifugation techniques or by flocculation techniques [128–130]. The energy input in the cultivation processes of algae is lower than the energy input of edible and nonedible oil cultivation. The main nutrients needed for the growth of algae is obtained from the atmospheric carbon dioxide and nitrogen oxides. Hence, third-generation biodiesel produces a very low amount of greenhouse gases [131–134]. Algae are containing FFA in traces, but upon storage of the algal biomass, the FFA content increases from traces to a large quantity by the internal enzymatic hydrolysis of triglycerides. The algal biomass containing a large amount of FFAs was effectively utilized in biodiesel production using a two-step catalytic conversion [102]. Also the large-scale production and the commercialization of third-generation biodiesel is throttled by the need for large investment for algae cultivation and the difficulties in the extraction of oil. Despite these difficulties, low sulfur content, water content, saponification value, phosphorus content, lipid content, wide productivity, and low affordable price make the algae-derived biodiesel a sustainable one [135, 136].

## 19.8 Waste Oils, Grease, and Animal Fats in Biodiesel Production

In continuation to the usage of algae oil for the production of biodiesel, researchers attempted to utilize waste oil and grease in the production of biodiesel [137–140]. The biodiesel produced by using the waste oils as the feedstock is also termed as the third-generation biodiesel. Usage of the waste oil and grease increases the sustainability by multifold, as it facilitates the development of renewable energy carriers along with waste management. Thus, it benefits environmental protection by reducing greenhouse gas emissions and waste management [141]. The common sources of waste oil are the food industry [142], households, and restaurants [139, 143]. These waste oils were obtained while using soybean, rapeseed, coconut, palm oil, and other edible oils. Waste oil samples obtained from several sources have been used in the production of biodiesel.

The oil feedstocks extracted from plants find the foremost application in cooking and the food industry. During the usage, quite a lot amount of the plant-derived oil is getting wasted and mixed into the waste water lines. This results in the accumulation of greasy materials in the municipal wastewater treatment plants. These greases contain a large amount of FFA content. Waste grease containing <15% FFA content is termed yellow grease. The other type of waste grease is brown grease containing >15% FFA content. Both yellow grease [144] and brown grease [145] contain a larger lipid content. Hence, they can be used in the production of biodiesel. The conversion of waste grease to biodiesel is carried out in two stages: (i) the pretreatment process and (ii) the traditional alkaline catalytic process.

In addition to the waste cooking oil and waste greases, animal fats are also used as feedstocks in the production of biodiesel [146–150]. The animal fat is sourced from chicken, cattle, hogs, and fish. These animal fats provide several advantages like stability to oxidation, shorter delay in the ignition, and a higher calorific value. The major problem associated with the use of animal fat in the production of biodiesel is due to the physical state of the animal fat. Animal fat remains solid or congealed at room temperature [147, 151]. Also animal fat-derived biodiesel has a high cloud point and causes problems while using in very cold conditions.

## 19.9 Technical Impact by the Biodiesel Usage

In the previous sections, a discussion on the various feedstocks utilized in the production of biodiesel is provided. In this section, the technical impact of biodiesel usage will be discussed. To reduce the negative impact on the environment by the intense use of fossil fuels, various alternate fuels have been suggested and developed [152–154]. In the development of alternate fuels, many of the proposed alternate fuels could not compete with and replace conventional fossil fuels due to their inability to meet the technical needs of an IC engine. These technical needs include energy content, burning rate, energy-releasing rate, etc. Among the various alternate fuels developed, very few fuels including biodiesel [79] and 5-hydroxymethylfurfural-derived 2,5-dimethylfuran [155] meet several of the technical needs required to use them in the IC engines. Biodiesel can be used alone in the IC engines by making a few minor modifications in the engines [156]. Else, biodiesel can be mixed with petroleum fuels in a wide ratio and can be applied to the engines. The use of biodiesel/petroleum fuel mixture does not require any modification in the IC engines [157, 158].

## 19.10 Socioeconomic Impacts

The economic feasibility of feedstock production is also the main factor to be considered in attaining sustainability in the energy sector. In the technical needs, biodiesel competes well with conventional fuels. In the production and processing of conventional fuels, processing needs a large energy input, while production does not need it, whereas, in biodiesel production, both the production and processing need a large energy input. Hence, the overall production cost is higher for biodiesel. This is a barrier to the commercial development of biodiesel. The cost of production of biodiesel keeps changing depending on the cost of the feedstock and the scale of operation. The feedstock cost can be reduced by utilizing the feedstocks such as waste cooking oils, grease, etc. [137, 159]. But biodiesel production is limited by the availability of these waste materials as feedstocks. The processing cost can be reduced by selecting an appropriate method of production. Microbial biodiesel production gives a higher yield and reduces the processing cost [31]. Alkali-mediated processes are less expensive compared to enzyme-catalyzed biodiesel production [63]. In the enzyme-catalyzed processes, very expensive enzyme catalysts were used, and the process is slower than the alkali-mediated biodiesel production. Recently immobilized catalysts were utilized in biodiesel production. The usage of immobilized catalysts facilitates the easy and efficient recovery of the catalysts. So the cost of solid acid catalyst-mediated biodiesel production is much lower. In the production of biodiesel, glycerol is obtained as a by-product. In most of the biodiesel production process, the obtained glycerol is highly impure as all the impurities formed during biodiesel production stays with the glycerol by its viscosity. The development of value-added products from crude glycerol will result in the best cost efficiency in the production of biodiesel.

## 19.11 Toxicological Impact

Depending on the feedstocks used, three generations of biodiesels are being developed. In the production of first-generation biodiesel, edible oils were used as feedstocks. Nonedible oils were used as feedstocks in the production of second-generation biodiesel. In the first- and second-generation biodiesel, plant-derived oils were used as feedstocks. Hence, they do not impart any negative impacts on human health and the environment. The algae are the potential source for the third-generation biodiesel. The undefined chemical composition of biodiesel from different feedstock can be a barrier to find its toxic effects; therefore, the toxicity assessment has been made by chemical and biochemical analysis. Biodiesel is a healthier and safer substitute for petroleum fuels, but biodiesel and its blend too cause negative toxicological impacts like carcinogenicity, phenotypic, and genotypic anomalies in living organisms through its mutagenic and genotoxic pollutants [160, 161]. The mutagenicity in biodiesel has been explored by a bioassay on Salmonella species by exposing different particulate matter extracts of low sulfur petroleum diesel and biodiesel [162]. It has shown no difference between biodiesel and petroleum fuel. According to the tests that have been made on the effects of soluble methanolic and ethanolic fractions of biodiesel, it has been found that these fractions imparted changes in the reactive oxygen species (ROS), disturbance in the DNA, and the activation of glutathione S-transferase (GST) in the cells, which behaves as the stressors for the cells.

## 19.12 Sustainability Challenges in the Biodiesel Production and Use

All the sustainable energy sources are renewable. The vice versa is not true. Various types of renewable feedstocks have been utilized in the preparation of biodiesel. The development of various feedstocks is mainly aimed at introducing sustainability in renewable feedstocks. It is evident that the first-generation biodiesel is a renewable energy carrier, but it is not a sustainable one due to its interruption in the food chain and food web. Also the sustainability of biodiesel is compromised in terms of land use changes as the land area is limited, and increasing the use of biofuels will lead to more competition. In the cultivation of biodiesel feedstocks, substituting the fossil-derived energy inputs with the biodiesel-based energy inputs and using organic fertilizers favors sustainability.

The main aspects of sustainability in biodiesel are to ensure energy security, increase the economic viability toward easy access for domestic use, mitigating the exploitation of natural resources, low greenhouse gas emission, and reduce water usage for cultivation. The assessment of environmental sustainability and risk analysis is measured using a tool named life cycle analysis (LCA). Based on LCA, biodiesel is a sustainable one, since economic and social sustainability has not been included in LCA [134, 163]. Efforts were made to attain sustainability via improvising the production process. In terms of the production process, the catalyst plays a vital role. Among the several catalysts explored, solid acid catalysts offer more advantages, because the presence of water and FFA does not affect the reactions mediated by the solid acid catalysts. Solid acid catalysts are capable of carrying out the esterification of FFAs and the transesterification at the same time and hence reduces the production cost. The environmentally greener enzyme catalysis is also preferred to improvise the biodiesel production process. But the enzyme catalysts are costlier than the solid acid catalysts. Efforts were made to increase sustainability by using supercritical methanol as a solvent in the transesterification process. The use of supercritical methanol can tolerate with high FFA and water content [33]. In this process, pretreatment of the feedstocks can be avoided.

In terms of emission characteristics, biodiesel is resulting in less environmental impact than petroleum diesel. In comparison with petroleum fuel, biodiesel shows less particulate matter emission (53–69%), CO emission (4–16%), and hydrocarbons (45–67%). Biodiesel is obtained from plant-derived oils, which contain more nitrogen compounds; a slight increase of 10–23% in the emission of nitric oxide is observed [164]. These variations in the emission of various compounds from the burning of biodiesel are the result of its oxygen content and viscosity.

Considering the sustainability in the feedstocks, the edible oil feedstocks are not sustainable as they compete with the food crops in the usage of water and natural resources. Hence the best alternate for edible oil feedstock is nonedible oils. Among the various nonedible oils, the *J. curcas* oil gained more attention as it is a drought-resistant plant that can withstand and survive in abandoned and fallowed agricultural land [86]. This will gain more income as it does not exploit the farming land. Algae-based biodiesel has gained more interest because it has a higher yield per hectare than other oil crops, its nonedible oil production, and does not compete with the food crops farmlands. But the main barrier to the usage of algae-based biodiesel is its high production cost, the need for high-yielding algae strains, and large bioreactor requirement as it can only grow in the bioreactor and farm. Waste cooking oil and greases are affordable and demonstrated as effective feedstocks in the production of biodiesel. But the collection of the waste cooking oil and grease might be difficult as they are scattered in different areas. For the production of the fourth-generation biodiesel [165, 166], the genetically engineered plants noncompetitive to food crops are being developed, for example, miscanthus, poplar, and switchgrass.

## 19.13 Concluding Remarks

In the energy-dependent and technologically driven world, various forms of energy carriers have been developed and utilized. The scarcity of fossil resources directed the energy research toward the development of sustainable energy carriers to be used in IC engines. In the race of sustainable energy carriers, plant-derived energy carriers are the prospective candidates. Among the plant-derived fuels, triglycerides are known forever to humankind and utilized in the production of heat energy and light energy, across the human race in all parts of the world from the poles to the equators. The use of triglycerides in IC engines is limited by their inherent viscosity. This issue has been addressed by the transesterification of the triglycerides to the corresponding methyl ester, which is known as biodiesel. To attain sustainability in the production and use of biodiesel, different forms of feedstocks have been developed and utilized in the production of biodiesel. Many countries have adopted the policy of using biodiesel to reduce greenhouse gas emissions. The incentives adopted by the government also play an important role in moving toward the sustainable use of biodiesel. These incentives are provided for the crop plantation in abandoned and fallowed agricultural lands, water management improvement, conservation practices, etc.

## References

1 Beerling, D.J. (2005). Evolutionary responses of land plants to atmospheric $CO_2$. In: *A History of Atmospheric $CO_2$ and Its Effects on Plants, Animals, and Ecosystems* (ed. J.R. Ehleringer, T. Cerling and M.D. Dearing), 114–132. Springer-Verlag.

2 Ghosh, P., Bhattacharya, S.K. and Ghosh, P. (2005). Atmospheric $CO_2$ during the late paleozoic and mesozoic: estimates from Indian soils. In: *A History of Atmospheric $CO_2$ and Its Effects on Plants, Animals, and Ecosystems* (ed. J.R. Ehleringer, T. Cerling and M.D. Dearing), 8–34. Springer-Verlag.

3 Berner, R.A. (2005). The rise of trees and how they changed paleozoic atmospheric $CO_2$, climate, and geology. In: *A History of Atmospheric $CO_2$ and Its Effects on Plants, Animals, and Ecosystems* (ed. J.R. Ehleringer, T. Cerling and M.D. Dearing), 1–7. Springer-Verlag.

4 Etheridge, D.M., Steele, L.P., Langenfelds, R.L. et al. (2021). Historical $CO_2$ records from the Law Dome DE08, DE08-2, and DSS ice cores [Internet]. https://cdiac.ess-dive.lbl.gov/trends/co2/lawdome.html.

5 Rambhiya, S.J., Magar, C.S. and Deodhar, M.A. (2021). Using seawater-based $Na_2CO_3$ medium for scrubbing the $CO_2$ released from Bio-CNG plant for enhanced biomass production of *Pseudanabaena limnetica*. *SN Appl Sci.* 3 (2): 1–17.

6 Feulner, G. (2017). Formation of most of our coal brought Earth close to global glaciation. *Proc Natl Acad Sci* 114 (43): 11333–11337.

7 Skevis, G. (2016). Liquid biofuels: bioalcohols, biodiesel and biogasoline and algal biofuels. In: *Handbook of Combustion* (ed. M. Lackner, F. Winter, A.K. Agarwal and J.S. Lighty), 1–43. Wiley-VCH Verlag GmbH & Co. KGaA.

8 Jansen, R.A. (2013). First- and second-generation biofuels. In: *Second Generation Biofuels and Biomass* (ed. R.A. Jansen), 21–29. Wiley-VCH Verlag GmbH & Co. KGaA.

9 Knothe, G. (2010). History of vegetable oil-based diesel fuels. In: *The Biodiesel Handbook*, 2e, 5–19. Elsevier Inc.

10 Ribeiro, N., Pinto, A.C., Quintella, C.M. et al. (2007). The role of additives for diesel and diesel blended (ethanol or biodiesel) fuels: a review. *Energy Fuels* 21: 2433–2445.

11 Salmani, M.H., Rehman, S., Zaidi, K. and Hasan, A.K. (2015). Study of ignition characteristics of microemulsion of coconut oil under off diesel engine conditions. *Eng Sci Technol Int J.* 18 (3): 318–324.

12 De, B. and Panua, R.S. (2014). An experimental study on performance and emission characteristics of vegetable oil blends with diesel in a direct injection variable compression ignition engine. *Procedia Eng.* 90: 431–438.

13 Murthi, M.K., Velliyangiri, B., Sengottaiyan, M. and Eswaran, S. (2020). Effect of additives on the performance characteristics of VCR engine fuelled with Karanja oil and blends by varying injection pressure. *Int J Eng Trends Technol.* 68 (2): 134–141.

14 Rajesh, K.V. and Mahla, S.K. (2020). Influence of EGR on a CI engine running on 20% blend of jatropha biodiesel. *Mater Today: Proc.* 43: 273–280.

15 Ravi, S., Karthikeyan, A. and Jayaprabakar, J. (2020). Experimental study of oxygenated additive in diesel-waste plastic oil-propanol blend operated in a single-cylinder diesel engine. *Dig J Nanomater Biostructures.* 15 (3): 757–767.

16 Melo-Espinosa, E.A., Piloto-Rodríguez, R., Goyos-Pérez, L. et al. (2015). Emulsification of animal fats and vegetable oils for their use as a diesel engine fuel: an overview. *Renew Sustain Energy Rev.* 47: 623–633.

17 Najjar, R., Zarei-Gharehbaba, L. and Tazerout, M. (2021). The exhaust emission characteristics of a water-emulsified diesel/sunflower oil blend fuel by ionic liquid surfactants. *Int J Environ Sci Technol.* DOI: 10.1007/s13762-021-03191-8.

**18** Kumar, H., Sarma, A.K. and Kumar, P. (2020). A comprehensive review on preparation, characterization, and combustion characteristics of microemulsion based hybrid biofuels. *Renew Sustai Energy Rev.* 117: 109498.

**19** Dimawarnita, F., Kartika, I.A. and Hambali, E. (2021). Sustainability of biodiesel B30, B40, and B50 in Indonesia with addition of emulsifier. *IOP Conf Ser Earth Environ Sci.* 749: 012026.

**20** Maroa, S. and Inambao, F. (2020). Introduction. In: *Green Energy and Technology* (ed. A. Doyle), 1–6. Springer Science and Business Media Deutschland GmbH.

**21** Helwani, Z., Othman, M.R., Aziz, N. et al. (2009). Technologies for production of biodiesel focusing on green catalytic techniques: a review. *Fuel Process Technol.* 90 (12): 1502–1514.

**22** Changmai, B., Vanlalveni, C., Ingle, A.P. et al. (2020). Widely used catalysts in biodiesel production: a review. *RSC Adv* 10: 41625–41679.

**23** Lin, X., Li, M., Chen, Z. et al. (2021). One-step fabrication of polymeric self-solidifying ionic liquids as the efficient catalysts for biodiesel production. *J Clean Prod.* 292: 125967.

**24** Bian, Y., Shan, Q., Guo, C. et al. (2021). Biodiesel production over esterification catalyzed by a novel poly(acidic ionic liquid)s. *Catal Lett.* 151: 3523–3531.

**25** Shah, S., Sharma, S. and Gupta, M.N. (2004). Biodiesel preparation by lipase-catalyzed transesterification of *Jatropha Oil*. *Energy Fuels.* 18 (1): 154–159.

**26** Arumugam, A. and Ponnusami, V. (2013). Enzymatic transesterification of *Calophyllum inophyllum* oil by lipase immobilized on functionalized SBA-15 synthesized from low-cost precursor. *Biomass Convers Biorefin.* 4 (1): 35–41.

**27** Arumugam, A. and Ponnusami, V. (2014). Biodiesel production from *Calophyllum inophyllum* oil using lipase producing *Rhizopus oryzae* cells immobilized within reticulated foams. *Renew Energy.* 64: 276–282.

**28** Amini, Z., Ong, H.C., Harrison, M.D. et al. (2017). Biodiesel production by lipase-catalyzed transesterification of *Ocimum basilicum* L. (sweet basil) seed oil. *Energy Convers Manag.* 132: 82–90.

**29** Okoro, O.V., Sun, Z. and Birch, J. (2018). Lipases for biofuel production. In: *Encyclopedia of Food Chemistry* (ed. L. Melton, F. Shahidi and P. Varelis), 150–157. Elsevier.

**30** Arumugam, A. and Ponnusami, V. (2017). Production of biodiesel by enzymatic transesterification of waste sardine oil and evaluation of its engine performance. *Heliyon.* 3 (12): e00486.

**31** Bhan, C. and Singh, J. (2020). Role of microbial lipases in transesterification process for biodiesel production. *Environ Sustain.* 3 (3): 257–266.

**32** Cao, X., Xu, H., Li, F. et al. (2021). One-step direct transesterification of wet yeast for biodiesel production catalyzed by magnetic nanoparticle-immobilized lipase. *Renew Energy.* 171: 11–21.

**33** Demirbas, A. (2006). Biodiesel production via non-catalytic SCF method and biodiesel fuel characteristics. *Energy Convers Manag.* 47 (15–16): 2271–2282.

**34** Tan, K.T. and Lee, K.T. (2011). A review on supercritical fluids (SCF) technology in sustainable biodiesel production: potential and challenges. *Renew Sustain Energy Rev.* 15: 2452–2456.

**35** Zhang, J., Liu, J., Huang, X. et al. (2020). Heterogeneous catalytic esterification of oleic acid under sub/supercritical methanol over $\gamma$-$Al_2O_3$. *Fuel.* 268: 117359.

**36** Akkarawatkhoosith, N., Kaewchada, A. and Jaree, A. (2020). Continuous catalyst-free biodiesel synthesis from rice bran oil fatty acid distillate in a microreactor. *Energy Rep.* 6: 545–549.

37 Omar, S., Yang, Y. and Wang, J. (2021). A review on catalytic & non-catalytic bio-oil upgrading in supercritical fluids. *Front Chem Sci Eng.* 15: 4–17.

38 Qadeer, M.U., Ayoub, M., Komiyama, M. et al. (2021). Review of biodiesel synthesis technologies, current trends, yield influencing factors and economical analysis of supercritical process. *J Clean Prod.* 309: 127388.

39 Alalwan, H.A., Alminshid, A.H. and HAS, A. (2019). Promising evolution of biofuel generations. Subject review. *Renew Energy Focus.* 28: 127–139.

40 Peterson, C.L. and Hustrulid, T. (1998). Carbon cycle for rapeseed oil biodiesel fuels. *Biomass Bioenergy.* 14 (2): 91–101.

41 Kusdiana, D. and Saka, S. (2001). Kinetics of transesterification in rapeseed oil to biodiesel fuel as treated in supercritical methanol. *Fuel.* 80 (5): 693–698.

42 Ma, H., Li, S., Wang, B. et al. (2008). Transesterification of rapeseed oil for synthesizing biodiesel by K/KOH/$\gamma$-$Al_2O_3$ as heterogeneous base catalyst. *J Am Oil Chem Soc.* 85 (3): 263–270.

43 Konur, O. (2021). Rapeseed oil-based biodiesel fuels. In: *Biodiesel Fuels Based on Edible and Nonedible Feedstocks, Wastes, and Algae* (ed. O. Konur), 497–514. CRC Press.

44 Antolín, G., Tinaut, F.V., Briceo, Y. et al. (2002). Optimisation of biodiesel production by sunflower oil transesterification. *Bioresour Technol.* 83 (2): 111–114.

45 Demirbas, A. (2007). Biodiesel from sunflower oil in supercritical methanol with calcium oxide. *Energy Convers Manag.* 48 (3): 937–941.

46 Tutunea, D., Dumitru, I., Racila, L. et al. (2019). Characterization of sunflower oil biodiesel as alternative for diesel fuel. In: *Proceedings of the 4th International Congress of Automotive and Transport Engineering (AMMA 2018). AMMA2018 2018. Proceedings in Automotive Engineering* (ed. N. Burnete and B. Varga), 172–180. Cham: Springer.

47 Liu, X., He, H., Wang, Y. et al. (2008). Transesterification of soybean oil to biodiesel using CaO as a solid base catalyst. *Fuel.* 87 (2): 216–221.

48 Celante, D., Schenkel, J.V.D. and de Castilhos, F. (2018). Biodiesel production from soybean oil and dimethyl carbonate catalyzed by potassium methoxide. *Fuel.* 212: 101–107.

49 Elkelawy, M., Alm-Eldin Bastawissi, H., Esmaeil, K.K. et al. (2019). Experimental studies on the biodiesel production parameters optimization of sunflower and soybean oil mixture and DI engine combustion, performance, and emission analysis fueled with diesel/biodiesel blends. *Fuel.* 255: 115791.

50 Soares Dias, A.P., Catarino, M. and Gomes, J. (2021). Co-processing lard/soybean oil over Ca-based catalysts to greener biodiesel. *Environ Technol Innov.* 21: 101220.

51 Kaya, C., Hamamci, C., Baysal, A. et al. (2009). Methyl ester of peanut (*Arachis hypogea* L.) seed oil as a potential feedstock for biodiesel production. *Renew Energy.* 34 (5): 1257–1260.

52 Jazie, A.A., Pramanik, H. and Sinha, A.S.K. (2013). Transesterification of peanut and rapeseed oils using waste of animal bone as cost effective catalyst. *Mater Renew Sustain Energy.* 2 (2): 1–10.

53 Silveira Junior, E.G., Simionatto, E., Perez, V.H. et al. (2016). Potential of Virginia-type peanut (*Arachis hypogaea* L.) as feedstock for biodiesel production. *Ind Crops Prod.* 89: 448–454.

54 Rupilius, W. and Ahmad, S. (2007). Palm oil and palm kernel oil as raw materials for basic oleochemicals and biodiesel. *Eur J Lipid Sci Technol.* 109 (4): 433–439.

**55** Aladetuyi, A., Olatunji, G.A., Ogunniyi, D.S. et al. (2014). Production and characterization of biodiesel using palm kernel oil; fresh and recovered from spent bleaching earth. *Biofuel Res J.* 1 (4): 134–138.

**56** Ajala, E.O., Ajala, M.A., Okedere, O.B. et al. (2021). Synthesis of solid catalyst from natural calcite for biodiesel production: case study of palm kernel oil in an optimization study using definitive screening design. *Biofuels.* 12 (6): 703–714.

**57** Einloft, S., Magalhães, T.O., Donato, A. et al. (2008). Biodiesel from rice bran oil: transesterification by Tin compounds. *Energy Fuels.* 22 (1): 671–674.

**58** Zaidel, D.N.A., Muhamad, I.I., Daud, N.S.M. et al. (2019). Production of biodiesel from rice bran oil. In: *Biomass, Biopolymer-Based Materials, and Bioenergy: Construction, Biomedical, and Other Industrial Applications* (ed. D. Verma, E. Fortunati, S. Jain and X. Zhang), 409–447. Elsevier.

**59** Hoang, A.T., Tabatabaei, M., Aghbashlo, M. et al. (2021). Rice bran oil-based biodiesel as a promising renewable fuel alternative to petrodiesel: a review. *Renew Sustain Energy Rev.* 135: 110204.

**60** Kumar, D., Kumar, G., Poonam, and Singh, C.P. (2010). Fast, easy ethanolysis of coconut oil for biodiesel production assisted by ultrasonication. *Ultrason Sonochem.* 17 (3): 555–559.

**61** Ahmad, M., Khan, M.A., Zafar, M. and Sultana, S. (2010). Environment-friendly renewable energy from sesame biodiesel. *Energy Sources, Part A Recover Util Environ Eff.* 32 (2): 189–196.

**62** Betiku, E. and Adepoju, T.F. (2013). Methanolysis optimization of sesame (*Sesamum indicum*) oil to biodiesel and fuel quality characterization. *Int J Energy Environ Eng.* 4 (1): 1–8.

**63** Azcan, N. and Danisman, A. (2007). Alkali catalyzed transesterification of cottonseed oil by microwave irradiation. *Fuel.* 86 (17–18): 2639–2644.

**64** Nabi, M.N., Rahman, M.M. and Akhter, M.S. (2009). Biodiesel from cotton seed oil and its effect on engine performance and exhaust emissions. *Appl Therm Eng.* 29 (11–12): 2265–2270.

**65** Shu, Q., Zhang, Q., Xu, G. et al. (2009). Synthesis of biodiesel from cottonseed oil and methanol using a carbon-based solid acid catalyst. *Fuel Process Technol.* 90 (7–8): 1002–1008.

**66** Shankar, A.A., Pentapati, P.R. and Prasad, R.K. (2017). Biodiesel synthesis from cottonseed oil using homogeneous alkali catalyst and using heterogeneous multi walled carbon nanotubes: characterization and blending studies. *Egypt J Pet.* 26 (1): 125–133.

**67** Rashid, U., Anwar, F., Moser, B.R. and Knothe, G. (2008). *Moringa oleifera* oil: a possible source of biodiesel. *Bioresour Technol.* 99 (17): 8175–8179.

**68** Rashid, U., Anwar, F., Ashraf, M. et al. (2011). Application of response surface methodology for optimizing transesterification of *Moringa oleifera* oil: biodiesel production. *Energy Convers Manag.* 52 (8–9): 3034–3042.

**69** Fernandes, D.M., Sousa, R.M.F., De Oliveira, A. et al. (2015). *Moringa oleifera*: a potential source for production of biodiesel and antioxidant additives. *Fuel.* 146: 75–80.

**70** Balamurugan, T., Arun, A. and Sathishkumar, G.B. (2018). Biodiesel derived from corn oil – a fuel substitute for diesel. *Renew Sustain Energy Rev.* 94: 772–778.

**71** Veljković, V.B., Biberdžić, M.O., Banković-Ilić, I.B. et al. (2018). Biodiesel production from corn oil: a review. *Renew Sustain Energy Rev.* 91: 531–548.

**72** He, B.B., Van Gerpen, J.H. and Thompson, J.C. (2009). Sulfur content in selected oils and fats and their corresponding methyl esters. *Appl Eng Agric.* 25 (2): 223–226.

**73** Hill, J., Nelson, E., Tilman, D. et al. (2006). Environmental, economic, and energetic costs and benefits of biodiesel and ethanol biofuels. *Proc Natl Acad Sci.* 103 (30): 11206–11210.

**74** Fedoroff, N.V. and Cohen, J.E. (1999). Plants and population: is there time? *Proc Natl Acad Sci.* 96 (11): 5903–5907.

**75** Tilman, D., Balzer, C., Hill, J. and Befort, B.L. (2011). Global food demand and the sustainable intensification of agriculture. *Proc Natl Acad Sci.* 108 (50): 20260–20264.

**76** Dunn, R.O. and Bagby, M.O. (1995). Low-temperature properties of triglyceride-based diesel fuels: transesterified methyl esters and petroleum middle distillate/ester blends. *J Am Oil Chem Soc.* 72 (8): 895–904.

**77** Chiu, C.W., Schumacher, L.G. and Suppes, G.J. (2004). Impact of cold flow improvers on soybean biodiesel blend. *Biomass Bioenergy.* 27 (5): 485–491.

**78** Bhale, P.V., Deshpande, N.V. and Thombre, S.B. (2009). Improving the low temperature properties of biodiesel fuel. *Renew Energy.* 34 (3): 794–800.

**79** Altaie, M.A.H., Janius, R.B., Rashid, U. et al. (2015). Cold flow and fuel properties of methyl oleate and palm-oil methyl ester blends. *Fuel.* 160: 238–244.

**80** Hazrat, M.A., Rasul, M.G., Mofijur, M. et al. (2020). A mini review on the cold flow properties of biodiesel and its blends. *Front Energy Res.* 8: 326.

**81** Sharifi-Rad, J., Sureda, A., Tenore, G.C. et al. (2017). Biological activities of essential oils: from plant chemoecology to traditional healing systems. *Molecules.* 22 (1): 70.

**82** Balat, M. (2011). Potential alternatives to edible oils for biodiesel production – a review of current work. *Energy Convers Manag.* 52 (2): 1479–1492.

**83** No, S.Y. (2011). Inedible vegetable oils and their derivatives for alternative diesel fuels in CI engines: a review. *Renew Sustain Energy Rev.* 15 (1): 131–149.

**84** Banković-Ilić, I.B., Stamenković, O.S. and Veljković, V.B. (2012). Biodiesel production from non-edible plant oils. *Renew Sustain Energy Rev.* 16 (6): 3621–3247.

**85** Bhuiya, M.M.K., Rasul, M.G., Khan, M.M.K. et al. (2014). Second generation biodiesel: potential alternative to-edible oil-derived biodiesel. *Energy Procedia.* 61: 1969–1972.

**86** Tiwari, A.K., Kumar, A. and Raheman, H. (2007). Biodiesel production from jatropha oil (*Jatropha curcas*) with high free fatty acids: an optimized process. *Biomass Bioenergy.* 31: 569–575.

**87** Koh, M.Y., Idaty, T. and Ghazi, M. (2011). A review of biodiesel production from *Jatropha curcas* L. oil. *Renew Sustain Energy Rev.* 15: 2240–2251.

**88** Ali, M.H., Mashud, M., Rubel, M.R. and Ahmad, R.H. (2013). Biodiesel from neem oil as an alternative fuel for diesel engine. *Procedia Eng.* 56: 625–630.

**89** Ramasubramanian, S., Manavalan, S., Gnanavel, C. and Balakrishnan, G. (2017). Processing of neem and Jatropha methyl esters – alternative fuels from vegetable oil. *IOP Conf Ser Mater Sci Eng.* 183 (1): 012036.

**90** Canoira, L., Alcántara, R., Jesús García-Martínez, M. and Carrasco, J. (2006). Biodiesel from Jojoba oil-wax: transesterification with methanol and properties as a fuel. *Biomass Bioenergy.* 30 (1): 76–81.

**91** Bouaid, A., Bajo, L., Martinez, M. and Aracil, J. (2007). Optimization of biodiesel production from Jojoba oil. *Process Saf Environ Prot.* 85 (5): 378–382.

**92** Patel, R.L. and Sankhavara, C.D. (2017). Biodiesel production from Karanja oil and its use in diesel engine: a review. *Renew Sustain Energy Rev.* 71: 464–474.

**93** Harreh, D., Saleh, A.A., Reddy, A.N.R. and Hamdan, S. (2018). An experimental investigation of Karanja biodiesel production in Sarawak, Malaysia. *J Eng.* 2018: 4174205.

**94** Ghadge, S.V. and Raheman, H. (2006). Process optimization for biodiesel production from mahua (*Madhuca indica*) oil using response surface methodology. *Bioresour Technol.* 97 (3): 379–384.

**95** Ghadge, S.V. and Raheman, H. (2005). Biodiesel production from mahua (*Madhuca indica*) oil having high free fatty acids. *Biomass Bioenergy.* 28 (6): 601–605.

**96** Baskar, G., Gurugulladevi, A., Nishanthini, T. et al. (2017). Optimization and kinetics of biodiesel production from Mahua oil using manganese doped zinc oxide nanocatalyst. *Renew Energy.* 103: 641–646.

**97** Silitonga, A.S., Masjuki, H.H., Ong, H.C. et al. (2016). Synthesis and optimization of *Hevea brasiliensis* and *Ricinus communis* as feedstock for biodiesel production: a comparative study. *Ind Crops Prod.* 85: 274–286.

**98** Soudagar, M.E.M., Mujtaba, M.A., Safaei, M.R. et al. (2021). Effect of Sr@ZnO nanoparticles and *Ricinus communis* biodiesel-diesel fuel blends on modified CRDI diesel engine characteristics. *Energy.* 215: 119094.

**99** Arumugam, A., Jegadeesan, G.B. and Ponnusami, V. (2017). Comparative studies on catalytic properties of immobilized lipase on low-cost support matrix for transesterification of pinnai oil. *Biomass Convers Biorefinery.* 8 (1): 69–77.

**100** Arumugam, A. and Sankaranarayanan, P. (2020). Biodiesel production and parameter optimization: an approach to utilize residual ash from sugarcane leaf, a novel heterogeneous catalyst, from *Calophyllum inophyllum* oil. *Renew Energy.* 153: 1272–1282.

**101** Demirbas, A. and Fatih Demirbas, M. (2011). Importance of algae oil as a source of biodiesel. *Energy Convers Manag.* 52 (1): 163–170.

**102** Chen, L., Liu, T., Zhang, W. et al. (2012). Biodiesel production from algae oil high in free fatty acids by two-step catalytic conversion. *Bioresour Technol.* 111: 208–214.

**103** Brownbridge, G., Azadi, P., Smallbone, A. et al. (2014). The future viability of algae-derived biodiesel under economic and technical uncertainties. *Bioresour Technol.* 151: 166–173.

**104** White, K., Lorenz, N., Potts, T. et al. (2011). Production of biodiesel fuel from tall oil fatty acids via high temperature methanol reaction. *Fuel.* 90 (11): 3193–3199.

**105** Adewale, P., Vithanage, L.N. and Christopher, L. (2017). Optimization of enzyme-catalyzed biodiesel production from crude tall oil using Taguchi method. *Energy Convers Manag.* 154: 81–91.

**106** Lawer-Yolar, G., Dawson-Andoh, B. and Atta-Obeng, E. (2021). Synthesis of biodiesel from tall oil fatty acids by homogeneous and heterogeneous catalysis. *Sustain Chem.* 2 (1): 206–221.

**107** Uddin, M.N., Techato, K., Rasul, M.G. et al. (2019). Waste coffee oil: a promising source for biodiesel production. *Energy Procedia.* 160: 677–682.

**108** Tarigan, J.B., Ginting, M., Mubarokah, S.N. et al. (2019). Direct biodiesel production from wet spent coffee grounds. *RSC Adv.* 9 (60): 35109–35116.

**109** Marx, S., Venter, R., Karmee, S.K. et al. (2020). Biofuels from spent coffee grounds: comparison of processing routes. *Biofuels.* DOI: 10.1080/17597269.2020.1793538.

**110** Roick, C., Otun, K.O., Diankanua, N. and Joshua, G. (2021). Non-edible feedstock for biodiesel production. In: *Biodiesel Technology and Applications* (ed. A.M.I. Inamuddin, R. Boddula and M. Rezakazemi), 285–309.

**111** Gui, M.M., Lee, K.T. and Bhatia, S. (2008). Feasibility of edible oil vs. non-edible oil vs. waste edible oil as biodiesel feedstock. *Energy.* 33 (11): 1646–1653.

**112** Dukare, M.B., Awasthi, S.K., Sapkal, V.S. and Sapkal, R.S. (2010). Two step process for biodiesel production using non-edible oils. *Indian Chem Eng.* 52 (3): 254–262.

**113** Karthikeyan, M., Renganathan, S. and Govindhan, P. (2017). Production of biodiesel via two-step acid-base catalyzed transesterification reaction of Karanja oil by $BaMoO_4$ as a catalyst. *Energy Sources, Part A Recover Util Environ Eff.* 39 (14): 1504–1510.

**114** Alam, F., Mobin, S. and Chowdhury, H. (2015). Third generation biofuel from algae. *Procedia Eng.* 105: 763–768.

**115** Chowdhury, H. and Loganathan, B. (2019). Third-generation biofuels from microalgae: a review. *Curr Opin Green Sustain Chem.* 20: 39–44.

**116** Behera, S., Singh, R., Arora, R. et al. (2015). Scope of algae as third generation biofuels. *Front Bioeng Biotechnol.* 2: 90.

**117** Xu, H., Miao, X. and Wu, Q. (2006). High quality biodiesel production from a microalga *Chlorella protothecoides* by heterotrophic growth in fermenters. *J Biotechnol.* 126 (4): 499–507.

**118** Shen, Y., Yuan, W., Pei, Z. and Mao, E. (2010). Heterotrophic culture of *Chlorella protothecoides* in various nitrogen sources for lipid production. *Appl Biochem Biotechnol.* 160 (6): 1674–1684.

**119** Krzemińska, I., Piasecka, A., Nosalewicz, A. et al. (2015). Alterations of the lipid content and fatty acid profile of *Chlorella protothecoides* under different light intensities. *Bioresour Technol.* 196: 72–77.

**120** Liu, J., Huang, J., Fan, K.W. et al. (2010). Production potential of *Chlorella zofingiensis* as a feedstock for biodiesel. *Bioresour Technol.* 101 (22): 8658–8663.

**121** Liu, J., Huang, J., Sun, Z. et al. (2011). Differential lipid and fatty acid profiles of photoautotrophic and heterotrophic *Chlorella zofingiensis*: assessment of algal oils for biodiesel production. *Bioresour Technol.* 102 (1): 106–110.

**122** Feng, P., Deng, Z., Fan, L. and Hu, Z. (2012). Lipid accumulation and growth characteristics of *Chlorella zofingiensis* under different nitrate and phosphate concentrations. *J Biosci Bioeng.* 114 (4): 405–410.

**123** Zhu, L., Zhang, X., Ji, L. et al. (2007). Changes of lipid content and fatty acid composition of *Schizochytrium limacinum* in response to different temperatures and salinities. *Process Biochem.* 42 (2): 210–214.

**124** Zhang, T., Lou, F., Tao, G. et al. (2016). Composition and structure of single cell oil produced by *Schizochytrium limacinum* SR31. *J Am Oil Chem Soc.* 93 (10): 1337–1346.

**125** Li, Z., Ling, X., Zhou, H. et al. (2019). Screening chemical modulators of benzoic acid derivatives to improve lipid accumulation in *Schizochytrium limacinum* SR21 with metabolomics analysis. *Biotechnol Biofuels.* 12 (1): 209.

**126** Jerney, J. and Spilling, K. (2018). Large scale cultivation of microalgae: open and closed systems. *Methods Mol Biol.* 1980: 1–8.

**127** Ruangsomboon, S., Dimak, J., Jongput, B. et al. (2020). Outdoor open pond batch production of green microalga *Botryococcus braunii* for high hydrocarbon production: enhanced production with salinity. *Sci Rep.* 10 (1): 1–12.

**128** Schlesinger, A., Eisenstadt, D., Bar-Gil, A. et al. (2012). Inexpensive non-toxic flocculation of microalgae contradicts theories; overcoming a major hurdle to bulk algal production. *Biotechnol Adv.* 30 (5): 1023–1030.

**129** Gerde, J.A., Yao, L., Lio, J.Y. et al. (2014). Microalgae flocculation: impact of flocculant type, algae species and cell concentration. *Algal Res.* 3 (1): 30–35.

**130** Najjar, Y.S.H. and Abu-Shamleh, A. (2020). Harvesting of microalgae by centrifugation for biodiesel production: a review. *Algal Res.* 51: 102046.

**131** Shirvani, T., Yan, X., Inderwildi, O.R. et al. (2011). Life cycle energy and greenhouse gas analysis for algae-derived biodiesel. *Energy Environ Sci.* 4 (10): 3773–3778.

**132** Resurreccion, E.P., Colosi, L.M., White, M.A. and Clarens, A.F. (2012). Comparison of algae cultivation methods for bioenergy production using a combined life cycle assessment and life cycle costing approach. *Bioresour Technol.* 126: 298–306.

**133** Khoo, H.H., Koh, C.Y., Shaik, M.S. and Sharratt, P.N. (2013). Bioenergy co-products derived from microalgae biomass via thermochemical conversion – life cycle energy balances and $CO_2$ emissions. *Bioresour Technol.* 143: 298–307.

**134** Gnansounou, E. and Kenthorai Raman, J. (2016). Life cycle assessment of algae biodiesel and its co-products. *Appl Energy.* 161: 300–308.

**135** Thomassen, G., Van Dael, M., Lemmens, B. and Van Passel, S. (2017). A review of the sustainability of algal-based biorefineries: towards an integrated assessment framework. *Renew Sustain Energy Rev.* 68: 876–887.

**136** Kumar, D. and Singh, B. (2019). Algal biorefinery: an integrated approach for sustainable biodiesel production. *Biomass Bioenergy.* 131: 105398.

**137** Li, M., Zheng, Y., Chen, Y. and Zhu, X. (2014). Biodiesel production from waste cooking oil using a heterogeneous catalyst from pyrolyzed rice husk. *Bioresour Technol.* 154: 345–348.

**138** Tshizanga, N., Aransiola, E.F. and Oyekola, O. (2017). Optimisation of biodiesel production from waste vegetable oil and eggshell ash. *South African J Chem Eng.* 23: 145–156.

**139** Degfie, T.A., Mamo, T.T. and Mekonnen, Y.S. (2019). Optimized biodiesel production from waste cooking oil (WCO) using calcium oxide (CaO) nano-catalyst. *Sci Rep.* 9 (1): 1–8.

**140** Samanta, S. and Sahoo, R.R. (2021). Waste cooking (palm) oil as an economical source of biodiesel production for alternative green fuel and efficient lubricant. *Bioenergy Res.* 14 (1): 163–174.

**141** Hajjari, M., Tabatabaei, M., Aghbashlo, M. and Ghanavati, H. (2017). A review on the prospects of sustainable biodiesel production: a global scenario with an emphasis on waste-oil biodiesel utilization. *Renew Sustain Energy Rev.* 72: 445–464.

**142** Samuel, O.D., Waheed, M.A., Taheri-Garavand, A. et al. (2021). Prandtl number of optimum biodiesel from food industrial waste oil and diesel fuel blend for diesel engine. *Fuel.* 285: 119049.

**143** Park, S.H., Khan, N., Lee, S. et al. (2019). Biodiesel production from locally sourced restaurant waste cooking oil and grease: synthesis, characterization, and performance evaluation. *ACS Omega.* 4 (4): 7775–7784.

**144** Panchal, B.M., Deshmukh, S.A. and Sharma, M.R. (2017). Production and kinetic transesterification of biodiesel from yellow grease with dimethyl carbonate using methanesulfonic acid as a catalyst. *Environ Prog Sustain Energy.* 36 (3): 802–807.

**145** Kolet, M., Zerbib, D., Nakonechny, F. and Nisnevitch, M. (2020). Production of biodiesel from brown grease. *Catalysts.* 10 (10): 1189.

**146** Alptekin, E. and Canakci, M. (2011). Optimization of transesterification for methyl ester production from chicken fat. *Fuel.* 90 (8): 2630–2638.

147 Shi, W., Li, J., He, B. et al. (2013). Biodiesel production from waste chicken fat with low free fatty acids by an integrated catalytic process of composite membrane and sodium methoxide. *Bioresour Technol.* 139: 316–322.

148 Banković-Ilić, I.B., Stojković, I.J., Stamenković, O.S. et al. (2014). Waste animal fats as feedstocks for biodiesel production. *Renew Sustain Energy Rev.* 32: 238–254.

149 Adewale, P., Dumont, M.J. and Ngadi, M. (2015). Recent trends of biodiesel production from animal fat wastes and associated production techniques. *Renew Sustain Energy Rev.* 45: 574–588.

150 Sander, A., Antonije Košćak, M., Kosir, D. et al. (2018). The influence of animal fat type and purification conditions on biodiesel quality. *Renew Energy.* 118: 752–760.

151 Canoira, L., Rodríguez-Gamero, M., Querol, E. et al. (2008). Biodiesel from low-grade animal fat: production process assessment and biodiesel properties characterization. *Ind Eng Chem Res.* 47 (21): 7997–8004.

152 Cooney, C.P., Yeliana, Worm, J.J. and Naber, J.D. (2009). Combustion characterization in an internal combustion engine with ethanol-gasoline blended fuels varying compression ratios and ignition timing. *Energy Fuels.* 23 (5): 2319–2324.

153 Lanzafame, P., Temi, D.M., Perathoner, S. et al. (2011). Etherification of 5-hydroxymethyl-2-furfural (HMF) with ethanol to biodiesel components using mesoporous solid acidic catalysts. *Catal Today.* 175 (1): 435–441.

154 Shi, J., Wang, Y., Yu, X. et al. (2016). Production of 2,5-dimethylfuran from 5-hydroxymethylfurfural over reduced graphene oxides supported Pt catalyst under mild conditions. *Fuel.* 163: 74–79.

155 Zhang, F., Liu, Y., Yuan, F. et al. (2017). Efficient production of the liquid fuel 2,5-dimethylfuran from 5-hydroxymethylfurfural in the absence of acid additive over bimetallic PdAu supported on graphitized carbon. *Energy Fuels.* 31 (6): 6364–6373.

156 Gulzar, M., Masjuki, H.H., Kalam, M.A. et al. (2015). Oil filter modification for biodiesel-fueled engine: a pathway to lubricant sustainability and exhaust emissions reduction. *Energy Convers Manag.* 91: 168–175.

157 Schumacher, L.G., Soylu, S., Van Gerpen, J. and Wetherell, W. (2005). Fueling direct injection diesel engines with 2% biodiesel blends. *Appl Eng Agric Trans ASAE.* 21 (2): 149–152.

158 Raheman, H., Jena, P.C. and Jadav, S.S. (2013). Performance of a diesel engine with blends of biodiesel (from a mixture of oils) and high-speed diesel. *Int J Energy Environ Eng.* 4 (1): 1–9.

159 Spiller, R., Knoshaug, E.P., Nagle, N. et al. (2020). Upgrading brown grease for the production of biofuel intermediates. *Bioresour Technol Rep.* 9: 100344.

160 Nogueira, L., Sanches, A.L.M., da Silva, D.G.H. et al. (2011). Biochemical biomarkers in Nile tilapia (*Oreochromis niloticus*) after short-term exposure to diesel oil, pure biodiesel and biodiesel blends. *Chemosphere.* 85 (1): 97–105.

161 Nogueira, L., da Silva, D.G.H., Oliveira, T.Y.K. et al. (2013). Biochemical responses in armored catfish (*Pterygoplichthys anisitsi*) after short-term exposure to diesel oil, pure biodiesel and biodiesel blends. *Chemosphere.* 93 (2): 311–319.

162 Leme, D.M., Grummt, T., de Oliveira, D.P. et al. (2012). Genotoxicity assessment of water soluble fractions of biodiesel and its diesel blends using the *Salmonella* assay and the in vitro MicroFlow® kit (Litron) assay. *Chemosphere.* 86 (5): 512–520.

163 Amouri, M., Mohellebi, F., Zaïd, T.A. and Aziza, M. (2017). Sustainability assessment of *Ricinus communis* biodiesel using LCA approach. *Clean Technol Environ Policy.* 19: 749–760.

**164** Hoekman, S.K. and Robbins, C. (2012). Review of the effects of biodiesel on NOx emissions. *Fuel Process Technol.* 96: 237–249.

**165** Lü, J., Sheahan, C. and Fu, P. (2011). Metabolic engineering of algae for fourth generation biofuels production. *Energy Environ Sci.* 4 (7): 2451–2466.

**166** Abdullah, B., Muhammad, S.A.F.S., Shokravi, Z. et al. (2019). Fourth generation biofuel: a review on risks and mitigation strategies. *Renew Sustain Energy Rev.* 107: 37–50.

# 20

## Advanced Practices in Biodiesel Production

*Trinath Biswal[1], Krushna Prasad Shadangi[2], and Rupam Kataki[3]*

[1] *Department of Chemistry, Veer Surendra Sai University of Technology, Burla, Sambalpur, Odisha, India*
[2] *Department of Chemical Engineering, Veer Surendra Sai University of Technology, Burla, Sambalpur, Odisha, India*
[3] *Department of Energy, Tezpur University, Tezpur, Assam, India*

## 20.1 Introduction

The major cause of environmental degradation is massive use of fossil fuel in the industrialized world, which threaten the survival of biological community in the world. Hence, the awareness for production of clean and sustainable fossil fuel by using alternative technologies is developed. Now it was estimated that in the European Union (EU) countries, more than 30% of the consumption of total energy is due to transportation, which was obtained mainly from petroleum products. This is the major source of emission of pollutants to the troposphere and cause of degradation in air quality. Therefore, there is an urgent need for development of clean, renewable, and environmental fuel, which should be a proper substitute of the presently used fossil fuel [1, 2]. The biofuel in the liquid state is treated as the most renewable and economically viable option of petroleum diesel. The production and use of biodiesel is almost 79% of total biofuels, and it was mainly produced from palm and soybean oil. According to 2012 data the total use of biofuel in the consumption of fuel sector in EU countries is almost 4.7% among which the countries like Germany, Italy, the United Kingdom, France, Spain, and Poland utilize 70% of the biodiesel. Biodiesel is a mixture of variety kinds of monoalkyl esters having long chain fatty acids and is extracted from natural resources such as vegetable oils, algae oil, animal fats, and waste cooking oil. There are so many advantages of biodiesel, which includes nontoxicity, biodegradability almost four times quicker than normal fossil petroleum oil, no sulfur, higher flash point, low rate of emission, and more efficiency of combustion. In addition to that its use is renewable, safe, and used effectively without any change in the engine. Now, the major difficulty is the high cost and required amount of production and commercialization of the biodiesel. Therefore, to save our environment from drastic pollution and survival of biological community, the only alternative is restriction in the use of fossil fuel and find out a suitable alternative of fossil fuel. Hence, massive research work is now carried out to

*Biodiesel Production: Feedstocks, Catalysts, and Technologies*, First Edition. Edited by Samuel Lalthazuala Rokhum, Gopinath Halder, Suttichai Assabumrungrat, and Kanokwan Ngaosuwan.

develop a suitable technology for production of biodiesel in massive rate in eco-friendly pathway [3, 4]. There are four basic processes of the application of biodiesel such as use of vegetable oil directly, microemulsion, transesterification, and thermal cracking. But the most widely used method in the biofuel industry is the method of transesterification. This process of transesterification is the cleavage of $C_2H_5OH$, $CH_3OH$ molecule chemically from the natural feedstock of lipids into ethyl or methyl esters (serves as biodiesel) along with glycerol (by-product) [5]. The transesterification reaction may be catalyzed or non-catalyzed through heterogeneous or homogeneous enzymatic catalysts. The use of homogeneous catalysts is suitable both economic and kinetic points of view, but the problems related with it are consumption, higher amount of energy, the separation of the catalyst from the reaction mixture [1] is costly, and production of huge quantity of wastewater effluents through the process of biodiesel purification. Hence, the major challenge of the traditional biodiesel production is in the process of purification and the by-product generation. Nowadays some advanced novel technology was developed regarding the decrease in biodiesel cost with increase in efficiency of separation of biodiesel and glycerol [6, 7].

*Example:* NaCl-assisted gravitational settling method is an advantageous strategy to increase the rate of separation between the glycerol and biodiesel. The major focus in this chapter is to develop an advanced sustainable technology for biodiesel industry such as enzymatic, supercritical transesterification, and heterogeneous instead of traditional technique. In addition to that more emphasis is given in the modern technologies like ultrasonic technology, cosolvent use, microwave irradiation, and membrane reactor.

## 20.2 Mechanism of Transesterification

The uprooting of alcohol by additional in an interaction likes hydrolysis, then again, actually alcohol is utilized rather than water from an ester, which is known as transesterification. This interaction is generally employed to decrease the thickness of fatty oils. The overall reaction of transesterification is addressed by Figure 20.1.

It is a reversible reaction. In any case, the occurrence of an impetus (a solid corrosive or base) quickens the transformation, and somewhat abundance of liquor is utilized to move the balance toward the development of unsaturated esters and glycerol. The glycerol level separated from the bottom of the reaction chamber. The intermediate products included

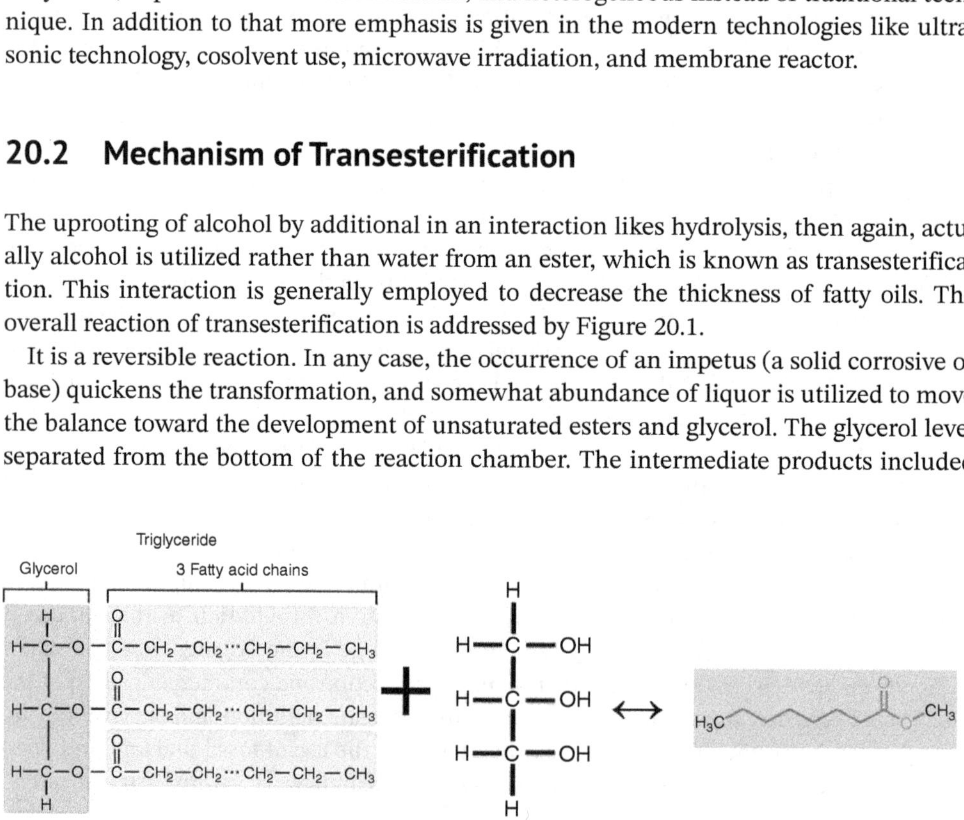


**Figure 20.1** Overall reaction of transesterification process.

with monoglycerides and diglycerides. The fatty oil transesterification through methanol comprises of the accompanying three sequential reactions. Right off the bat the fatty substance responds with the liquor to deliver a diglyceride and the unsaturated fat ester. At that point the diglyceride shaped in the principal response responds further with the liquor to frame another atom of the unsaturated fat ester and a monoglyceride. This at that point responds with the liquor to get glycerol and the unsaturated fat ester [8]. The stepwise responses are reversible, and somewhat abundance of alcohol is utilized to move the harmony toward the arrangement of esters. Within the sight of abundance liquor, the forward response is pseudo-first request, and the converse response is discovered to be second request. This reaction proceeds very fast in presence of salt catalyst [9]. The salt-catalyzed transesterification process is well discussed depicted somewhere else [9]. The initial step includes the assault of the alkoxide particle to the carbonyl carbon [10] of the fatty substance atom, which brings about the development of a tetrahedral middle. The response of this transitional with liquor creates the alkoxide particle in the subsequent advance. In the last advance the revamp of the tetrahedral halfway offers ascend to an ester and a diglyceride [10]. At the point when the transesterification response happened within the sight of the strong base impetus, for example, BaO, SrO, CaO, and MgO, the synergist responses occur on the outside of strong base impetuses. The component is as per the following: in the initial step, surface $O^{2-}$ extricates $H^+$ from $CH_3OH$ to frame surface $CH_3O^-$, which is unequivocally fundamental and has high synergist movement in the transesterification response. In the subsequent advance, the carbonyl carbon particle of the fatty oil atom pulls in a methoxide anion from the outside of the SrO to shape a tetrahedral halfway. In the third step, the tetrahedral transitional gets $H^+$ from the outside of the SrO. The tetrahedral moderate likewise can respond with methanol to produce methoxide anions [9]. Transesterification can be catalyzed by acids impetuses, ideally by sulfonic and sulfuric acids. These impetuses give extremely exceptional returns in alkyl esters; however these responses are moderate, requiring regularly temperature over 100 °C and in excess of 3 h to finish the change [9]. As per the system of corrosive catalyzed transesterification of vegetable oil, carbonyl protonation and gathering of the ester prompt the carbocation, which later a nucleophilic assault of the liquor creates a tetrahedral halfway. This middle of the road disposes of glycerol to structure another organic compound (ester) and to recover the energy. This instrument of corrosive catalyzed transesterification identified with a monoglyceride. Notwithstanding, this response can be reached out to di- and fatty oils [9].

## 20.3 Advanced Biodiesel Production Technologies

### 20.3.1 Production of Biodiesel Using Membrane Reactor

#### 20.3.1.1 Principle

Membrane technique enables the membrane for the discriminating conveyance of liquids by varying the mass transfer speed. Here, the reaction and isolation shall occur inside the reaction chamber. It functions because of the interfacial adhesion of alcohol, oil, and also the varying surface forces. Use of such membranes in transesterification process has gained substantial interest in recent years [11]. This knowledge participates a reaction and separation phase that decreases regeneration and segregation expenses and increases transformation due to improved suppressed product responses. It also increases the efficiency of biodiesel by changing the balance onto the product aspect and precise isolation [12]. In the manufacture of biodiesel by extracting glycerol and maintaining triglycerides from the reaction mixture, membrane reactors play a vital role [13]. $CH_3OH$ is almost immiscible in triglycerides or oils due to the polarity variation and, thus, defines a two-phase system. The phase separation of such two is the key explanation for the restriction of the transition of mass throughout the process of transesterification, however, is beneficial mostly where a membrane reactor is used. In general, biodiesel produced along with catalyst. The glycerol is dissolvable in natural solvents, while fatty substances stay adjourned and transported in a blend as retentate [14, 15]. Since the molecular sizes of alcohol, biodiesel, catalyst, and glycerol are minimal, the microporous membrane reaches the permeate side after building up the transmembrane pressure (Figure 20.2) [16], whereas the bigger emulsified blend of triglycerides is contained inside the layer and persistently changed over to biodiesel. This membrane may be natural or inorganic. The inorganic films may be much more secure than natural membranes since they are ordinarily more thermally steady [12, 13]. The use of alcohol and catalyst is comparatively less in the membrane process [15]. The product includes water, alcohol (methanol), glycerol, and fatty acid ethyl ester (FAAE), which require further processing. Higher purity products can be obtained in antiacid environment [15]. The use of less amount of catalyst with a higher alcohol-to-oil ratio is less in membrane reactor compared to clump reactor for producing a better yield of biodiesel [16]. The ratio of alcohol to oil during the biodiesel production can be reduced if the alcohol and glycerol is reused [15]. The concentration of biodiesel can be enhanced to about 92.4 wt% from 85.7% once 100% alcohol and glycerol is reused rather than 50 or 75%. The yield of biodiesel rate can be maintained by optimizing the appropriate ratio of alcohol

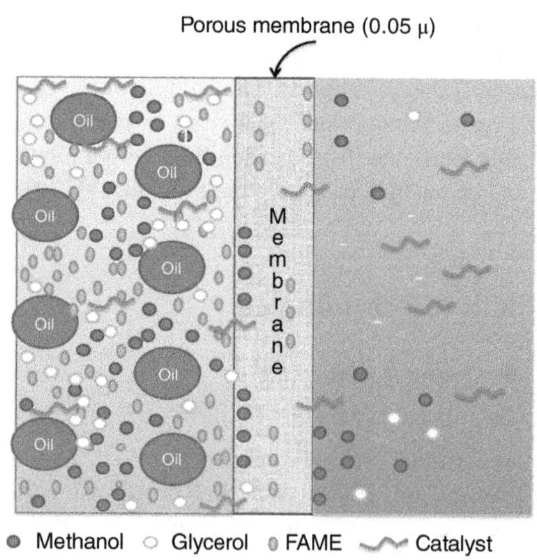


**Figure 20.2** Biodiesel separation using membrane process..

and oil. It was reported that ceramic membrane results better yield at lower alcohol-to-oil ratio [17].

***Advantages***

- Proficient partition and less contamination.
- Well-organized separation and cleaning.
- Less use of water washing.
- Miscibility of glycerin is less compared to other reactors (batch).
- Less yield of glycerin.
- Working pressure and temperature is low.
- Ester conversion rate is higher.
- Cost of operating is low.

***Disadvantages***

- The formation of soap in the reactor causes the fouling.
- Cleaning is expensive in membrane reactor.
- Controlling of operating parameters is must to enhance the yield.
- Required higher control of operating parameters to avoid fouling in the membrane.

***Application***

- It is applicable for pilot scale.

## 20.3.2 Microwave-Assisted Transesterification Technology

### 20.3.2.1 Principle

In a microwave, the heat is created by the interaction of particles, owing to polar liquor atoms arrangement by the continually shifting of attractive ground of microwave, which are the electromagnetic irradiations (0.3–300 GHz) (Figure 20.3) [18]. The impacts of microwave light on responses are both warm and nonthermal. The heating is produced by the interaction of microwaves without changing the atomic structure of the chemicals at the atomic level [19, 20]. The microwave light strategy has numerous preferences over the ordinary warming transesterification strategy, such as decreased warm slope, less

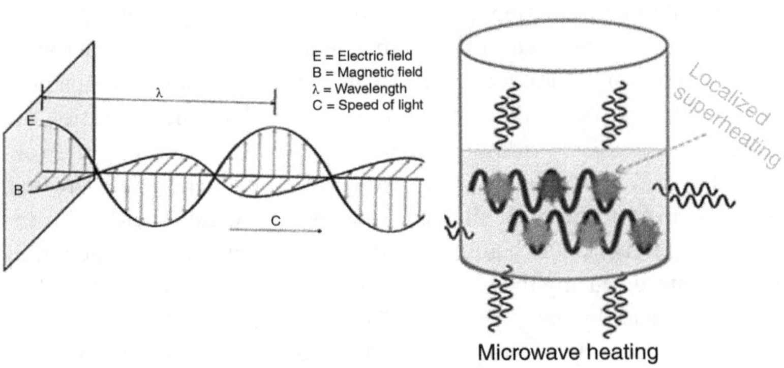


**Figure 20.3** Biodiesel production using microwave heating.

overheating of surfaces due to contactless warming, vitality exchange instead of warm exchange, invert warm impact, i.e. warm produces from interior of the fabric, etc. Atomic grinding actuated by microwaves produces a heat response by utilizing the attractive field on polar particles. In this manner, microwave-assisted transesterification response is proficiently completed in exceptionally brief response time since it is of localized superheating that improves the response rate [21–23]. Separated from the various points of interest, there are some disadvantages of production of biodiesel in presence of microwave. The most important is scale-up to industrial scale and its maintenance. Some of them are also discussed in the following paragraphs [24].

***Advantages***

- Condensed thermal gradient.
- Less hotness of outsides surface.
- Restricted superheating.
- Small reaction time with high efficiency.
- Improved reaction kinetics.
- It take less time for the separation and purification.
- Ecologically friendly.
- Energy efficient.

***Disadvantages***

- Less depth of penetration microwaves.
- Internal mixing is required.
- Localized superheating is required to decrease.
- Mass transfer confines.
- More safety aspect required.
- Scale-up is difficult.

***Application***

- Laboratory scale.

### 20.3.3 Ultrasonic-Assisted Transesterification Techniques

The ultrasonic-assisted biodiesel production process improves the miscibility between stages by mixing the reactants in presence of microturbulence (Figure 20.4). Cavitation could be recently inquired about proficient technique to progress mass exchange in a non-miscible liquid–liquid framework. Acoustic cavitation and hydrodynamic cavitation are the two categories of ultrasonic-assisted transesterification techniques. By using the venturi plate, perforated plate, and throttling valve, hydrodynamic cavitation was created. However, the ultrasound is used to produce the acoustic cavitation. The increase in the temperature, along with the energy density and pressure acoustic cavitation, is generally produced in the reaction volume. During the transesterification process the emulsification of one liquid with other takes place by the generation and collapsing the bubbles due to the acoustic cavitation at the phase boundary. The generated bubbles get collapsed as it reaches the phase boundary. Due to the cavitation proper mixing in the reaction mixture happens,

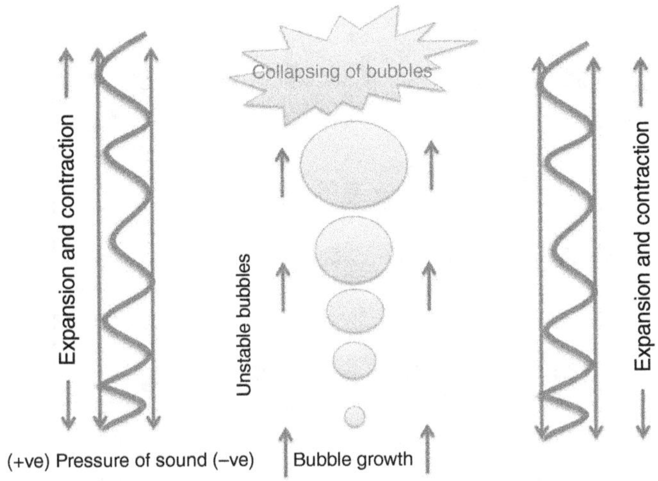


**Figure 20.4** Formation of ultrasonic sound and formation and growth of bubbles.

which do not require the addition of heat and external mixing. The presence of ultrasonic cavitation rises the temperature (5000 K) and pressure (1000 atm), catalyst surface area, and catalyst activity, which speed up the rate of transesterification reaction [25–27]. However, more studies are required to find out the minimum energy required, efficient reactor configuration to enhance the process of transesterification process for the production of biodiesel. Table 20.1 shows the yield of biodiesel using ultrasonic techniques.

***Advantages***

- It increases the contact area between two immiscible fluids.
- It enhances the rate of mass transfer.
- Decreases the time of reaction.
- Decreases the requirement of catalyst.
- This process is able to produce good result from low quality of raw materials (high fatty acid content) including the waste cooking oil and animal fats.
- It decreases about 50% of the need of alcohol and catalyst.
- It generated less waste.

***Disadvantages***

- Easy damage of the ultrasonic probe.
- Not easy to control the temperature during the reaction process.
- It generates some undesirable by-products.

***Application***

- It is generally used for industrial scale.

**Table 20.1** Synthesis of biodiesel using ultrasonic techniques.

| Raw material | Catalyst | Quantity of catalyst (wt%) | Alcohol | Alcohol:oil (molar ratio) | Reaction time (min) | Temperature (°C) | Ultrasonic frequency (kHz) | Yield, % |
|---|---|---|---|---|---|---|---|---|
| Soybean | CaO [28] | 6 | Methanol | 10 : 1 | 60 | 62 | 35 | 90 |
| | magnetic $NaSiO_3$ @$Fe_3O_4$/C [29] | 7 | | 7 : 1 | 80 | 54 | 20–25 | 97.7 |
| | Novozym 435 [30] | 6 | | 6 : 1 | 240 | 40 | 40 | 96 |
| | NaOH [31] | 1 | | | 10 | 45 | 197 | 99 |
| | $Ti(Pr)_4$ $Al(Pr)_3$ [32] | 3 | | | 120 | 60 | — | 64 |
| | Na or K [33] | 0.15 | | | 60 | 45 | 21.5 | 100 |
| | KOH [34] | 0.5 | | | 30 | 26–45 | 611 | 90 |
| Triolein | NaOH [35] | 1 | Ethanol | 6 : 1 | <20 | 25 | 40 | 92 |
| | KOH [36] | 1 | Methanol | 6 : 1 | 10 | 25 | | >90 |
| Palm oil | KOH [37] | — | | 6 : 1 | 20 | 38–40 | 45 | 95 |
| Palm oil | KOH [38] | 1.5–1.7 | Ethanol | 9–11 : 1 | | 35–40 | 20 | 92 |
| Palm oil | CaO [39] | 8 | Methanol | 9 : 1 | 60 | 60 | | 92.7 |
| Jatropha oil | CaO [40] | 5.5 | | 11 : 1 | — | 64 | 35 | 95 |
| Frying Fish oil | KOH [41] | 0.5–1 | Ethanol | 6 : 1 | 60 | 60 | 20 | 98.2 |
| Jatropha oil | KOH [42] | 0.5 | Methanol | 5 : 1 | 7 | 25–35 | | 97.6 |
| Canola oil | CaO [43] | 5.3 | | 7.5 : 1 | 150 | 60 | | 99.4 |
| Sesame oil | $Ba(OH)_2$ [44] | 1.8 | | 6.7 : 1 | 40.3 | 31.9 | | 98.6 |
| Muskmelon oil | KOH [45] | 1.15 | | 6.6 : 1 | 3.56 | 41.9 | | 97.9 |
| *Silybum marianum* oil | KOH [46] | 1.5 | Methanol/ ethanol | 8 : 1 | 20 | 60 | 40 | 95.7 |

## 20.3.4 Production of Biodiesel Using Cosolvent Method

### 20.3.4.1 Principle

The methanol and oil are immiscible, hence affect the rate of mass transfer during the process of transesterification. To prepare a proper mixing of these two liquids, an additional solvent is required, which is known as cosolvent. The introduction of third component decreases the mass transfer limitations by reducing the resistances and hence enhances the rate of mass transfer at the interphase. Till date few researches have provided the optimized conditions for the production of biodiesel using cosolvent methods. So more research is required in the use of cosolvent and the other essential operating parameters such as pressure, temperature, and oil-to-alcohol molar ratio [47, 48]. Literatures report about the use of several cosolvents for the production of biodiesel using diethyl ether, carbon dioxide, acetone, heptane, tetrahydrofuran, propane, dimethyl ether, and mixture of solvents [49, 50]. This process requires extra cost in the form of cosolvent for the synthesis of biodiesel and their purification [51]. Some other cosolvents are used for the production of biodiesel from rapeseed oil which includes, DEE, diisopropyl ether, tertbutyl methyl ether, di-butyl ether, and tetrahydrofuran [52]. However, the highest yield about 97.6% of biodiesel was obtained when methanol and DEE are mixed in 1 : 1 in addition to the molar ratio of methanol to oil of 9 : 1 and 0.7% of catalyst KOH. The other operating conditions were 303 K temperature, 700 rpm stirrer speed, and 2 h of reaction [52]. On this behalf, some researchers report that the acetone is a good cosolvent [49].One of the good point of the cosolvent method is that it requires less methanol compared with other process to produce about more 20% higher yield of biodiesel through batch and continuous reaction process [48]. The yield and operating parameters conditions required for the biodiesel production from various source and cosolvent are reported in Table 20.2.

***Advantages***

- Increases rate of reaction
- Reduces oil and alcohol molar ratio
- Decreases the operating pressure and temperature

***Disadvantages***

- Difficult to separate the cosolvent from the reaction solution
- Generation of more amount of wastewater
- Higher the biodiesel synthesis cost because of the cosolvent purification

***Application***
Industrial scale

## 20.3.5 In Situ Biodiesel Production Technology

### 20.3.5.1 Principle

Transesterification followed by extraction process is collectively known as in situ process. First the oil is extracted by ordinary techniques, and further esterification or transesterification process is monitored for the production of biodiesel. In this process simultaneous production and conversion takes place. Apart from the regular extraction process, alcohol and other suitable solvent can also be used for the better yield of oil, whereas the alcohol

**Table 20.2** Yield of biodiesel using cosolvent techniques.

| Feed | Catalyst type | Reaction temperature, °C | Reaction time, min | Cosolvent type | Alcohol type | Alcohol-to-oil molar ratio | Yield, % |
|---|---|---|---|---|---|---|---|
| Castor oil [53] | Sodium methoxide | 60 | 120 | Hexane | Methanol | 6 : 1 | 95.5 |
| Colza oil [47] | KOH | 23 | 10 | Tetra hydro furan | | 1 : 1 | 99.89 |
| Waste cooking oil [54] | Sulfated tin oxide | 150 | 90 | Biodiesel | | 15 | 88.2 |
| Sunflower oil [55] | KOH | 25 | 20 | DME, DEE, TBME, THF | | 6 : 1 | 78 |
| Crude palm kernel oil [56] | — | 290–350 | 10 | THF and hexane | Supercritical methanol | 12 : 1–42 : 1 | 96.3 |
| Palm oil [57] | Catalyst-free process | 280 | — | Heptane | | | 66 |
| Soybean oil [58–62] | Catalyst-free process | 280 | 10 | Propane | | 24 : 1 | 97 |
| | KOH | 160 | 20 | Hexane, carbon dioxide | | 24 : 1 | 98 |
| | Methoxide base catalyzed | 40 | — | Tetra hydro furan (THF) | Methanol, butanol | | — |
| | Catalyst-free process | 250–325 | Continuous process | Carbon dioxide | Supercritical ethanol | 20 : 1–40 : 1 | 80% |
| | KOH | 60 | 60 | Biodiesel | Methanol | 6 : 1 | >90 |

used for the transesterification process [63]. The alcohol used for the extraction process is regenerated and reused again for the transesterification process. If the similar alcohol is used for the extraction and production biodiesel, then the processing time get reduced, which decreases the cost of production. Else the requirement of alcohol increases to produce biodiesel by this process [64, 65]. The use of 5% by volume of catalyst H2SO4 and oil-to-methanol ratio at 2 : 1; the yield of biodiesel was highest where the time of reaction was very less [66]. It was reported that the use of pore solvent having low selectivity decreases the yield for the production of biodiesel from microalgae; however, such reaction rate can be enhanced with the application of acid catalyst to produce higher yield of biodiesel [67]. Another specialist explores that a two-step of in situ process is more suitable to produce biodiesel from a high free fatty acids content biomass such as microalgae [68]. The yield was intensively high about 94.87 $\pm$0.86% by using Amberlyst-15 catalyst when used at the pre-esterification stage and at the transesterification stage also [69–72]. This process of in situ process is helpful for the extraction and production of biodiesel from nonedible seeds.

***Advantages***

- It reduces the processing time.
- Reduces the quantity of solvent.
- Less cost of production.

***Disadvantages***

- Complicated refinement system of the biodiesel.
- Purification cost is higher.

***Application***

- Laboratory scale

### 20.3.6 Production of Biodiesel Through Reactive Distillation Process

#### 20.3.6.1 Principle

The reactive distillation biodiesel production process includes two operations such as reaction and separation that occur simultaneously inside a partial refining mechanical assembly. In this process simultaneous production and extraction of products takes place. As soon as the products form, it is separated through distillation process. However, the product formation and distillation is somehow complex due to the formation of azeotrope [73]. This kind of method is further reasonable for esterification responses, where the amount of free unsaturated fats is high in the reactant [74]. Both pressed and plate kind refining segments can be utilized for receptive refining; however, plate segments are recommended on behalf of homogeneous responses because of the better fluid holdup notwithstanding the drawn out maintenance time [75]. Higher transformation, decreased energy utilization, lesser principal expense, improved selectivity, evacuation of solvents all through the response, and shirking of azeotropes are a portion of the likely points of interest of responsive refining. The constant expulsion of items from the response framework not just builds the change, yet additionally assists with lessening capital and venture costs [76]. Henceforth, it gives various focal points along limits over the regular response methods joined by refining or some other partition procedures. Table 20.3 shows the yield of biodiesel by the process of reactive distillation.

**Table 20.3** Yield of biodiesel by the process of reactive distillation.

| Feed | Reaction types | Catalyst | Catalyst quantity | Alcohol types | Quantity | Temperature, °C | Yield, % |
|---|---|---|---|---|---|---|---|
| Oleic acid [77] | Esterification | Amberlyst-15 | 5 wt% | n-butanol | 4 : 1 mol% | 125 | 72 |
| Triolein: oleic acid (86 : 14 kmol $h^{-1}$) [78] | | Amberlyst-15 | 800 kg | Methanol | 70 kmol $h^{-1}$ | 99.98 | 99.926 |
| Triolein 85.9 kmol $h^{-1}$ [78] | Transesterification | MgO | 10 080 kg | 2 streams of methanol | 400 + 500 kmol $h^{-1}$ | 64.45 | 99.9 |
| Soybean oil [79] | | NaOH | 0.65 wt% | Bioethanol | 8 : 1 mol% | 50 | 98.18 |
| Palm oil [73] | | KOH | 1 wt% | Methanol | 4 : 1 mol% | 64 | 92.75 |
| Palm oil [80] | | Potassium (K)=based activated carbon(K/AC) | — | | 6 : 1 mol% | 60 | 82.69 |
| Soybean oil [79] | | NaOH | 1.5 wt% | Bioethanol | | 50 | 94.54 |
| Palm oil [81] | Esterification and transesterification | $H_2SO_4$ and KOH | 0.8 wt% | Methanol | | 61–65 | 94 |
| Oleic acid and triolein [82] | | Amberlyst-15 & $CaO/Al_2O_3$ | 2.5 wt% | | | — | 97 |
| Waste cooking oil [83] | | Amberlyst-15 & $CaO/Al_2O_3$ | 2.5 wt% & 3 wt% | | 4 : 1 mol% | — | 98.87 |

***Advantages***

- Generation of less waste.
- Reduces the prerequisite of external reprocess streams.
- Requirement of low amount of catalyst.
- The time of reaction is less.
- Higher conversion rate.
- Higher yield of biodiesel.
- Separation product is easy.

***Disadvantages***

- Higher energy is required.
- The conversion is directly affected by the efficiency of catalyst.

***Application***

- Pilot scale.

## 20.4 Conclusion

To save our environment and biological community from drastic fossil fuel pollution, the only alternative is biofuel, which is eco-friendly and suitable to our society. Biodiesel is considered as one of a significant biofuels used practically in various sectors, but still now it is not commercialized because of lack of proper technology of production in a cost effective way. Therefore, scientists are always searching a new advanced novel method of production of biodiesel in a cost effective pathway. Now the different vegetable-based oils such as sunflower oil, soybean oil, rapeseed oil, and palm oil are considered as the major source or biomass in the production of biodiesel. Because of high cost of production, hap hazard manageability, and improper technique of extraction, it was made unsuitable for industrial production of biodiesel. Microalgae are treated as an amazing source of biodiesel, owing to its higher percentage of lipid, no requirement of agricultural land, and less carbon impression. Some genetically developed plants such as switch grass, enormous bluestem, miscanthus, and poplar are considered as a potential feedstock for production of biodiesel. The quality of the biodiesel produced depends upon the steadiness of oxidation, cetane number, cloud point, pour point, octane number, and antiknocking properties. Hence, before production of biodiesel we have to analyze the quality of the feedstock. Therefore, to make the process of biodiesel production in industrial scale or commercialization, two factors are considered as important. The first one is the source of biomass feedstock, and the second one is the technology of production. More recently some new advanced technology was developed such as reactive distillation process, use of microwave technology, ultrasonication, and membrane technology, which is although capable of producing biodiesel in comparatively less cost and sustainable way, but still it is not sufficient for industrialization of this project. Therefore, extensive research is needed for commercialization of this project and save our future generation not only for drastic fossil fuel contamination but also save our valuable stock natural resources.

## 20.5 Future Perspectives

The biodiesel research is now expanding very rapidly worldwide because of limitation of our stock resources, excessive environmental contamination, energy scarcity, and some important ecological problems. The major challenge of biodiesel production is the availability of raw materials including oilseeds, higher percentage of oil containing crop cultivation, and use of suitable advanced cost-effective technology. The combination of fossil energy resources with biodiesel is also some extent sustainable and cost effective, but its industrialization is not permissible. If the percentage of oil in oil seeds like soybean oil, sunflower oil, and other vegetable oils has to be increased by modification of the crop seeds, then the quantity of production of biodiesel is noticeably increased. Another important challenge is the choice of suitable feedstock biomass raw material for production of biodiesel. The scarcity of suitable agricultural land for cultivation of oilseed crops is a great challenge; therefore, suitable time frame should have to be maintained without affecting normal agriculture for food production. Hence, in future microalgae, nonpalatable oil, wastes oil, and genetically modified plant species may serve as a future alternative for production of biodiesel. Hence, the nonfood source plants such as microalgae, jatropha, and some genetically modified grass/plant species are considered as potential feedstocks of biodiesel production. The use of heterogeneous enzymatic stimulates the process of biodiesel synthesis than homogeneous enzymes. The heterogeneous stimuli enzymes can be easily separated, efficiently reused, and offer excellent effectiveness. The use of enzymatic impetuses is highly beneficial and is treated as future prospective of the production of biodiesel. The requirement of energy can be solved by using the advanced technology such as microwave warming, unmoving blender, ultrasonic light, film reactor, and responsive refining. Therefore, we can finally conclude that biodiesel is an alternative fuel of high importance in future generation.

## References


1 G Lourinho, P Brito, Advanced biodiesel production technologies: novel developments. *Rev. Environ. Sci. Biotechnol.* 14 (2015). 287–316. https://doi.org/10.1007/s11157-014-9359-x.

2 Oyetola Ogunkunle, Noor A. Ahmed, A review of global current scenario of biodiesel adoption and combustion in vehicular diesel engines, *Energy Rep.*, 5 (2019) 1560–1579. https://doi.org/10.1016/j.egyr.2019.10.028.

3 K Pikula, A Zakharenko, A Stratidakis, M Razgonova, A Nosyrev, Y Mezhuev, K Golokhvast, The advances and limitations in biodiesel production: feedstock's, oil extraction methods, production, and environmental life cycle assessment. *Green Chem. Lett. Rev.*, 13(4), (2020)11–30. doi:https://doi.org/10.1080/17518253.2020.1829099.

4 Zahan, K., & Kano, M. Biodiesel production from palm oil, its by-products, and mill effluent: a review. *Energies*, 11(8), (2018) 2132–2156. doi:https://doi.org/10.3390/en11082132.

5 Bhuiya, M.K., Rasul, M.G., Khan, M.M.K. et al. (2016). Prospects of 2nd generation biodiesel as a sustainable fuel – Part: 1 selection of feedstocks, oil extraction techniques and conversion technologies. *Renew. Sustain. Energy Rev.* 55: 1109–1128.

6 S N Gebremariam, J M Marchetti, Biodiesel production technologies: review, *AIMS Energy*, 5 (3) (2017) 425–457. https://doi.org/10.3934/energy.2017.3.425.

7 A Talebian-Kiakalaieh, N. A. S. Amin, & H Mazaheri. A review on novel processes of biodiesel production from waste cooking oil. *Appl. Energy*, 104(2013) 683–710. https://doi.org/10.1016/j.apenergy.2012.11.061.

8 Koberg, M. and Gedanken, A. (2013). Using microwave radiation and SrO as a catalyst for the complete conversion of oils, cooked oils, and microalgae to biodiesel. In: *New and Future Developments in Catalysis. Catalytic Biomass Conversion* (ed. S.L. Suib), 209–227. Elsevier https://doi.org/10.1016/B978-0-444-53878-9.00010-2.

9 Athar, M. and Zaidi, S. A review of the feedstocks, catalysts, and intensification techniques for sustainable biodiesel production. *J. Environ. Chem. Eng.* https://doi.org/10.1016/j.jece.2020.104523.

10 Patil, P.D., Gude, V.G., Mannarswamy, A. et al. (2011). Optimization of microwave-assisted transesterification of dry algal biomass using response surface methodology. *Bioresour. Technol.* 102 (2): 1399–1405.

11 M.A. Dubé, A.Y. Tremblay, J. Liu, Biodiesel production using a membrane reactor, *Bioresour. Technol.* 98 (2007) 639–647. https://doi.org/10.1016/j.biortech.2006.02.019.

12 P. Cao, A.Y. Tremblay, M.A. Dubé, K. Morse, Effect of membrane pore size on the performance of a membrane reactor for biodiesel production, *Ind. Eng. Chem. Res.* 46 (2007) 52–58. https://doi.org/10.1021/ie060555o.

13 J. Saleh, A.Y. Tremblay, M.A. Dubé, Glycerol removal from biodiesel using membrane separation technology, *Fuel* 89 (2010) 2260–2266. https://doi.org/10.1016/j.fuel.2010.04.025.

14 S.H. Shuit, K.T. Lee, A.H. Kamaruddin, S. Yusup, Reactive extraction of *Jatropha curcas* L. seed for production of biodiesel: process optimization study, *Environ. Sci. Technol.* 44 (2010) 4361–4367. https://doi.org/10.1021/es902608v.

15 P. Cao, M.A. Dubé, A.Y. Tremblay, High-purity fatty acid methyl ester production from canola, soybean, palm, and yellow grease lipids by means of a membrane reactor, *Biomass Bioenergy* 32 (2008) 1028–1036. https://doi.org/10.1016/j.biombioe.2008.01.020.

16 S. Baroutian, M.K. Aroua, A.A.A. Raman, N.M.N. Sulaiman, A packed bed membrane reactor for production of biodiesel using activated carbon supported catalyst, *Bioresour. Technol.* 102 (2011) 1095–1102. https://doi.org/10.1016/j.biortech.2010.08.076.

17 I. Reyes, G. Ciudad, M. Misra, A. Mohanty, D. Jeison, R. Navia, Novel sequential batch membrane reactor to increase fatty acid methyl esters quality at low methanol to oil molar ratio, *Chem. Eng. J.* 197 (2012) 459–467. https://doi.org/10.1016/j.cej.2012.05.038.

18 V. Gude, P. Patil, E. Martinez-Guerra, S. Deng, N. Nirmalakhandan, Microwave energy potential for biodiesel production, *Sustain. Chem. Process.* (2013). https://doi.org/10.1186/2043-7129-1-5.

19 B. Karmakar, G. Halder, Progress and future of biodiesel synthesis: advancements in oil extraction and conversion technologies, *Energy Convers. Manag.* 182 (2019) 307–339. https://doi.org/10.1016/j.enconman.2018.12.066.

20 J.Z. Yin, M. Xiao, J. Bin Song, Biodiesel from soybean oil in supercritical methanol with co-solvent, *Energy Convers. Manag.* 49 (2008) 908–912. https://doi.org/10.1016/j.enconman.2007.10.018.

21 V. Lertsathapornsuk, R. Pairintra, K. Aryusuk, K. Krisnangkura, Microwave assisted in continuous biodiesel production from waste frying palm oil and its performance in a 100 kW diesel generator, *Fuel Process. Technol.* 89 (2008) 1330–1336. https://doi.org/10.1016/j.fuproc.2008.05.024.

22 A.A. Refaat, S.T. Sheltawy, K.U. Sadek, Optimum reaction time, performance and exhaust emissions of biodiesel produced by microwave irradiation, *Int. J. Environ. Sci. Technol.* 5 (2008) 315–322. https://doi.org/10.1007/BF03326026.

23 Azcan, N. and Danisman, A. (2007). Alkali catalyzed transesterification of cottonseed oil by microwave irradiation. *Fuel* 86: 2639–2644. https://doi.org/10.1016/j.fuel.2007.05.021.390-392.

24 Groisman, Y. and Gedanken, A. (2008). Continuous flow, circulating microwave system and its application in nanoparticle fabrication and biodiesel synthesis. *J. Phys. Chem. C.* 112: 8802–8808.

25 P.R. Gogate, A.M. Kabadi, A review of applications of cavitation in biochemical engineering/biotechnology, *Biochem. Eng. J.* 44 (2009) 60–72. https://doi.org/10.1016/j.bej.2008.10.006.

26 P. Patil, S. Deng, J. Isaac Rhodes, P.J. Lammers, Conversion of waste cooking oil to biodiesel using ferric sulfate and supercritical methanol processes, *Fuel* 89 (2010) 360–364. https://doi.org/10.1016/j.fuel.2009.05.024.

27 Qiu, Z., Zhao, L., and Weatherley, L. (2010). Process intensification technologies in continuous biodiesel production. *Chem. Eng. Process. Process Intensif.* 49: 323–330. https://doi.org/10.1016/j.cep.2010.03.005.108.

28 H.A. Choudhury, S. Chakma, V.S. Moholkar, Mechanistic insight into sonochemical biodiesel synthesis using heterogeneous base catalyst, *Ultrason. Sonochem.* 21 (2014) 169–181. https://doi.org/10.1016/j.ultsonch.2013.04.010.

29 F. Zhang, Z. Fang, Y.T. Wang, Biodiesel production directly from oils with high acid valueby magnetic $Na_2SiO_3@Fe_3O_4/C$ catalyst and ultrasound, *Fuel* 150 (2015) 370–377. https://doi.org/10.1016/j.fuel.2015.02.032.

30 D. Yu, L. Tian, H. Wu, S. Wang, Y. Wang, D. Ma, X. Fang, Ultrasonic irradiation with vibration for biodiesel production from soybean oil by Novozym 435, *Process Biochem.* 45 (2010) 519–525. https://doi.org/10.1016/j.procbio.2009.11.012.

31 J. Ji, J. Wang, Y. Li, Y. Yu, Z. Xu, Preparation of biodiesel with the help of ultrasonic and hydrodynamic cavitation, *Ultrasonics* 44 (2006) e411–e414. https://doi.org/10.1016/j.ultras.2006.05.020.

32 X. Ye, S. Fernando, W. Wilson, A. Singh, Application of amphiphilic catalysts, ultrasonication, and nanoemulsions for biodiesel production process, *Chem. Eng. Technol.* 30 (2007) 1481–1487. https://doi.org/10.1002/ceat.200700283.

33 P. Cintas, S. Mantegna, E.C. Gaudino, G. Cravotto, A new pilot flow reactor for high-intensity ultrasound irradiation. Application to the synthesis of biodiesel, *Ultrason. Sonochem.* 17 (2010) 985–989. https://doi.org/10.1016/j.ultsonch.2009.12.003.

34 N.N. Mahamuni, Y.G. Adewuyi, Optimization of the synthesis of biodiesel via ultrasound-enhanced base-catalyzed transesterification of soybean oil using a multifrequency ultrasonic reactor, *Energy and Fuels* 23 (2009) 2757–2766. https://doi.org/10.1021/ef900047j.

35 A.P. Vyas, J.L. Verma, N. Subrahmanyam, Effects of molar ratio, alkali catalyst concentration and temperature on transesterification of Jatropha oil with methanol under ultrasonic irradiation, *Adv. Chem. Eng. Sci.* 1 (2011) 45–50. https://doi.org/10.4236/aces.2011.12008.

36 H.D. Hanh, N.T. Dong, K. Okitsu, R. Nishimura, Y. Maeda, Biodiesel production through transesterification of triolein with various alcohols in an ultrasonic field, *Renew. Energy* 34 (2009) 766–768. https://doi.org/10.1016/j.renene.2008.04.007.

**37** C. Stavarache, M. Vinatoru, Y. Maeda, H. Bandow, Ultrasonically driven continuous process for vegetable oil transesterification, *Ultrason. Sonochem.* 14 (2007) 413–417. https://doi.org/10.1016/j.ultsonch.2006.09.014.

**38** K. Noipin, S. Kumar, Optimization of ethyl ester production assisted by ultrasonic irradiation, *Ultrason. Sonochem.* 22 (2015) 548–558. https://doi.org/10.1016/j.ultsonch.2014.07.019.

**39** G. Chen, R. Shan, J. Shi, B. Yan, Ultrasonic-assisted production of biodiesel from transesterification of palm oil over ostrich eggshell-derived CaO catalysts, *Bioresour. Technol.* 171 (2014) 428–432. https://doi.org/10.1016/j.biortech.2014.08.102.

**40** H.A. Choudhury, P.P. Goswami, R.S. Malani, V.S. Moholkar, Ultrasonic biodiesel synthesis from crude *Jatropha curcas* oil with heterogeneous base catalyst: mechanistic insight and statistical optimization, *Ultrason. Sonochem.* 21 (2014) 1050–1064. https://doi.org/10.1016/j.ultsonch.2013.10.023.

**41** R.E. Armenta, M. Vinatoru, A.M. Burja, J.A. Kralovec, C.J. Barrow, Transesterification of fish oil to produce fatty acid ethyl esters using ultrasonic energy, *J. Am. Oil Chem. Soc.* 84 (2007) 1045–1052. https://doi.org/10.1007/s11746-007-1129-2.

**42** A.S. Badday, A.Z. Abdullah, K.T. Lee, Transesterification of crude Jatropha oil by activated carbon-supported heteropolyacid catalyst in an ultrasound-assisted reactor system, *Renew. Energy* 62 (2014) 10–17. https://doi.org/10.1016/j.renene.2013.06.037.

**43** I. Korkut, M. Bayramoglu, Selection of catalyst and reaction conditions for ultrasound assisted biodiesel production from Canola oil, *Renew. Energy* 116 (2017) 543–551. https://doi.org/10.1016/j.renene.2017.10.010.

**44** A. Sarve, S.S. Sonawane, M.N. Varma, Ultrasound assisted biodiesel production from sesame (*Sesamum indicum* L.) oil using barium hydroxide as a heterogeneous catalyst: comparative assessment of prediction abilities between response surface methodology (RSM) and artificial neural network, *Ultrason. Sonochem.* (2015). https://doi.org/10.1016/j.ultsonch.2015.01.013.

**45** J.P. Maran, B. Priya, Ultrasonics sonochemistry comparison of response surface methodology and artificial neural network approach towards efficient ultrasound-assisted biodiesel production from muskmelon oil, *Ultrason. Sonochem.* 23 (2015) 192–200. https://doi.org/10.1016/j.ultsonch.2014.10.019.

**46** M. Takase, W. Feng, W. Wang, X. Gu, Y. Zhu, T. Li, L. Yang, X. Wu, *Silybum marianum* oil as a new potential non-edible feedstock for biodiesel: a comparison of its production using conventional and ultrasonic assisted method, *Fuel Process. Technol.* 123 (2014) 19–26. https://doi.org/10.1016/j.fuproc.2014.01.032.

**47** Çağlar, E.(2007). Biodiesel production using co-solvent. In: *Book of Abstracts European Congress Chemical Engineering* Copenhagen (16–20 September, 2007).

**48** S. Sakthivel, S. Halder, P.D. Gupta, Influence of co-solvent on the production of biodiesel in batch and continuous process, *Int. J. Green Energy* 10 (2013) 876–884. https://doi.org/10.1080/15435075.2012.727365.

**49** L.T. Thanh, K. Okitsu, Y. Sadanaga, N. Takenaka, Y. Maeda, H. Bandow, A new co-solvent method for the green production of biodiesel fuel – optimization and practical application, *Fuel* 103 (2013) 742–748. https://doi.org/10.1016/j.fuel.2012.09.029.

**50** C. Calgaroto, S. Calgaroto, M.A. Mazutti, D. De Oliveira, S. Pergher, J. Vladimir De Oliveira, Production of biodiesel from soybean and *Jatropha curcas* oils with KSF and amberlyst 15 catalysts in the presence of co-solvents, *Sustain. Chem. Process.* 1:17 (2013) 1–6. https://doi.org/10.1186/2043-7129-1-17.

**51** G.R. Kumar, R. Ravi, A. Chadha, Kinetic studies of base-catalyzed transesterification reactions of non-edible oils to prepare biodiesel: the effect of co-solvent and temperature, *Energy and Fuels* 25 (2011) 2826–2832. https://doi.org/10.1021/ef200469u.

**52** Encinar, J.M., González, J.F., Pardal, A., and Martínez, G. (n.d.). Transesterification of rapeseed oil with methanol in the presence of various co-solvents. *Third International Symposium on Energy from Biomass Waste*, Venice, Italy, (8–11 November 2010).

**53** R. Pen, R. Romero, S. Luz, M.J. Ramos, A. Marti, R. Natividad, S.L. Martí, A. Martí, Transesterification of castor oil : effect of catalyst and co-solvent, *Ind. Eng. Chem. Res.* 48 (2009) 1186–1189. https://doi.org/10.1021/ie8005929.

**54** M.K. Lam, K.T. Lee, Accelerating transesterification reaction with biodiesel as co-solvent: a case study for solid acid sulfated tin oxide catalyst, *Fuel* 89 (2010) 3866–3870. https://doi.org/10.1016/j.fuel.2010.07.005.

**55** G. Guan, N. Sakurai, K. Kusakabe, Synthesis of biodiesel from sunflower oil at room temperature in the presence of various cosolvents, *Chem. Eng. J.* 146 (2009) 302–306. https://doi.org/10.1016/j.cej.2008.10.009.

**56** R. Sawangkeaw, K. Bunyakiat, S. Ngamprasertsith, Effect of co-solvents on production of biodiesel via transesterification in supercritical methanol, *Green Chem.* 9 (2007) 679–685. https://doi.org/10.1039/b614064e.

**57** K.T. Tan, K.T. Lee, A.R. Mohamed, Effects of free fatty acids, water content and co solvent on biodiesel production by supercritical methanol reaction, *J. Supercrit. Fluids.* 53 (2010) 88– 91. https://doi.org/10.1016/j.supflu.2010.01.012.

**58** C.M. Trentin, A.P. Lima, I.P. Alkimim, C. Da Silva, F. De Castilhos, M.A. Mazutti, J.V. Oliveira, Continuous production of soybean biodiesel with compressed ethanol in a microtube reactor using carbon dioxide as co-solvent, *Fuel Process. Technol.* 92 (2011) 952–958. https://doi.org/10.1016/j.fuproc.2010.12.016.

**59** W. Cao, H. Han, J. Zhang, Preparation of biodiesel from soybean oil using supercriticalmethanol and co-sol vent, *Fuel* 84 (2005) 347–351. https://doi.org/10.1016/j.fuel.2004.10.001.

**60** K. Bunyakiat, S. Makmee, R. Sawangkeaw, S. Ngamprasertsith, Continuous production of biodiesel via transesterification from vegetable oils in supercritical methanol, *Energy & Fuels* 20 (2006) 812–817. https://doi.org/10.1021/ef050329b.

**61** D.G.B. Boocock, S.K. Konar, V. Mao, H. Sidi, Fast one-phase oil-rich processes for the preparation of vegetable oil methyl esters, *Biomass Bioenergy.* 11 (1996) 43–50. https://doi.org/10.1016/0961-9534(95)00111-5.

**62** J.Y. Park, D.K. Kim, Z.M. Wang, J.S. Lee, Fast biodiesel production with one-phase reaction, *Appl. Biochem. Biotechnol.* 154 (2009) 246–252. https://doi.org/10.1007/s12010-008-8421-y.

**63** Knothe, G., Van Gerpen, J., and Krahl, J. (2005). *The Biodiesel Handbook*, 1e. AOCS Press.

**64** Ivana B. Banković-Ilić, Olivera S. Stamenković, Vlada B. Veljković. Biodiesel production from nonedible plant oils, *Renew. Sustain. Energy Rev.* 16 (2012) 3621–3647. https://doi.org/10.1016/j.rser.2012.03.002.

**65** A. Mondala, K. Liang, H. Toghiani, R. Hernandez, T. French, Biodiesel production by in situ transesterification of municipal primary and secondary sludges, *Bioresour. Technol.* 100 (2009) 1203–1210. https://doi.org/10.1016/j.biortech.2008.08.020.

**66** P.-J. Shiu, S. Gunawana, W.-H. Hsieh, N.S. Kasim, Y.-H. Ju, Biodiesel production from rice bran by a two-step in-situ process, *Bioresour. Technol.* 101 (2010) 984–989. https://doi.org/10.1016/j.biortech.2009.09.011.

**67** M.J. Haas, K. Wagner, Simplifying biodiesel production: the direct or in situ transesterification of algal biomass, *Eur. J. Lipid Sci. Technol.* 113 (2011) 1219–1229. https://doi.org/10.1002/ejlt.201100106.

**68** T. Dong, J. Wang, C. Miao, Y. Zheng, S. Chen, Two-step in situ biodiesel production from microalgae with high free fatty acid content, *Bioresour. Technol.* 136 (2013) 8–15. https://doi.org/10.1016/j.biortech.2013.02.105.

**69** Sastry, S.V.A.R. and Murthy, C.V.R. (2014). Synthesis of biodiesel by in-situ transesterification of Karanja oil. *Bangladesh J. Sci. Ind. Res.* 49: 211–218.

**70** V.T. Wyatt, M.J. Haas, Production of fatty acid methyl esters via the in situ transesterification of soybean oil in carbon dioxide-expanded methanol, *JAOCS, J. Am. Oil Chem. Soc.* 86 (2009) 1009–1016. https://doi.org/10.1007/s11746-009-1438-8.

**71** J. Zeng, X. Wang, B. Zhao, J. Sun, Y. Wang, Rapid in situ transesterification of sunflower oil, *Ind. Eng. Chem. Res.* 48 (2009) 850–856. https://doi.org/10.1021/ie8008956.

**72** Prabaningrum, N., Subbarao, D., and Ismail, L. (2011). In-situ transesterification of *Jatropha curcas* seeds using the mixture of methanol and isopropanol. *National Postgraduate Conferences,* Perak, Malaysia (19–20 September 2011). IEEE. pp. 1–6. https://doi.org/10.1109/NatPC.2011.6136282.

**73** C. Mueanmas, K. Prasertsit, C. Tongurai, Feasibility study of reactive distillation column for transesterification of palm oils, *Int. J. Chem. Eng. Appl.* 1 (2010) 77–83. https://doi.org/10.7763/IJCEA.2010.V1.13.

**74** Dimian, A.C. and Rothenberg, G. (2010). Process for manufacturing acid esters through reactive distillation. EP Patent 2,154,226. https://doi.org/10.1371/joumal.pone.0010853.

**75** He, B.B., Singh, A.P., and Thompson, J.C. (2006). A novel continuous-flow reactor using reactive distillation for biodiesel production. *Trans. ASABE (American Soc. Agric. Biol. Eng.)* 49: 107–112.

**76** Shinde, G.B., Sapkal, V.S., Sapkal, R.S., and Raut, N.B. (2011). Transesterification by reactive distillation for synthesis and characterization of biodiesel. In: *Biodiesel – Feedstocks Processing. Technologies* (ed. M. Stoytcheva), 29. InTech.

**77** Mahajani, P.S.M., Anekar, P.S.V., Jadhav, P.A.L., and Patil, P.A.V. (2017). Production of bio-diesel (*Butyl Oleate*) by reactive distillation technique. *Int. Res. J. Eng. Technol.* 4: 770–778. https://irjet.net/archives/V4/i1/IRJET-V4I1135.pdf.

**78** E.S. Pérez-Cisneros, X. Mena-Espino, V. Rodríguez-López, M. Sales-Cruz, T. Viveros-García, R. Lobo-Oehmichen, An integrated reactive distillation process for biodiesel production, *Comput. Chem. Eng.* 91 (2016) 233–246 https://doi.org/10.1016/j.compchemeng.2016.01.008.

**79** N. De Lima Da Silva, C.M.G. Santander, C.B. Batistella, R.M.I. Filho, M.R.W. MacIel, Biodiesel production from integration between reaction and separation system: reactive distillation process, *Appl. Biochem. Biotechnol.* 161 (2010) 245–254. https://doi.org/10.1007/s12010-009-8882-7.

**80** Y.S. Pradana, A. Hidayat, A. Prasetya, A. Budiman, Biodiesel production in a reactive distillation column catalyzed by heterogeneous potassium catalyst, *Energy Procedia* 143 (2017) 742–747. https://doi.org/10.1016/j.egypro.2017.12.756.

**81** Dos Santos, M., Barros, A.A.C., Gabriel, K., and Correia, M.J.N. (2017). Strategy to biodiesel production through the reactive distilation column using palm oil as feedstock. *Int. J. Adv. Eng. Manag. Res.* 2: 2202–2216.

**82** A. Anantapinitwatna, K. Ngaosuwan, W. Kiatkittipong, D. Wongsawaeng, S. Assabumrungrat, Effect of water content in waste cooking oil on biodiesel production via ester-transesterification in a single reactive distillation, *IOP Conf. Ser. Mater. Sci. Eng.* 559 (2019). https://doi.org/10.1088/1757-899X/559/1/012014.

**83** N. Petchsoongsakul, K. Ngaosuwan, W. Kiatkittipong, F. Aiouache, S. Assabumrungrat, Process design of biodiesel production: hybridization of ester-and transesterification in a single reactive distillation, *Energy Convers. Manag.* 153 (2017) 493–503. https://doi.org/10.1016/j.enconman.2017.10.013.

# Index



*Biodiesel Production: Feedstocks, Catalysts, and Technologies*, First Edition. Edited by Samuel Lalthazuala Rokhum, Gopinath Halder, Suttichai Assabumrungrat, and Kanokwan Ngaosuwan.

## C

## d

www.ingramcontent.com/pod-product-compliance
Lightning Source LLC
LaVergne TN
LVHW062204200926
848484LV00006B/1318